NOTES

NOTES

Theory of Vibration

with Applications

Theory of Vibration with Applications

WILLIAM T. THOMSON

Professor of Mechanical Engineering
University of California, Santa Barbara

Prentice-Hall, Inc., Englewood Cliffs, New Jersey

10 9 8 7 6 5 4

ISBN: 0–13–914549–4

Library of Congress Catalog Card Number: 72–9039

Printed in the United States of America.

PRENTICE-HALL INTERNATIONAL, INC. *London*
PRENTICE-HALL OF AUSTRALIA PTY. LTD., *Sydney*
PRENTICE-HALL OF CANADA, LTD., *Toronto*
PRENTICE-HALL OF INDIA PRIVATE LIMITED, *New Delhi*
PRENTICE-HALL OF JAPAN, INC., *Tokyo*

Contents

Preface

The subject of vibrations has a unique fascination. It is a logical subject explainable by basic principles of mechanics. Unlike some subjects, its mathematical concepts are all associated with physical phenomena which can be experienced and measured. It is a satisfying subject to teach and to share with students. From the first elementary text, *Mechanical Vibrations*, published in 1948, the author has attempted to improve its presentation in keeping with technological advances and experience gained by teaching and practice. In this respect, many teachers and students have contributed with suggestions and interactions over the years.

This new text, which has been almost entirely rewritten, is again a desire on the author's part towards clearer presentation with modern techniques which have become commonplace. In the first five chapters, which deal with single degree of freedom systems and with two degrees of freedom systems, the simplicity of the previous text has been adhered to and, hopefully, improved upon. Since the digital computer is now a commonly available facility, its use in the vibration field is encouraged by some simple examples. In spite of the versatility of the digital computer, the analog computer is still a useful tool, and, in many cases, its use is fully justified. The first five chapters, which keep two degrees of freedom systems on a simple and physical basis, form a background for the understanding of the basic subject of vibrations, which can be covered in a quarter or a semester in a first course on vibration.

In Chapter 6 the concepts of the two degrees of freedom systems are generalized to those of multidegree of freedom systems. The emphasis of this chapter is theory, and, with the aid of matrix algebra, the extension to multidegree of freedom systems can be presented elegantly. All of the basis for coordinate decoupling becomes clear with matrices. Some uncommon ideas of normal modes in forced vibration and the method of state space used commonly in control theory are introduced.

There are many analytical approaches to vibration analysis of complex structures of many degrees of freedom. Chapter 7 presents some of the more useful procedures, and, although most multidegree of freedom systems are today solved on the digital computer, one still needs to know how to formulate such problems for efficient computation and to know some of the approximations which can be made to check the calculations. All of the problems here can be programmed for the computer, but the theory behind the computations must be understood. A digital computation of a Holzer-type problem is illustrated.

Chapter 8 deals with continuous sytems, or those problems associated with partial differential equations. A finite difference approach to beam problems offers an opportunity to solve such problems on the digital computer.

Lagrange's equations, covered in Chapter 9, strengthen again the understanding of dynamical systems presented earlier and broaden one's view for other extensions. For example, the important concepts of the mode summation procedure is a natural consequence of the Lagrangian generalized coordinates. The meaning of constraint equations as physical boundary conditions for modal synthesis is again logically understood through Lagrange's theory.

Chapter 10 treats dynamical systems excited by random forces or displacements. Such problems must be examined from a statistical point of view and, in many cases, the probability density of the random excitation is normally distributed. The point of view taken here is that, given a random record, an autocorrelation can be easily determined from which the spectral density and mean square response can be calculated. The digital computer is again essential for the numerical work.

In Chapter 11 the treatment of nonlinear systems is introduced with emphasis on the phase plane method. When the nonlinearities are small, the methods of perturbation or iteration offer an analytical approach. Results of machine computations for a nonlinear system illustrate what can be done.

Chapters 6 through 11 represent subject matter appropriate for a second course in vibration, which may be covered at the graduate level.

WILLIAM T. THOMSON

Theory of Vibration

with Applications

1

Oscillatory Motion

1.1 INTRODUCTION

The study of vibration is concerned with the oscillatory motions of bodies and the forces associated with them. All bodies possessing mass and elasticity are capable of vibration. Thus most engineering machines and structures experience vibration to some degree, and their design generally requires consideration of their oscillatory behavior.

Oscillatory systems can be broadly characterized as *linear* or *nonlinear*. For linear systems the principle of superposition holds, and the mathematical techniques available for their treatment are well-developed. In contrast, techniques for the analysis of nonlinear systems are less well known, and difficult to apply. However, some knowledge of nonlinear systems is desirable, since all systems tend to become nonlinear with increasing amplitude of oscillation.

There are two general classes of vibrations—free and forced. *Free vibration* takes place when a system oscillates under the action of forces inherent in the system itself, and when external impressed forces are absent. The system under free vibration will vibrate at one or more of its *natural frequencies*, which are properties of the dynamical system established by its mass and stiffness distribution.

Vibration that takes place under the excitation of external forces is called *forced vibration*. When the excitation is oscillatory, the system is forced to vibrate at the excitation frequency. If the frequency of excitation coincides with one of the natural frequencies of the system, a condition of *resonance* is encountered, and dangerously large oscillations may result. The failure of major structures, such as bridges, buildings, or airplane wings, is an awesome possibility under resonance. Thus, the calculation of the natural frequencies is of major importance in the study of vibrations.

Vibrating systems are all subject to *damping* to some degree because energy is dissipated by friction and other resistances. If the damping is small, it has very little influence on the natural frequencies of the system, and hence the calculations for the natural frequencies are generally made on the basis of no damping. On the other hand, damping is of great importance in limiting the amplitude of oscillation at resonance.

The number of independent coordinates required to describe the motion of a system is called the *degrees of freedom* of the system. Thus a free particle undergoing general motion in space will have three degrees of freedom, while a rigid body will have six degrees of freedom, i.e., three components of position and three angles defining its orientation. Furthermore, a continuous elastic body will require an infinite number of coordinates (three for each point on the body) to describe its motion; hence its degrees of freedom must be infinite. However, in many cases, parts of such bodies may be assumed to be rigid, and the system may be considered to be dynamically equivalent to one having finite degrees of freedom. In fact, a surprisingly large number of vibration problems can be treated with sufficient accuracy by reducing the system to one having a single degree of freedom.

1.2 HARMONIC MOTION

Oscillatory motion may repeat itself regularly, as in the balance wheel of a watch, or display considerable irregularity, as in earthquakes. When the motion is repeated in equal intervals of time τ, it is called *periodic motion*. The repetition time τ is called the *period* of the oscillation, and its reciprocal, $f = 1/\tau$, is called the *frequency*. If the motion is designated by the time function $x(t)$, then any periodic motion must satisfy the relationship $x(t) = x(t + \tau)$.

Irregular motions, which appear to possess no definite period, can be considered to be the sum of a very large number of regular motions of different frequencies. Properties of such motion can be described statistically; the discussion of these properties will be deferred to a later section.

The simplest form of periodic motion is *harmonic motion.* It can be demonstrated by a mass suspended from a light spring, as shown in Fig. 1.2-1. If the mass is displaced from its rest position and released, it will oscillate up and down. By placing a light source on the oscillating mass, its motion can be recorded on a light-sensitive film strip which is made to move past it at constant speed.

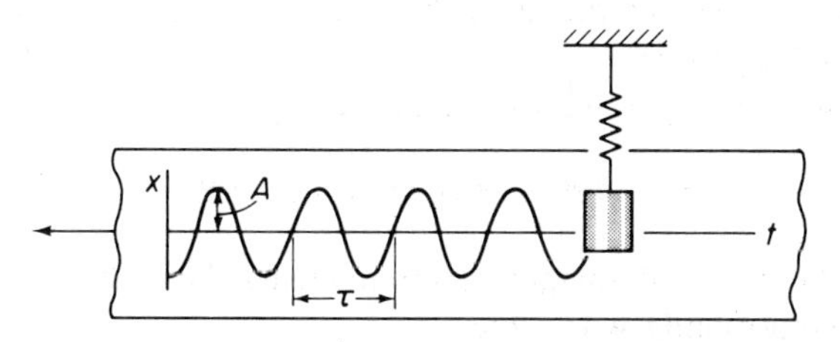

Figure 1.2-1. Recording of harmonic motion.

The motion recorded on the film strip can be expressed by the equation

$$x = A \sin 2\pi \frac{t}{\tau} \tag{1.2-1}$$

where A is the amplitude of oscillation, measured from the equilibrium position of the mass, and τ is the period. The motion is repeated when $t = \tau$.

Harmonic motion is often represented as the projection on a straight line of a point that is moving on a circle at constant speed, as shown in Fig. 1.2-2. With the angular speed of the line op designated by ω, the displacement x can be written as

$$x = A \sin \omega t \tag{1.2-2}$$

The quantity ω is generally measured in radians per second, and is referred

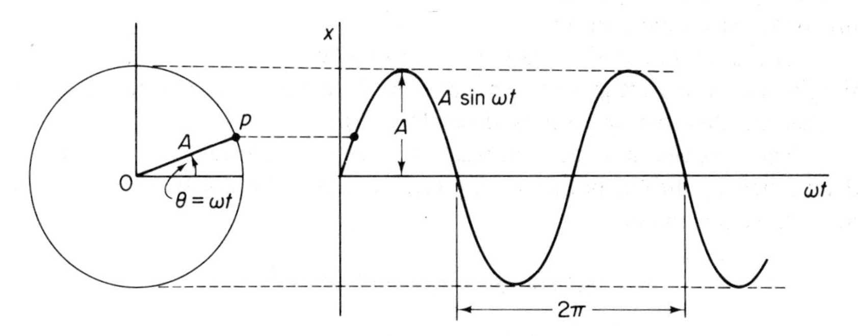

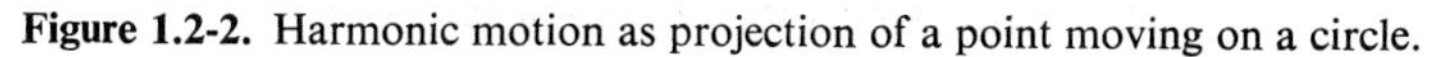

Figure 1.2-2. Harmonic motion as projection of a point moving on a circle.

to as the *circular frequency*. Since the motion repeats itself in 2π radians, we have the relationship

$$\omega = \frac{2\pi}{\tau} = 2\pi f \tag{1.2-3}$$

where τ and f are the period and frequency of the harmonic motion, usually measured in seconds and cycles per second respectively.

For the motion of a point around a circle, it is convenient to use an imaginary axis i and let the radius of the circle be represented by a complex quantity z called a *phasor*.

The phasor z is expressed by the equation

$$z = Ae^{i\theta} = A\cos\theta + iA\sin\theta \tag{1.2-4}$$

which define the real and imaginary components. With $\theta = \omega t$, the components vary sinusoidally with time

$$\text{Re } z = A\cos\omega t$$
$$\text{Im } z = A\sin\omega t$$

It is often necessary to consider two harmonic motions of the same frequency but differing in phase by ϕ. The two motions may be expressed by the phasors

$$z_1 = A_1 e^{i\omega t}$$
$$z_2 = A_2 e^{i(\omega t + \phi)}$$

where A_1 and A_2 are real numbers. The second phasor can be further rewritten as

$$z_2 = A_2 e^{i\phi} e^{i\omega t} = \bar{A}_2 e^{i\omega t} \tag{1.2-5}$$

where $\bar{A}_2$ is now a complex number. This form is often useful in problems involving harmonic motion.

The addition, multiplication, or raising to powers of phasors follow simple rules which are given in Appendix A. With harmonic motion expressed as phasors, their manipulations are easily carried out.

The velocity and acceleration of harmonic motion can be simply determined by differentiation of Eq. (1.2-2). Using the dot notation for the derivative, we obtain

$$\dot{x} = \omega A\cos\omega t = \omega A\sin\left(\omega t + \frac{\pi}{2}\right) \tag{1.2-6}$$

$$\ddot{x} = -\omega^2 A\sin\omega t = \omega^2 A\sin(\omega t + \pi) \tag{1.2-7}$$

Thus the velocity and acceleration are also harmonic with the same frequency of oscillation, but lead the displacement by $\pi/2$ and π radians respectively, as shown in Fig. 1.2-3.

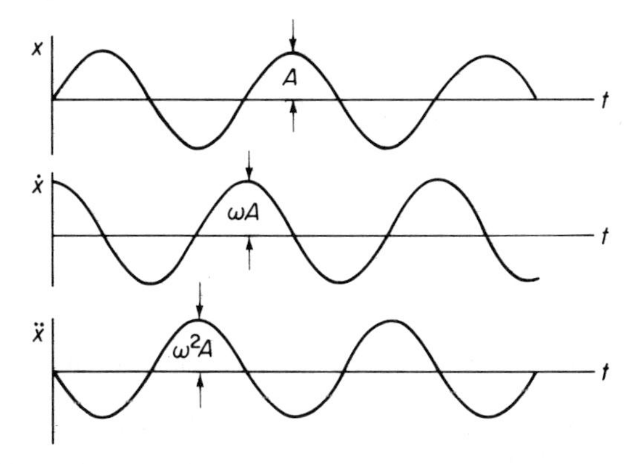

Figure 1.2-3. In harmonic motion, the velocity and acceleration lead the displacement by $\pi/2$ and π.

Examination of Eqs. (1.2-2) and (1.2-7) reveals that

$$\ddot{x} = -\omega^2 x \tag{1.2-8}$$

so that in harmonic motion the acceleration is proportional to the displacement and is directed towards the origin. Since Newton's second law of motion states that the acceleration is proportional to the force, harmonic motion can be expected for systems with linear springs with force varying as kx.

1.3 HARMONIC ANALYSIS

It is quite common for vibrations of several different frequencies to exist simultaneously. For example, the vibration of a violin string is composed of the fundamental frequency f and all its harmonics $2f$, $3f$, etc. Another example is the free vibration of a multidegree-of-freedom system, to which the vibrations at each natural frequency contribute. Such vibrations result in a complex waveform which is repeated periodically as shown in Fig. 1.3-1.

The French mathematician J. Fourier (1768–1830) showed that any periodic motion can be represented by a series of sines and cosines which are harmonically related. If $x(t)$ is a periodic function of the period τ, it is represented by the Fourier series

$$\begin{aligned} x(t) = \frac{a_0}{2} &+ a_1 \cos \omega_1 t + a_2 \cos 2\omega_1 t + \cdots \\ &+ b_1 \sin \omega_1 t + b_2 \sin 2\omega_1 t + \cdots \end{aligned} \tag{1.3-1}$$

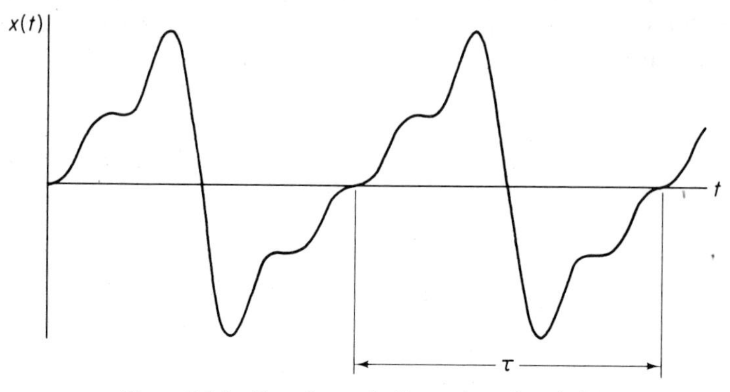

Figure 1.3-1. Complex periodic motion of period τ.

where $\omega_1 = 2\pi/\tau$ is the fundamental frequency. To determine the coefficients a_n and b_n, we multiply both sides of Eq. (1.3-1) by $\cos n\omega_1 t$ or $\sin n\omega_1 t$, and integrate each term over the period τ. Recognizing the following relations,

$$\int_{-\tau/2}^{\tau/2} \cos n\omega_1 t \cos m\omega_1 t \, dt = \begin{cases} 0 & \text{if } m \neq n \\ \dfrac{\pi}{\omega_1} & \text{if } m = n \end{cases}$$

$$\int_{-\tau/2}^{\tau/2} \sin n\omega_1 t \sin m\omega_1 t \, dt = \begin{cases} 0 & \text{if } m \neq n \\ \dfrac{\pi}{\omega_1} & \text{if } m = n \end{cases}$$

$$\int_{-\tau/2}^{\tau/2} \cos n\omega_1 t \sin m\omega_1 t \, dt = \begin{cases} 0 & \text{if } m \neq n \\ 0 & \text{if } m = n \end{cases}$$

all terms except one on the right side of the equation will be zero, and we obtain the results

$$a_n = \frac{\omega_1}{\pi} \int_{-\tau/2}^{\tau/2} x(t) \cos n\omega_1 t \, dt \tag{1.3-2}$$

$$b_n = \frac{\omega_1}{\pi} \int_{-\tau/2}^{\tau/2} x(t) \sin n\omega_1 t \, dt \tag{1.3-3}$$

Returning to Eq. (1.3-1) and examining the two terms at one of the frequencies, $n\omega_1$, their sum can be written as

$$\begin{aligned} &a_n \cos n\omega_1 t + b_n \sin n\omega_1 t \\ &= \sqrt{a_n^2 + b_n^2}\left\{\frac{a_n}{\sqrt{a_n^2 + b_n^2}} \cos n\omega_1 t + \frac{b_n}{\sqrt{a_n^2 + b_n^2}} \sin n\omega_1 t\right\} \\ &= c_n \cos(n\omega_1 t - \phi_n) \end{aligned}$$

where

$$c_n = \sqrt{a_n^2 + b_n^2} \tag{1.3-4}$$

and

$$\tan \phi = \frac{b_n}{a_n} \tag{1.3-5}$$

Thus c_n and ϕ_n (or a_n and b_n) completely define the harmonic contribution of the periodic wave.

When c_n and ϕ_n are plotted against the frequency $n\omega_1$ for all n, the result is a series of discrete lines at ω_1, $2\omega_1$, $3\omega_1$, etc., as shown in Fig. 1.3-2. Such plots are called the *Fourier spectrum* of the waveform.

With the aid of the digital computer, harmonic analysis today is efficiently carried out in minimum time. A new computer algorithm introduced recently, known as the Fast Fourier Transform,* further reduces the computational time.

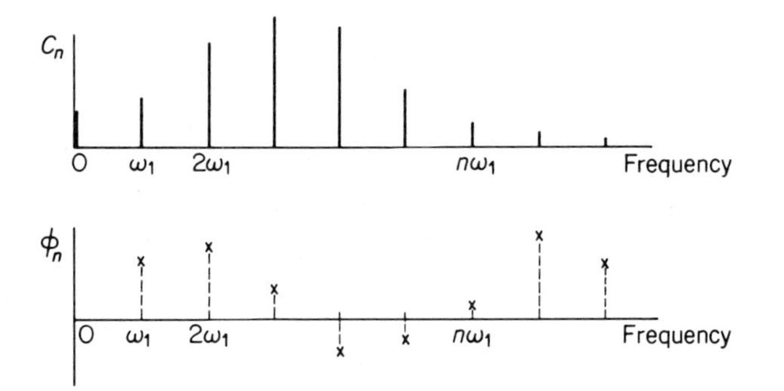

Figure 1.3-2. Fourier spectrum for a periodic time function.

1.4 TRANSIENT TIME FUNCTION

A function that exists only for a limited time and is zero at all other times is called a *transient time function.* Such functions are not periodic. Figure 1.4-1 shows a typical pressure variation of a sonic boom that is a transient time function. The force of impact during a collision of two bodies is another example.

*J. W. Cooley and J. W. Tukey, "An Algorithm for the Machine Calculation of Complex Fourier Series," *Mathematics of Computation* 19; 90 (April 1965), pp. 297–301.

See also—"Special Issue on Fast Fourier Transform," *IEEE Trans. on Audio & Electroacoustics*, Vol. AU-15, No. 2 (1967).

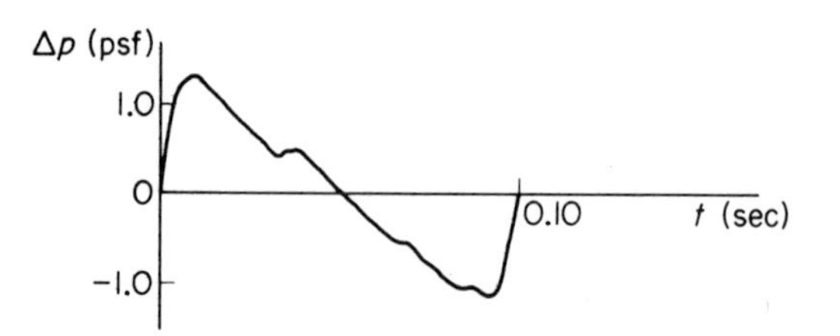

Figure 1.4-1. The sonic boom (N-Wave) is a transient time function.

The response of a mechanical system to an impulse or shock is generally referred to as a *transient response*. Due to the presence of damping, such vibrations will die down after the excitation is over.

Since transient waves are not periodic, the method of Fourier series is *not* applicable. However, nonperiodic functions can be analyzed for their frequency content by the method of Fourier Transforms (see Chapter 10). In contrast to the discrete frequency spectrum of the periodic function, the frequency spectrum of a transient time function is continuous.

1.5 RANDOM TIME FUNCTION

The types of functions we have considered up to now can be classified as deterministic, i.e., mathematical expressions can be written which will determine their instantaneous values at any time t. There are, however, a number of physical phenomena that result in nondeterministic data where future instantaneous values cannot be predicted in a deterministic sense. As examples, we can mention the output of a noise generator, the heights of waves in a choppy sea, ground motion during an earthquake, and pressure gusts encountered by an airplane in flight. These phenomena all have one thing in common: the unpredictability of their instantaneous value at any future time. Nondeterministic data of this type are referred to as *random time functions*.

A sample of a typical random time function is shown in Fig. 1.5-1. In spite of the irregular character of the function, certain averaging proce-

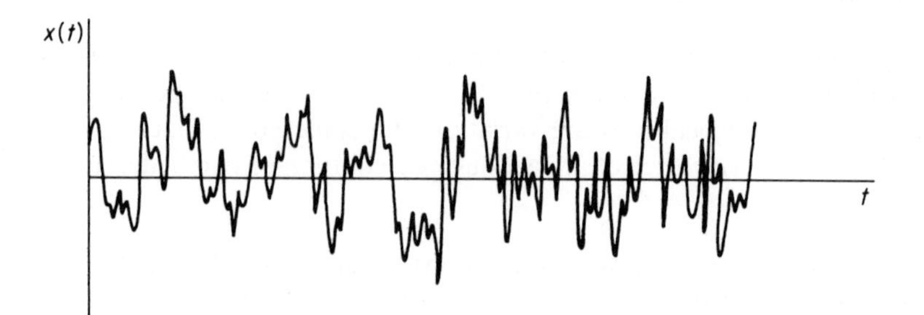

Figure 1.5-1. A record of random time function.

dures can be applied to such data to establish gross characteristics useful in engineering design. These are discussed in detail in Chapter 10. Briefly, it can be mentioned that, as in the periodic and transient vibrations, the concepts of amplitude and their frequency distribution are of paramount importance. These quantities for the random vibration are characterized by statistically evaluated average quantities such as the *root mean square* and the *mean square spectral density*.

1.6 PROPERTIES OF OSCILLATORY MOTION

Certain properties of oscillatory motion are of interest in the measurement of vibration. The simplest of these are the *peak value* and the *average value*.

The peak value will generally indicate the maximum stress which the vibrating part is undergoing. It also places a limitation on the "rattle space" requirement.

The average value indicates a steady or static value somewhat like the DC level of an electrical current. It can be found by the time integral

$$\bar{x} = \lim_{T \to \infty} \frac{1}{T} \int_0^T x(t)\, dt \tag{1.6-1}$$

For example, the average value for a complete cycle of a sine wave, $A \sin t$, is zero; whereas its average value for a half cycle is

$$\bar{x} = \frac{A}{\pi} \int_0^{\pi} \sin t\, dt = \frac{2A}{\pi}$$

It is evident that this is also the average value of the rectified sine wave shown in Fig. 1.6-1.

The square of the displacement generally is associated with the energy of the vibration for which the mean square value is a measure. The *mean square value* of a time function $x(t)$ is found from the average of the squared values, integrated over some time interval T:

$$\overline{x^2} = \lim_{T \to \infty} \frac{1}{T} \int_0^T x^2(t)\, dt$$

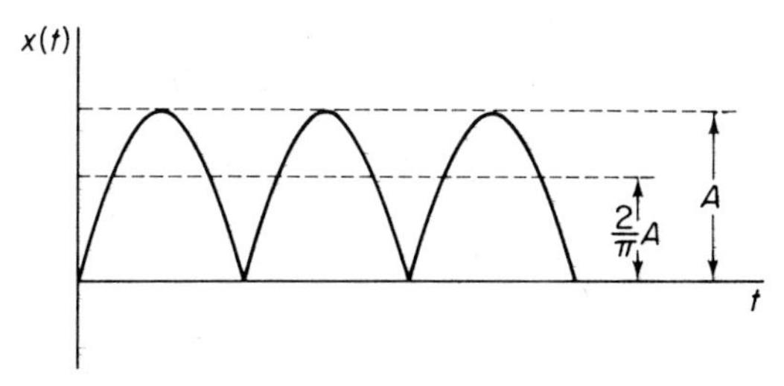

Figure 1.6-1. Average value of a rectified sine wave.

For example, if $x(t) = A \sin \omega t$, its mean square value is

$$\overline{x^2} = \lim_{T\to\infty} \frac{A^2}{T} \int_0^T \frac{1}{2}(1 - \cos 2\omega t)\, dt = \frac{1}{2} A^2$$

The *root mean square* (rms) value is the square root of the mean square value. From the previous example, the rms of the sine wave of amplitude A is $A/\sqrt{2}$.

Frequency Spectrum. The frequency content of an oscillatory motion is of importance in characterizing vibration. For a single sine wave, the frequency content is represented by a line equal in length to its amplitude, drawn at the frequency of its motion.

For a periodic motion, the frequency spectrum is a series of lines that occur at integral multiples of the fundamental frequency as defined by its Fourier series. The phase of each component with respect to the fundamental can also be presented so that its complete description may appear as in Fig. 1.6-2.

Transient motion, although limited in time, can be viewed as periodic motion of infinite period by including the zero value regions to infinity. With $\tau = 2\pi/\omega_1 \to \infty$, or $\omega_1 \to 0$, the spectral lines are crowded together to approach a continuous curve.

Random time functions are not periodic, and their frequency spectra are determined by the Fourier integral rather than by the Fourier series. This subject is treated in Chapter 10; for the present, we will merely state that its spectrum is a presentation of its mean square density plotted against frequency, as shown in Fig. 1.6-3. Such curves are continuous and may be

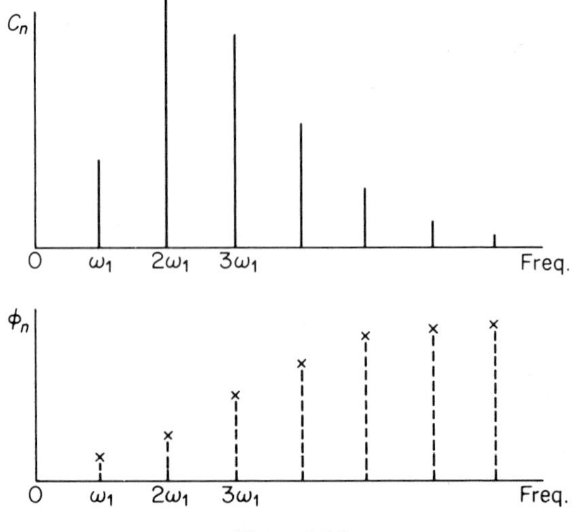

Figure 1.6-2

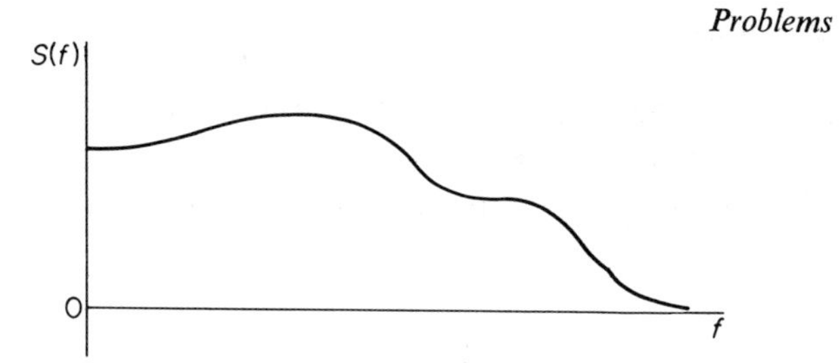

Figure 1.6-3. Frequency spectrum for a random time function.

determined by electronic instruments designed for this specific purpose. In general, the phase of a random time function is of no interest and is disregarded.

PROBLEMS

1-1 A harmonic motion has an amplitude of 0.20 inch and a period of 0.15 sec. Determine the maximum velocity and acceleration.

1-2 An accelerometer indicates that a structure is vibrating harmonically at 82 cps with a maximum acceleration of 50 g. Determine the amplitude of vibration.

1-3 A harmonic motion has a frequency of 10 cps and its maximum velocity is 180 in./sec. Determine its amplitude, its period, and its maximum acceleration.

1-4 Find the sum of two harmonic motions of equal amplitude but of slightly different frequencies. Discuss the beating phenomena that results from this sum.

1-5 Express the complex number $4 + 3i$ in the exponential form $Ae^{i\theta}$.

1-6 Add two complex numbers $(2 + 3i)$ and $(4 - i)$ expressing the result as $A \angle \theta$.

1-7 Show that the multiplication of a phasor by i rotates it by 90°.

1-8 Determine the sum of two phasors $5e^{i\pi/6}$ and $4e^{i\pi/3}$ and find the angle between the resultant and the first phasor.

1-9 Determine the Fourier series for the rectangular wave shown in Fig. P1-9.

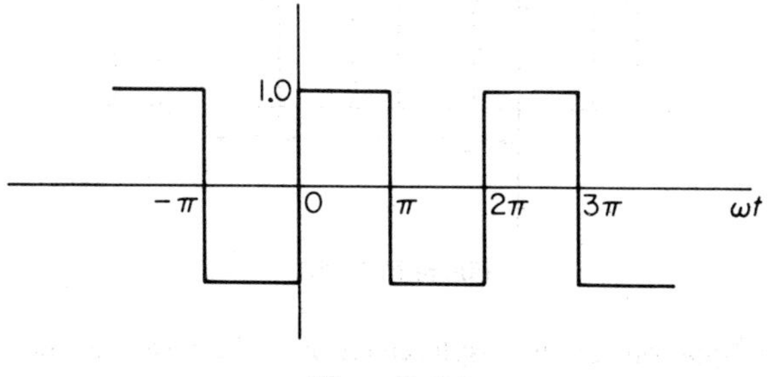

Figure P. 1-9.

1-10 If the origin of the square wave of Prob. 1-9 is shifted to the right by $\pi/2$, determine the Fourier series.

1-11 Determine the Fourier series for the triangular wave shown in Fig. P1-11.

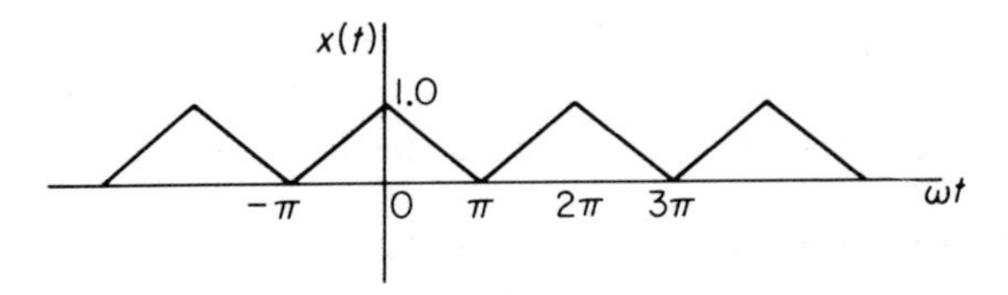

Figure P. 1-11.

1-12 Determine the Fourier series for the saw tooth curve shown in Fig. P1-12.

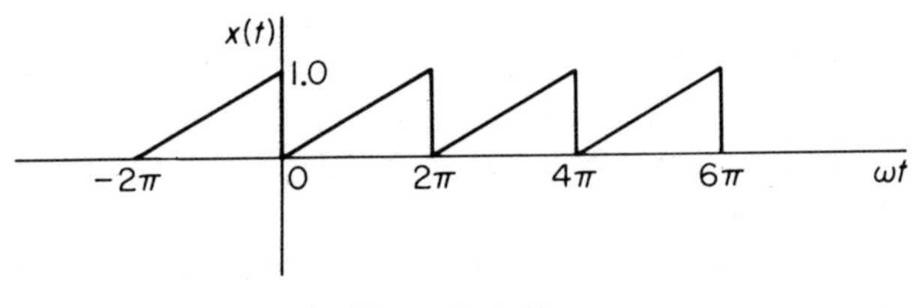

Figure P. 1-12.

1-13 Determine the *rms* value of a wave consisting of the positive portions of a sine wave.

1-14 Determine the mean square value of the saw tooth wave of Prob. 1-12. Do this two ways, from the squared curve and from the Fourier series.

1-15 Plot the frequency spectrum for the triangular wave of Prob. 1-11.

1-16 Determine the Fourier series and the frequency spectrum of a series of rectangular pulses shown in Fig. P1-16.

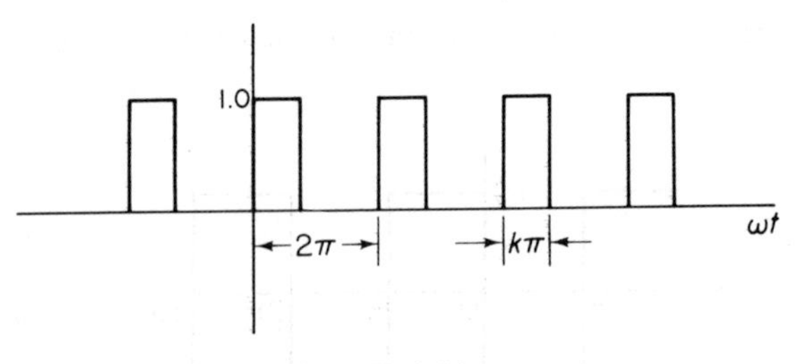

Figure P. 1-16.

1-17 Write the equation for the displacement s of the piston in the crank-piston mechanism shown in Fig. P1-17, and determine the harmonic components and their relative magnitudes.

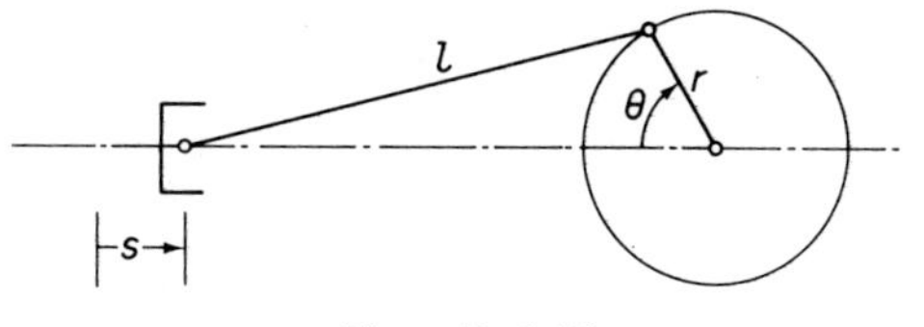

Figure P. 1-17.

2

Free Vibration

2.1 FORCE SUMMATION METHOD

Any system possessing mass and elasticity is capable of vibration. The simplest oscillatory system consists of a mass and a spring as shown in Fig. 2.1-1. The spring supporting the mass is assumed to be negligible in weight with a stiffness k lb per unit deflection. The system possesses one degree of freedom since its motion is described by a single coordinate x.

When placed into motion, oscillation will take place at the natural frequency f_n, which is a property of the system. We now examine some of the basic concepts associated with the free vibration of systems with one degree of freedom.

Newton's second law is the first basis for examining the motion of the system. As shown in Fig. 2.1-1 the deformation of the spring in the static equilibrium position is Δ, and the spring force $k\Delta$ is equal to the gravitational force w acting on the mass m:

$$k\Delta = w \tag{2.1-1}$$

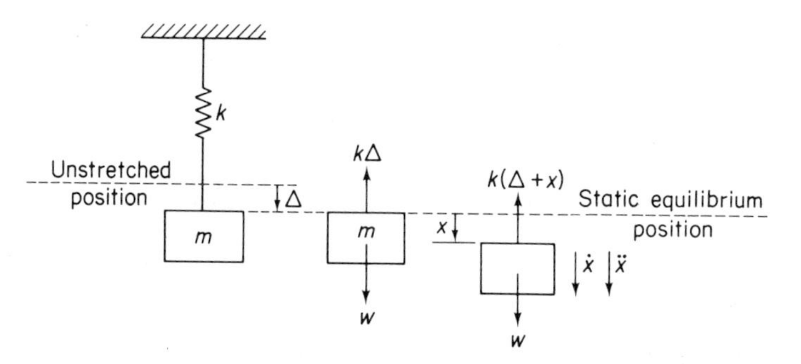

Figure 2.1-1. Spring-mass system and free-body diagram.

Measuring the displacement x from the static equilibrium position, the forces acting on m are $k(\Delta + x)$ and w. With x chosen to be positive in the downward direction, all quantities—force, velocity, and acceleration—are also positive in the downward direction.

We now apply Newton's second law of motion to the mass m

$$m\ddot{x} = \Sigma F = w - k(\Delta + x)$$

and since $k\Delta = w$, we obtain

$$m\ddot{x} = -kx \tag{2.1-2}$$

It is evident that the choice of the static equilibrium position as reference for x has eliminated the weight w and the static spring force $k\Delta$ from the equation of motion, and the resultant force on m is simply the spring force due to the displacement x.

Defining the circular frequency ω_n by the equation

$$\omega_n^2 = \frac{k}{m} \tag{2.1-3}$$

Eq. (2.1-2) may be written as

$$\ddot{x} + \omega_n^2 x = 0 \tag{2.1-4}$$

and we conclude by comparison with Eq. (1.2-8) that the motion is harmonic. Equation (2.1-4), a homogeneous second order linear differential equation, has the following general solution

$$x = A \sin \omega_n t + B \cos \omega_n t \tag{2.1-5}$$

where A and B are the two necessary constants. These constants are evaluated from initial conditions $x(0)$ and $\dot{x}(0)$, and Eq. (2.1-5) can be shown to reduce to

$$x = \frac{\dot{x}(0)}{\omega_n} \sin \omega_n t + x(0) \cos \omega_n t \tag{2.1-6}$$

The natural period of the oscillation is established from $\omega_n \tau = 2\pi$, or

$$\tau = 2\pi \sqrt{\frac{m}{k}} \tag{2.1-7}$$

and the natural frequency is

$$f_n = \frac{1}{\tau} = \frac{1}{2\pi} \sqrt{\frac{k}{m}} \tag{2.1-8}$$

These quantities may be expressed in terms of the statical deflection Δ by observing Eq. (2.1-1), $k\Delta = mg$. Using $g = 386$ in./sec² and Δ in inches, the natural frequency in terms of Δ becomes

$$\begin{aligned} f_n &= \frac{1}{2\pi} \sqrt{\frac{g}{\Delta''}} = \frac{3.127}{\sqrt{\Delta''}} \text{ cps (Hertz)} \\ &= \frac{187.6}{\sqrt{\Delta''}} \text{ c.p.m.} \end{aligned} \tag{2.1-9}$$

Thus the natural frequency of a single degree of freedom system is uniquely determined by the statical deflection Δ. A logarithmic plot of Eq. (2.1-9) is shown in Fig. 2.1-2.

Although oscillatory systems may differ in appearance, the discussion in this section applies to all single degree of freedom systems undergoing undamped free vibration. In some cases the oscillations may be rotational,

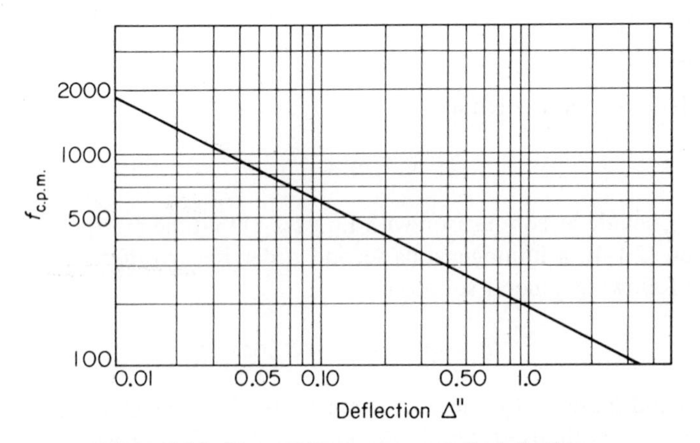

Figure 2.1-2. Natural frequency vs. static deflection.

as with the torsional pendulum, in which case Newton's second law is replaced by its rotational counterpart

$$J\ddot{\theta} = \Sigma M \tag{2.1-10}$$

where M is the moment, J the mass moment of inertia, and $\ddot{\theta}$ the angular acceleration, all referenced to a fixed inertial axis of rotation. The above equation is also valid with respect to the center of mass axis which may be in motion.

2.2 THE ENERGY METHOD

In a conservative system the total energy is constant, and the differential equation of motion can be established by the principle of conservation of energy. For the free vibration of an undamped system, the energy is partly kinetic and partly potential. The kinetic energy T is stored in the mass by virtue of its velocity, whereas the potential energy U is stored in the form of strain energy in elastic deformation or work done in a force field such as gravity. The total energy being constant, its rate of change is zero as illustrated by the following equations

$$T + U = \text{constant} \tag{2.2-1}$$

$$\frac{d}{dt}(T + U) = 0 \tag{2.2-2}$$

If our interest is only in the natural frequency of the system, it can be determined by the following considerations. From the principle of conservation of energy we can write

$$T_1 + U_1 = T_2 + U_2 \tag{2.2-3}$$

where $_1$ and $_2$ represent two instances of time. Let $_1$ be the time when the mass is passing through its static equilibrium position and choose $U_1 = 0$ as reference for the potential energy. Let $_2$ be the time corresponding to the maximum displacement of the mass. At this position, the velocity of the mass is zero, and hence $T_2 = 0$. We then have

$$T_1 + 0 = 0 + U_2 \tag{2.2-4}$$

However, if the system is undergoing harmonic motion, then T_1 and U_2 are maximum values, and hence

$$T_{\max} = U_{\max} \tag{2.2-5}$$

The above equation leads directly to the natural frequency.

torsional stiffness K in. lb/rad
mass moment of inertia J lb in. $\sec^2$

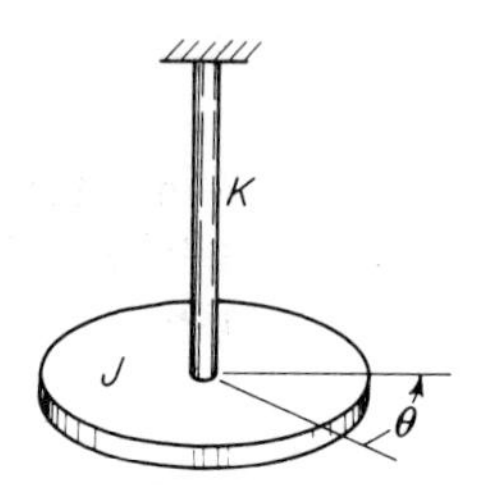

Figure 2.2-1. Torsional pendulum.

Example 2.2-1

Determine the natural frequency of the torsional pendulum shown in Fig. 2.2-1.

Solution: Assume the oscillatory motion to be harmonic and expressible by the equation

$$\theta = A \sin \omega_n t$$

The maximum kinetic and potential energies are

$$T_{\max} = \tfrac{1}{2} J \dot{\theta}^2_{\max} = \tfrac{1}{2} J \omega_n^2 A^2$$

and

$$U_{\max} = \tfrac{1}{2} K \theta^2_{\max} = \tfrac{1}{2} K A^2$$

Equating the two energies, we arrive at the expression for its natural frequency, which is

$$\omega_n = \sqrt{\frac{K}{J}}.$$

Example 2.2-2

A cylinder of weight w and radius r rolls without slipping on a cylindrical surface of radius R, as shown in Fig. 2.2-2. Determine its differential equation of motion for small oscillations about the lowest point. For no slipping $r\phi = R\theta$.

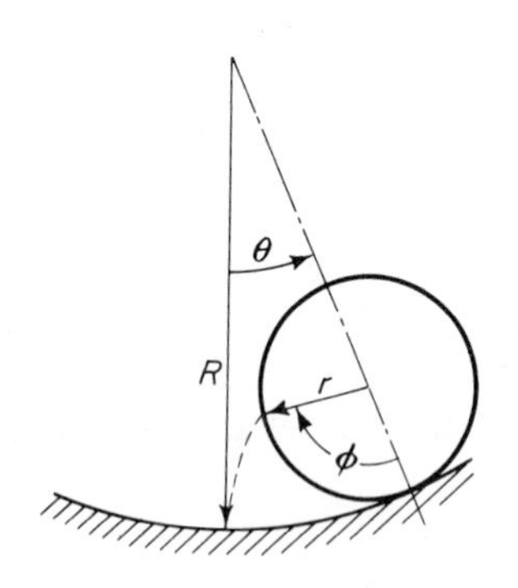

Figure 2.2-2.

Solution: In determining the kinetic energy of the cylinder, it must be noted that both translation and rotation take place. The translational velocity of the center of the cylinder is $(R - r)\dot{\theta}$, whereas the rotational velocity is $(\dot{\phi} - \dot{\theta}) = (R/r - 1)\dot{\theta}$, since $\dot{\phi} = (R/r)\dot{\theta}$ for no slipping. The kinetic energy may now be written as

$$T = \frac{1}{2}\frac{w}{g}[(R - r)\dot{\theta}]^2 + \frac{1}{2}\frac{w}{g}\frac{r^2}{2}\left[\left(\frac{R}{r} - 1\right)\dot{\theta}\right]^2$$

$$= \frac{3}{4}\frac{w}{g}(R - r)^2\dot{\theta}^2$$

where $(w/g)(r^2/2)$ is the moment of inertia of the cylinder about its mass center.

The potential energy referred to its lowest position is

$$U = w(R - r)(1 - \cos\theta)$$

which is equal to the negative of the work done by the gravity force in lifting the cylinder through the vertical height $(R - r)(1 - \cos\theta)$.

Substituting into Eq. (2.2-2)

$$\left[\frac{3}{2}\frac{w}{g}(R - r)^2\ddot{\theta} + w(R - r)\sin\theta\right]\dot{\theta} = 0,$$

and letting $\sin\theta = \theta$ for small angles, we obtain the familiar equation for harmonic motion

$$\ddot{\theta} + \frac{2g}{3(R - r)}\theta = 0$$

By inspection, the circular frequency of oscillation is

$$\omega_n = \sqrt{\frac{2g}{3(R - r)}}.$$

2.3 EFFECTIVE MASS

So far in the calculation of the natural frequency, we have assumed the spring to be without mass. Often the spring and other moving elements may represent a sizable fraction of the total mass of the system, and their neglect may result in natural frequencies which are too high.

To obtain a better estimate of the natural frequency, we can compute the additional kinetic energy of the moving elements which were previously neglected. This, of course, requires a guess as to the motion of the distributed elements. The integrated result of the additional kinetic energy can then be expressed in terms of the velocity $\dot{x}$ of the lumped mass in the form

$$T_{\text{add}} = \tfrac{1}{2}m_{\text{eff}}\dot{x}^2$$

where m_{eff} is the effective mass to be added to the lumped mass.

Example 2.3-1

Determine the effect of the mass of the spring on the natural frequency of the system shown in Fig. 2.3-1.

Solution: With $\dot{x}$ equal to the velocity of the lumped mass m, we will assume the velocity of a spring element located a distance y from the fixed

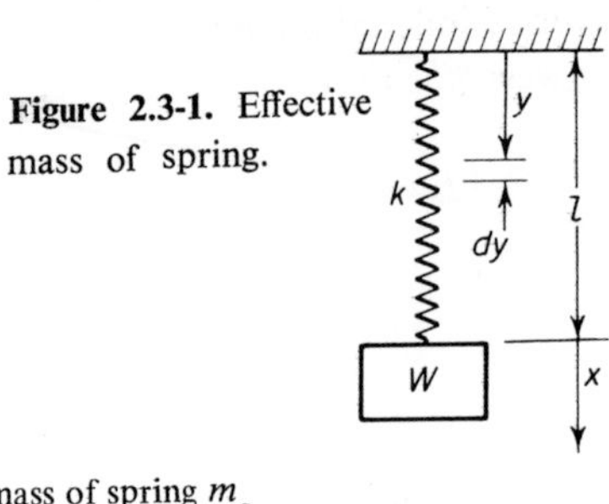

Figure 2.3-1. Effective mass of spring.

mass of spring m_s

mass of spring element $m_s \frac{dy}{l}$

velocity of spring element $\dot{x}\frac{y}{l}$

end to vary linearly with y as follows

$$\dot{x}\frac{y}{l}$$

The kinetic energy of the spring may then be integrated to

$$T_{\text{add}} = \frac{1}{2}\int_0^l \left(\dot{x}\frac{y}{l}\right)^2 \frac{m_s}{l}\,dy = \frac{1}{2}\frac{m_s}{3}\dot{x}^2$$

and the effective mass is found to be one-third the mass of the spring. Adding this to the lumped mass, the revised natural frequency is

$$\omega_n = \sqrt{\frac{k}{m + \frac{1}{3}m_s}}$$

Example 2.3-2

Oscillatory systems are often composed of levers, gears, and other linkages that seemingly complicate the analysis. The engine valve system of Fig. 2.3-2 is an example of such a system. The reduction

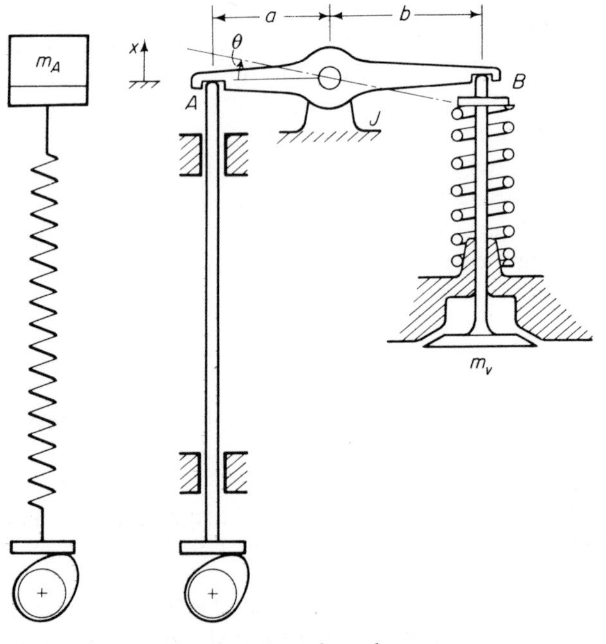

Figure 2.3-2. Engine valve system.

of such a system to a simpler equivalent system is generally desirable.

The rocker arm of moment of inertia J, the valve of mass m_v, and the spring of mass m_s can be reduced to a single mass at A by writing the kinetic energy equation as follows

$$T = \frac{1}{2}J\dot{\theta}^2 + \frac{1}{2}m_v(b\dot{\theta})^2 + \frac{1}{2}\left(\frac{m_s}{3}\right)(b\dot{\theta})^2$$
$$= \frac{1}{2}\left(J + m_v b^2 + \frac{1}{3}m_s b^2\right)\dot{\theta}^2$$

Recognizing that the velocity at A is $\dot{x} = a\dot{\theta}$, the above equation becomes

$$T = \frac{1}{2}\left(\frac{J + m_v b^2 + \frac{1}{3}m_s b^2}{a^2}\right)\dot{x}^2$$

Thus the effective mass at A is

$$m_A = \left(\frac{J + m_v b^2 + \frac{1}{3}m_s b^2}{a^2}\right)$$

If the push rod is now reduced to a spring and an additional mass at the end A, the entire system is reduced to a single spring and a mass as shown in Fig. 2.3-2.

2.4 DAMPED FREE VIBRATION

When a linear system of one degree of freedom is excited, its response will depend on the type of excitation and the damping which is present. The equation of motion will in general be of the form

$$m\ddot{x} + F_d + kx = F(t) \tag{2.4-1}$$

where $F(t)$ is the excitation and F_d the damping force. Although the actual description of the damping force is difficult, ideal damping models can be assumed that will often result in satisfactory prediction of the response. Of these models, the viscous damping force, proportional to the velocity, leads to the simplest mathematical treatment.

Viscous damping force is expressed by the equation

$$F_d = c\dot{x} \tag{2.4-2}$$

where c is a constant of proportionality. Symbolically it is designated by a

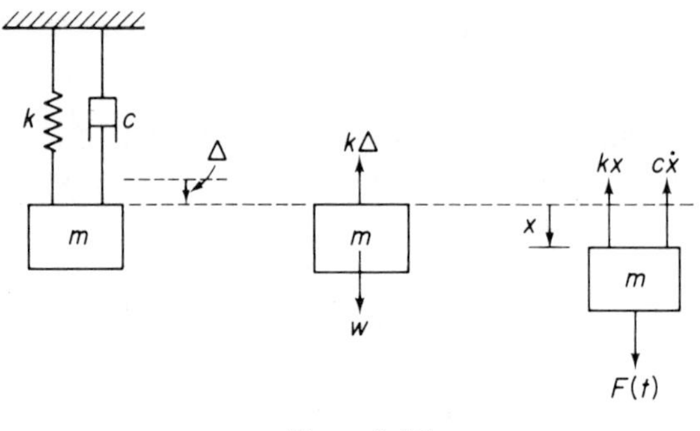

Figure 2.4-1.

dashpot as shown in Fig. 2.4-1. From the free-body diagram the equation of motion is seen to be

$$m\ddot{x} + c\dot{x} + kx = F(t) \tag{2.4-3}$$

The solution of the above equation has two parts. If $F(t) = 0$, we have the homogeneous differential equation whose solution corresponds physically to that of *free-damped vibration.* With $F(t) \neq 0$, we obtain the particular solution that is due to the excitation irrespective of the homogeneous solution. We will first examine the homogeneous equation that will give us some understanding of the role of damping.

With the homogeneous equation

$$m\ddot{x} + c\dot{x} + kx = 0 \tag{2.4-4}$$

the traditional approach is to assume a solution of the form

$$x = e^{st} \tag{2.4-5}$$

where s is a constant. Upon substitution into the differential equation, we obtain

$$(ms^2 + cs + k)e^{st} = 0$$

which is satisfied for all values of t when

$$s^2 + \frac{c}{m}s + \frac{k}{m} = 0 \tag{2.4-6}$$

Equation (2.4-6), which is known as the *characteristic equation*, has two roots

$$s_{1,2} = -\frac{c}{2m} \pm \sqrt{\left(\frac{c}{2m}\right)^2 - \frac{k}{m}} \tag{2.4-7}$$

Hence, the general solution is given by the equation

$$x = Ae^{s_1 t} + Be^{s_2 t} \tag{2.4-8}$$

where A and B are constants to be evaluated from the initial conditions $x(0)$ and $\dot{x}(0)$.

Equation (2.4-7) substituted into (2.4-8) gives

$$x = e^{-(c/2m)t}\left(Ae^{\sqrt{(c/2m)^2 - k/m}\,t} + Be^{-\sqrt{(c/2m)^2 - k/m}\,t}\right) \tag{2.4-9}$$

The first term $e^{-(c/2m)t}$ is simply an exponentially decaying function of time. The behavior of the terms in the parenthesis, however, depends on whether the numerical value within the radical is positive, zero, or negative.

When the damping term $(c/2m)^2$ is larger than k/m, the exponents in the above equation are real numbers and no oscillations are possible. We refer to this case as *overdamped.*

When the damping term $(c/2m)^2$ is less than k/m, the exponent becomes an imaginary number, $\pm i\sqrt{k/m - (c/2m)^2}\,t$. Since

$$e^{\pm i\sqrt{k/m-(c/2m)^2}\,t} = \cos\sqrt{\frac{k}{m} - \left(\frac{c}{2m}\right)^2}\,t \pm i \sin\sqrt{\frac{k}{m} - \left(\frac{c}{2m}\right)^2}\,t$$

the terms of Eq. (2.4-9) within the parenthesis are oscillatory. We refer to this case as *underdamped.*

As a limiting case between the oscillatory and nonoscillatory motion, we define *critical damping* as the value of c which reduces the radical to zero.

It is now advisable to examine these three cases in detail, and in terms of quantities used in practice. We begin with the critical damping.

Critical Damping. For critical damping c_c the radical in Eq. (2.4-9) is zero.

$$\left(\frac{c_c}{2m}\right)^2 = \frac{k}{m} = \omega_n^2$$

or

$$c_c = 2\sqrt{km} = 2m\omega_n \tag{2.4-10}$$

It is convenient to express the value of any damping in terms of the critical damping by the nondimensional ratio

$$\zeta = \frac{c}{c_c} \tag{2.4-11}$$

which is called the *damping ratio.* We now express the roots of Eq. (2.4-7) in terms of ζ by noting that

$$\frac{c}{2m} = \zeta\frac{c_c}{2m} = \zeta\omega_n$$

Equation (2.4-7) then becomes

$$s_{1,2} = (-\zeta \pm \sqrt{\zeta^2 - 1})\omega_n \tag{2.4-12}$$

and the three cases of damping previously discussed now depend on whether ζ is greater than, less than, or equal to unity.

Figure 2.4-2 shows Eq. (2.4-12) plotted in a complex plane with ζ along the horizontal axis. If $\zeta = 0$, Eq. (2.4-12) reduces to $s_{1,2}/\omega_n = \pm i$ so that the roots on the imaginary axis correspond to the undamped case. For $0 \leq \zeta \leq 1$, Eq. (2.4-12) can be rewritten as

$$\frac{s_{1,2}}{\omega_n} = -\zeta \pm i\sqrt{1 - \zeta^2}.$$

The roots s_1 and s_2 are then conjugate complex points on a circular arc converging at the point $s_{1,2}/\omega_n = -1.0$. As ζ increases beyond unity, the roots separate along the horizontal axis and remain real numbers. With this diagram in mind, we are now ready to examine the solution given by Eq. (2.4-9).

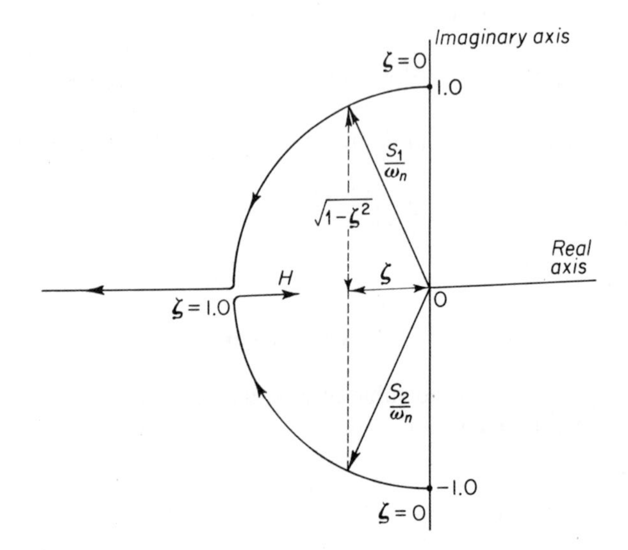

Figure 2.4-2.

Oscillatory Motion. [$\zeta < 1.0$ (Underdamped Case).] Substituting Eq. (2.4-12) into (2.4-8), the general solution becomes

$$x = e^{-\zeta\omega_n t}(Ae^{i\sqrt{1-\zeta^2}\,\omega_n t} + Be^{-i\sqrt{1-\zeta^2}\,\omega_n t}) \qquad (2.4\text{-}13)$$

The above equation can also be written in either of the following two forms

$$x = Xe^{-\zeta\omega_n t}\sin(\sqrt{1-\zeta^2}\,\omega_n t + \phi) \qquad (2.4\text{-}14)$$

$$= e^{-\zeta\omega_n t}(C_1 \sin\sqrt{1-\zeta^2}\,\omega_n t + C_2 \cos\sqrt{1-\zeta^2}\,\omega_n t) \qquad (2.4\text{-}15)$$

where the arbitrary constants X, ϕ, or C_1, C_2 are determined from initial conditions. With initial conditions $x(0)$ and $\dot{x}(0)$, Eq. (2.4-15) can be shown to reduce to

$$x = e^{-\zeta\omega_n t}\left(\frac{\dot{x}(0) + \zeta\omega_n x(0)}{\omega_n\sqrt{1-\zeta^2}} \sin\sqrt{1-\zeta^2}\,\omega_n t + x(0)\cos\sqrt{1-\zeta^2}\,\omega_n t\right) \qquad (2.4\text{-}16)$$

The equation indicates that the *frequency of damped oscillation* is equal to

$$\omega_d = \frac{2\pi}{\tau_d} = \omega_n\sqrt{1-\zeta^2} \qquad (2.4\text{-}17)$$

Figure 2.4-3 shows the general nature of the oscillatory motion.

Nonoscillatory Motion. [$\zeta > 1.0$ (Overdamped Case).] As ζ exceeds unity, the two roots remain on the real axis of Fig. 2.4-2 and separate, one increasing and the other decreasing. The general solution then becomes

$$x = Ae^{(-\zeta+\sqrt{\zeta^2-1})\omega_n t} + Be^{(-\zeta-\sqrt{\zeta^2-1})\omega_n t} \qquad (2.4\text{-}18)$$

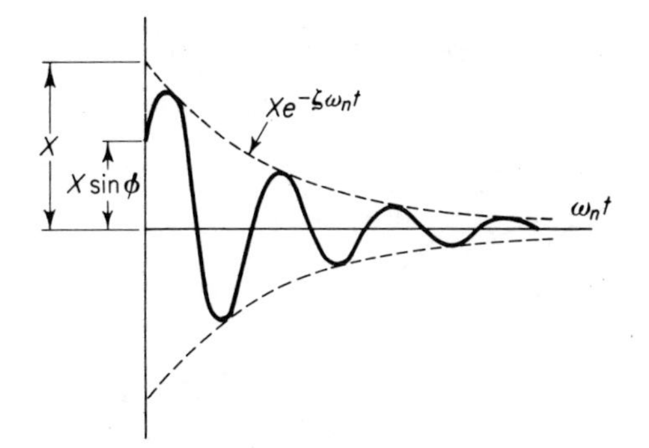

Figure 2.4-3. Damped oscillation $\zeta < 1.0$.

where

$$A = \frac{\dot{x}(0) + (\zeta + \sqrt{\zeta^2 - 1})\omega_n x(0)}{2\omega_n\sqrt{\zeta^2 - 1}}$$

and

$$B = \frac{-\dot{x}(0) - (\zeta - \sqrt{\zeta^2 - 1})\omega_n x(0)}{2\omega_n\sqrt{\zeta^2 - 1}}$$

The motion is an exponentially decreasing function of time as shown in Fig. 2.4-4, and is referred to as *aperiodic*.

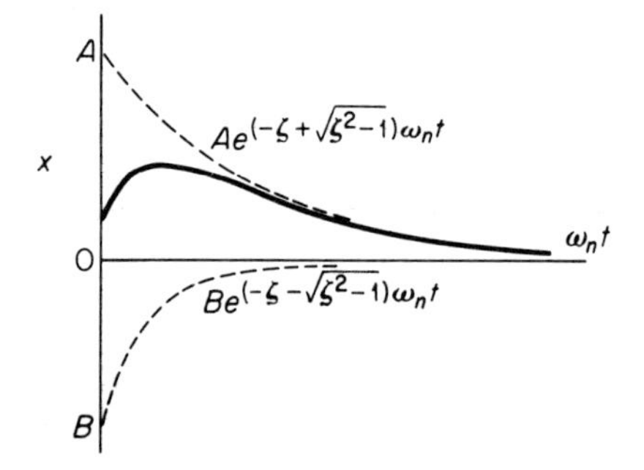

Figure 2.4-4. Aperiodic motion $\zeta > 1.0$.

Critically Damped Motion. [$\zeta = 1.0$] For $\zeta = 1$, we obtain a double root $s_1 = s_2 = -\omega_n$, and the two terms of Eq. (2.4-8) combine to form a single term

$$x = (A + B)e^{-\omega_n t} = Ce^{-\omega_n t}$$

which is lacking in the number of constants required to satisfy the two initial conditions. The solution for the initial conditions $x(0)$ and $\dot{x}(0)$ can be found from Eq. (2.4-16) by letting $\zeta \longrightarrow 1$

$$x = e^{-\omega_n t}\{[\dot{x}(0) + \omega_n x(0)]t + x(0)\} \tag{2.4-19}$$

Figure 2.4-5 shows three types of response with initial displacement $x(0)$.

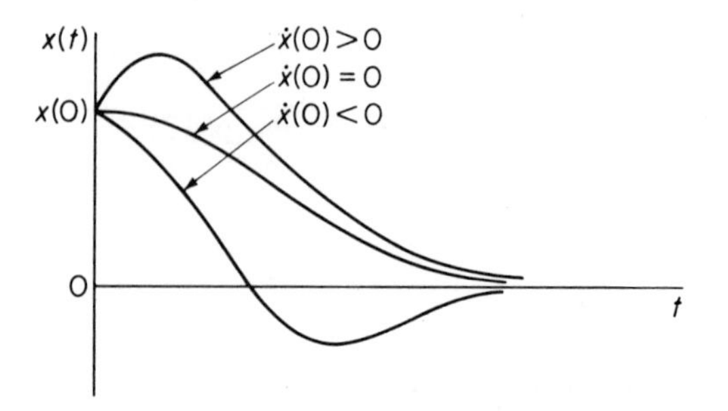

Figure 2.4-5. Critically damped motion $\zeta = 1.0$.

The moving parts of many electrical meters and instruments are critically damped to avoid overshoot and oscillation.

2.5 LOGARITHMIC DECREMENT

A convenient way to determine the amount of damping present in a system is to measure the rate of decay of free oscillations. The larger the damping, the greater will be the rate of decay.

Consider a damped vibration expressed by the general equation (2.4-14)

$$x = Xe^{-\zeta\omega_n t} \sin(\sqrt{1-\zeta^2}\,\omega_n t + \phi)$$

which is shown graphically in Fig. 2.5-1. We introduce here a term called *logarithmic decrement* which is defined as the natural logarithm of the ratio of any two successive amplitudes. The expression for the logarithmic decrement then becomes

$$\delta = \ln\frac{x_1}{x_2} = \ln\frac{e^{-\zeta\omega_n t_1}\sin(\sqrt{1-\zeta^2}\,\omega_n t_1 + \phi)}{e^{-\zeta\omega_n(t_1+\tau_d)}\sin(\sqrt{1-\zeta^2}\,\omega_n(t_1+\tau_d)+\phi)} \tag{2.5-1}$$

and since the values of the sines are equal when the time is increased by the damped period τ_d, the above relation reduces to

$$\delta = \ln\frac{e^{-\zeta\omega_n t_1}}{e^{-\zeta\omega_n(t_1+\tau_d)}} = \ln e^{\zeta\omega_n\tau_d} = \zeta\omega_n\tau_d \tag{2.5-2}$$

Substituting for the damped period, $\tau_d = 2\pi/\omega_n\sqrt{1-\zeta^2}$, the expression for the logarithmic decrement becomes

$$\delta = \frac{2\pi\zeta}{\sqrt{1-\zeta^2}} \tag{2.5-3}$$

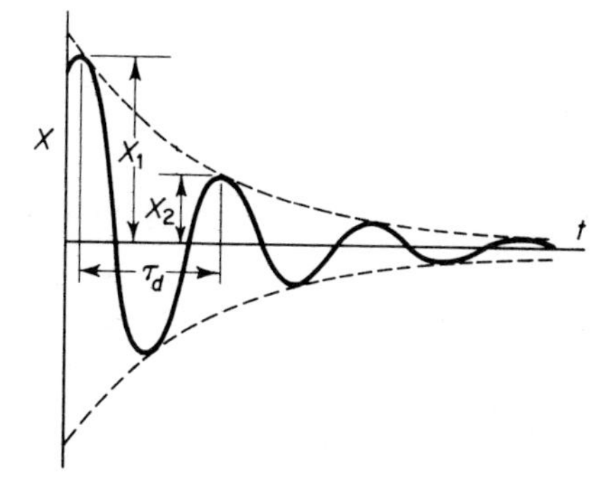

Figure 2.5-1. Rate of decay of oscillation measured by the logarithmic decrement.

which is an exact equation.

When ζ is small, $\sqrt{1-\zeta^2} \cong 1$, and an approximate equation

$$\delta \cong 2\pi\zeta \tag{2.5-4}$$

is obtained. Figure 2.5-2 shows a plot of the exact and approximate values of δ as a function of ζ.

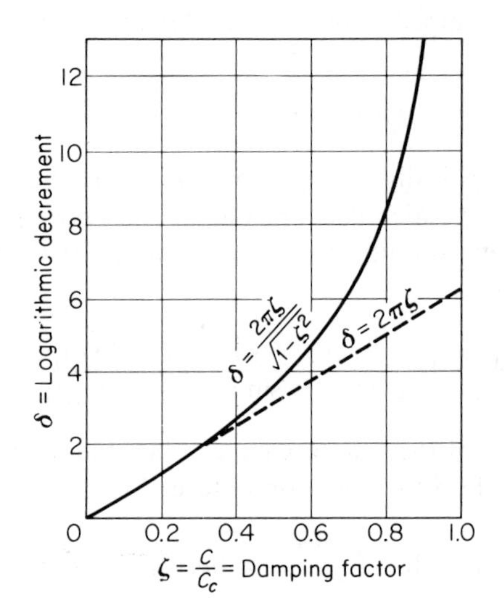

Figure 2.5-2. Logarithmic decrement as function of ζ.

Example 2.5-1

The following data are given for a vibrating system with viscous damping: $w = 10$ lb, $k = 30$ lb/in., and $c = 0.12$ lb/in. per sec. Determine the logarithmic decrement and the ratio of any two successive amplitudes.

Solution: The undamped natural frequency of the system in radians per second is

$$\omega_n = \sqrt{\frac{k}{m}} = \sqrt{\frac{30 \times 386}{10}} = 34.0 \text{ rad/sec}$$

The critical damping coefficient c_c and damping factor ζ are

$$c_c = 2m\omega_n = 2 \times \tfrac{10}{386} \times 34.0 = 1.76 \text{ lb/in. per sec}$$

$$\zeta = \frac{c}{c_c} = \frac{0.12}{1.76} = 0.0681$$

The logarithmic decrement, from Eq. (2.5-3), is

$$\delta = \frac{2\pi\zeta}{\sqrt{1-\zeta^2}} = \frac{2\pi \times 0.0681}{\sqrt{1-0.0681^2}} = 0.429$$

The amplitude ratio for any two consecutive cycles is

$$\frac{x_1}{x_2} = e^{\delta} = e^{0.429} = 1.54$$

Example 2.5-2

Show that the logarithmic decrement is also given by the equation

$$\delta = \frac{1}{n}\ln\frac{x_0}{x_n}$$

where x_n represents the amplitude after n cycles have elapsed. Plot a curve giving the number of cycles elapsed against ζ for the amplitude to diminish by 50 per cent.

Solution: The amplitude ratio for any two consecutive amplitudes is

$$\frac{x_0}{x_1} = \frac{x_1}{x_2} = \frac{x_2}{x_3} = \cdots = \frac{x_{n-1}}{x_n} = e^{\delta}$$

The ratio x_0/x_n can be written as

$$\frac{x_0}{x_n} = \left(\frac{x_0}{x_1}\right)\left(\frac{x_1}{x_2}\right)\left(\frac{x_2}{x_3}\right)\cdots\left(\frac{x_{n-1}}{x_n}\right) = (e^{\delta})^n = e^{n\delta}$$

from which the required equation is obtained as

$$\delta = \frac{1}{n}\ln\frac{x_0}{x_n}$$

To determine the number of cycles elapsed for 50 per cent reduction in amplitude, we obtain the following relation from the above equation

$$\delta \cong 2\pi\zeta = \frac{1}{n}\ln 2 = \frac{0.693}{n}$$

$$n\zeta = \frac{0.693}{2\pi} = 0.110$$

The last equation is that of a rectangular hyperbola, and is plotted in Fig. 2.5-3.

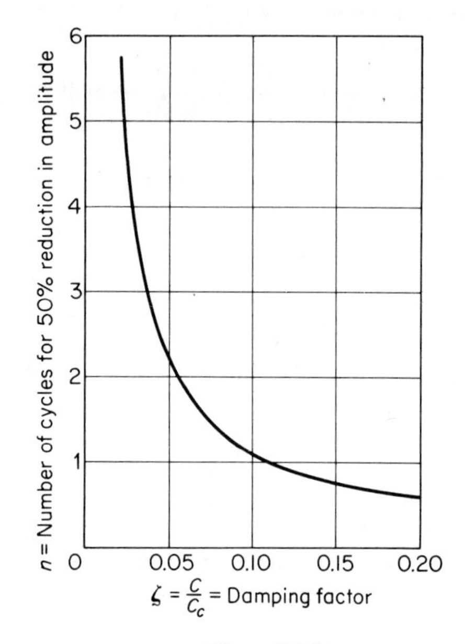

Figure 2.5-3.

Example 2.5-3

For small damping, show that the logarithmic decrement is expressible in terms of the vibrational energy U and the energy dissipated per cycle ΔU.

Solution: Figure 2.5-4 shows a damped vibration with consecutive amplitudes $x_1, x_2, x_3, \ldots$. From the definition of the logarithmic decrement

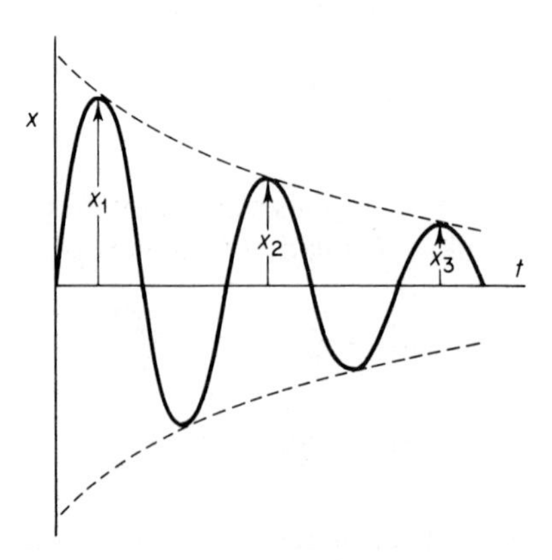

Figure 2.5-4.

$\delta = \ln x_1/x_2$, we can write the ratio of amplitudes in exponential form:

$$\frac{x_2}{x_1} = e^{-\delta} = 1 - \delta + \frac{\delta^2}{2!} - \cdots$$

The vibrational energy of the system is that stored in the spring at maximum displacement, or

$$U_1 = \tfrac{1}{2}kx_1^2, \qquad U_2 = \tfrac{1}{2}kx_2^2$$

The loss of energy divided by the original energy is

$$\frac{U_1 - U_2}{U_1} = 1 - \frac{U_2}{U_1} = 1 - \left(\frac{x_2}{x_1}\right)^2 = 1 - e^{-2\delta} = 2\delta - \frac{(2\delta)^2}{2!} + \cdots$$

Thus for small δ we obtain the relationship

$$\frac{\Delta U}{U} = 2\delta$$

2.6 COULOMB DAMPING

Coulomb damping results from the sliding of two dry surfaces. The damping force is equal to the product of the normal force and the coefficient of friction μ and is assumed to be independent of the velocity, once the motion is initiated. Since the sign of the damping force is always opposite to that of the velocity, the differential equation of motion for each sign is valid only for half cycle intervals.

To determine the decay of amplitude, we resort to the work-energy principle of equating the work done to the change in kinetic energy. Choosing a half cycle starting at the extreme position with velocity equal to zero and the amplitude equal to X_1, the change in the kinetic energy is zero and the work done on m is also zero.

$$\tfrac{1}{2}k(X_1^2 - X_{-1}^2) - F_d(X_1 + X_{-1}) = 0$$

or

$$\tfrac{1}{2}k(X_1 - X_{-1}) = F_d$$

where X_{-1} is the amplitude after the half cycle as shown in Fig. 2.6-1. Repeating this procedure for the next half cycle, a further decrease in amplitude of $2F_d/k$ will be found, so that the decay in amplitude per cycle is a constant and equal to

$$X_1 - X_2 = \frac{4F_d}{k} \tag{2.6-1}$$

The motion will cease, however, when the amplitude becomes less than Δ, at which position the spring force is insufficient to overcome the static

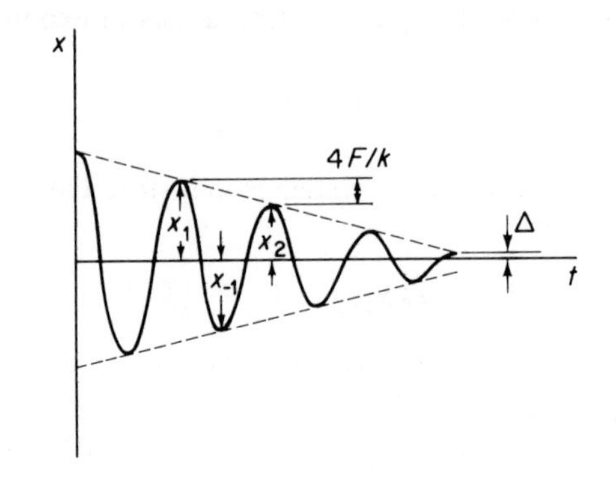

Figure 2.6-1. Free vibration with coulomb damping.

friction force, which is generally greater than the kinetic friction force. It can also be shown that the frequency of oscillation is $\omega_\mu = \sqrt{k/m}$, which is the same as that of the undamped system.

Figure 2.6-1 shows the free vibration of a system with Coulomb damping. It should be noted that the amplitudes decay linearly with time.

2.7 STIFFNESS AND FLEXIBILITY

In the natural frequency calculations of single degree of freedom systems, the mass and stiffness measurements are required. Effective mass may be calculated using any appropriate point in the system as a reference; however, the stiffness for that point must also be determined. Stiffness is defined as the force required to produce a unit displacement in the direction specified. If x is the specified displacement under a force F, the stiffness is determined from the ratio

$$k = \frac{F}{x} \tag{2.7-1}$$

Flexibility is the reciprocal of the stiffness. It is designated by the letter "a" and is defined by the equation

$$a = \frac{1}{k} = \frac{x}{F} \tag{2.7-2}$$

In a later section we will need to determine the stiffness and flexibility in terms of two points i and j of the system. The flexibility a_{ij} is then defined as the deflection at i produced by a unit force at j. The stiffness k_{ij} is the force required at i for a unit deflection at j with all other deflections equal to zero. In terms of these quantities, the k and "a" of Eqs. (2.7-1) and (2.7-2) are k_{ii} and a_{ii}. The table at the end of this section presents stiffnesses for various types of springs.

Example 2.7-1

Determine the spring stiffness for the system of springs shown in Fig. 2.7-1.

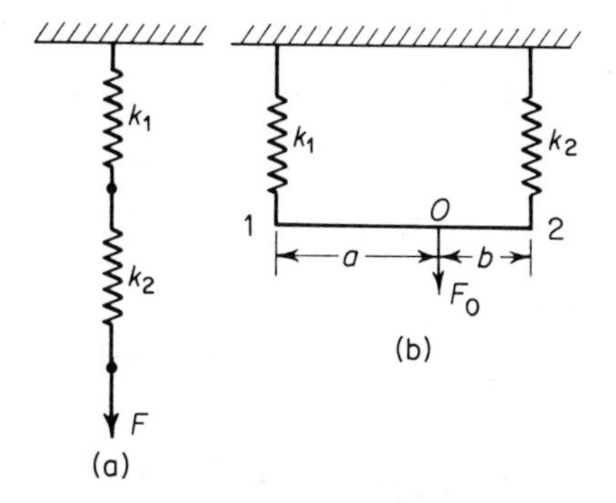

Figure 2.7-1. Stiffness of spring combinations.

Solution: System (a): Applying a force F at the lower end of the second spring, each spring will stretch by an amount F/k_1 and F/k_2, and the total displacement of the lower end becomes $x = F/k_1 + F/k_2$. The stiffness from Eq. (2.7-1) is then

$$k = \frac{F}{\dfrac{F}{k_1} + \dfrac{F}{k_2}} = \frac{k_1 k_2}{k_1 + k_2}$$

System (b): A force F_0 applied at 0 divides into $F_0 b/(a + b)$ and $F_0 a/(a + b)$ respectively. The deflections of 1 and 2 are $F_0 b/(a + b)k_1$ and $F_0 a/(a + b)k_2$, and that of point 0 is

$$x_0 = F_0\left\{\frac{b}{(a + b)k_1} + \frac{a}{(a + b)}\left[\frac{a}{(a + b)k_2} - \frac{b}{(a + b)k_1}\right]\right\}$$
$$= \frac{F_0}{(a + b)^2}\left(\frac{a^2}{k_2} + \frac{b^2}{k_1}\right)$$

The spring stiffness at 0 is then

$$k_0 = \frac{F_0}{x_0} = \frac{(a + b)^2}{\left(\dfrac{a^2}{k_2} + \dfrac{b^2}{k_1}\right)}$$

If $k_1 = k_2 = k$ and $a = b$, the above equation reduces to $k_0 = 2k$.

Table of Spring Stiffness.

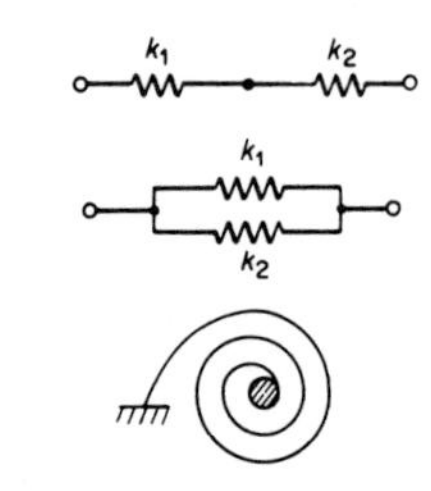

$$k = \frac{1}{1/k_1 + 1/k_2}$$

$$k = k_1 + k_2$$

$$k = \frac{EI}{l}, \quad I = \text{moment of inertia of cross-sectional area}$$

$$l = \text{total length}$$

$$k = \frac{EA}{l} \quad A = \text{cross-sectional area}$$

$$k = \frac{GJ}{l} \quad J = \text{torsion constant of cross section}$$

$$k = \frac{Gd^4}{64nR^3} \quad n = \text{number of turns}$$

$$k = \frac{3EI}{l^3}$$

$$k = \frac{48EI}{l^3}$$

$$k = \frac{192\,EI}{l^3}$$

$$k = \frac{768EI}{7l^3}$$

$$k = \frac{3EIl}{a^2b^2}$$

PROBLEMS

2-1 A 1 lb weight attached to a light spring elongates it 0.31 in. Determine the natural frequency of the system.

2-2 In a spring-mass system k_1, m has a natural frequency of f_1. If a second spring k_2 is added in series with the first spring, the natural frequency is lowered to $\frac{1}{2}f_1$. Determine k_2 in terms of k_1.

2-3 A 10 lb weight, attached to the lower end of a spring whose upper end is fixed, vibrates with a natural period of 0.45 sec. Determine the natural period when a 5 lb weight is attached to the mid-point of the same spring with the upper and lower ends fixed.

2-4 An unknown weight W lb attached to the end of an unknown spring k has a natural frequency of 94 cpm. When a 1 lb weight is added to W, the natural frequency is lowered to 76.7 cpm. Determine the unknown weight W lb and the spring constant k lb/in.

2-5 A weight w_1 hangs from a spring k and is in static equilibrium. A second weight w_2 drops through a height h and sticks to w_1 without rebound, as shown in Fig. P2-5. Determine the subsequent motion.

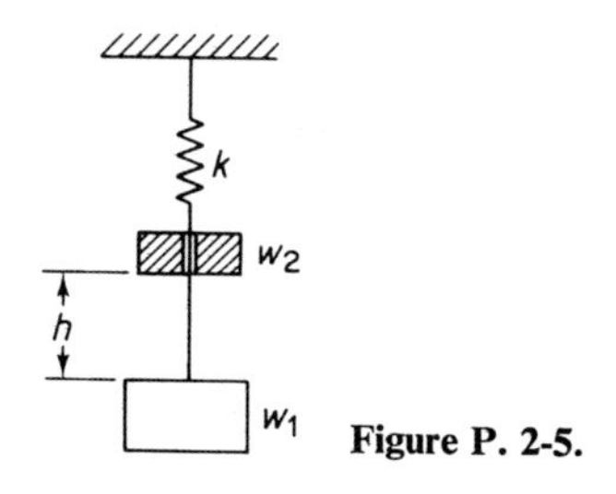

Figure P. 2-5.

2-6 In the torsional pendulum of Fig. 2.2-1, explain how the natural frequency depends on (a) the length of wire, (b) the diameter of wire, (c) the material of the wire, (d) the suspended weight, and (e) the radius of gyration of the suspended weight.

2-7 A flywheel weighing 70 lb was allowed to swing as a pendulum about a knife-edge at the inner side of the rim as shown in Fig. P2-7. If the measured period of oscillation was 1.22 sec, determine the moment of inertia of the flywheel about its geometric axis.

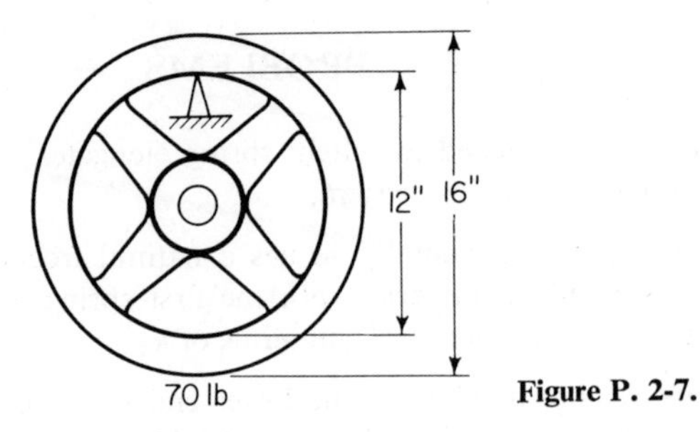

Figure P. 2-7.

2-8 A connecting rod weighing 4.80 lb oscillates 53 times in 1 min when suspended as shown in Fig. P2-8. Determine its moment of inertia about its center of gravity, which is located 10.0 in. from the point of support.

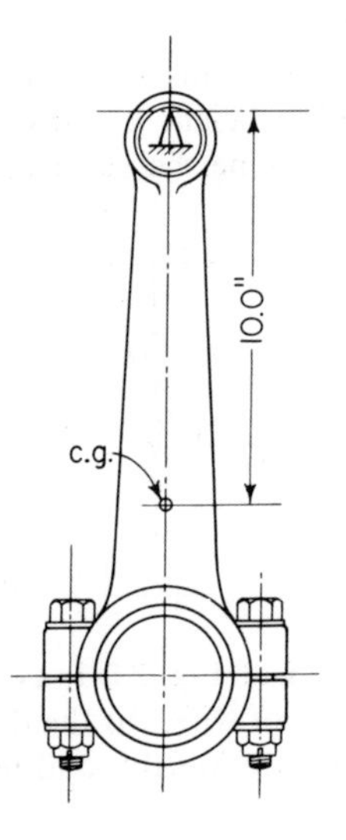

Figure P. 2-8.

2-9 A flywheel of weight W is suspended in the horizontal plane by three wires of 6 ft length, equally spaced around a circle of 10 in. radius. If the period of oscillation about a vertical axis through the center of the wheel is 2.17 sec, determine its radius of gyration.

2-10 A wheel and axle assembly of moment inertia J is inclined from the vertical by an angle α as shown in Fig. P2-10. Determine the frequency of oscillation due to a small unbalance weight w lb at a distance a in. from the axle.

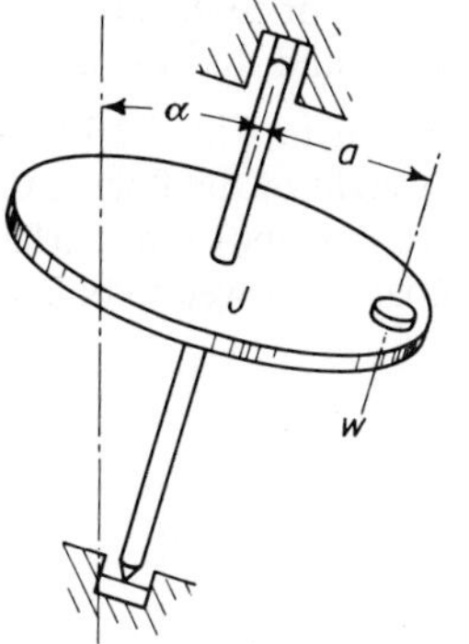

Figure P. 2-10.

2-11 A cylinder of mass m and mass moment of inertia J_0 is free to roll without slipping but is restrained by the spring k as shown in Fig. P2-11. Determine the natural frequency of oscillation.

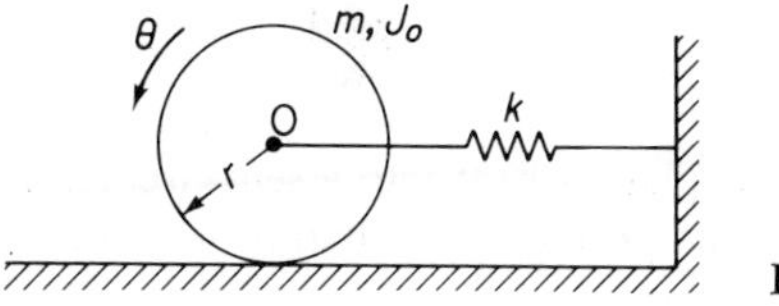

Figure P. 2-11.

2-12 A chronograph is to be operated by a 2 sec pendulum of length L shown in Fig. P2-12. A platinum wire attached to the bob completes the electric timing circuit through a drop of mercury as it swings through the lowest point. (a) What should be the length L of the pendulum? (b) If the platinum wire is in contact with the mercury for $\frac{1}{8}$ in. of the swing, what must be the ampli-

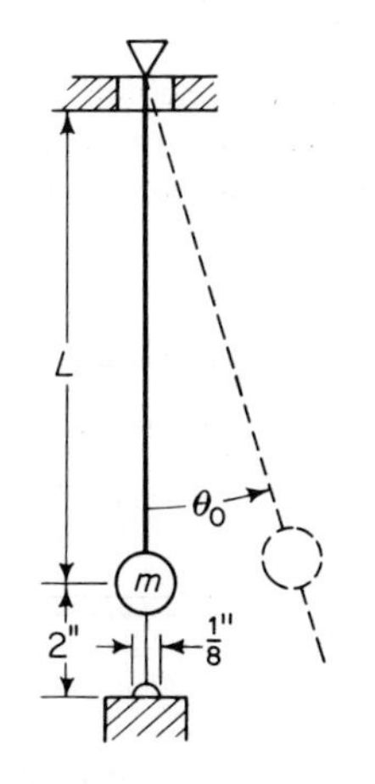

Figure P. 2-12.

tude θ_0 to limit the duration of contact to 0.01 sec? (Assume that the velocity during contact is constant, and that the amplitude of oscillation is small.)

2-13 A hydrometer float, shown in Fig. P2-13, is used to measure the specific gravity of liquids. The weight of the float is 0.082 lb, and the diameter of the cylindrical section protruding above the surface is $\frac{1}{4}$ in. Determine the period of vibration when the float is allowed to bob up and down in a fluid of specific gravity 1.20.

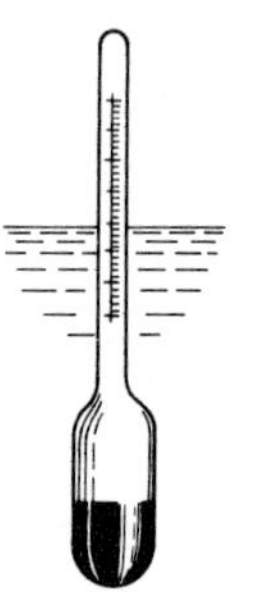

Figure P. 2-13.

2-14 A thin rectangular plate is bent into a semicircular cylinder as shown in Fig. P2-14. Determine its period of oscillation if it is allowed to rock on a horizontal surface.

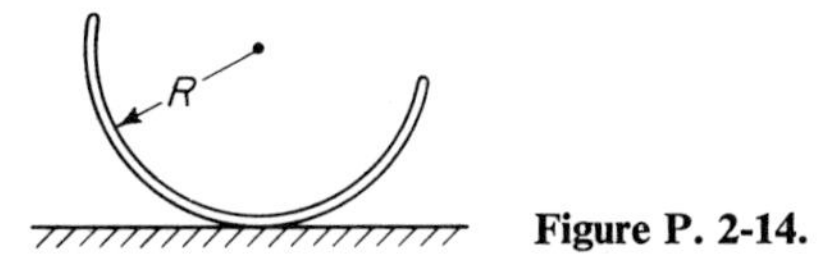

Figure P. 2-14.

2-15 A uniform bar of length L and weight W is suspended symmetrically by two strings as shown in Fig. P2-15. Set up the differential equation of motion for small angular oscillations of the bar about the vertical axis 0–0, and determine its period.

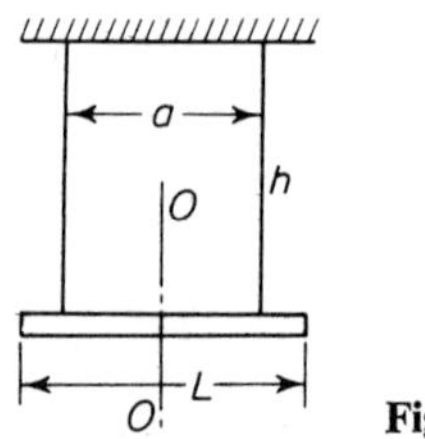

Figure P. 2-15.

2-16 A uniform bar of length L is suspended in the horizontal position by two vertical strings of equal length attached to the ends. If the period of oscillation in the plane of the bar and strings is t_1, and the period of oscillation about a vertical line through the center of gravity of the bar is t_2, show that the radius of gyration of the bar about the center of gravity is given by the expression

$$k = \left(\frac{t_2}{t_1}\right)\frac{L}{2}$$

2-17 A uniform bar of radius of gyration k about its center of gravity is suspended horizontally by two vertical strings of length h, at distances a and b from the mass center. Prove that the bar will oscillate about the vertical line through the mass center, and determine the frequency of oscillation.

2-18 A steel shaft 50 in. long and $1\frac{1}{2}$ in. in diameter is used as a torsion spring for the wheels of a light automobile as shown in Fig. P2-18. Determine the natural frequency of the system if the weight of the wheel and tire assembly is 38 lb and its radius of gyration about its axle is 9.0 in. Discuss the difference in the natural frequency with the wheel locked and unlocked to the arm.

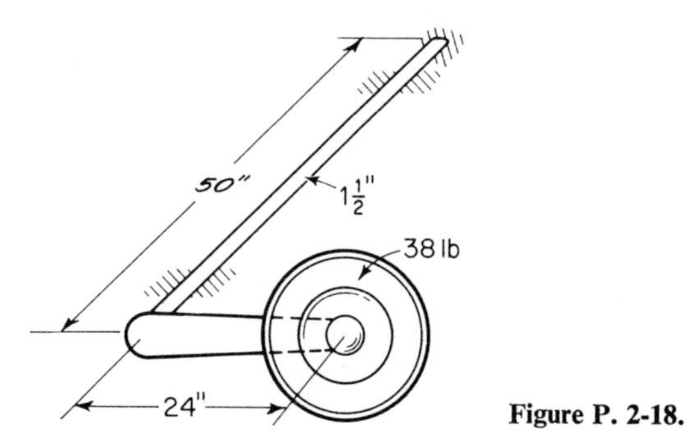

Figure P. 2-18.

2-19 Tachometers are a reed type of frequency-measuring instrument consisting of small cantilever beams with weights attached at the ends. When the frequency of vibration corresponds to the natural frequency of one of the reeds, the tachometer will vibrate, thereby indicating the frequency. How large a weight must be placed on the end of a reed made of spring steel 0.04 in. thick, 0.25 in. wide, and 3.50 in. long, for a natural frequency of 20 cps?

2-20 Determine the effective mass at point O of a uniform rod of mass m and length l pivoted at a distance nl from O, as shown in Fig. P2-20.

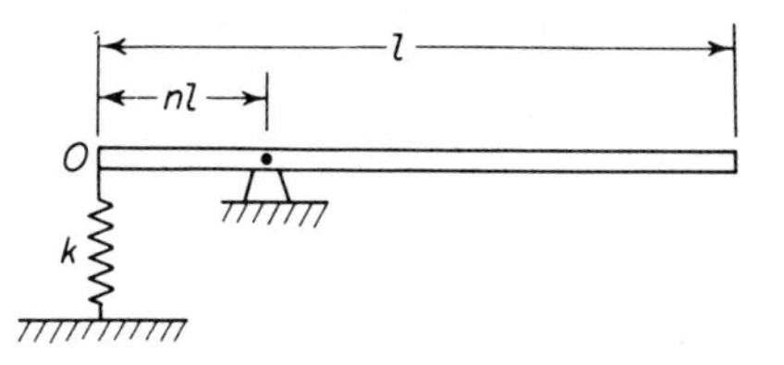

Figure P. 2-20.

2-21 Determine the effective mass of the rocket engine shown in Fig. P2-21 to be added to the actuator mass m_1.

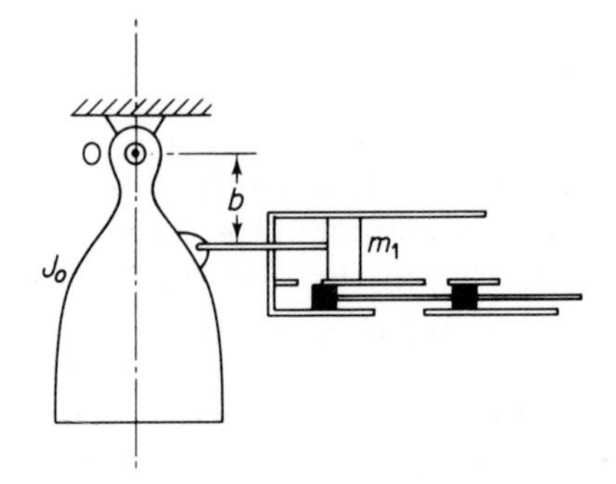

Figure P. 2-21.

2-22 A uniform cantilever beam is to be replaced by an effective mass at its free end. Assume a static deflection curve for a uniform load. Find the effective mass.

2-23 Determine the effective mass moment of inertia for shaft 1 in the system shown in Fig. P2-23.

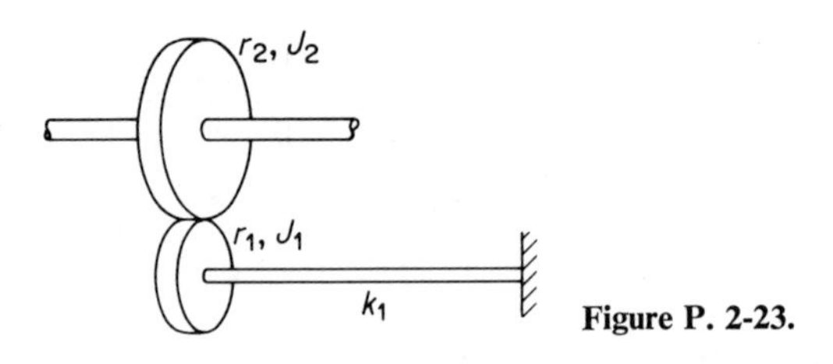

Figure P. 2-23.

2-24 Determine the kinetic energy of the system shown in Fig. P2-24 in terms of $\dot{x}$.

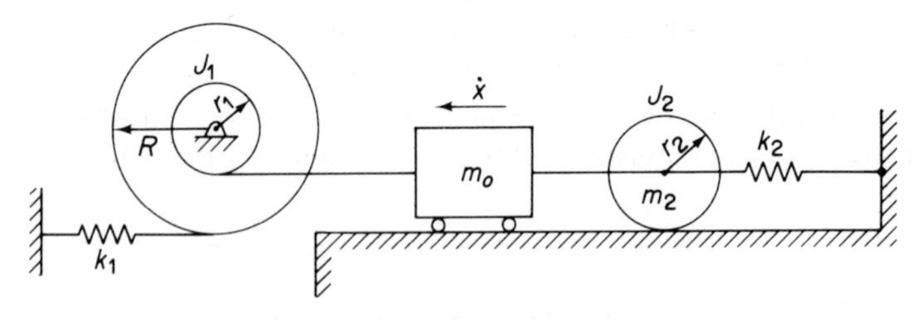

Figure P. 2-24.

2-25 Determine the effective mass at point n for the system shown in Fig. P2-25.

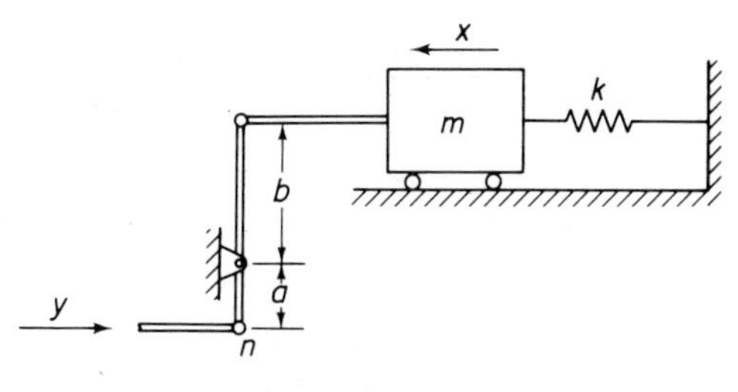

Figure P. 2-25.

2-26 A weight of 2 lb is attached to the end of a spring with a stiffness of 4 lb/in. Determine the critical damping coefficient.

2-27 To calibrate a dashpot, the velocity of the plunger was measured when a given force was applied to it. If a $\frac{1}{2}$ lb weight produced a constant velocity of 1.20 in./sec, determine the damping factor ζ when used with the system of Problem 2-26.

2-28 A vibrating system is started under the following initial conditions: $x = 0$, $\dot{x} = v_0$. Determine the equation of motion when (a) $\zeta = 2.0$, (b) $\zeta = 0.50$, (c) $\zeta = 1.0$. Plot non-dimensional curves for the three cases with $\omega_n t$ as abscissa and $x\omega_n/v_0$ as ordinate.

2-29 A vibrating system consisting of a weight of 5 lb and a spring of stiffness 10 lb/in. is viscously damped such that the ratio of any two consecutive amplitudes is 1.00 to 0.98. Determine (a) the natural frequency of the damped system, (b) the logarithmic decrement, (c) the damping factor, and (d) the damping coefficient.

2-30 A vibrating system consists of a weight of 10 lb, a spring of stiffness 20 lb/in., and a dashpot with a damping coefficient of 0.071 lb/in. per sec. Find (a) the damping factor, (b) the logarithmic decrement, and (c) the ratio of any two consecutive amplitudes.

2-31 A vibrating system has the following constants: $w = 38.6$ lb, $k = 40$ lb/in., and $c = 0.40$ lb/in. per sec. Determine (a) the damping factor, (b) the natural frequency of damped oscillation, (c) the logarithmic decrement, and (d) the ratio of any two consecutive amplitudes.

2-32 Set up the differential equation of motion for the system shown in Fig. P2-32. Determine the expression for (a) the critical damping coefficient, and (b) the natural frequency of damped oscillation.

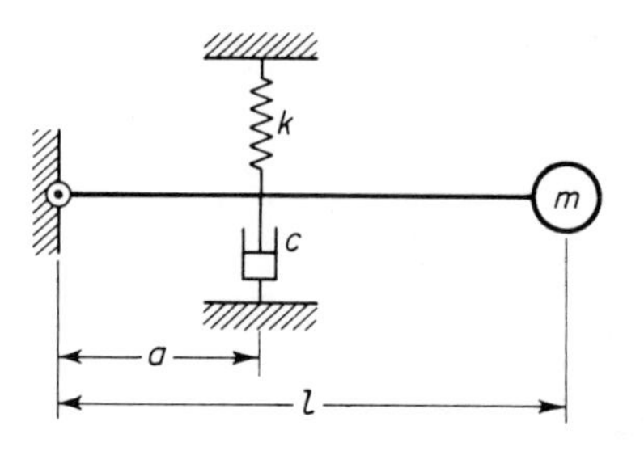

Figure P. 2-32.

2-33 Write the differential equation of motion for the system shown in Fig. P2-33, and determine the natural frequency of damped oscillation and the critical damping coefficient.

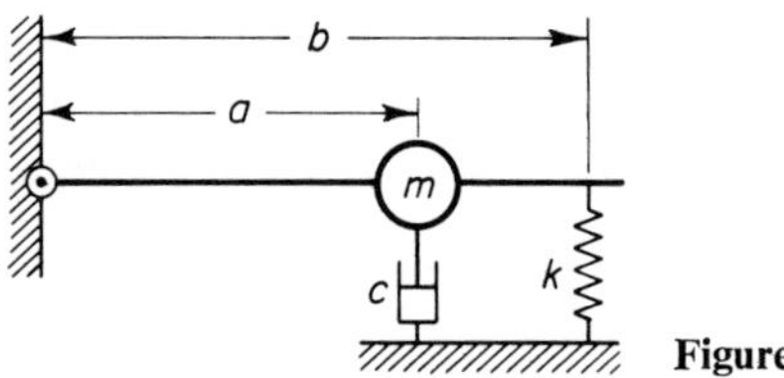

Figure P. 2-33.

2-34 A spring-mass system with viscous damping is displaced from the equilibrium position and released. If the amplitude diminished by 5% each cycle, what fraction of the critical damping does the system have?

2-35 A rigid uniform bar of mass m and length l is pinned at O and supported by a spring and viscous damper as shown in Fig. P2-35. Measuring θ from the static equilibrium position, determine (a) the equation for small θ (the moment of inertia of the bar about 0 is $ml^2/3$), (b) the equation for the undamped natural frequency, and (c) the expression for critical damping.

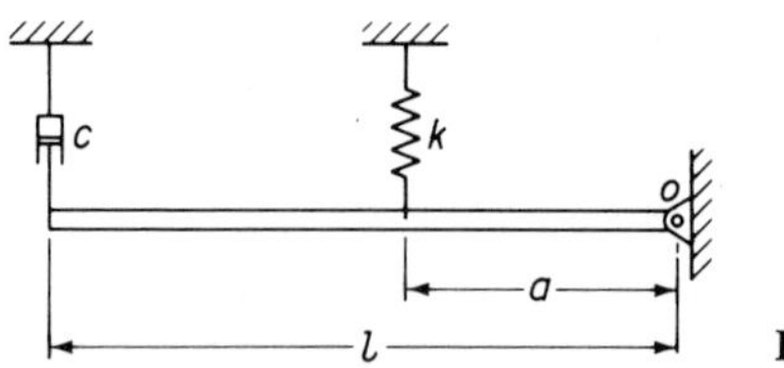

Figure P. 2-35.

2-36 A thin plate of area A and weight W is attached to the end of a spring and is allowed to oscillate in a viscous fluid as shown in Fig. P2-36. If τ_1 is the natural period of undamped oscillation (that is, with the system oscillating in air), and τ_2 the damped period with the plate immersed in the fluid, show that

$$\mu = \frac{2\pi W}{gA\tau_1\tau_2}\sqrt{\tau_2^2 - \tau_1^2}$$

where the damping force on the plate is $F_d = \mu 2Av$, $2A$ is the total surface area of the plate, and v is its velocity.

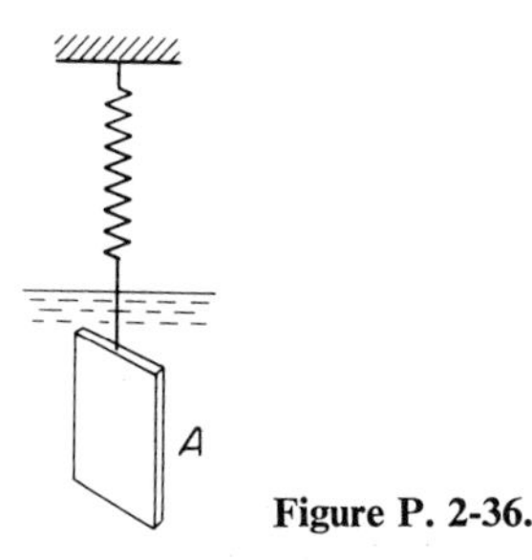

Figure P. 2-36.

2-37 A gun barrel weighing 1200 lb has a recoil spring of stiffness 20,000 lb/ft. If the barrel recoils 4 ft on firing, determine (a) the initial recoil velocity of the barrel, (b) the critical damping coefficient of a dashpot which is engaged at the end of the recoil stroke, and (c) the time required for the barrel to return to a position 2 in. from its initial position.

2-38 A piston weighing 10 lbs is traveling in a tube with a velocity of 50 ft/sec and engages a spring and damper as shown in Fig. P2-38. Determine the maximum displacement of the piston after engaging the spring-damper. How many seconds does it take?

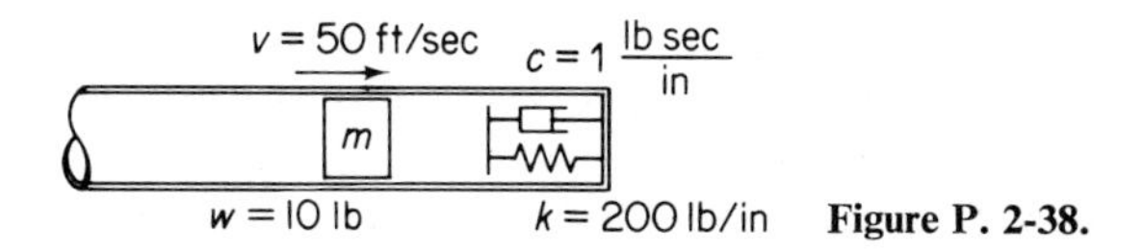

Figure P. 2-38.

2-39 A shock absorber is to be designed so that its overshoot is 10% of the initial displacement when released. Determine ζ_1. If ζ is made equal to $\frac{1}{2}\zeta_1$, what will be the overshoot?

2-40 Discuss the limitations of the equation $\Delta U/U = 2\delta$ by considering the case where $x_2/x_1 = \frac{1}{2}$.

2-41 Determine the effective stiffness of the springs shown in Fig. P2-41.

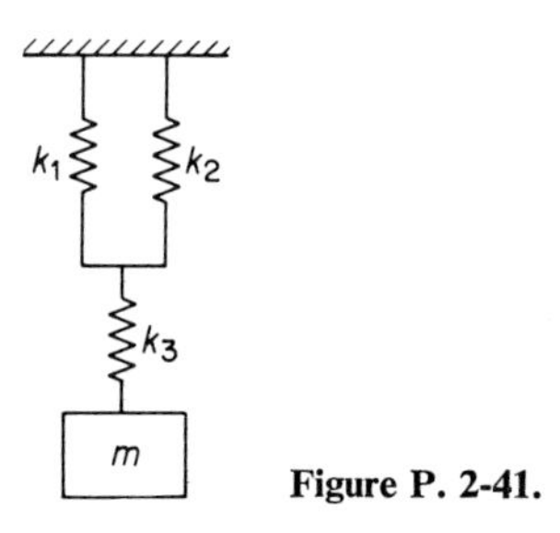

Figure P. 2-41.

2-42 Determine the flexibility of a simply supported uniform beam of length L at a point $\frac{1}{3}L$ from the end.

2-43 Determine the effective stiffness of the system shown in Fig. P2-43, in terms of the displacement x.

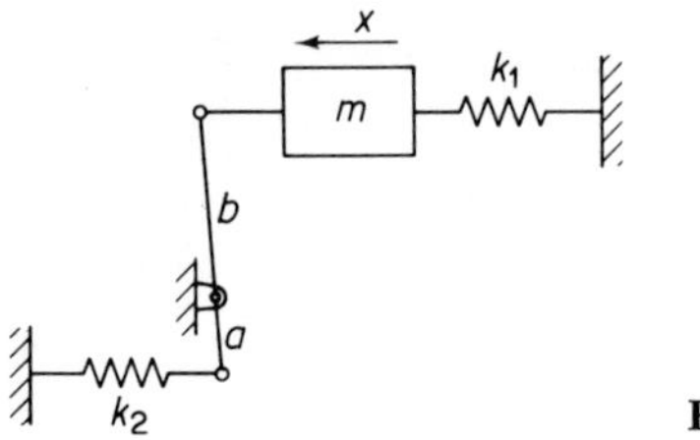

Figure P. 2-43.

2-44 Determine the effective stiffness of the torsional system shown in Fig. P2-44. The two shafts in series have torsional stiffnesses of k_1 and k_2.

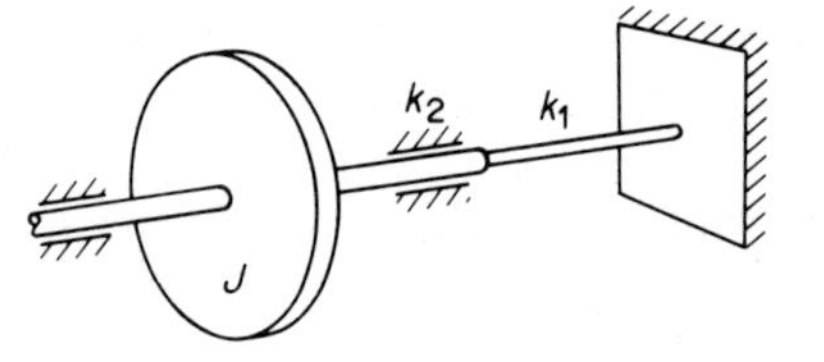

Figure P. 2-44.

2-45 A spring-mass system m, k, is started with an initial displacement of unity and an initial velocity of zero. Plot $ln X$ vs. n where X is amplitude at cycle n for (a) viscous damping with $\zeta = 0.05$, and (b) Coulomb damping with damping force $F_d = 0.05\,k$. When will the two amplitudes be equal?

3

Harmonically Excited Motion

3.1 INTRODUCTION

Harmonic excitation is often encountered in engineering systems. It is commonly produced by the unbalance in rotating machinery. Although pure harmonic excitation is less likely to occur than periodic or other types of excitation, understanding the behavior of a system undergoing harmonic excitation is essential in order to comprehend how the system will respond to more general types of excitation. Harmonic excitation may be in the form of a force or displacement of some point in the system.

3.2 FORCED HARMONIC VIBRATION

We will first consider a single degree of freedom system with viscous damping, excited by a harmonic force $F_0 \sin \omega t$ as shown in Fig. 3.2-1. Its differential

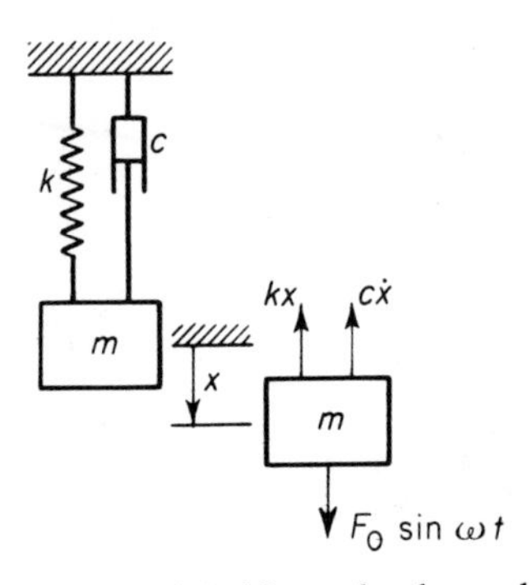

Figure 3.2-1. Viscously damped system with harmonic excitation.

equation of motion is found from the free-body diagram to be

$$m\ddot{x} + c\dot{x} + kx = F_0 \sin \omega t \qquad (3.2\text{-}1)$$

The solution to this equation consists of two parts, the *complimentary function*, which is the solution of the homogeneous equation, and the *particular integral*. The complimentary function, in this case, is a damped free vibration that was discussed in Chapter 2.

The particular solution to the above equation is a steady-state oscillation of the same frequency ω as that of the excitation. We can assume the particular solution to be of the form

$$x = X \sin (\omega t - \phi) \qquad (3.2\text{-}2)$$

where X is the amplitude of oscillation and ϕ is the phase of the displacement with respect to the exciting force.

The amplitude and phase in the above equation are found by substituting Eq. (3.2-2) into the differential equation Eq. (3.2-1). Remembering that in harmonic motion the phases of the velocity and acceleration are ahead of the displacement by 90° and 180° respectively, the terms of the differential equation can also be displayed graphically as in Fig. 3.2-2. It is easily seen from this diagram that

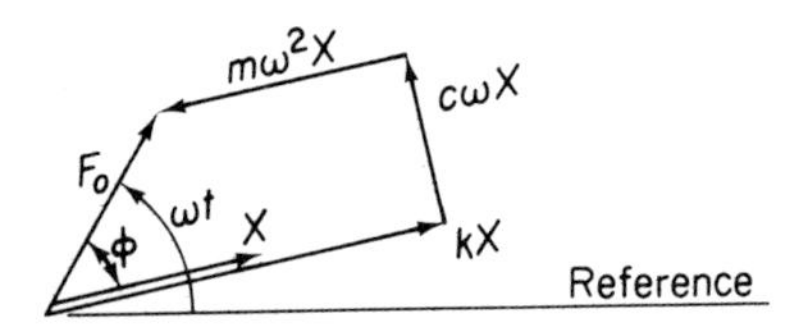

Figure 3.2-2. Vector relationship for forced vibration with damping.

$$X = \frac{F_0}{\sqrt{(k - m\omega^2)^2 + (c\omega)^2}} \qquad (3.2\text{-}3)$$

and

$$\phi = \tan^{-1} \frac{c\omega}{k - m\omega^2} \qquad (3.2\text{-}4)$$

We will now express Eqs. (3.2-3) and (3.2-4) in nondimensional form that enables a concise graphical presentation of these results. Dividing the numerator and denominator of Eqs. (3.2-3) and (3.2-4) by k, we obtain

$$X = \frac{\frac{F_0}{k}}{\sqrt{\left(1 - \frac{m\omega^2}{k}\right)^2 + \left(\frac{c\omega}{k}\right)^2}} \tag{3.2-5}$$

and

$$\tan\phi = \frac{\frac{c\omega}{k}}{1 - \frac{m\omega^2}{k}} \tag{3.2-6}$$

The above equations may be further expressed in terms of the following quantities:

$$\omega_n = \sqrt{\frac{k}{m}} = \text{natural frequency of undamped oscillation}$$

$$c_c = 2m\omega_n = \text{critical damping}$$

$$\zeta = \frac{c}{c_c} = \text{damping factor}$$

$$\frac{c\omega}{k} = \frac{c}{c_c}\frac{c_c\omega}{k} = 2\zeta\frac{\omega}{\omega_n}$$

The nondimensional expressions for the amplitude and phase then become

$$\frac{Xk}{F_0} = \frac{1}{\sqrt{\left[1 - \left(\frac{\omega}{\omega_n}\right)^2\right]^2 + \left[2\zeta\left(\frac{\omega}{\omega_n}\right)\right]^2}} \tag{3.2-7}$$

and

$$\tan\phi = \frac{2\zeta\left(\frac{\omega}{\omega_n}\right)}{1 - \left(\frac{\omega}{\omega_n}\right)^2} \tag{3.2-8}$$

These equations indicate that the nondimensional amplitude Xk/F_0 and the phase ϕ are functions only of the frequency ratio ω/ω_n and the damping factor ζ and can be plotted as shown in Fig. 3.2-3. These curves show that the damping factor has a large influence on the amplitude and phase angle in the frequency region near resonance. Further understanding of the behavior of the system may be obtained by studying the force diagram corresponding to Fig. 3.2-2 in the regions ω/ω_n small, $\omega/\omega_n = 1$, and ω/ω_n large.

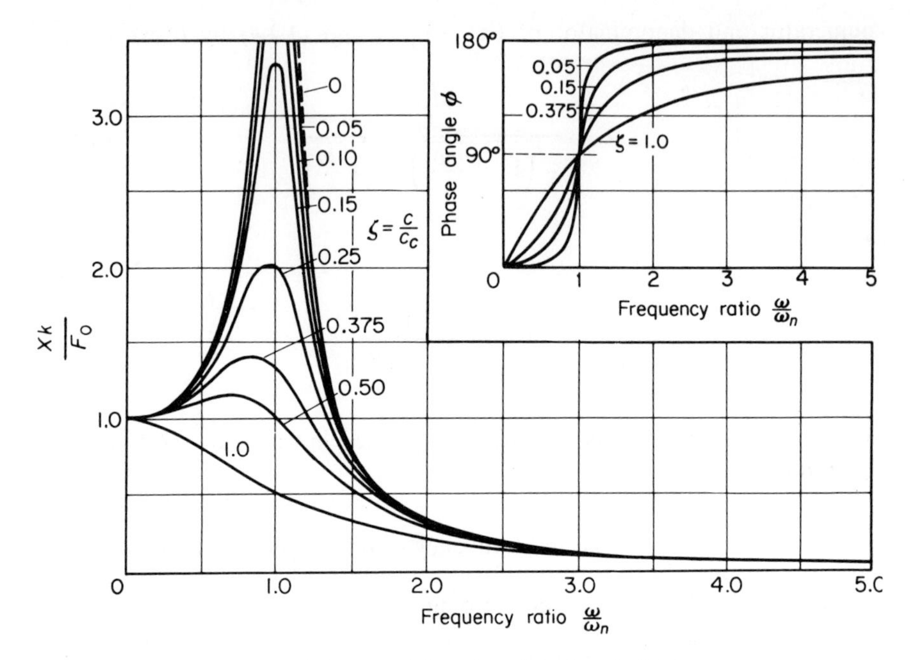

Figure 3.2-3. Plot of Equations (3.2-7) and (3.2-8).

For small values of $\omega/\omega_n \ll 1$, both the inertia and damping forces are small, which results in a small phase angle ϕ. The magnitude of the impressed force is then nearly equal to the spring force as shown in Fig. 3.2-4a.

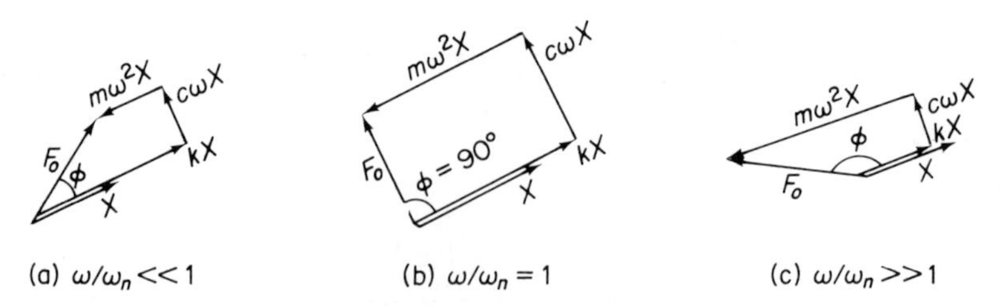

Figure 3.2-4. Vector relationship in forced vibration.

For $\omega/\omega_n = 1.0$, the phase angle is 90° and the force diagram appears as in Fig. 3.2-4b. The inertia force, which is now larger, is balanced by the spring force; whereas the impressed force overcomes the damping force. The amplitude at resonance can be found, either from Eqs. (3.2-5) or (3.2-7) or from Fig. 3.2-4b, to be

$$X = \frac{F_0}{c\omega_n} = \frac{F_0}{2\zeta k} \tag{3.2-9}$$

At large values of $\omega/\omega_n \gg 1.00$, ϕ approaches 180°, and the impressed force is expended almost entirely in overcoming the large inertia force as shown in Fig. 3.2-4c.

In summary, we can write the differential equation and its complete solution, including the transient term as

$$\ddot{x} + 2\zeta\omega_n\dot{x} + \omega_n^2 x = \frac{F_0}{m}\sin\omega t \tag{3.2-10}$$

$$x(t) = \frac{F_0}{k}\frac{\sin(\omega t - \phi)}{\sqrt{\left[1 - \left(\frac{\omega}{\omega_n}\right)^2\right]^2 + \left[2\zeta\frac{\omega}{\omega_n}\right]^2}} + X_1 e^{-\zeta\omega_n t}\sin(\sqrt{1-\zeta^2}\,\omega_n t + \phi_1) \tag{3.2-11}$$

3.3 ROTATING UNBALANCE

Unbalance in rotating machines is a common source of vibration excitation. We consider here a spring mass system constrained to move in the vertical direction and excited by a rotating machine that is unbalanced, as shown in Fig. 3.3-1. The unbalance is represented by an eccentric mass m with ec-

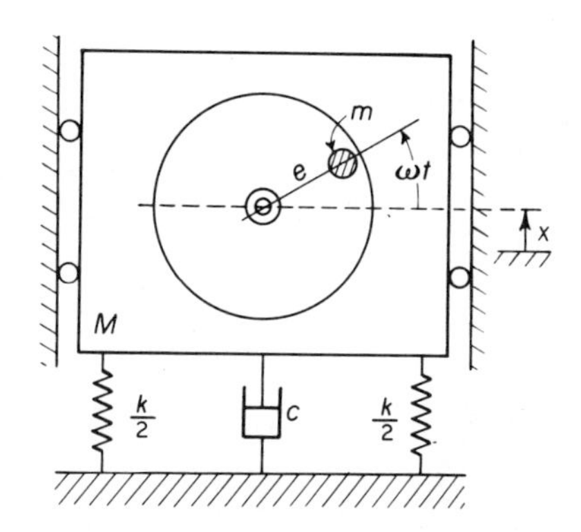

Figure 3.3-1. Harmonic disturbing force resulting from rotating unbalance.

centricity e which is rotating with angular velocity ω. Letting x be the displacement of the nonrotating mass $(M - m)$ from the static equilibrium position, the displacement of m is

$$x + e\sin\omega t$$

The equation of motion is then

$$(M - m)\ddot{x} + m\frac{d^2}{dt^2}(x + e\sin\omega t) = -kx - c\dot{x}$$

which can be rearranged to

$$M\ddot{x} + c\dot{x} + kx = (me\omega^2)\sin\omega t \tag{3.3-1}$$

It is evident, then, that the above equation is identical to Eq. (3.2-1), where F_0 is replaced by $me\omega^2$, and hence the steady state solution of the previous section can be replaced by

$$X = \frac{me\omega^2}{\sqrt{(k - M\omega^2)^2 + (c\omega)^2}} \tag{3.3-2}$$

and

$$\tan\phi = \frac{c\omega}{k - M\omega^2} \tag{3.3-3}$$

These can be further reduced to nondimensional form

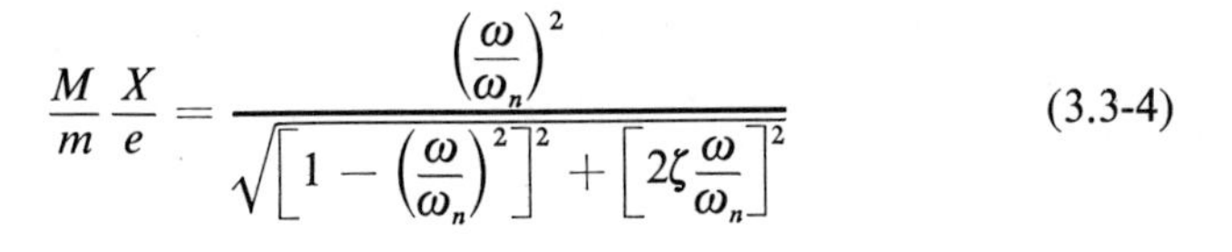

$$\frac{M}{m}\frac{X}{e} = \frac{\left(\frac{\omega}{\omega_n}\right)^2}{\sqrt{\left[1 - \left(\frac{\omega}{\omega_n}\right)^2\right]^2 + \left[2\zeta\frac{\omega}{\omega_n}\right]^2}} \tag{3.3-4}$$

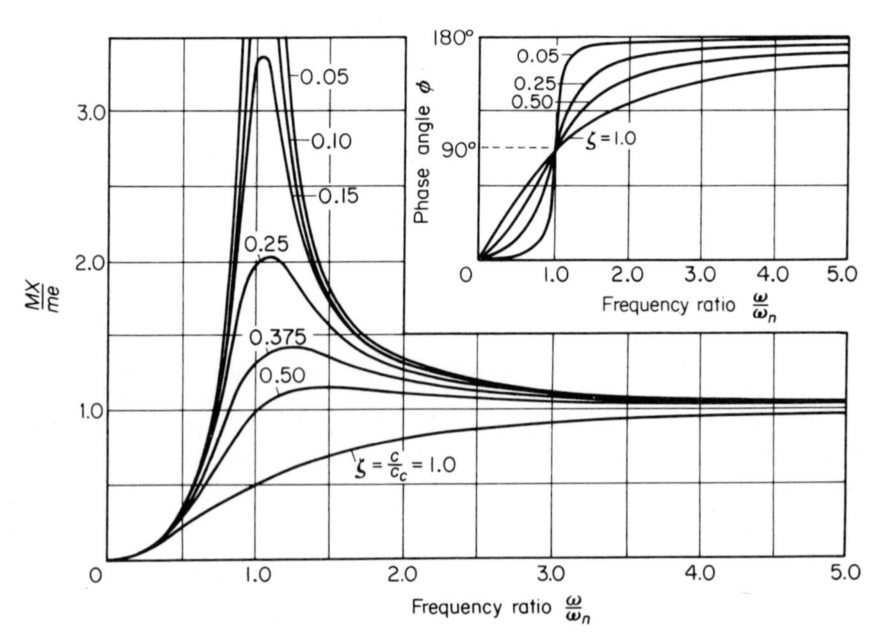

Figure 3.3-2. Plot of equations (3.3-4) and (3.3-5) for forced vibration with rotating unbalance.

and

$$\tan\phi = \frac{2\zeta\left(\frac{\omega}{\omega_n}\right)}{1-\left(\frac{\omega}{\omega_n}\right)^2} \tag{3.3-5}$$

and presented graphically as in Fig. 3.3-2. The complete solution is given by

$$x(t) = X_1 e^{-\zeta\omega_n t}\sin(\sqrt{1-\zeta^2}\,\omega_n t+\phi_1) + \frac{me\omega^2}{\sqrt{(k-M\omega^2)^2+(c\omega)^2}}\sin(\omega t-\phi) \tag{3.3-6}$$

Example 3.3-1

A counterrotating eccentric weight exciter is used to produce forced oscillation of a spring-supported mass, as shown in Fig. 3.3-3. By varying the speed of rotation, a resonant amplitude of 0.60 in. was recorded. When the speed of rotation was increased considerably beyond the resonant frequency, the amplitude appeared to approach a fixed value of 0.08 in. Determine the damping factor of the system.

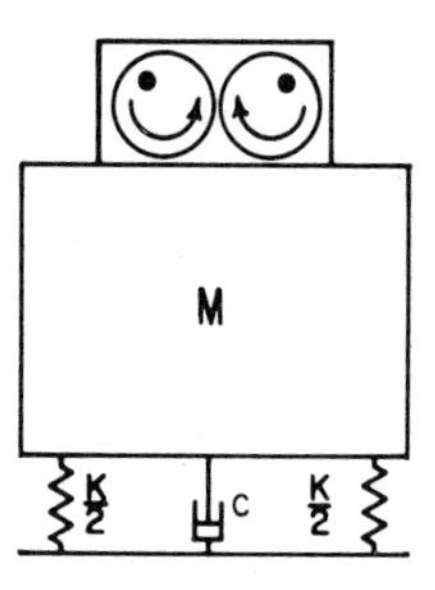

Figure 3.3-3

Solution: From Eq. (3.3-4), the resonant amplitude is

$$X = \frac{\frac{me}{M}}{2\zeta} = 0.60 \text{ in.}$$

When ω is very much greater than ω_n, the same equation becomes

$$X = \frac{me}{M} = 0.08 \text{ in.}$$

Solving the two equations simultaneously, the damping factor of the system is

$$\zeta = \frac{0.08}{2\times 0.60} = 0.0666$$

We have shown that a mass m at a radial distance e from the axis of rotation results in a centrifugal force of $me\omega^2$. Such forces produce static unbalance or dynamic unbalance depending on their distribution in the rotor.

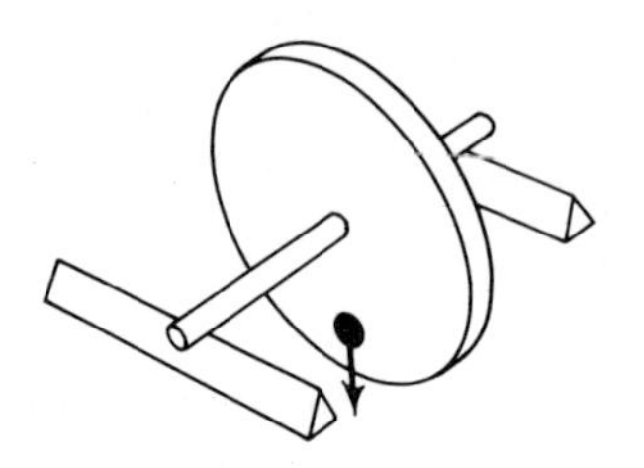

Figure 3.3-4. System with static unbalance.

Static Unbalance. When the unbalanced masses all lie in a single plane, as in the case of a thin rotor disk, the resultant unbalance is a single radial force. As shown in Fig. 3.3-4, such unbalance can be detected by a static test in which the wheel-axle assembly is placed on a pair of horizontal rails. The wheel will roll to a position where the heavy point is directly below the axle. Since such unbalance can be detected without spinning the wheel, it is called static unbalance.

Dynamic Unbalance. When the unbalance appears in more than one plane, the resultant is a force and a rocking moment which is referred to as dynamic unbalance. As previously described, a static test may detect the resultant force but the rocking moment cannot be detected without spinning the rotor. For example, consider a shaft with two disks as shown in Fig. 3.3-5. If the two

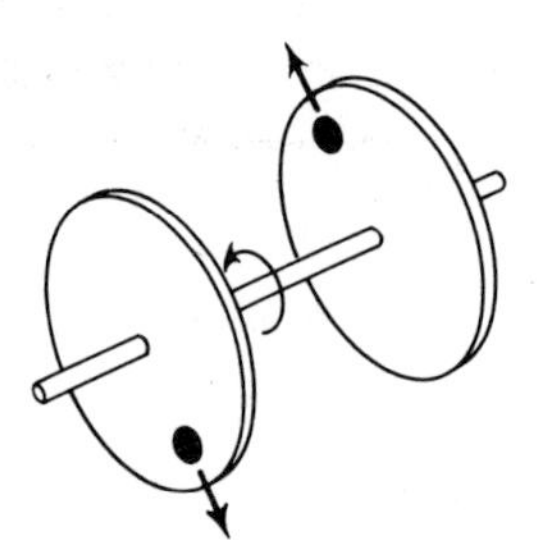

Figure 3.3-5. System with dynamic unbalance.

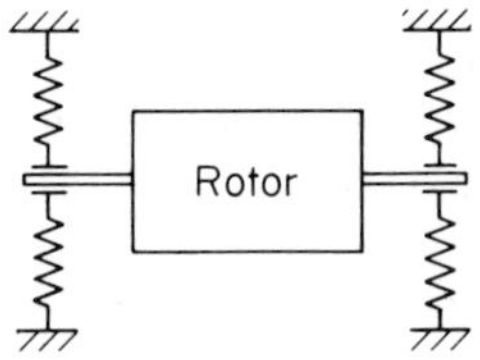

Figure 3.3-6. A rotor balancing machine.

unbalanced masses are equal and 180° apart, the rotor will be statically balanced about the axis of the shaft. However, when the rotor is spinning, each unbalanced disk would set up a rotating centrifugal force, tending to rock the shaft on its bearings.

In general, a long rotor, such as a motor armature or an automobile engine crankshaft, can be considered to be a series of thin disks, each with some unbalance. Such rotors must be spun in order to detect the unbalance. Machines to detect and correct the rotor unbalance are called balancing machines. Essentially the balancing machine consists of supporting bearings which are spring mounted so as to detect the unbalanced forces by their motion, as shown in Fig. 3.3-6. Knowing the amplitude of each bearing and their relative phase, it is possible to determine the unbalance of the rotor and correct for them.

Example 3.3-2

Although a thin disk can be balanced statically, it can also be balanced dynamically. We describe one such test which can be simply performed.

The disk is supported on spring restrained bearings that can move horizontally as shown in Fig. 3.3-7. Running at any

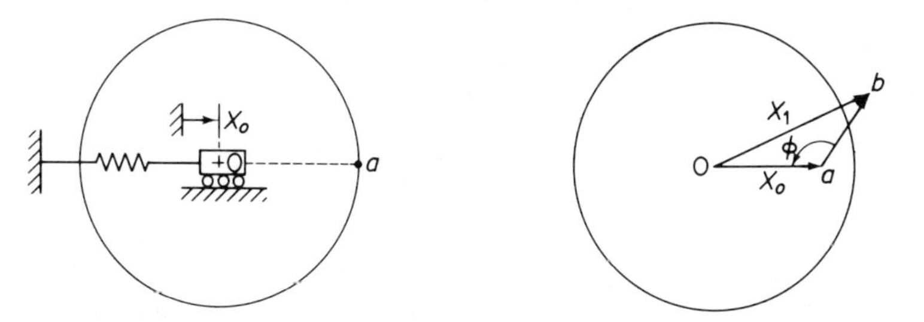

Figure 3.3-7. Experimental balancing of thin disk.

predetermined speed, the amplitude X_0 and the wheel position "*a*" at maximum excursion are noted. An accelerometer on the bearing and a stroboscope can be used for this observation. The amplitude X_0, due to the original unbalance w_0, is drawn to scale on the wheel in the direction from *o* to *a*.

Next a trial weight w_1 is added at any point on the wheel and the procedure is repeated at the same speed. The new amplitude X_1 and wheel position "*b*", which are due to the original unbalance w_0 and the trial weight w_1, are represented by the vector ob. The difference vector ab is then the effect of the trial weight w_1 alone. If the position of w_1 is now advanced by the angle ϕ shown in the vector diagram, and the magnitude of w_1 is increased to w_1 (oa/ab), the vector *ab* will become equal and opposite to the vector *oa*. The wheel is now balanced since X_1 is zero.

Example 3.3-3

A long rotor can be balanced by the addition or removal of correction weights in any two parallel planes. Generally, the correction is made by drilling holes in the two end planes; i.e., each radial inertia force $me\omega^2$ is replaced by two parallel forces, one in each end plane. With several unbalanced masses treated similarly, the correction to be made is found from their resultant in the two end planes.

Consider the balancing of a 4 in. long rotor shown in Fig. 3.3-8. It has a 3 oz in. unbalance in a plane 1 in. from the left end,

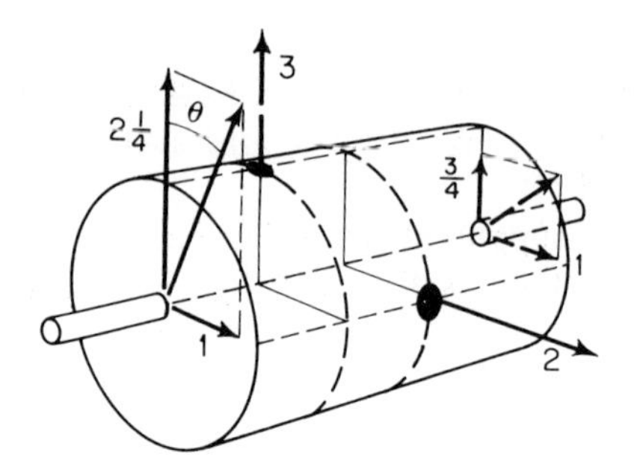

Figure 3.3-8. Correcting a long rotor unbalance in two end planes.

and a 2 oz in. unbalance in the middle plane angularly displaced 90° from the first unbalance.

The 3 oz in. unbalance is equivalent to $2\frac{1}{4}$ oz in. at the left end and $\frac{3}{4}$ oz in. at the right end, as shown. The 2 oz in. at the middle is obviously equal to 1 oz in. at the ends. Combining the two unbalances at each end, the corrections are

Left end:

$$C_1 = \sqrt{1^2 + (2.25)^2} = 2.47 \text{ oz in. to be removed}$$

$$\theta_1 = \tan^{-1}\frac{1}{2.25} = 24°0' \text{ clockwise from plane of first unbalance}$$

Right end:

$$C_2 = \sqrt{(\tfrac{3}{4})^2 + 1^2} = 1.25 \text{ oz in. to be removed}$$

$$\theta_2 = \tan^{-1}\frac{1}{(\frac{3}{4})} = 53° \text{ clockwise from plane of first unbalance}$$

3.4 WHIRLING OF ROTATING SHAFTS

Rotating shafts tend to bow out at certain speeds, and whirl in a complicated manner. *Whirling* is defined as the rotation of the plane made by the bent shaft and the line of centers of the bearings. The phenomenon results from various causes such as mass unbalance, hysteresis damping in the shaft, gyroscopic forces, fluid friction in bearings, etc. The whirling of the shaft may take place in the same or opposite direction as that of the rotation of the shaft, and the whirling speed may or may not be equal to the rotation speed.

The subject of shaft whirl is a subtle topic, and its general motion comes under the classification of self-excited motion in which the exciting forces inducing the motion are controlled by the motion itself. Consideration of the general motion of shaft whirl is beyond the scope of this text; those interested are referred to an excellent report of a study on the subject by Edgar J. Gunter, Jr.*

In this section we will consider the simplest case of *synchronous whirl*, where the whirling speed of the shaft is equal to the rotation speed of the shaft. For this purpose we will assume an idealized system consisting of a single disk of mass m symmetrically located on a shaft supported by two bearings as shown in Fig. 3.4-1. The center of mass G of the disk is at a radial distance

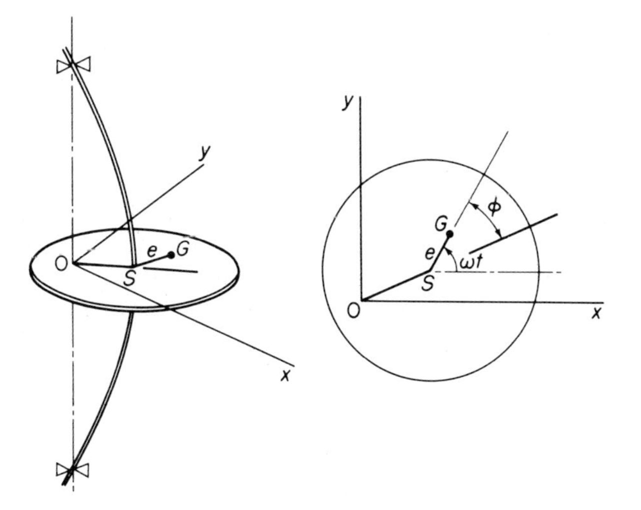

Figure 3.4-1. Synchronous whirl of a shaft due to mass unbalance.

e from the geometric center S of the disk. The center line of the bearings intersects the plane of the disk at O, and the shaft center is deflected by OS. In a synchronous whirl, O, S, and G remain fixed relative to each other, and the shaft and disk rotate at a constant speed ω. With x_s and y_s defining the position of the shaft center S, the coordinates of the mass center G are $(x_s + e\cos\omega t)$ and $(y_s + e\sin\omega t)$. Assuming viscous damping proportional to the velocity of S, the equations of motion in the x and y directions are

$$m\frac{d^2}{dt^2}(x_s + e\cos\omega t) = -kx_s - c\dot{x}_s$$

$$m\frac{d^2}{dt^2}(y_s + e\sin\omega t) = -ky_s - c\dot{y}_s$$

or

$$\begin{aligned} m\ddot{x}_s + c\dot{x}_s + kx_s &= me\omega^2\cos\omega t \\ m\ddot{y}_s + c\dot{y}_s + ky_s &= me\omega^2\sin\omega t \end{aligned} \qquad (3.4\text{-}1)$$

*Edgar J. Gunter, Jr., "Dynamic Stability of Rotor-Bearing Systems," *NASA SP-113*, 1966, U. S. Government Printing Office, Washington, D.C. 20402.

These equations are similar to Eq. (3.3-1), and we can write the solution by inspection as

$$x_s = \frac{me\omega^2 \cos(\omega t - \phi)}{\sqrt{(k - m\omega^2)^2 + (c\omega)^2}}$$
$$y_s = \frac{me\omega^2 \sin(\omega t - \phi)}{\sqrt{(k - m\omega^2)^2 + (c\omega)^2}} \tag{3.4-2}$$

$$OS = r = \sqrt{x_s^2 + y_s^2} = \frac{me\omega^2}{\sqrt{(k - m\omega^2)^2 + (c\omega)^2}} \tag{3.4-3}$$

$$\tan \phi = \frac{c\omega}{k - m\omega^2} \tag{3.4-4}$$

It is evident then that the line $SG = e$ leads the displacement line $OS = r$ by a phase angle ϕ, which depends on the amount of damping and the rotation speed ω. When the rotation speed ω is equal to the critical speed $\omega_n = \sqrt{k/m}$, or the natural frequency of the shaft in lateral vibration, a condition of resonance is encountered in which the amplitude is restrained only by the damping. Fig. 3.4-2 shows the disk-shaft system under three different speed conditions.

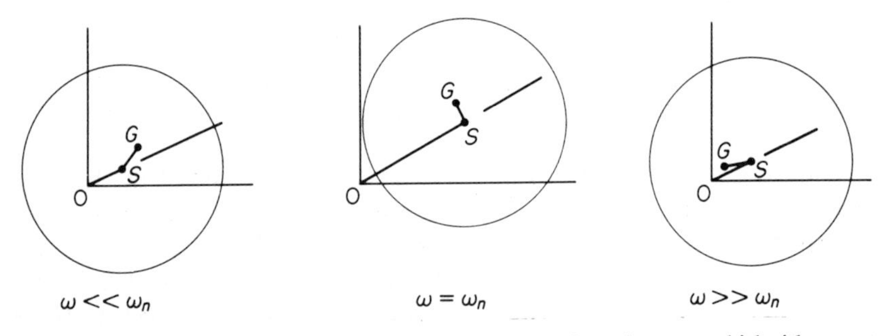

Figure 3.4-2. Amplitude and phase relationship of synchronous whirl with viscous damping.

3.5 SUPPORT MOTION

In many cases the dynamical system is excited by the motion of the support point, as shown in Fig. 3.5-1. We let y be the harmonic displacement of the support point and measure the displacement x of the mass m from a fixed reference.

In the displaced position, the unbalanced forces are due to the damper and the springs, and the differential equation of motion becomes

$$m\ddot{x} = -k(x - y) - c(\dot{x} - \dot{y}) \tag{3.5-1}$$

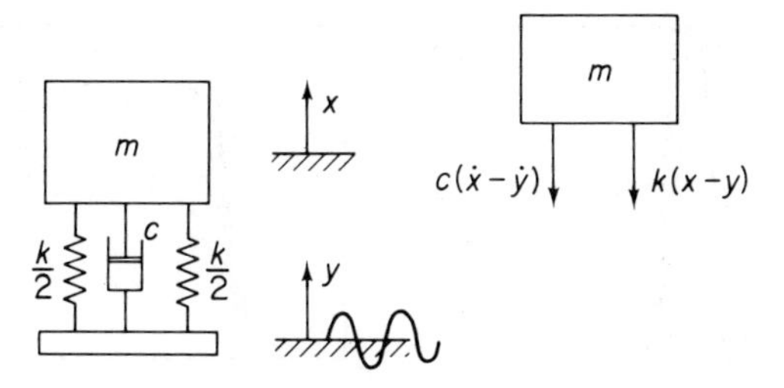

Figure 3.5-1. System excited by motion of support point.

which may be rearranged to

$$m\ddot{x} + c\dot{x} + kx = ky + c\dot{y} \tag{3.5-2}$$

Using complex algebra, we let

$$y = Ye^{i\omega t}$$

and

$$x = Xe^{i(\omega t - \phi)} = Xe^{-i\phi}e^{i\omega t} \tag{3.5-3}$$

which allows the displacement x to differ in phase from the displacement y by the angle ϕ. Substituting these into Eq. (3.5-2), we obtain

$$(-m\omega^2 + i\omega c + k)Xe^{-i\phi} = (k + i\omega c)Y$$

or

$$\frac{Xe^{-i\phi}}{Y} = \frac{k + i\omega c}{(k - m\omega^2) + i\omega c} \tag{3.5-4}$$

The absolute value of the amplitude ratio is then

$$\left|\frac{X}{Y}\right| = \sqrt{\frac{k^2 + (\omega c)^2}{(k - m\omega^2)^2 + (\omega c)^2}} = \sqrt{\frac{1 + \left(2\zeta\frac{\omega}{\omega_n}\right)^2}{\left[1 - \left(\frac{\omega}{\omega_n}\right)^2\right]^2 + \left[2\zeta\frac{\omega}{\omega_n}\right]^2}} \tag{3.5-5}$$

To find the phase angle ϕ, we equate the real and imaginary parts in Eq. (3.5-4) to determine sin ϕ and cos ϕ. The ratio then results in the equation for the phase angle, which is

$$\tan\phi = \frac{mc\omega^3}{k^2\left[1 - \left(\frac{\omega}{\omega_n}\right)^2\right] + [c\omega]^2} = \frac{2\zeta\left(\frac{\omega}{\omega_n}\right)^3}{1 - \left(\frac{\omega}{\omega_n}\right)^2 + \left(2\zeta\frac{\omega}{\omega_n}\right)^2} \tag{3.5-6}$$

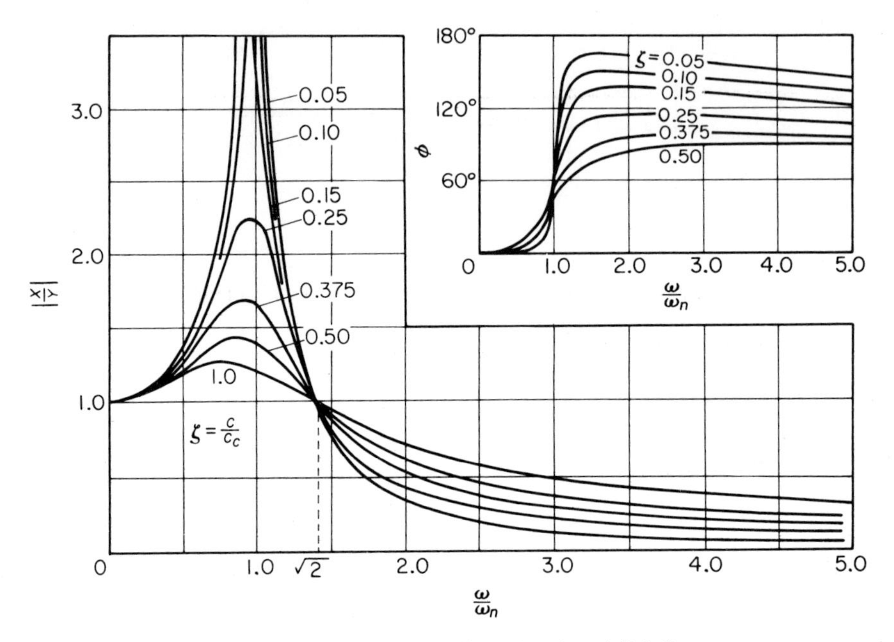

Figure 3.5-2. Plot of equations (3.5-5) and (3.5-6).

Equations (3.5-5) and (3.5-6) for the steady state amplitude and phase are plotted in Fig. 3.5-2. It should be observed that the amplitude curves for different damping all have the same value of $|X/Y| = 1.0$ at the frequency ratio $\omega/\omega_n = \sqrt{2}$.

3.6 VIBRATION MEASURING INSTRUMENTS

Figure 3.6-1 shows the essential elements of a vibration measuring instrument. It consists of a seismic mass m supported by springs inside a case which is to be fastened to the vibrating body. The motion to be measured is y and the relative motion $(x - y)$ between the mass m and the supporting case is sensed.

To determine the behavior of such instruments we consider the equation of motion of m, which is

$$m\ddot{x} = -c(\dot{x} - \dot{y}) - k(x - y) \tag{3.6-1}$$

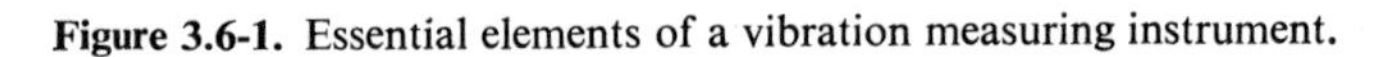

Figure 3.6-1. Essential elements of a vibration measuring instrument.

Letting the relative displacement of the mass and the case be

$$z = (x - y) \tag{3.6-2}$$

the above equation becomes

$$m\ddot{z} + c\dot{z} + kz = -m\ddot{y} \tag{3.6-3}$$

Assuming sinusoidal motion $y = Y \sin \omega t$ for the vibrating body, we obtain the equation

$$m\ddot{z} + c\dot{z} + kz = mY\omega^2 \sin \omega t \tag{3.6-4}$$

which is identical in form to Eq. (3.3-1) with z and $mY\omega^2$ replacing x and $me\omega^2$ respectively. The steady state solution $z = Z \sin (\omega t - \phi)$ is then available from inspection to be

$$Z = \frac{mY\omega^2}{\sqrt{(k - m\omega^2)^2 + (c\omega)^2}} = \frac{Y\left(\frac{\omega}{\omega_n}\right)^2}{\sqrt{\left[1 - \left(\frac{\omega}{\omega_n}\right)^2\right]^2 + \left[2\zeta\frac{\omega}{\omega_n}\right]^2}} \tag{3.6-5}$$

and

$$\tan \phi = \frac{\omega c}{k - m\omega^2} = \frac{2\zeta\left(\frac{\omega}{\omega_n}\right)}{1 - \left(\frac{\omega}{\omega_n}\right)^2} \tag{3.6-6}$$

Figure 3.6-2 shows a plot of Eq. (3.6-5) which is identical to Fig. 3.3-2 with Z/Y replacing MX/me. The type of the instrument is determined by the useful range of frequencies with respect to the natural frequency.

Seismometer. A *seismometer* is a low natural frequency instrument. Thus the range of frequencies for which it is intended is characterized by a large value of ω/ω_n. An examination of Eq. (3.6-5) shows that as $\omega/\omega_n \rightarrow \infty$, the relative displacement Z becomes equal to Y, or $|Z/Y| = 1.0$. The mass m then remains stationary while the supporting case moves with the vibrating body.

One of the disadvantages of the seismometer is its large size. Since $|Z| = |Y|$, the relative motion of the seismic mass must be of the same order of magnitude as the vibration to be measured.

The relative motion is generally converted to an electrical voltage by making the seismic mass a magnet moving relative to coils fixed in the case. Since the voltage generated is proportional to the rate of cutting of the mag-

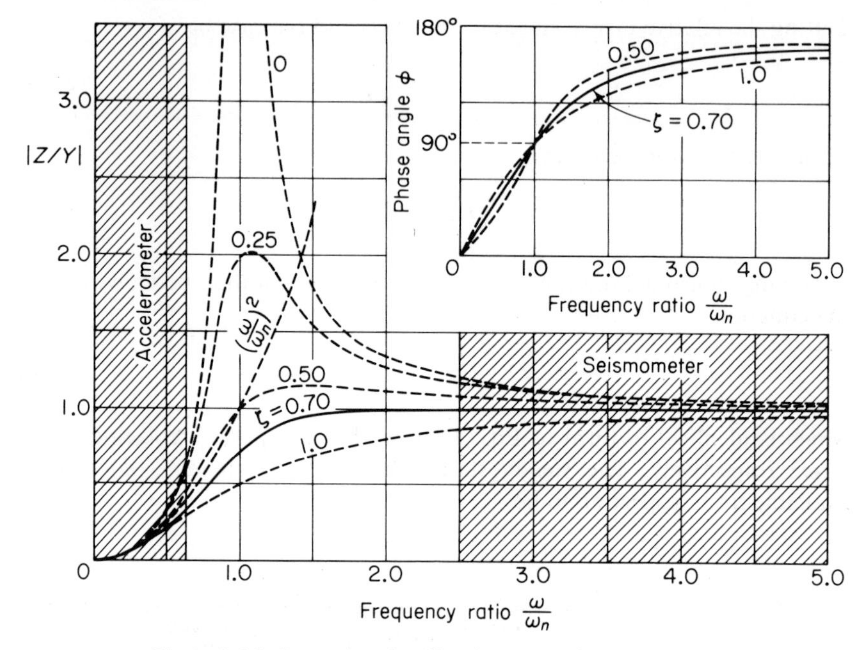

Figure 3.6-2. Response of a vibration measuring instrument.

netic field, the output of the instrument will be proportional to the velocity of the vibrating body. A typical instrument of this kind may have a natural frequency of 2 to 5 cps and a useful range of 10 to 500 cps. The sensitivity of such instruments ranges around 100 mv per in./sec with displacement limited to 0.2 in.

Accelerometer. Most vibration measurements today are made with accelerometers. Even earthquakes are recorded by accelerometers, and the velocity and displacement are obtained by integration. Accelerometers are preferred as vibration measuring instruments because of their small size and high sensitivity.

Accelerometers are high natural frequency instruments, and the useful range of their operation is ω/ω_n from zero to about 0.4. Examination of Eq. (3.6-5) for $\omega/\omega_n \longrightarrow 0$ leads to

$$Z = \frac{\omega^2 Y}{\omega_n^2} = \frac{\text{acceleration}}{\omega_n^2} \tag{3.6-7}$$

and hence Z becomes proportional to the acceleration of the motion to be measured. The sensitivity decreases, however, as ω_n increases, so that ω_n should be no higher than necessary. For example, accelerometers used extensively for earthquake measurements have a natural frequency of 20 cps, which allows ground motions of frequencies less than 8 cps to be faithfully reproduced. Actually, motions up to 16 cps can be measured by applying a correction from the instrument calibration.

For a greater range of frequencies, the piezoelectric crystal accelerometer is extensively used. Its natural frequency is generally very high, and the crystal accelerometer can be used up to frequencies of 1000 cps or more.

The useful frequency range of the undamped accelerometer is somewhat limited because the denominator $1 - (\omega/\omega_n)^2$ drops off rapidly as ω increases. However with damping in the range of $\zeta = 0.65$ to 0.70, the reduction in the term $1 - (\omega/\omega_n)^2$ is compensated for by the additional term $(2\zeta\, \omega/\omega_n)^2$ to greatly extend the useful range of the instrument.

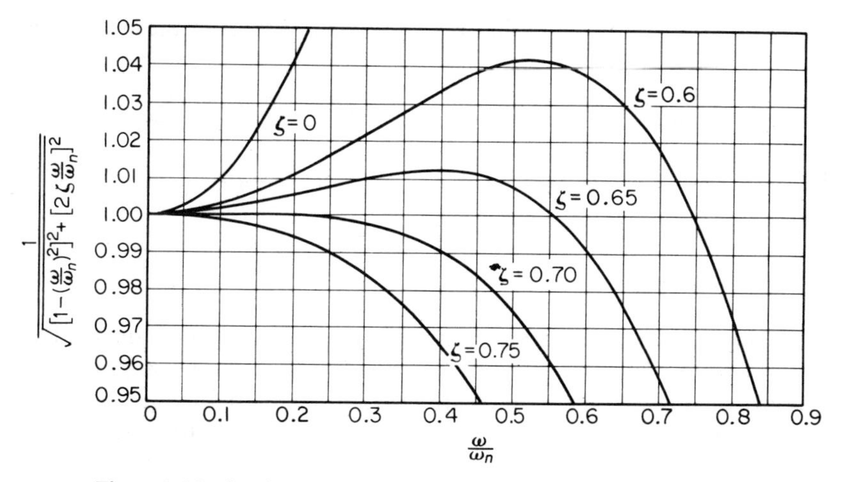

Figure 3.6-3. Accelerometer error vs. frequency with ζ as parameter.

Fig. 3.6-3 shows the factor

$$\frac{1}{\sqrt{\left[1 - \left(\frac{\omega}{\omega_n}\right)^2\right]^2 + \left[2\zeta\frac{\omega}{\omega_n}\right]^2}} \tag{3.6-8}$$

for various damping, plotted on a magnified scale. Most accelerometers utilize damping near $\zeta = 0.7$, which not only extends the useful frequency range but prevents phase distortion.

Phase Distortion. To reproduce a complex wave without changing its shape, the phase of all harmonic components must be shifted equally along the time axis. This can be accomplished if the phase angle ϕ of the accelerometer output increases linearly with frequency. For example, if $\phi = \pi/2 \times \omega/\omega_n$, which is nearly satisfied for $\zeta = 0.70$, the phase distortion is practically eliminated.

Example 3.6-1

Investigate the output of an accelerometer with damping $\zeta = 0.70$ when used to measure a periodic motion with the displacement given by the equation

$$y = Y_1 \sin \omega_1 t + Y_2 \sin \omega_2 t$$

Solution: For $\zeta = 0.70$, $\phi \cong \pi/2 \times \omega/\omega_n$, so that $\phi_1 = \pi/2 \times \omega_1/\omega_n$ and $\phi_2 = \pi/2 \times \omega_2/\omega_n$. The output of the accelerometer is then

$$z = Z_1 \sin(\omega_1 t - \phi_1) + Z_2 \sin(\omega_2 t - \phi_2)$$

Substituting for Z_1 and Z_2 from Eq. (3.6-7), the output of the instrument is

$$z = \frac{1}{\omega_n^2}\left\{\omega_1^2 Y_1 \sin \omega_1\left(t - \frac{\pi}{2\omega_n}\right) + \omega_2^2 Y_2 \sin \omega_2\left(t - \frac{\pi}{2\omega_n}\right)\right\}$$

Thus the instrument faithfully reproduces the acceleration $\ddot{y}$ without distortion.

3.7 VIBRATION ISOLATION

Vibratory forces generated by machines and engines are often unavoidable; however, their effect on a dynamical system can be reduced substantially by properly designed springs, which are referred to as isolators.

In Fig. 3.7-1, let $F_0 \sin \omega t$ be the exciting force acting on the single

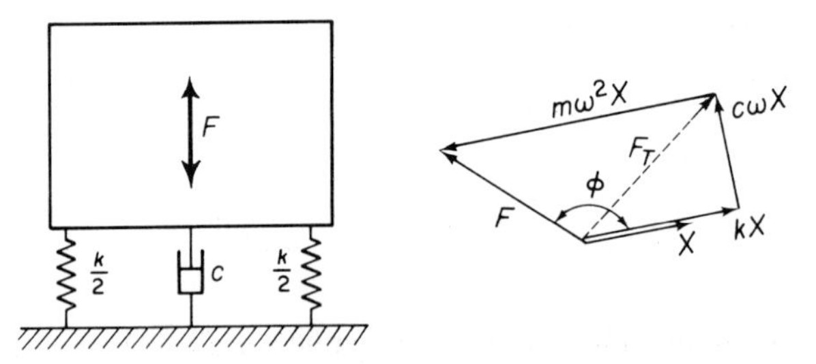

Figure 3.7-1. Disturbing force transmitted through springs and damper.

degree of freedom system. The transmitted force through the springs and damper is

$$F_T = \sqrt{(kX)^2 + (c\omega X)^2} = kX\sqrt{1 + \left(\frac{c\omega}{k}\right)^2} \tag{3.7-1}$$

Since the amplitude X developed under the force $F_0 \sin \omega t$ is given by Eq.

(3.2-5), the above equation reduces to

$$F_T = \frac{F_0\sqrt{1 + \left(\frac{c\omega}{k}\right)^2}}{\sqrt{\left[1 - \frac{m\omega^2}{k}\right]^2 + \left(\frac{c\omega}{k}\right)^2}} = \frac{F_0\sqrt{1 + \left(2\zeta\frac{\omega}{\omega_n}\right)^2}}{\sqrt{\left[1 - \left(\frac{\omega}{\omega_n}\right)^2\right]^2 + \left(2\zeta\frac{\omega}{\omega_n}\right)^2}} \tag{3.7-2}$$

Comparison of Eqs. (3.7-2) and (3.5-5) indicates that $|F_T/F_0|$ is identical to $|X/Y| = |\omega^2 X/\omega^2 Y|$. Thus the problem of isolating a mass from the motion of the support point is identical to that of isolating disturbing forces. Each of these ratios is referred to as transmissibility, and the ordinate of Fig. 3.5-2 can equally represent transmissibility of force or of displacement. These curves show that the transmissibility is less than unity only for $\omega/\omega_n > \sqrt{2}$, thereby establishing the fact that vibration isolation is possible only for $\omega/\omega_n > \sqrt{2}$. As seen from Fig. 3.5-2, in the region $\omega/\omega_n > \sqrt{2}$, an undamped spring is superior to a damped spring in reducing the transmissibility. Some damping is desirable when it is necessary for ω to vary through the resonant region, although the large amplitude at resonance can be limited by stops.

When the damping is negligible, the transmissibility equation reduces to

$$TR = \frac{1}{\left(\frac{\omega}{\omega_n}\right)^2 - 1} \tag{3.7-3}$$

where it is understood that the value of ω/ω_n to be used is always greater than $\sqrt{2}$. On further replacing ω_n^2 with g/Δ'', where $g = 386$ in./sec^2 and Δ'' = statical deflection in inches, Eq. (3.7-3) may be expressed as

$$TR = \frac{1}{\frac{(2\pi f)^2\Delta}{g} - 1} \tag{3.7-4}$$

Solving for f and converting it to cycles per minute, we obtain the following equation

$$f = 188\sqrt{\frac{1}{\Delta''}\left(\frac{1}{TR} + 1\right)}, \qquad f = 188\sqrt{\frac{1}{\Delta''}\left(\frac{2 - R}{1 - R}\right)} \tag{3.7-5}$$

where the percent reduction in the transmissibility is defined as $R = (1 - TR)$. Fig. 3.7-2 displays Eq. (3.7-5) for f vs. Δ'' with R as parameter.

This discussion has been limited to bodies with translation along a single coordinate. In general, a rigid body has six degrees of freedom; namely, translation along and rotation about the three perpendicular coordinate axes. For these more advanced cases the reader is referred to the excellent text on vibration isolation by C. Crede.*

*C. E. Crede, *Vibration and Shock Isolation* (New York: John Wiley & Sons, 1951).

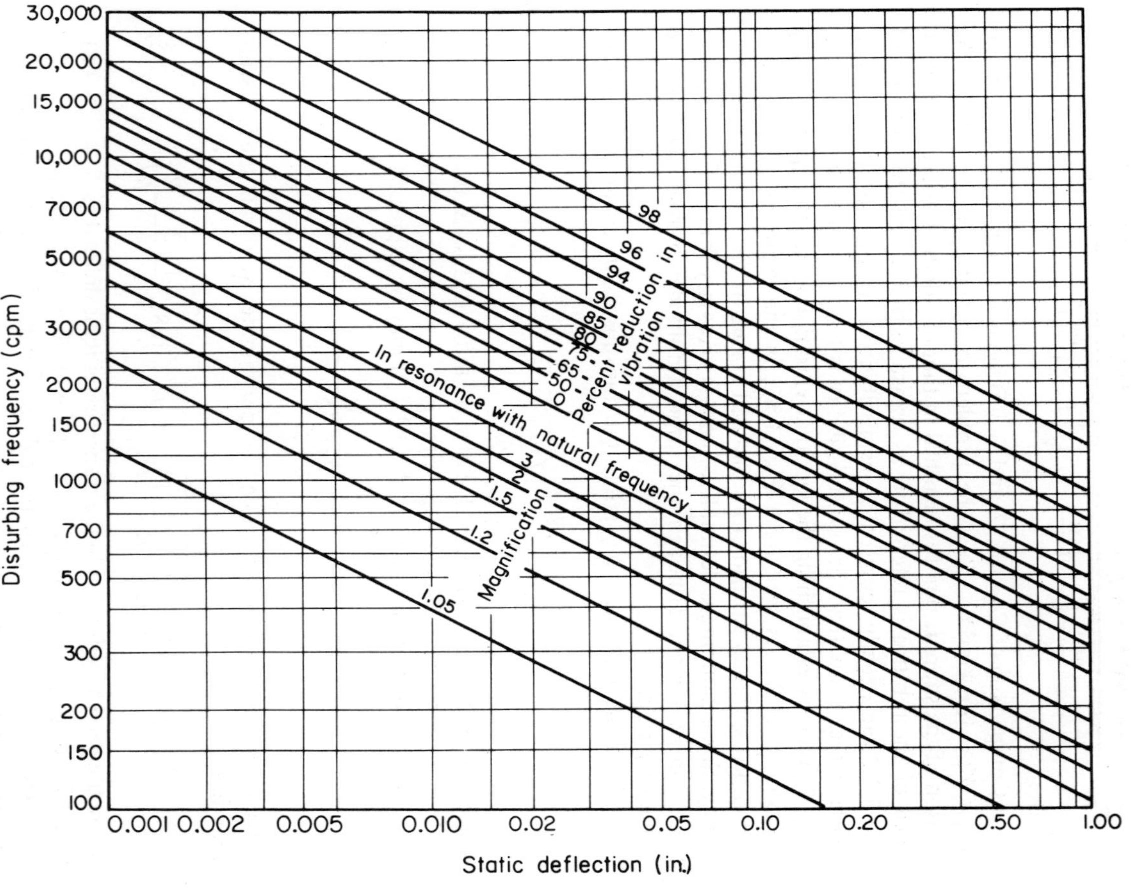

Figure 3.7-2. Isolation efficiency for flexibly mounted system.

Example 3.7-1

A machine weighing 200 lb and supported on springs of total stiffness 4000 lb/in. has an unbalanced rotating element which results in a disturbing force of 80 lb at a speed of 3000 rpm. Assuming a damping factor of $\zeta = 0.20$, determine (a) its amplitude of motion due to the unbalance, (b) the transmissibility, and (c) the transmitted force.

Solution: The statical deflection of the system is $\frac{200}{4000} = 0.05$ in., and its natural frequency is

$$f_n = \frac{60}{2\pi}\sqrt{\frac{386}{0.05}} = 841 \text{ cpm.}$$

(a) Substituting into Eqs. (3.2-7), the amplitude of vibration becomes

$$X = \frac{(\frac{80}{4000})}{\sqrt{[1 - (\frac{3000}{841})^2]^2 + [2 \times 0.20 \times \frac{3000}{841}]^2}} = 0.00169 \text{ in.}$$

(b) The transmissibility, from Eq. (3.7-2), is

$$TR = \frac{\sqrt{1 + (2 \times 0.20 \times \frac{3000}{841})^2}}{\sqrt{[1 - (\frac{3000}{841})^2]^2 + (2 \times 0.20 \times \frac{3000}{841})^2}}$$

(c) The transmitted force is the disturbing force multiplied by the transmissibility

$$F_{TR} = 80 \times 0.148 = 11.8 \text{ lb}$$

3.8 DAMPING

Damping is present in all oscillatory systems. Its effect is to remove energy from the system. Energy in a vibrating system is either dissipated into heat or radiated away. Dissipation of energy into heat can be experienced simply by bending a piece of metal back and forth a number of times. We are all aware of the sound which is radiated from an object given a sharp blow. When a buoy is made to bob up and down in the water, waves radiate out and away from it, thereby resulting in its loss of energy.

In vibration analysis, we are generally concerned with damping in terms of system response. The loss of energy from the oscillatory system results in the decay of amplitude of free vibration. In steady-state forced vibration, the loss of energy is balanced by the energy which is supplied by the excitation.

A vibrating system may encounter many different types of damping forces, from internal molecular friction to sliding friction and fluid resistance. Generally their mathematical description is quite complicated and not suitable for vibration analysis. Thus simplified damping models have been developed that in many cases are found to be adequate in evaluating the system response. For example, we have already used the viscous damping model, designated by the dashpot, which leads to manageable mathematical solutions.

Energy dissipation is usually determined under conditions of cyclic oscillations. Depending on the type of damping present, the force-displacement relationship when plotted may differ greatly. In all cases, however, the force-displacement curve will enclose an area, referred to as the *hysteresis loop*, that is proportional to the energy lost per cycle. The energy lost per cycle due to a damping force F_d is computed from the general equation

$$W_d = \oint F_d \, dx \tag{3.8-1}$$

In general, W_d will depend on many factors, such as temperature, frequency, or amplitude.

We will consider in this section the simplest case of energy dissipation, that of a spring-mass system with viscous damping. The damping force in this case is $F_d = c\dot{x}$. With the steady state displacement and velocity

$$x = X \sin(\omega t - \phi)$$
$$\dot{x} = \omega X \cos(\omega t - \phi)$$

the energy dissipated per cycle, from Eq. (3.8-1) becomes

$$\begin{aligned} W_d &= \oint c\dot{x} \, dx = \oint c\dot{x}^2 \, dt \\ &= c\omega^2 X^2 \int_0^{2\pi/\omega} \cos^2(\omega t - \phi) \, dt = \pi c \omega X^2 \end{aligned} \tag{3.8-2}$$

Of particular interest is the energy dissipated in forced vibration at resonance. Substituting $\omega_n = \sqrt{k/m}$ and $c = 2\zeta\sqrt{km}$, the above equation at resonance becomes

$$W_d = 2\zeta\pi k X^2 \tag{3.8-3}$$

The energy dissipated per cycle by the damping force can be represented graphically as follows. Writing the velocity in the form

$$\begin{aligned} \dot{x} &= \omega X \cos(\omega t - \phi) = \pm \omega X \sqrt{1 - \sin^2(\omega t - \phi)} \\ &= \pm \omega \sqrt{X^2 - x^2} \end{aligned}$$

the damping force becomes

$$F_d = c\dot{x} = \pm c\omega\sqrt{X^2 - x^2} \tag{3.8-4}$$

Rearranging the above equation to

$$\left(\frac{F_d}{c\omega X}\right)^2 + \left(\frac{x}{X}\right)^2 = 1 \tag{3.8-5}$$

we recognize it as that of an ellipse with F_d and x plotted along the vertical and horizontal axes, as shown in Fig. 3.8-1(a). The energy dissipated per cycle is then given by the area enclosed by the ellipse. If we add to F_d the force kx of the lossless spring, the hysteresis loop is rotated as shown in Fig. 3.8-1(b). This representation then conforms to the *Voigt model*, which consists of a dashpot in parallel with a spring.

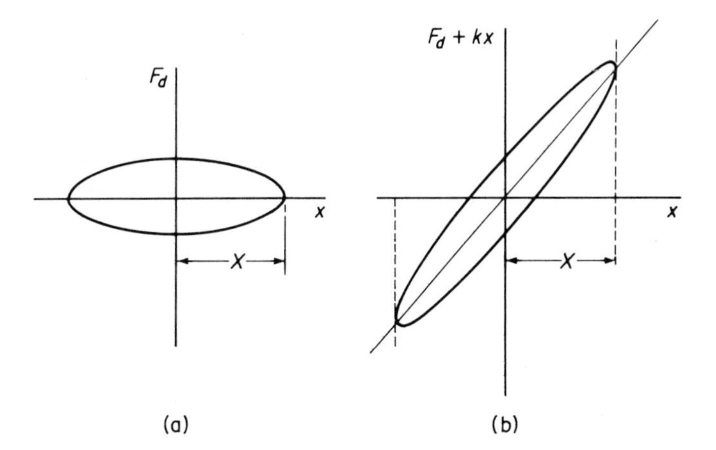

Figure 3.8-1. Energy dissipated by viscous damping.

Damping properties of materials are listed in many different ways depending on the technical areas to which they are applied. Of these we list two relative energy units which have wide usage. First of these is *specific damping capacity* defined as the energy loss per cycle W_d divided by the peak potential energy U.

$$\frac{W_d}{U} \tag{3.8-6}$$

The second quantity is the *loss coefficient* defined as the ratio of damping energy loss per radian $W_d/2\pi$ divided by the peak potential or strain energy U.

$$\eta = \frac{W_d}{2\pi U} \tag{3.8-7}$$

For the case of linear damping where the energy loss is proportional to the square of the strain or amplitude, the hysteresis curve is an ellipse.

The loss coefficient for most materials varies between point 0.001 to unity depending on the material and conditions under which the tests are performed. When the damping loss is not a quadratic function of the strain or amplitude, the hysteresis curve is no longer an ellipse. Again, the loss coefficient may vary from 0.001 to approximately 0.2.

Example 3.8-1

Determine the expression for the power developed by a force $F = F_0 \sin(\omega t + \phi)$ acting on a displacement $x = X_0 \sin \omega t$.

Solution: Power is the rate of doing work which is the product of the force and velocity.

$$\begin{aligned} P = F\frac{dx}{dt} &= (\omega X_0 F_0) \sin(\omega t + \phi) \cos \omega t \\ &= (\omega X_0 F_0)[\cos \phi \cdot \sin \omega t \cos \omega t + \sin \phi \cdot \cos^2 \omega t] \\ &= \tfrac{1}{2}\omega X_0 F_0[\sin \phi + \sin(2\omega t + \phi)] \end{aligned}$$

The first term is a constant, representing the steady flow of work per unit time. The second term is a sine wave of twice the frequency that represents the fluctuating component of power, the average value of which is zero over any interval of time that is a multiple of the period.

Example 3.8-2

A force $F = 10 \sin \pi t$ lb acts on a displacement of $x = 2 \sin(\pi t - \pi/6)$. Determine (a) the work done during the first 6 sec; (b) the work done during the first $\frac{1}{2}$ sec.

Solution: Rewriting Eq. (3.8-1) as $W = \int F\dot{x}\, dt$, and substituting $F = F_0 \sin \omega t$ and $x = X \sin(\omega t - \phi)$, the work done per cycle becomes

$$W = \pi F_0 X \sin \phi$$

For the force and displacement given in this problem, $F_0 = 10$, $X = 2$, $\phi = \pi/6$, and the period $\tau = 2$ seconds. Thus in the 6 seconds specified in (a), three complete cycles take place, and the work done is

$$W = 3(\pi F_0 X \sin \phi) = 3\pi \times 10 \times 2 \times \sin 30^\circ = 94.2 \text{ in. lb}$$

The work done in part (b) is determined by integrating the expression for work between the limits 0 and $\frac{1}{2}$ sec.

$$W = \omega F_0 X_0 \left[\cos 30^\circ \int_0^{1/2} \sin \pi t \cos \pi t \, dt + \sin 30^\circ \int_0^{1/2} \sin^2 \pi t \, dt \right]$$

$$= \pi \times 10 \times 2 \left[-\frac{0.866}{4\pi} \cos 2\pi t + 0.50 \left(\frac{t}{2} - \frac{\sin 2\pi t}{4\pi} \right) \right]_0^{1/2}$$

$$= 16.51 \text{ in. lb}$$

3.9 EQUIVALENT VISCOUS DAMPING

The primary influence of damping on oscillatory systems is that of limiting the amplitude of response at resonance. As seen from the response curves of Fig. 3.2-3, damping has little influence on the response in the frequency regions away from resonance.

In the case of viscous damping, the amplitude at resonance, Eq. (3.2-9), was found to be

$$X = \frac{F_0}{c\omega_n} \tag{3.9-1}$$

For other types of damping, no such simple expression exists. It is possible, however, to approximate the resonant amplitude by substituting an equivalent damping c_{eq} in the above equation.

The equivalent damping c_{eq} is found by equating the energy dissipated by the viscous damping to that of the nonviscous damping force with assumed harmonic motion. From Eq. (3.8-2)

$$\pi c_{eq} \omega X^2 = W_d \tag{3.9-2}$$

where W_d must be evaluated from the particular type of damping force.

Example 3.9-1

Bodies moving with moderate speed (10 to 50 ft/sec) in fluids such as water or air, are resisted by a damping force which is proportional to the square of the speed. Determine the equivalent damping for such forces acting on an oscillatory system, and find its resonant amplitude.

Solution: Let the damping force be expressed by the equation

$$F_d = \pm a\dot{x}^2$$

where the negative sign must be used when $\dot{x}$ is positive, and vice versa. Assuming harmonic motion with the time measured from the position of extreme negative displacement

$$x = -X \cos \omega t$$

the energy dissipated per cycle is

$$W_d = 2\int_{-X}^{X} a\dot{x}^2 dx = 2a\omega^2 X^3 \int_0^{\pi} \sin^3 \omega t d(\omega t)$$
$$= \tfrac{8}{3} a\omega^2 X^3$$

The equivalent viscous damping from Eq. (3.9-2) is then

$$c_{eq} = \frac{8}{3\pi} a\omega X$$

The amplitude at resonance is found by substituting $c = c_{eq}$ in Eq. (3.9-1) with $\omega = \omega_n$

$$X = \sqrt{\frac{3\pi F_0}{8a\omega_n^2}}$$

Example 3.9-2

> An oscillatory system forced to vibrate by an exciting force $F_0 \sin \omega t$ is known to be acted upon by several different forms of damping. Develop the equation for the equivalent damping and indicate the procedure for determining the amplitude at resonance.

Solution: Let U_1, U_2, U_3, etc., be the energy dissipated per cycle by the various damping forces. Equating the total energy dissipated to that of equivalent viscous damping

$$\pi c_{eq} \omega X^2 = U_1 + U_2 + U_3 + \cdots$$

The equivalent viscous damping coefficient is found to be

$$c_{eq} = \frac{\sum U}{\pi \omega X^2}$$

To determine the amplitude, it is necessary to obtain expressions for U_1, U_2, U_3, etc., which will contain X raised to various powers. Substituting c_{eq} into the expression

$$X = \frac{F_0}{c_{eq}\omega}$$

the equation with $\omega = \omega_n$ is solved for X.

3.10 STRUCTURAL DAMPING

When materials are cyclicly stressed, energy is dissipated internally within the material itself. Experiments by several investigators*,† indicate that for most structural metals, such as steel or aluminum, the energy dissipated per cycle is independent of the frequency over a wide frequency range, and proportional to the square of the amplitude of vibration. Internal damping fitting this classification is called *solid damping* or *structural damping*. With the energy dissipation per cycle proportional to the square of the vibration amplitude, the loss coefficient is a constant, and the shape of the hysteresis curve remains unchanged with amplitude and independent of the strain rate.

Energy dissipated by structural damping may be written as

$$W_d = \alpha X^2 \tag{3.10-1}$$

where α is a constant. Using the concept of equivalent viscous damping, Eq. (3.9-2) gives

$$\pi c_{eq} \omega X^2 = \alpha X^2$$

or

$$c_{eq} = \frac{\alpha}{\pi \omega} \tag{3.10-2}$$

Substitution of c_{eq} for c, the differential equation of motion for a system with structural damping may be written as

$$m\ddot{x} + \left(\frac{\alpha}{\pi\omega}\right)\dot{x} + kx = F(t) \tag{3.10-3}$$

Complex Stiffness. In the calculation of the flutter speeds of airplane wings and tail surfaces, the concept of *complex stiffness* is used. It is arrived at by assuming the oscillations to be harmonic, which enables Eq. (3.10-3) to be written as

$$m\ddot{x} + \left(k + i\frac{\alpha}{\pi}\right)x = F_0 e^{i\omega t}$$

By factoring out the stiffness k and letting $\gamma = \alpha/\pi k$, the above equation becomes

$$m\ddot{x} + k(1 + i\gamma)x = F_0 e^{i\omega t} \tag{3.10-4}$$

*Kimball, A. L. "Vibration damping, including the case of solid damping," Trans. ASME, APM 51–52, (1929).

†Lazan, B. J. "Damping of Materials and Members in Structural Mechanics" Pergamon Press, (1968).

The quantity $k(1 + i\gamma)$ is called the *complex stiffness* and γ is the *structural damping factor*.

Using the concept of complex stiffness for problems in structural vibrations is advantageous in that one needs only to multiply the stiffness terms in the system by $(1 + i\gamma)$. The method is justified, however, only for harmonic oscillations. With the solution $x = \bar{X}e^{i\omega t}$, the steady state amplitude from Eq. (3.10-4) becomes

$$\bar{X} = \frac{F_0}{(k - m\omega^2) + i\gamma k} \tag{3.10-5}$$

The amplitude at resonance is then

$$|X| = \frac{F_0}{\gamma k} \tag{3.10-6}$$

Comparing this with the resonant response of a system with viscous damping

$$|X| = \frac{F_0}{2\zeta k}$$

we conclude that with equal amplitudes at resonance, the structural damping factor is equal to twice the viscous damping factor.

3.11 SHARPNESS OF RESONANCE

In forced vibration there is a quantity Q related to damping that is a measure of the sharpness of resonance. To determine this quantity, we will assume viscous damping and start with Eq. (3.2-7).

When $\omega/\omega_n = 1$, the resonant amplitude is $X_{res} = (F_0/k)/2\zeta$. We now seek the two frequencies on either side of resonance (often referred to as *sidebands*), where X is 0.707 X_{res}. These points are also referred to as the *half-power points* and are shown in Fig. 3.11-1.

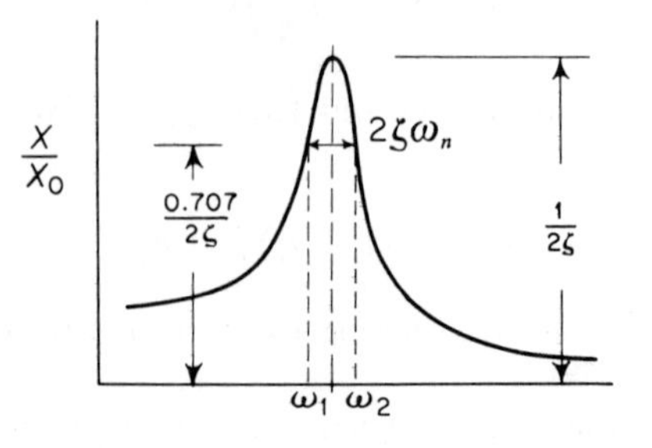

Figure 3.11-1.

Letting $X = 0.707\, X_{res}$ and squaring Eq. (3.2-7), we obtain

$$\frac{1}{2}\left(\frac{1}{2\zeta}\right)^2 = \frac{1}{\left[1 - \left(\frac{\omega}{\omega_n}\right)^2\right]^2 + \left[2\zeta\left(\frac{\omega}{\omega_n}\right)\right]^2}$$

or

$$\left(\frac{\omega}{\omega_n}\right)^4 - 2(1 - 2\zeta^2)\left(\frac{\omega}{\omega_n}\right)^2 + (1 - 8\zeta^2) = 0 \tag{3.11-1}$$

Solving for $(\omega/\omega_n)^2$ we have

$$\left(\frac{\omega}{\omega_n}\right)^2 = (1 - 2\zeta^2) \pm 2\zeta\sqrt{1 - \zeta^2} \tag{3.11-2}$$

Assuming $\zeta \ll 1$ and neglecting higher order terms of ζ, we arrive at the result

$$\left(\frac{\omega}{\omega_n}\right)^2 = 1 \pm 2\zeta \tag{3.11-3}$$

Letting the two frequencies corresponding to the roots of Eq. (3.11-3) be ω_1 and ω_2, we obtain

$$4\zeta = \frac{\omega_2^2 - \omega_1^2}{\omega_n^2} \cong 2\left(\frac{\omega_2 - \omega_1}{\omega_n}\right)$$

The quantity Q is then defined as

$$Q = \frac{\omega_n}{\omega_2 - \omega_1} = \frac{f_n}{f_2 - f_1} = \frac{1}{2\zeta} \tag{3.11-4}$$

Here, again, equivalent damping can be used to define Q for systems with other forms of damping. Thus, for structural damping, Q is equal to

$$Q = \frac{1}{\gamma} \tag{3.11-5}$$

PROBLEMS

3-1 A machine part weighing 4.3 lb vibrates in a viscous medium. Determine the damping coefficient when a harmonic exciting force of 5.5 lb results in a resonant amplitude of 0.50 in. with a period of 0.20 sec.

3-2 If the system of Prob. 3-1 is excited by a harmonic force of frequency 4 cps, what will be the percentage increase in the amplitude of forced vibration when the dashpot is removed?

3-3 A weight attached to a spring of stiffness 3.0 lb/in. has a viscous damping device. When the weight is displaced and released, the period of vibration is found to be 1.80 sec, and the ratio of consecutive amplitudes is 4.2 to 1.0. Determine the amplitude and phase when a force $F = 2 \cos 3t$ acts on the system.

3-4 Show that for the damped spring-mass system, the peak amplitude occurs at a frequency ratio given by the expression

$$\left(\frac{\omega}{\omega_n}\right)_p = \sqrt{1 - 2\zeta^2}$$

3-5 A spring-mass system is excited by a force $F_0 \sin \omega t$. At resonance the amplitude is measured to be 0.58 in. At 0.80 resonant frequency, the amplitude is measured to be 0.46 in. Determine the damping factor c of the system. (*Hint:* Assume damping term is negligible at 0.80 resonance.)

3-6 A circular disk rotating about its geometric axis has two holes A and B drilled through it. The diameter and position of the holes are $d_A = 1.0$ in., $r_A = 3.0$ in., $\theta_A = 0$ deg; $d_B = \frac{1}{2}$ in., $r_B = 2.0$ in., $\theta_B = 90$ deg. Determine the diameter and position of a third hole at 1 in. radius that will balance the disk.

3-7 The crank arm and pin of the two-cylinder crankshaft shown in Fig. P3-7 is equivalent to an eccentric weight of w lb at a radius of r in. Determine the counterweights necessary at the two flywheels if they are also placed at a radial distance of r in.

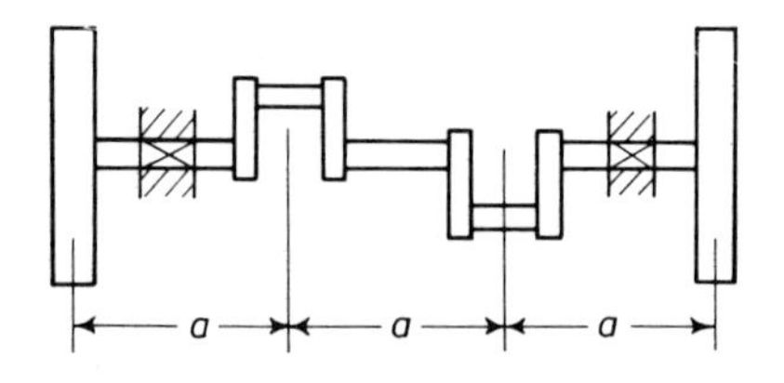

Figure P. 3-7.

3-8 For the system shown in Fig. P3-8, set up the equation of motion and solve for the steady state amplitude and phase angle by the use of complex algebra.

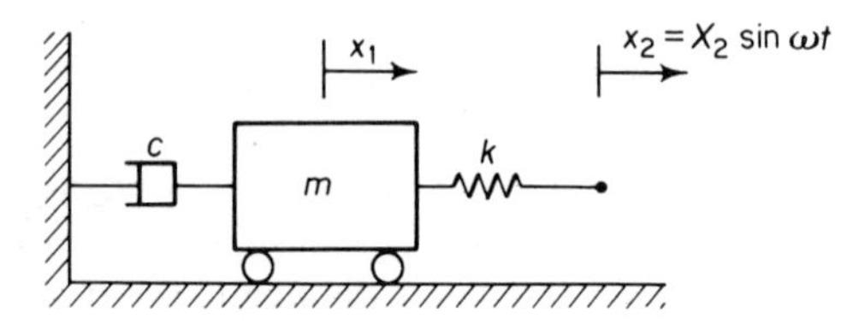

Figure P. 3-8.

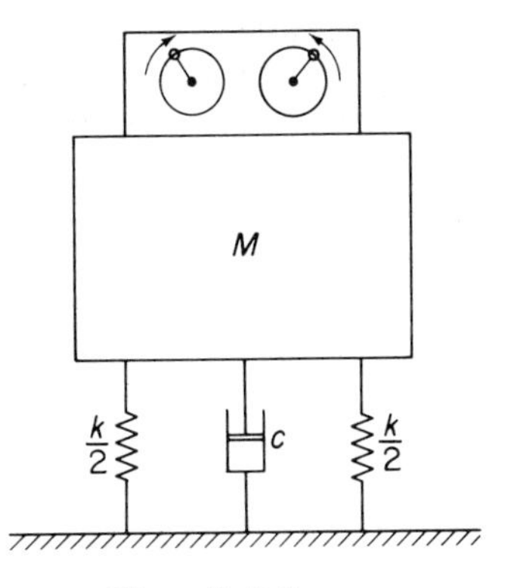

Figure P. 3-9.

3-9 A counter-rotating eccentric weight exciter shown in Fig. P3-9 is used to determine the vibrational characteristics of a structure weighing 400 lb. At a speed of 900 rpm, a stroboscope shows the eccentric weights to be at the top at the instant the structure is moving upward through its static equilibrium position, and the corresponding amplitude is 0.85 in. If the unbalance of each wheel of the exciter is 4 lb in., determine (a) the natural frequency of the structure, (b) the damping factor of the structure, (c) the amplitude at 1200 rpm, and (d) the angular position of the eccentrics at the instant the structure is moving upward through its equilibrium position.

3-10 A solid disk of weight 10 lb is keyed to the center of a $\frac{1}{2}$ in. steel shaft 2 ft between bearings. Determine the lowest critical speed. (Assume shaft to be simply supported at the bearings.)

3-11 The rotor of a turbine weighing 30 lb is supported at the midspan of a shaft with bearings 16 in. apart, as shown in Fig. P3-11. The rotor is known to have an unbalance of 4 oz in. Determine the forces exerted on the bearings at a speed of 6000 rpm if the diameter of the steel shaft is 1.0 in. Compare this result with that of the same rotor mounted on a steel shaft of diameter $\frac{3}{4}$ in. (Assume the shaft to be simply supported at the bearings.)

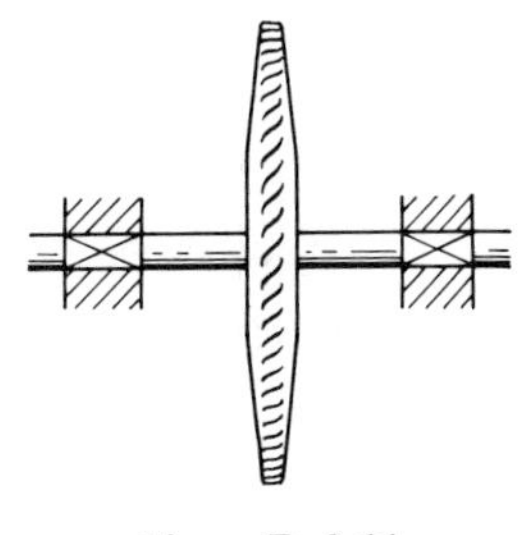
Figure P. 3-11.

3-12 Show that if damping is small, the amplitude of lateral vibration of a shaft at the critical speed builds up according to the equation

$$r = \frac{e}{2\zeta}(1 - e^{-\zeta\omega_n t})$$

where e is the eccentricity.

3-13 For turbines operating above the critical speed, stops are provided to limit the amplitude as it runs through the critical speed. In the turbine of Prob. 3-11, if the clearance between the 1-in. shaft and the stops is 0.02 in., and the eccentricity is $\frac{1}{120}$ in., determine the time required for the shaft to hit the stops, assuming that the critical speed is reached with zero amplitude.

3-14 Figure P3-14 represents a simplified diagram of a spring-supported vechicle traveling over a rough road. Determine the equation for the amplitude of W as a function of the speed and determine the most unfavorable speed.

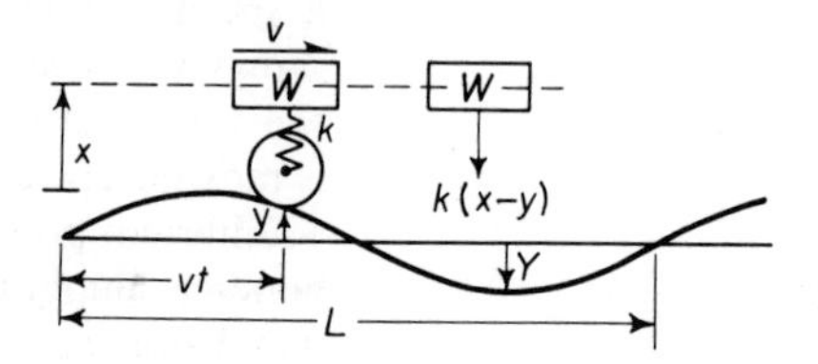

Figure P. 3-14.

3-15 The springs of an automobile trailer are compressed 4 in. under its weight. Find the critical speed when the trailer is traveling over a road with a profile approximated by a sine wave of amplitude 3.0 in. and wave length of 48 ft. What will be the amplitude of vibration at 40 mph? (Neglect damping.)

3-16 Shown in Fig. P.3-16 is a cylinder of mass m connected to a spring of stiffness k excited through viscous friction c to a piston with motion $y = A \sin \omega t$. Determine the amplitude of the cylinder motion and its phase with respect to the piston.

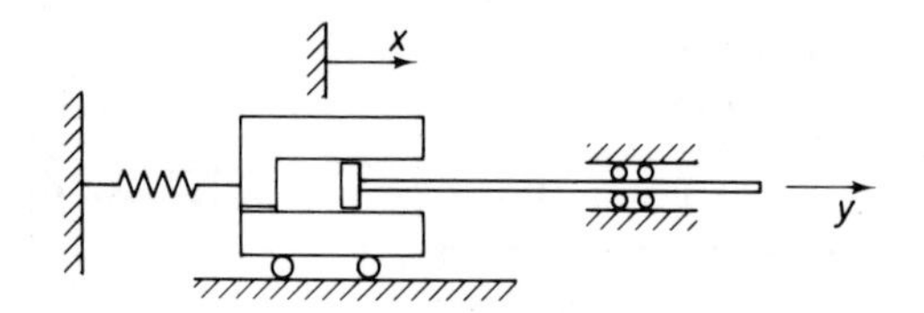

Figure P. 3-16.

3-17 The point of suspension of a simple pendulum is given a harmonic motion $x_0 = X_0 \sin \omega t$ along a horizontal line, as shown in Fig. P3-17. Write the differential equation of motion for small amplitude of oscillation, using the coordinates shown. Determine the solution for x/x_0 and show that when $\omega = \sqrt{2}\, \omega_n$, the node is found at the mid-point of l. Show that in general the distance h from the mass to the node is given by the relation $h = l(\omega_n/\omega)^2$ where $\omega_n = \sqrt{g/l}$.

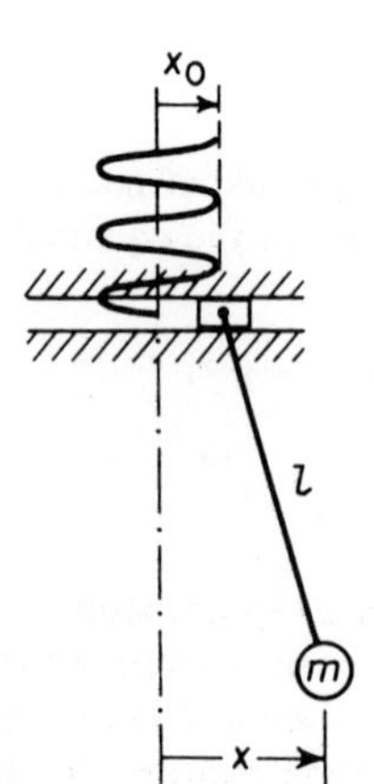

Figure P. 3-17.

3-18 A commercial-type vibration pickup has a natural frequency of 4.75 cps and a damping factor $\zeta = 0.65$. What is the lowest frequency that can be measured with (a) 1 percent error, (b) 2 percent error?

3-19 An undamped vibration pickup having a natural frequency of 1 cps is used to measure a harmonic vibration of 4 cps. If the amplitude indicated by the pickup (relative amplitude between pickup mass and frame) is 0.052 in., what is the correct amplitude?

3-20 The shaft of a torsiograph, shown in Fig. P3-20, undergoes harmonic torsional oscillation $\theta_0 \sin \omega t$. Determine the expression for the relative amplitude of the outer wheel with respect to (a) the shaft, (b) a fixed reference.

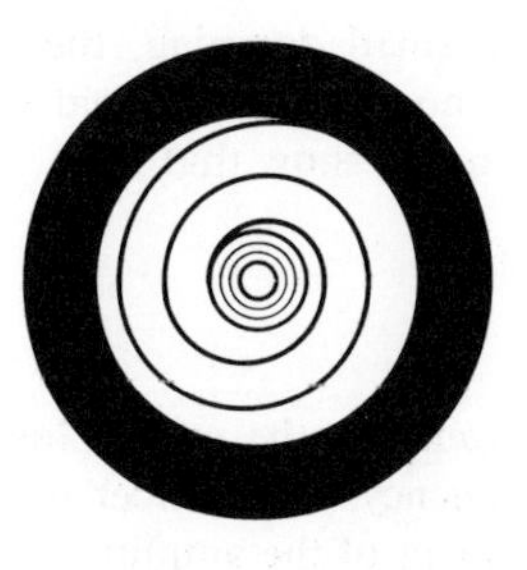

Figure P. 3-20.

3-21 Discuss the requirements of a seismic instrument from the standpoint of limiting phase distortion of complex waves.

3-22 A refrigerator unit weighing 65 lb is to be supported by three springs of stiffness k lb/in. each. If the unit operates at 580 rpm, what should be the value of the spring constant k if only 10 per cent of the shaking force of the unit is to be transmitted to the supporting structure?

3-23 An industrial machine weighing 1000 lb is supported on springs with a statical deflection of 0.20 in. If the machine has a rotating unbalance of 20 lb in., determine (a) the force transmitted to the floor at 1200 rpm, (b) the dynamical amplitude at this speed. (Assume damping to be negligible.)

3-24 If the machine of Prob. 3-23 is mounted on a large concrete block weighing 2500 lb and the stiffness of the springs or pads under the block is increased so that the statical deflection is still 0.20 in., what will be the dynamic amplitude?

3-25 An aircraft radio weighing 24 lb is to be isolated from engine vibrations ranging in frequencies from 1600 to 2200 cpm. What statical deflection must the isolators have for 85 per cent isolation?

3-26 Show that for viscous damping, the loss factor η is independent of the amplitude and proportional to the frequency.

3-27 Express the equation for the free vibration of a single degree of freedom system in terms of the loss factor η at resonance.

3-28 Show that τ_n/τ_d plotted against ζ is a quarter circle where τ_d = damped natural period and τ_n = undamped natural period.

3-29 Show that the energy dissipated per cycle for viscous friction can be expressed as

$$W_d = \frac{\pi F_0^2}{k} \frac{2\zeta}{[1 - (\omega/\omega_n)^2]^2 + [2\zeta(\omega/\omega_n)]^2}$$

3-30 For energy dissipation per cycle to be independent of the frequency ratio ω/ω_n find the required damping.

3-31 For small damping, the energy dissipated per cycle divided by the peak potential energy is equal to 2δ and also to $1/Q$. (See Eq. 3.7-6). For viscous damping show that

$$\delta = \frac{\pi c \omega_n}{k}$$

3-32 In general, the energy loss per cycle is a function of both amplitude and frequency. State under what condition the logarithmic decrement δ is independent of the amplitude.

3-33 Coulomb damping between dry surfaces is a constant D always opposed to the motion. Determine the equivalent viscous damping.

3-34 Using the result of Prob 3-33, determine the amplitude of motion of a spring mass system with Coulomb damping when excited by a harmonic force $F_0 \sin \omega t$. Under what condition can this motion be maintained?

3-35 For structural damping assume the stiffness to be a complex quantity in the form $k = ke^{i2\beta}$. Determine the equation for the response under harmonic excitation.

3-36 Show that $\delta = \pi(f_2 - f_1)/f_r$ where f_1 and f_2 are frequencies corresponding to the half power points of the resonance curve.

4

Transient Vibration

4.1 INTRODUCTION

When a dynamical system is excited by a suddenly applied nonperiodic excitation $F(t)$, such as the one shown in Fig. 4.1-1, the response to such excitation is called *transient response* since steady state oscillations are generally not produced. Such oscillations take place at the natural frequency of the system with the amplitude varying in a manner dependent on the type of excitation.

We first study the response of a spring-mass system to an impulse excitation because this case is important in the understanding of the more general problem of transients.

4.2 IMPULSE EXCITATION

Impulse is the time integral of the force, and we designate it by the notation $\hat{F}$

$$\hat{F} = \int F(t)\,dt \tag{4.2-1}$$

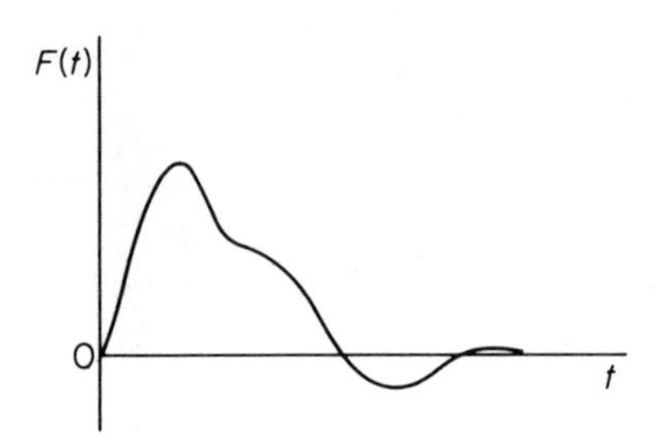

Figure 4.1-1. Nonperiodic excitation.

We frequently encounter a force of very large magnitude which acts for a very short time, but with a time integral which is finite. Such forces are called *impulsive*.

Figure 4.2-1 shows an impulsive force of magnitude $\hat{F}/\epsilon$ with a time duration of ϵ. As ϵ approaches zero, such forces tend to become infinite; however, the impulse defined by its time integral is $\hat{F}$ which is considered to be finite. When $\hat{F}$ is equal to unity, such force in the limiting case $\epsilon \rightarrow 0$ is called the *unit impulse* or the *delta function.* A delta function at $t = \xi$ is identified by the symbol $\delta(t - \xi)$ and has the following properties

$$\begin{aligned} \delta(t-\xi) &= 0 \qquad \text{for all } t \neq \xi \\ &= \text{greater than any assumed value for } t = \xi \qquad (4.2\text{-}2) \end{aligned}$$

$$\int_0^\infty \delta(t-\xi)\,dt = 1.0 \qquad 0 < \xi < \infty$$

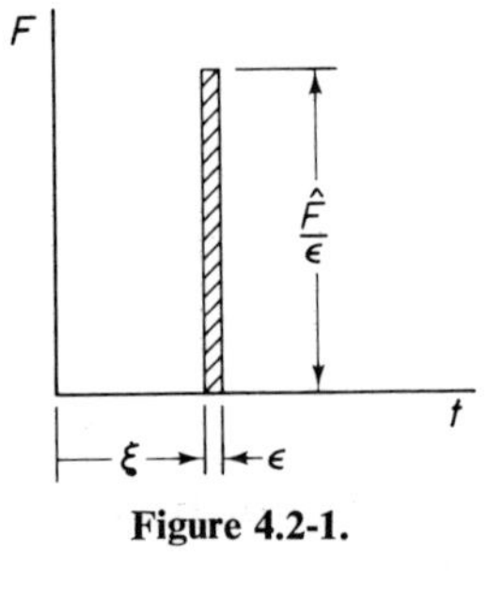

Figure 4.2-1.

If $\delta(t - \xi)$ is multiplied by any time function $f(t)$ as shown in Fig. 4.2-2, the product will be zero everywhere except at $t = \xi$, and its time integral will be

$$\int_0^\infty f(t)\delta(t-\xi)\,dt = f(\xi) \qquad 0 < \xi < \infty \tag{4.2-3}$$

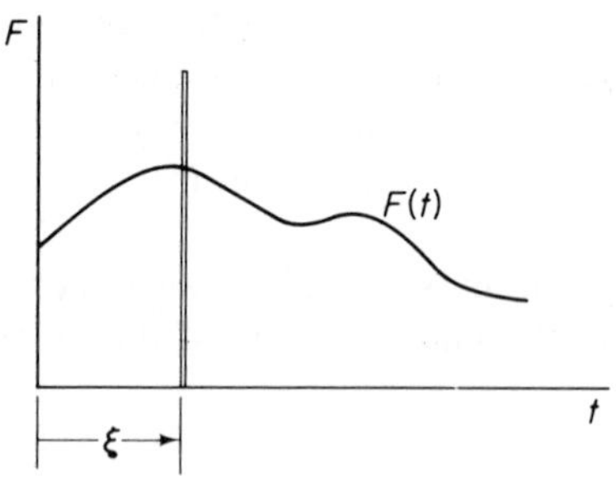

Figure 4.2-2.

Since $Fdt = mdv$, the impulse $\hat{F}$ acting on the mass will result in a sudden change in its velocity equal to $\hat{F}/m$ without an appreciable change in its displacement. Under free vibration we found that the undamped spring-mass system with initial conditions $x(0)$ and $\dot{x}(0)$ behaved according to the equation

$$x = \frac{\dot{x}(0)}{\omega_n} \sin \omega_n t + x(0) \cos \omega_n t$$

Hence the response of a spring-mass system initially at rest and excited by an impulse $\hat{F}$ is

$$x = \frac{\hat{F}}{m\omega_n} \sin \omega_n t \tag{4.2-4}$$

where

$$\omega_n = \sqrt{\frac{k}{m}}$$

When damping is present we can start with the free vibration equation

$$x = Xe^{-\zeta\omega_n t} \sin (\sqrt{1 - \zeta^2}\, \omega_n t - \phi)$$

and substituting the above initial conditions, we arrive at the equation

$$x = \frac{\hat{F}}{m\omega_n\sqrt{1 - \zeta^2}} e^{-\zeta\omega_n t} \sin \sqrt{1 - \zeta^2}\omega_n t \tag{4.2-5}$$

The response to the unit impulse is of importance to the problems of transients, and is identified by the special designation $g(t)$. Thus, in either the damped or undamped case, the equation for the impulsive response can be expressed in the form

$$x = \hat{F}g(t) \tag{4.2-6}$$

where the right side of the equation is given by either Eq. (4.2-4) or (4.2-5).

4.3 ARBITRARY EXCITATION

Having the response $g(t)$ to a unit impulse excitation, it is possible to establish the equation for the response of the system excited by an arbitrary force $f(t)$. For this development, we consider the arbitrary force to be a series of impulses as shown in Fig. 4.3-1. If we examine one of the impulses (shown crosshatched) at time $t = \xi$, its strength is

$$\hat{F} = f(\xi)\, \Delta\xi$$

and its contribution to the response at the time t is dependent upon the elapsed time $(t - \xi)$, or

$$f(\xi)\, \Delta\xi\, g(t - \xi)$$

Since the system we are considering is linear, the principle of superposition holds. Thus, by combining all such contributions, the response to the arbitrary

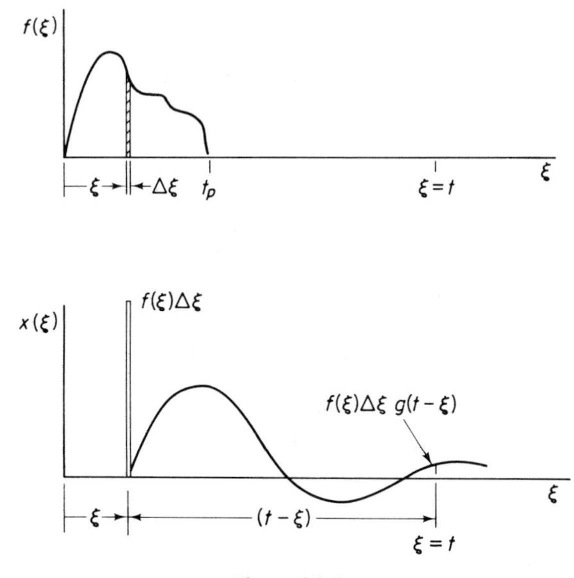

Figure 4.3-1.

excitation $f(t)$ is represented by the integral

$$x(t) = \int_0^t f(\xi)g(t - \xi)\,d\xi \tag{4.3-1}$$

The above integral is called the *Convolution integral*, or is sometimes referred to as the *superposition integral.*

Another form of this equation is found by letting $\tau = (t - \xi)$. Then $\xi = t - \tau$, $d\xi = -d\tau$, and we obtain

$$x(t) = \int_0^t f(t - \tau)g(\tau)\,d\tau \tag{4.3-2}$$

When t is greater than the pulse time, say t_p, the upper limit of the general equation, Eq. (4.3-1), remains at t_p because the integral can then be written as

$$\begin{aligned} x(t) &= \int_0^{t_p} f(\xi)g(t - \xi)\,d\xi + \int_{t_p}^t f(\xi)g(t - \xi)\,d\xi \\ &= \int_0^{t_p} f(\xi)g(t - \xi)\,d\xi, \qquad t > t_p \end{aligned} \tag{4.3-3}$$

Here the second integral is zero since $f(\xi) = 0$ for $\xi > t_p$.

Base Excitation. Often the support of the dynamical system is subjected to a sudden movement specified by its displacement, velocity, or acceleration.

The equation of motion can then be expressed in terms of the relative displacement $z = x - y$ as follows

$$\ddot{z} + 2\zeta\omega_n\dot{z} + \omega_n^2 z = -\ddot{y} \tag{4.3-4}$$

and hence all of the results for the force-excited system apply to the base-excited system for z when the term F_0/m is replaced by $-\ddot{y}$ or the negative of the base acceleration.

For an undamped system initially at rest, the solution for the relative displacement becomes

$$z = -\frac{1}{\omega_n}\int_0^t \ddot{y}(\xi)\sin\omega_n(t - \xi)\,d\xi \tag{4.3-5}$$

Example 4.3-1

Determine the response of a single degree of freedom system to the step excitation shown in Fig. 4.3-2.

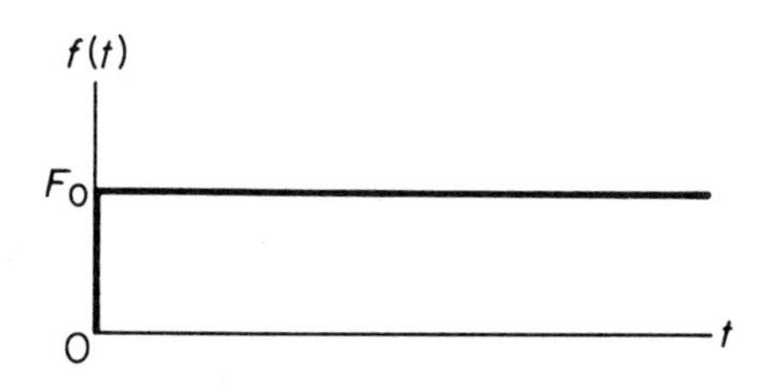

Figure 4.3-2. Step function excitation.

Solution: Considering the undamped system, we have

$$g(t) = \frac{1}{m\omega_n}\sin\omega_n t$$

Substituting into Eq. (4.3-1) the response of the undamped system is

$$\begin{aligned} x(t) &= \frac{F_0}{m\omega_n}\int_0^t \sin\omega_n(t - \xi)\,d\xi \\ &= \frac{F_0}{k}(1 - \cos\omega_n t) \end{aligned}$$

This result indicates that the peak response to the step excitation of magnitude F_0 is equal to twice the statical deflection.

For a damped system the procedure can be repeated with

$$g(t) = \frac{e^{-\zeta\omega_n t}}{m\omega_n\sqrt{1 - \zeta^2}}\sin\sqrt{1 - \zeta^2}\,\omega_n t$$

or, alternatively, we can simply consider the differential equation

$$\ddot{x} + 2\zeta\omega_n\dot{x} + \omega_n^2 x = \frac{F_0}{m}$$

whose solution is the sum of the solutions to the homogeneous equation and that of the particular solution, which for this case is $F_0/m\omega_n^2$. Thus the equation

$$x(t) = Xe^{-\zeta\omega_n t}\sin(\sqrt{1-\zeta^2}\,\omega_n t - \phi) + \frac{F_0}{m\omega_n^2}$$

fitted to the initial conditions of $x(0) = \dot{x}(0) = 0$ will result in the solution which is given as

$$x = \frac{F_0}{k}\left[1 - \frac{e^{-\zeta\omega_n t}}{\sqrt{1-\zeta^2}}\cos(\sqrt{1-\zeta^2}\,\omega_n t - \psi)\right]$$

where

$$\tan\psi = \frac{\zeta}{\sqrt{1-\zeta^2}}$$

Figure 4.3-3 shows a plot of xk/F_0 versus $\omega_n t$ with ζ as parameter, and it is evident that the peak response is less than $2F_0/k$ when damping is present.

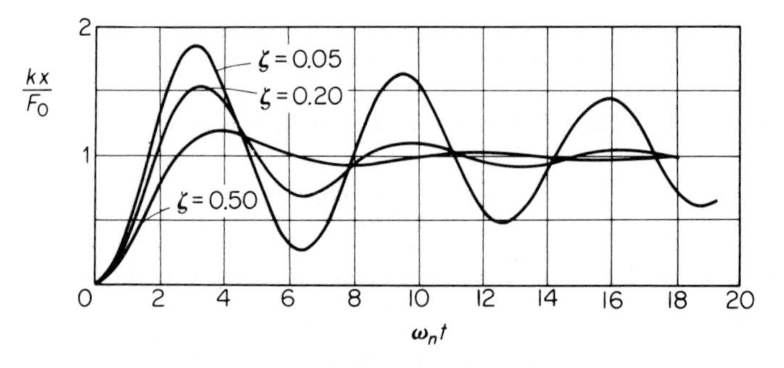

Fiture 4.3-3. Response to a unit step function.

Example 4.3-2

Consider an undamped spring-mass system where the motion of the base is specified by a velocity pulse of the form

$$\dot{y}(t) = v_0 e^{-t/t_0}$$

which is shown in Fig. 4.3-4 together with its time rate of change $a = \dot{v}$.

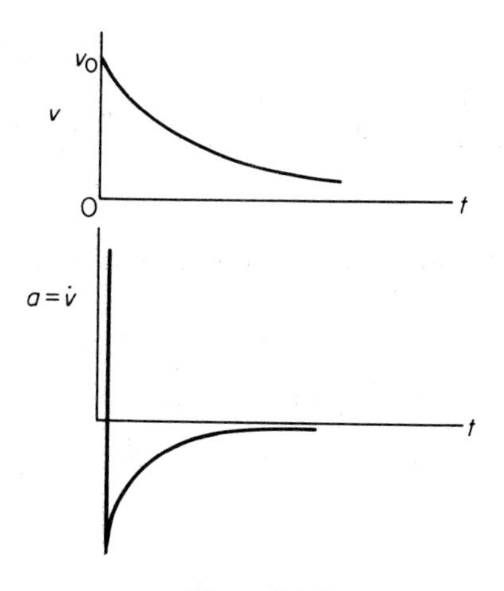

Figure 4.3-4.

Solution: The velocity pulse at $t = 0$ has a sudden jump from zero to v_0, and its rate of change (or acceleration) is infinite. Recognizing that $\int a\,dt = v$, the sudden change in the velocity at $t = 0$ is satisfied by

$$v_0 \int_0^{0+} \delta(t)\,dt = v_0$$

Thus the acceleration of the base becomes

$$\ddot{y}(t) = v_0 \delta(t) - \frac{v_0}{t_0} e^{-t/t_0}$$

Substitution of $\ddot{y}(t)$ into Eq. (4.3-5) yields

$$\begin{aligned} z(t) &= -\frac{v_0}{\omega_n} \int_0^t \left\{ \delta(\xi) - \frac{1}{t_0} e^{-\xi/t_0} \right\} \sin \omega_n (t - \xi)\, d\xi \\ &= \frac{v_0 t_0}{1 + (\omega_n t_0)^2} \{ e^{-t/t_0} - \omega_n t_0 \sin \omega_n t - \cos \omega_n t \} \end{aligned}$$

Example 4.3-3

A mass m attached to a spring of stiffness k is subjected to repeated impulse $\hat{F}$ of negligible duration at intervals of τ_i, as shown in Fig. 4.3-5. Determine the steady state response.

Solution: Between each impulse, the system is in free vibration at its natural frequency $\omega_n = \sqrt{k/m}$. Letting $t = 0$ immediately after the impulse,

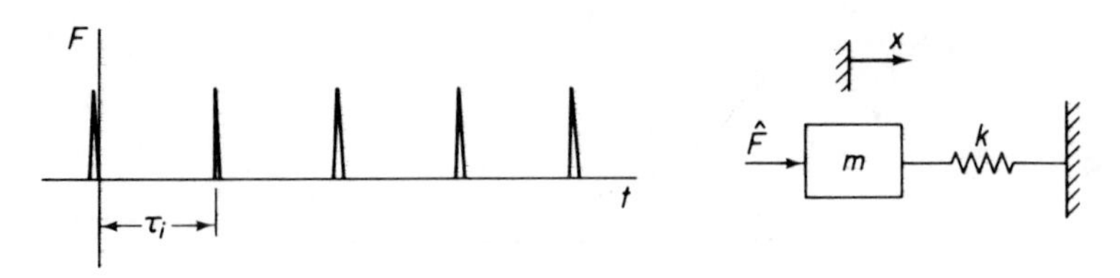

Figure 4.3-5. Repeated impulse on a spring-mass system.

the displacement and velocity may be expressed as

$$x = A \sin (\omega_n t + \phi) \tag{a}$$

$$\dot{x} = \omega_n A \cos (\omega_n t + \phi) \tag{b}$$

and hence at $t = 0$ we have

$$x(0) = A \sin \phi \tag{c}$$

$$\dot{x}(0) = \omega_n A \cos \phi \tag{d}$$

Just prior to the next impulse, the displacement and velocity are

$$x(\tau_i) = A \sin (\omega_n \tau_i + \phi) \tag{e}$$

$$\dot{x}(\tau_i) = \omega_n A \cos (\omega_n \tau_i + \phi) \tag{f}$$

where τ_i is the time interval of the impulses. The impulse acting at this time increases the velocity suddenly by $\hat{F}/m$ although the displacement remains essentially unchanged.

If steady state is attained, the displacement and velocity after each cycle must repeat themselves. Thus we can write

$$A \sin \phi = A \sin (\omega_n \tau_i + \phi) \tag{g}$$

$$\omega_n A \cos \phi = \omega_n A \cos (\omega_n \tau_i + \phi) + \frac{\hat{F}}{m} \tag{h}$$

Rearranging these equations to

$$\sin (\omega_n \tau_i + \phi) - \sin \phi = 0 \tag{i}$$

$$\cos (\omega_n \tau_i + \phi) - \cos \phi = -\frac{\hat{F}}{\omega_n m A} \tag{j}$$

we note that these equations may be rewritten as

$$\sin \frac{\omega_n \tau_i}{2} \cos \left(\frac{\omega_n \tau_i}{2} + \phi\right) = 0 \tag{i'}$$

$$\sin \frac{\omega_n \tau_i}{2} \sin \left(\frac{\omega_n \tau_i}{2} + \phi\right) = \frac{\hat{F}}{2\omega_n m A} \tag{j'}$$

Since $\sin \omega_n \tau_i/2$ cannot be zero for arbitrary τ_i, Eq. (i′) can be satisfied only if

$$\cos\left(\frac{\omega_n \tau_i}{2} + \phi\right) = 0$$

or

$$\sin\left(\frac{\omega_n \tau_i}{2} + \phi\right) = 1$$

Equation (j′) then becomes

$$\sin \frac{\omega_n \tau_i}{2} = \frac{\hat{F}}{2\omega_n m A} \tag{j''}$$

from which the amplitude is available as

$$A = \frac{\hat{F}}{2\omega_n m \sin \dfrac{\omega_n \tau_i}{2}} \tag{k}$$

The maximum spring force $F_s = kA$ may be of interest, in which case Eq. (k) takes the form

$$\frac{\tau_i F_s}{\hat{F}} = \frac{\dfrac{\omega_n \tau_i}{2}}{\sin \dfrac{\omega_n \tau_i}{2}} \tag{l}$$

Thus the amplitude or spring force becomes infinite when

$$\frac{\omega_n \tau_i}{2} = \frac{\pi f_n}{f_i} = 0, \pi, 2\pi, 3\pi, \ldots$$

Equation (l) also shows that the maximum spring force F_s is a minimum when

$$\frac{\pi f_n}{f_i} = \frac{\pi}{2}, \frac{3\pi}{2}, \frac{5\pi}{2}, \ldots$$

The time variation of the displacement and velocity may appear as in Fig. 4.3-6.

When damping is included, a similar procedure can be applied, although the numerical work is increased considerably.

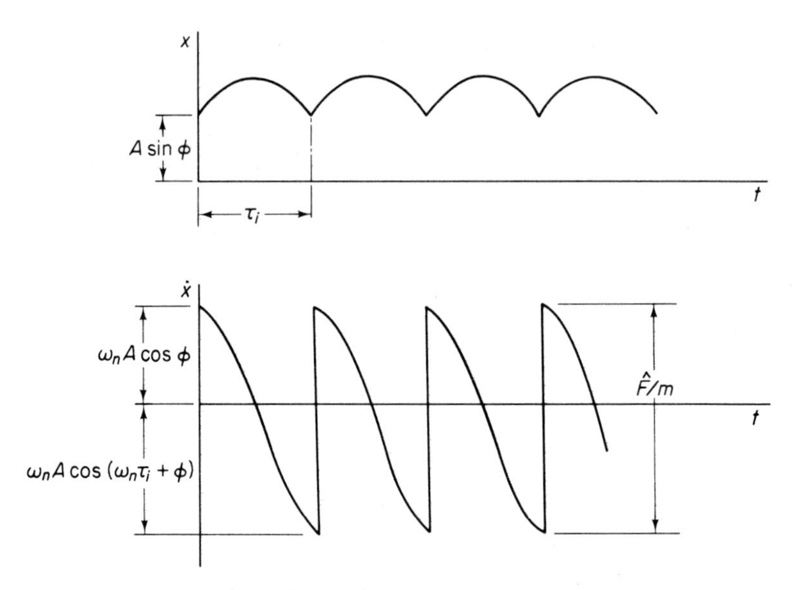

Figure 4.3-6. Displacement and velocity.

4.4 THE LAPLACE TRANSFORM FORMULATION

The Laplace transform method of solving the differential equation provides a complete solution, yielding both transient and forced vibration. For those unfamiliar with the method, a brief presentation of the Laplace transform theory is given in the Appendix. In this section we will illustrate its use by some simple examples.

Example 4.4-1

Formulate the Laplace transform solution of a viscously damped spring-mass system with initial conditions $x(0)$ and $\dot{x}(0)$.

Solution: The equation of motion for the system excited by an arbitrary force $F(t)$ is

$$m\ddot{x} + c\dot{x} + kx = F(t)$$

Taking its Laplace transform, we find

$$m[s^2\bar{x}(s) - x(0)s - \dot{x}(0)] + c[\bar{x}(s) + x(0)] + k\bar{x}(s) = \bar{F}(s)$$

Solving for $\bar{x}(s)$ we obtain the *subsidiary equation*

$$\bar{x}(s) = \frac{\bar{F}(s)}{ms^2 + cs + k} + \frac{(ms + c)x(0) + m\dot{x}(0)}{ms^2 + cs + k} \qquad (4.4\text{-}1)$$

The response $x(t)$ is found from the inverse of Eq. (4.4-1); the first term represents the forced vibration and the second term represents the transient solution due to the initial conditions.

For the more general case, the subsidiary equation can be written in the form

$$\bar{x}(s) = \frac{A(s)}{B(s)} \tag{4.4-2}$$

where $A(s)$ and $B(s)$ are polynomials and $B(s)$, in general, is of higher order than $A(s)$.

If only the forced solution is considered, we can define the *impedance transform* as

$$\frac{\bar{F}(s)}{\bar{x}(s)} = z(s) = ms^2 + cs + k \tag{4.4-3}$$

Its reciprocal is the admittance transform

$$H(s) = \frac{1}{z(s)} \tag{4.4-4}$$

Frequently a block diagram is used to denote input and output as shown in Fig. 4.4-1. The admittance transform $H(s)$ then can also be considered as the *system transfer function*, defined as the ratio in the subsidiary plane of the output over the input with all initial conditions equal to zero.

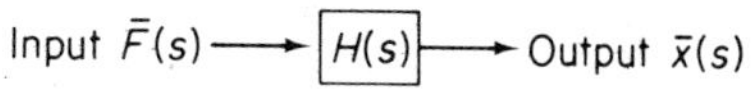

Input $\bar{F}(s)$ → $H(s)$ → Output $\bar{x}(s)$

Figure 4.4-1. Block diagram.

Example 4.4-2

A mass m is packaged in a box, as shown in Fig. 4.4-2, and dropped through a height h. It is desired to determine the maxi-

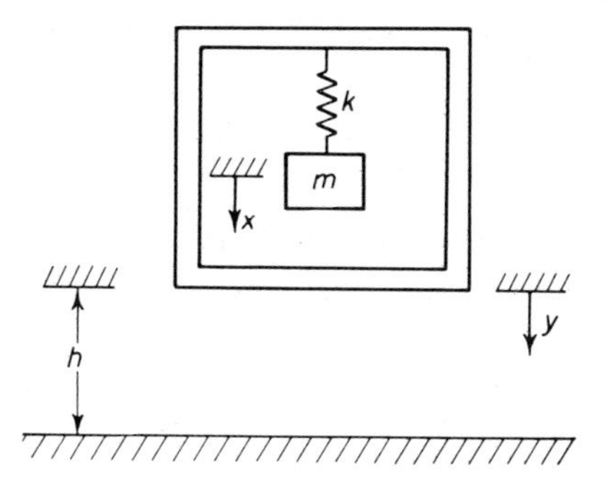

Figure 4.4-2. Drop test of a packaged mass.

mum force transmitted to the mass m and the required rattle space.*

Solution: We make the following idealized assumptions: (1) The mass m is supported within the box by a linear spring of stiffness k lb/in. (2) The mass of the box is large compared to that of the enclosed mass m, so that the free fall of the box is not influenced by the relative motion of the mass m. (3) On striking the floor, the box remains in contact with the floor.

Letting x be the displacement of m relative to the box, measured downward from the static equilibrium position, and y, the displacement of the box from the starting position, the general equation of motion of m is

$$m(\ddot{x} + \ddot{y}) = -kx \tag{4.4-5}$$

With $\omega_n^2 = k/m$, this equation becomes

$$\ddot{x} + \omega_n^2 x = -\ddot{y} \tag{4.4-6}$$

For initial conditions $x(0)$, $\dot{x}(0)$, $y(0)$, $\dot{y}(0)$, the Laplace transform of the above equation is

$$\bar{x}(s) = [x(0) + y(0)]\frac{s}{s^2 + \omega_n^2} + [\dot{x}(0) + \dot{y}(0)]\frac{1}{s^2 + \omega_n^2} - \frac{s^2\bar{y}(s)}{s^2 + \omega_n^2} \tag{4.4-7}$$

and its invese can be written as

$$x(t) = [x(0) + y(0)]\cos\omega_n t + \frac{1}{\omega_n}[\dot{x}(0) + \dot{y}(0)]\sin\omega_n t - \mathfrak{L}^{-1}\frac{s^2\bar{y}(s)}{s^2 + \omega_n^2} \tag{4.4-8}$$

We can now specialize this equation for the conditions of our problem. Of particular interest are the displacement $x(t_0)$ and the velocity $\dot{x}(t_0)$ of m at the time t_0 when the box strikes the floor.

The initial conditions for the free-fall interval are $x(0) = y(0) = \dot{x}(0) = \dot{y}(0) = 0$, and the motion of the box and its transform are

$$y(t) = \frac{1}{2}gt^2, \qquad \bar{y}(s) = \frac{g}{s^3} \tag{4.4-9}$$

The motion of m during free fall is then found from substituting Eq. (4.4-9) into (4.4-8) with the zero initial conditions, which gives

$$x(t) = -\mathfrak{L}^{-1}\frac{g}{s(s^2 + \omega_n^2)} = -\frac{g}{\omega_n^2}(1 - \cos\omega_n t) \tag{4.4-10}$$

*R. D. Mindlin, "Dynamics of Package Cushioning," *Bell Syst. Tech., Jour.*, 24, (July 1954) pp. 353–461.

Since the time to fall through the height h is $t_0 = \sqrt{2h/g}$, we arrive at the quantities of interest as

$$x(t_0) = -\frac{g}{\omega^2}(1 - \cos \omega_n t_0)$$

$$\dot{x}(t_0) = -\frac{g}{\omega_n} \sin \omega_n t_0$$

These quantities become the initial conditions for the second phase of the problem after impact of the box with the floor.

Redefining the time from the instant of impact, the initial conditions for the second phase of the problem are

$$y(0) = 0, \qquad x(0) = -\frac{g}{\omega_n^2}(1 - \cos \omega_n t_0)$$

$$\dot{y}(0) = g t_0, \qquad \dot{x}(0) = -\frac{g}{\omega_n} \sin \omega_n t_0$$

From the general equation, Eq. (4.4-8), the equation for the displacement of m after impact becomes

$$\begin{aligned} x(t) &= -\frac{g}{\omega_n^2}(1 - \cos \omega_n t) \cos \omega_n t + \left(\frac{g t_0}{\omega_n} - \frac{g}{\omega_n^2} \sin \omega_n t_0\right) \sin \omega_n t \\ &= \frac{g}{\omega_n^2}\sqrt{(1 - \cos \omega_n t_0)^2 + (\omega_n t_0 - \sin \omega_n t_0)^2} \sin (\omega_n t - \phi) \end{aligned} \tag{4.4-11}$$

where

$$\tan \phi = \frac{(1 - \cos \omega_n t_0)}{(\omega_n t_0 - \sin \omega_n t_0)} \tag{4.4-12}$$

Thus the maximum amplitude attained by m is

$$X_1 = \frac{g}{\omega_n^2}\sqrt{(1 - \cos \omega_n t_0)^2 + (\omega_n t_0 - \sin \omega_n t_0)^2} \tag{4.4-13}$$

which occurs at the time $(\omega_n t_1 - \phi) = \pi/2$. The maximum force is simply kX_1.

For any drop height h, or drop time $t_0 = \sqrt{2h/g}$, X_1 has a maximum value for $\omega_n \longrightarrow 0$. This can be shown by replacing $\sin \omega_n t_0$ and $\cos \omega_n t_0$ with their series form and allowing $\omega_n t_0$ to approach zero. Figure 4.4-3 shows a displacement response of X_1 versus frequency $f_n = (1/2\pi)\sqrt{k/m}$ for $h = 10, 5, 1$, and 0.15 in. Equation (4.4-13) however, indicates that $\omega_n^2 X_1/g = X_1/\delta_{st}$ is a function only of

$$\omega_n t_0 = \omega_n \sqrt{2h/g} = \sqrt{2h/\delta_{st}} \tag{4.4-14}$$

so that the curves of Fig. 4.4-3 plot as a single nondimensional curve as shown in Fig. 4.4-4.

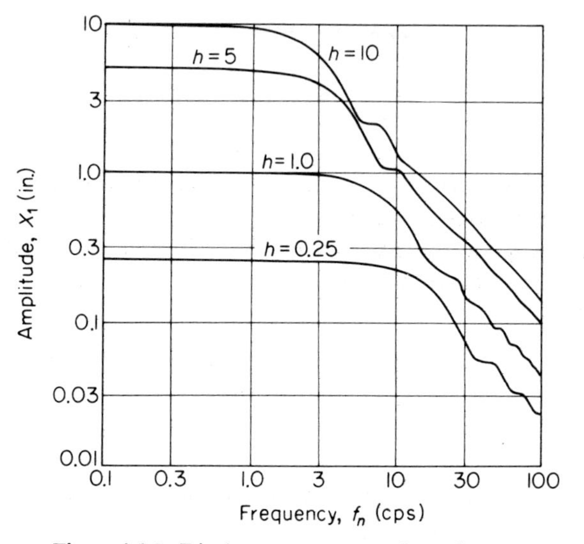

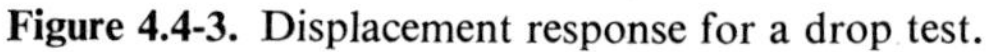

Figure 4.4-3. Displacement response for a drop test.

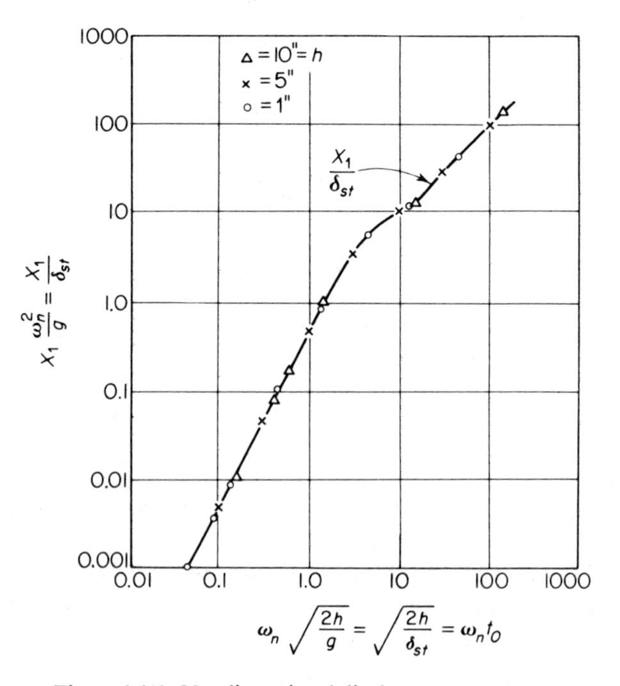

Figure 4.4-4. Nondimensional displacement response.

4.5 RESPONSE SPECTRUM

A shock represents a sudden application of a force or other form of disruption which results in a transient response of a system. The maximum value of the response is a good measure of the severity of the shock and is, of course, dependent upon the dynamic characteristics of the system. In order to categorize all types of shock excitations, a single degree of freedom undamped oscillator (spring-mass system) is chosen as a standard system.

Engineers have found the concept of the response spectrum to be useful in design. A *response spectrum* is a plot of the maximum peak response of the single degree of freedom oscillator as a function of the natural frequency of the oscillator. Different types of shock excitations will then result in different response spectra.

Since the response spectrum is determined from a single point on the time response curve, which is itself an incomplete bit of information, it does not uniquely define the shock input. In fact, it is possible for two different shock excitations to have very similar response spectra. In spite of this limitation, the response spectrum is a useful concept that is extensively used.

The response of a system to arbitrary excitation $f(t)$ was expressed in terms of the impulse response $g(t)$ by Eq. (4.3-1).

$$x(t) = \int_0^t f(\xi)g(t - \xi)\, d\xi \tag{4.5-1}$$

For the undamped single degree of freedom oscillator, we have

$$g(t) = \frac{1}{m\omega_n} \sin \omega_n t \tag{4.5-2}$$

so that the peak response to be used in the response spectrum plot is given by the equation

$$x(t)_{\max} = \left| \frac{1}{m\omega_n} \int_0^t f(\xi) \sin \omega_n (t - \xi)\, d\xi \right|_{\max} \tag{4.5-3}$$

In the case where the shock is due to the sudden motion of the support point, $f(t)$ in Eq. (4.5-3) is replaced by $-\ddot{y}(t)$, the acceleration of the support point, or

$$z(t)_{\max} = \left| \frac{-1}{\omega_n} \int_0^t \ddot{y}(\xi) \sin \omega_n (t - \xi)\, d\xi \right|_{\max} \tag{4.5-4}$$

Associated with the shock excitation $f(t)$ or $-\ddot{y}(t)$ is some characteristic time t_1, such as the duration of the shock pulse. With τ as the natural frequency of the oscillator, the maximum value of $x(t)$ or $z(t)$ is plotted as a function of t_1/τ.

Figures 4.5-1, 4.5-2 and 4.5-3 represent response spectra for three

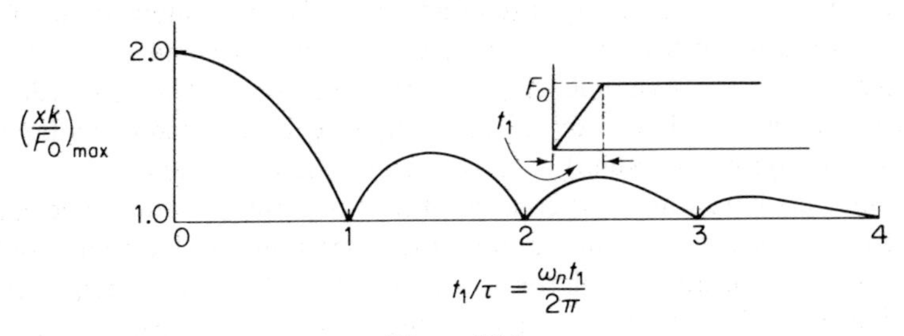

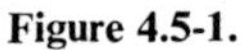

Figure 4.5-1.

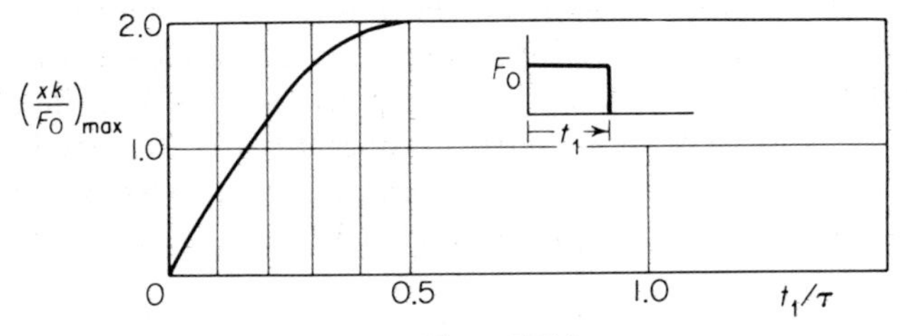

Figure 4.5-2.

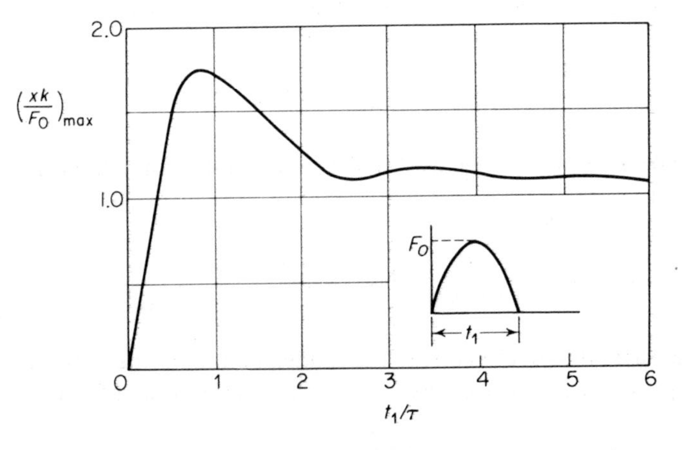

Figure 4.5-3.

different excitations. The horizontal scale is equal to the ratio t_1/τ, while the vertical scale is a nondimensional number which is a measure of the dynamic effect over that of a statically applied load. The dynamic factor of a shock is then generally less than two.

Pseudo Response Spectra. In ground shock situations, it is often convenient to express the response spectra in terms of the *velocity spectra.* The displacement and acceleration spectra then can be expressed in terms of the velocity spectra by dividing or multiplying by ω_n. Such results are called *pseudo spectra* since they are exact only if the peak response occurs after the shock pulse has passed, in which case the motion is harmonic.

Velocity spectra are used extensively in earthquake analysis, and damping is generally included. With relative displacement $z = x - y$, the equation for the damped oscillator is

$$\ddot{z} + 2\zeta\omega_n\dot{z} + \omega_n^2 = -\ddot{y} \tag{4.5-5}$$

and Eq. (4.5-4) is replaced by

$$z(t) = \frac{-1}{\omega_n\sqrt{1-\zeta^2}}\int_0^t \ddot{y}(\xi)e^{-\zeta\omega_n(t-\xi)} \sin\sqrt{1-\zeta^2}\,\omega_n(t-\xi)\,d\xi \tag{4.5-6}$$

Differentiating, using the equation

$$\frac{d}{dt}\int_0^t f(t_1\xi)\,d\xi = \int_0^t \frac{\partial f(t,\xi)}{dt}\,d\xi + f(t,\xi) \qquad \xi = t \tag{4.5-7}$$

we obtain for the velocity

$$\begin{aligned}\dot{z}(t) = \frac{-1}{\omega_n\sqrt{1-\zeta^2}}&\int_0^t \ddot{y}(\xi)e^{-\zeta\omega_n(t-\xi)} \\ &[-\zeta\omega_n \sin\sqrt{1-\zeta^2}\,\omega_n(t-\xi) \\ &+ \omega_n\sqrt{1-\zeta^2}\cos\sqrt{1-\zeta^2}\,\omega_n(t-\xi)]\,d\xi\end{aligned} \tag{4.5-8}$$

Letting

$$A = \int_0^t \ddot{y}(\xi)e^{\zeta\omega_n\zeta}\cos\sqrt{1-\xi^2}\,\omega_n\xi\,d\xi \tag{4.5-9}$$

$$B = \int_0^t \ddot{y}(\xi)e^{\zeta\omega_n\xi}\sin\sqrt{1-\xi^2}\,\omega_n\xi\,d\xi \tag{4.5-10}$$

Eq. (4.5-8) can be written as

$$\begin{aligned}\dot{z}(t) = \frac{e^{-\zeta\omega_n t}}{\sqrt{1-\zeta^2}}&\{[A\zeta - B\sqrt{1-\zeta^2}]\sin\sqrt{1-\zeta^2}\,\omega_n t \\ &+ [A\sqrt{1-\zeta^2} + B\zeta]\cos\sqrt{1-\zeta^2}\,\omega_n t\}\end{aligned} \tag{4.5-11}$$

or

$$\dot{z}(t) = \frac{e^{-\zeta\omega_n t}}{\sqrt{1-\zeta^2}}\sqrt{A^2+B^2}\sin(\sqrt{1-\zeta^2}\,\omega_n t - \phi) \tag{4.5-12}$$

If Eq. (4.5-12) is plotted against time, it would appear as an amplitude modulated wave, as shown in Fig. 4.5-4. Thus the peak velocity response S_v or the velocity spectrum is given with sufficient accuracy by the peak value of the envelope

$$S_v = |\dot{z}(t)|_{\max} = \left| \frac{e^{-\zeta\omega_n t}}{\sqrt{1-\zeta^2}} \sqrt{A^2 + B^2} \right|_{\max} \tag{4.5-13}$$

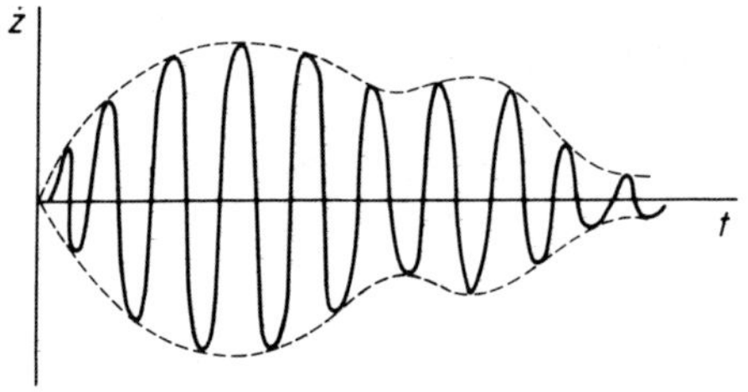

Figure 4.5-4.

Approximate relations for the peak displacement and acceleration, known as *pseudo spectra*, are then

$$|x - y|_{\max} = \frac{S_v}{\omega_n} \tag{4.5-14}$$

$$|\ddot{z}|_{\max} = \omega_n S_v \tag{4.5-15}$$

Example 4.5-1

Determine the undamped response spectrum for a step function with a rise time t_1, shown in Fig. 4.5-5.

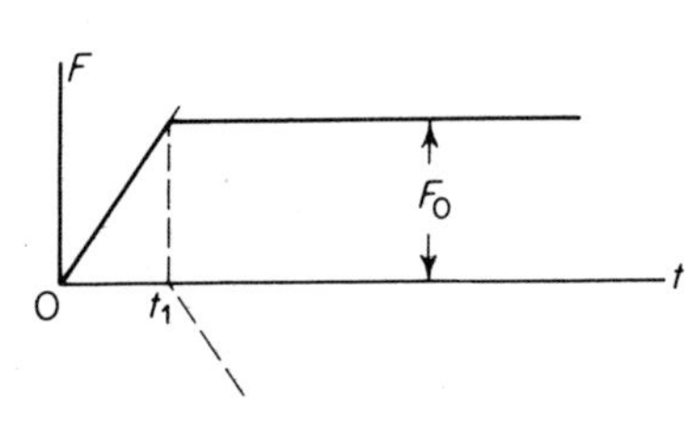

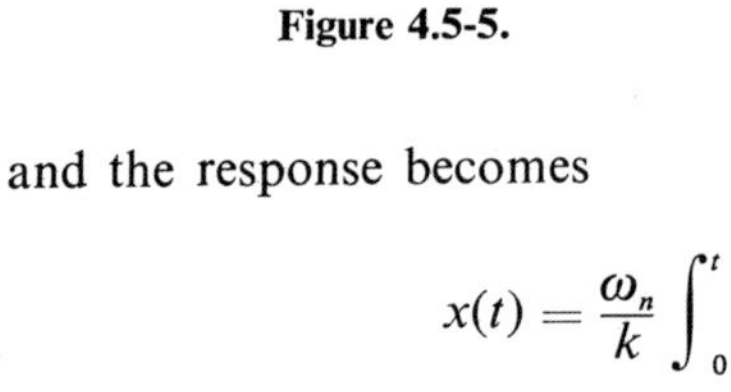

Figure 4.5-5.

Solution: The input can be considered to be the sum of two ramp functions $F_0(t/t_1)$, the second of which is negative and delayed by the time t_1. For the first ramp function the terms of the convolution integral are

$$f(t) = F_0(t/t_1)$$

$$g(t) = \frac{1}{m\omega_n} \sin \omega_n t = \frac{\omega_n}{k} \sin \omega_n t$$

and the response becomes

$$x(t) = \frac{\omega_n}{k} \int_0^t \frac{F_0 \xi}{t_1} \sin \omega_n (t - \xi)\, d\xi$$

$$= \frac{F_0}{k}\left(\frac{t}{t_1} - \frac{\sin \omega_n t}{\omega_n t_1}\right), \qquad t < t_1$$

For the second ramp function starting at t_1, the solution can be written down by inspection of the above equation as

$$x(t) = -\frac{F_0}{k}\left[\frac{(t-t_1)}{t_1} - \frac{\sin \omega_n(t-t_1)}{\omega_n t_1}\right]$$

By superimposing these two equations the response for $t > t_1$ becomes

$$x(t) = \frac{F_0}{k}\left[1 - \frac{\sin \omega_n t}{\omega_n t_1} + \frac{1}{\omega_n t_1}\sin \omega_n(t-t_1)\right] \qquad t > t_1$$

Differentiating and equating to zero, the peak time is obtained as

$$\tan \omega_n t_p = \frac{1 - \cos \omega_n t_1}{\sin \omega_n t_1}$$

Since $\omega_n t_p$ must be greater than π, we also obtain

$$\sin \omega_n t_p = -\sqrt{\tfrac{1}{2}(1 - \cos \omega_n t_1)}$$

$$\cos \omega_n t_p = \frac{-\sin \omega_n t_1}{\sqrt{2(1 - \cos \omega_n t_1)}}$$

Substituting these quantities into $x(t)$, the peak amplitude is found as

$$\left(\frac{xk}{F_0}\right)_{\max} = 1 + \frac{1}{\omega_n t_1}\sqrt{2(1 - \cos \omega_n t_1)}$$

Letting $\tau = 2\pi/\omega_n$ be the period of the oscillator, the above equation is plotted against t_1/τ as in Fig. 4.5-1.

Example 4.5-2

Determine the response spectrum for the base velocity input, $\dot{y}(t) = v_0 e^{-t/t_0}$ of Example 4.3-2.

Solution: The relative displacement $z(t)$ was found in Example 4.3-2 to be

$$z(t) = \frac{v_0 t_0}{1 + (\omega_n t_0)^2} \times (e^{-t/t_0} - \omega_n t_0 \sin \omega_n t - \cos \omega_n t)$$

To determine the peak value z_p, the usual procedure is to differentiate the equation with respect to time t, set it equal to zero, and substitute this time back into the equation for $z(t)$. It is evident that for this problem this results in a transcendental equation which must be solved by plotting. To avoid this numerical task, we will consider a different approach as follows.

For a very stiff system, which corresponds to large ω_n, the peak response will certainly occur at small t, and we would obtain for the time varying part of the equation the peak value

$$(1 - \omega_n t_0 - 1) = -\omega_n t_0$$

Thus for large ω_n the peak value will be nearly equal to

$$|z_p| \cong \frac{v_0 t_0}{1 + (\omega_n t_0)^2}(\omega_n t_0) \cong \frac{v_0 t_0}{\omega_n t_0}$$

so that $z_p/v_0 t_0$ plots against $\omega_n t_0$ as a rectangular hyperbola.

For small ω_n, or a very soft spring, the duration of the input would be small compared to the period of the system. Hence the input would appear as an impulsive doublet shown in Fig. 4.5-6 with the equation $v_0 t_0 \delta'(t)$. The solution for $z(t)$ is then

$$z(t) = v_0 t_0 \cos \omega_n t$$

and its peak value is

$$|z_p| \cong v_0 t_0$$

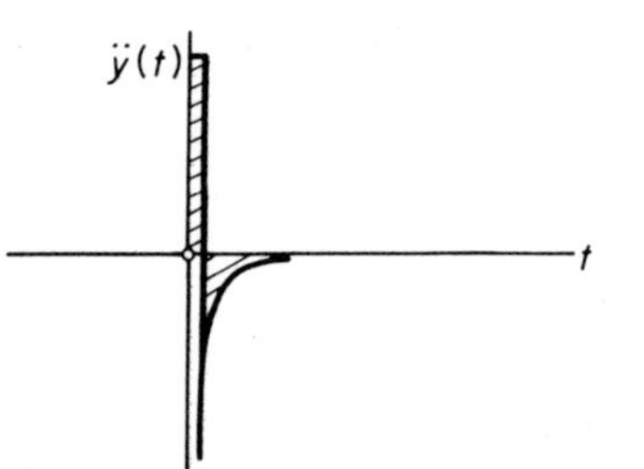

Figure 4.5-6. Impulsive doublet.

With these extreme conditions evaluated, we can now fill in the response spectrum which is shown in Fig. 4.5-7.

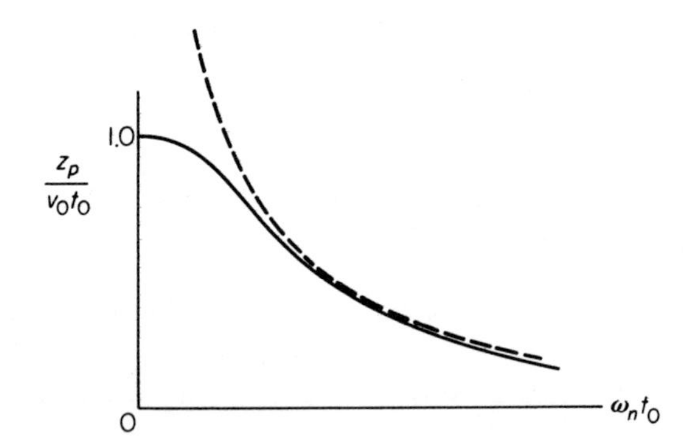

Figure 4.5-7. Response spectrum for the base velocity input $\dot{y}(t) = v_0 e^{-t/t_0}$.

4.6 THE ANALOG COMPUTER

Our brief encounter with transients and response spectra sufficiently indicates the algebraic difficulties confronted in even the very simple problems. Such problems are conveniently solved by the analog computer which is ideally suited for the solution of ordinary differential equations. These computers are also capable of solving nonlinear problems for which analytic techniques are either nonexistent or too complex for practical use.

The basic element of the analog computer is the high gain dc operational amplifier, shown symbolically in Fig. 4.6-1. It is not necessary, however, to know the details of its electronics. Such amplifiers are characterized by the equation

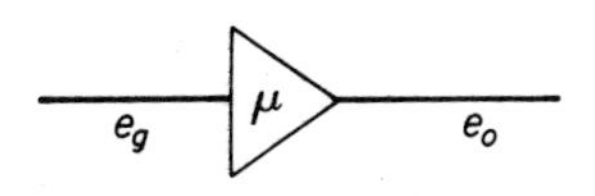

Figure 4.6-1. Operational amplifier.

$$e_0 = -\mu e_g \tag{4.6-1}$$

where μ is the amplification factor, and e_g and e_0 are the grid and output voltages. The amplification factor μ for the modern operational amplifiers is approximately 10^8. Since the output voltage is generally limited to ± 100 volts, the order of magnitude of the grid voltage e_g is $\pm 10^{-6}$ volts. The current drawn by the grid is also quite small with a representative value of approximately 10^{-7} amps.

By connecting the operational amplifier to different types of impedances the various operations of differentiation, integration, summing, etc., can be performed. Figure 4.6-2 indicates a general hookup of the amplifier with an

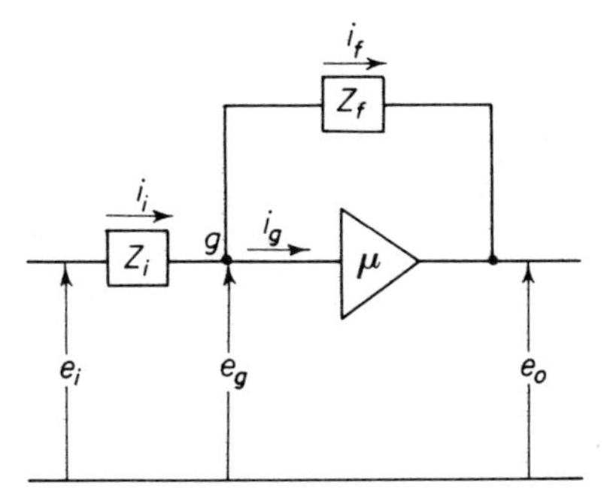

Figure 4.6-2. Operational amplifier circuit.

input impedance Z_i and a feedback impedance Z_f. The following equations can be written for the above circuit.

$$\begin{aligned} e_i - e_g &= i_i Z_i \\ e_g - e_0 &= i_f Z_f \\ i_i &= i_f + i_g \\ e_0 &= -\mu e_g \end{aligned}$$

In these equations, e_g is negligible in comparison to e_0 and e_i, and i_g is a negligible quantity in comparison to i_i and i_f. Thus with $e_g = i_g = 0$, the above equations become

$$e_i = i_i Z_i$$
$$-e_0 = i_f Z_f$$
$$i_i = i_f$$

From these, we obtain the output-input relationship

$$\frac{e_0}{e_i} = -\frac{Z_f}{Z_i} \tag{4.6-2}$$

which is fundamental to the analog computer.

(*a*) *Sign Change.* The simplest operation is that of changing sign. By letting $Z_i = R_i$ and $Z_f = R_f$, Eq. (4.6-2) becomes

$$e_0 = -\frac{R_f}{R_i} e_i \tag{4.6-3}$$

so that if $R_f = R_i$ we simply get

$$e_0 = -e_i$$

The circuit for the sign change is shown in Fig. 4.6-3.

(*b*) *Summation.* If more than one input is connected to point g, as shown in Fig. 4.6-4, then i_f is the sum of the input currents

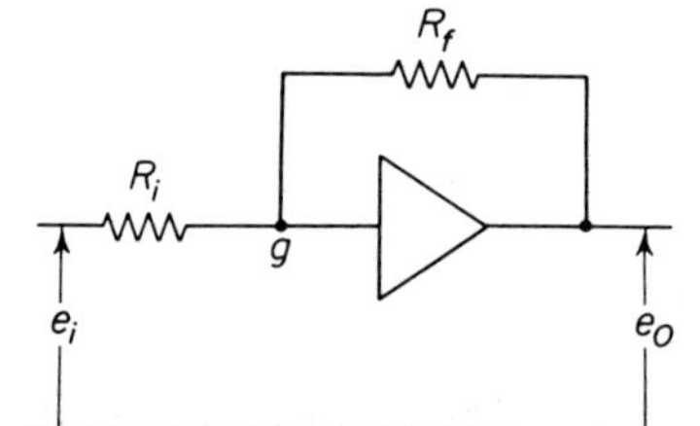

Figure 4.6-3. Circuit for scale factor, R_f/R_i.

$$i_1 + i_2 + i_3 = i_f$$

$$\frac{e_1}{R_1} + \frac{e_2}{R_2} + \frac{e_3}{R_3} = -\frac{e_0}{R_f}$$

and the output voltage is given by the sum

$$e_0 = -\sum_i \frac{R_f}{R_i} e_i \tag{4.6-4}$$

If all the resistances are equal, this then leads to the sum of the input voltages.

$$e_0 = -\sum_i e_i$$

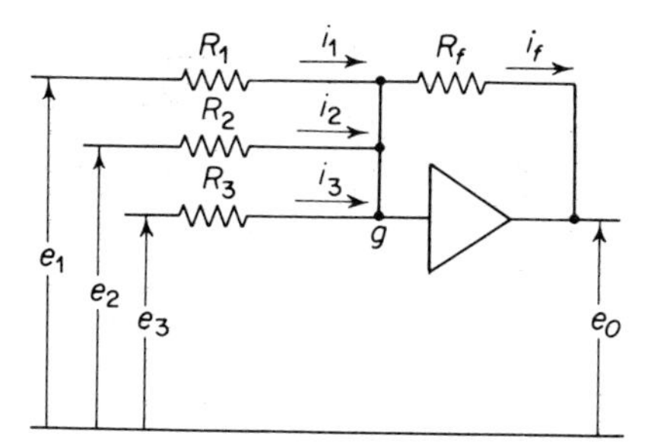

Figure 4.6-4. Circuit for summation.

(*c*) *Integration.* If the feedback impedance is a capacitor C, as shown in Fig. 4.6-5, the circuit will perform the function of integration. With initial voltage $e(0)$ across the capacitor, the capacitor voltage at any time t is (remembering that $e_g \cong 0$)

$$e_0 = -\frac{1}{C}\int_0^t i\,dt + e(0)$$

But $i = e_i/R$, so that the above equation becomes

$$e_0 = -\frac{1}{RC}\int_0^t e_i\,dt + e(0) \qquad (4.6\text{-}5)$$

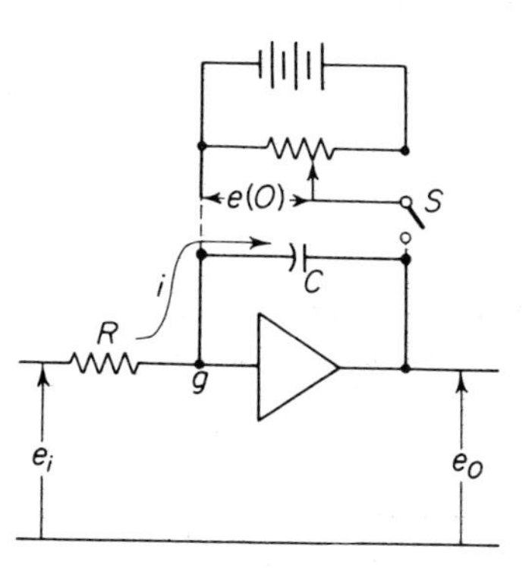

Figure 4.6-5. Integrating circuit with initial conditions.

With $R = 1$ megohm and $C = 1$ microfarad, $RC = 1$ second, and the computer time is directly in seconds.

The initial voltage $e(0)$ is obtained from the circuit shown in dotted lines by closing switch S prior to starting the computation. When the computation is initiated, the switch S is simultaneously opened by a relay.

(*d*) *Differentiation.* Differentiation is avoided in analog computers because the unavoidable noise signal in the input is amplified by μ, thereby causing the amplifier to saturate. Instead, it is usually possible to rearrange the equations for integration.

(*e*) *Voltage Division.* The potentiometer, often used to obtain a fraction k times the input voltage

$$e_2 = ke_1$$

is designated symbolically in Fig. 4.6-6(a). The fractional setting k on potentiometers holds only when the output is open circuited. When a load resis-

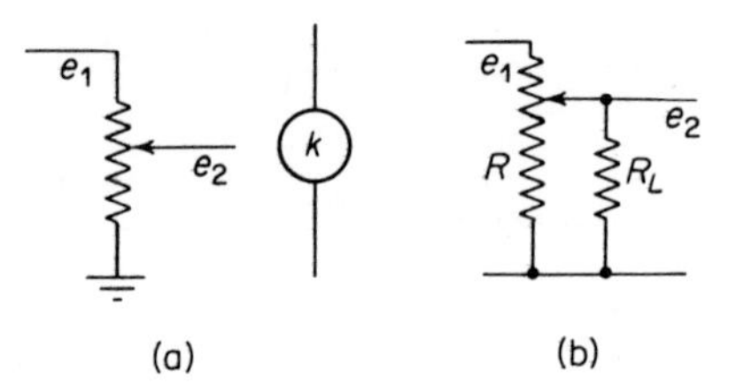

Figure 4.6-6. Voltage division.

tance R_L is placed across the output, as in Fig. 4.6-6(b), the output voltage can be shown to be equal to

$$e_2 = ke_1\left[\frac{1}{1 + \frac{R}{R_L}k(1 - k)}\right] \tag{4.6-6}$$

where R is the resistance of the potentiometer. It is evident that this equation approaches ke_1 when $R/R_L \longrightarrow 0$.

(*f*) *Multiplication.* Multiplication is one of the most difficult operations for the analogue computer. In one method, the principle of the unloaded potentiometer is exploited by using a servo-drive together with a ganged potentiometer, as shown in Fig. 4.6-7. The first potentiometer is connected to ± 100 volts, whereas the second potentiometer is connected to $\pm e_2$, the voltage to be multiplied by e_1.

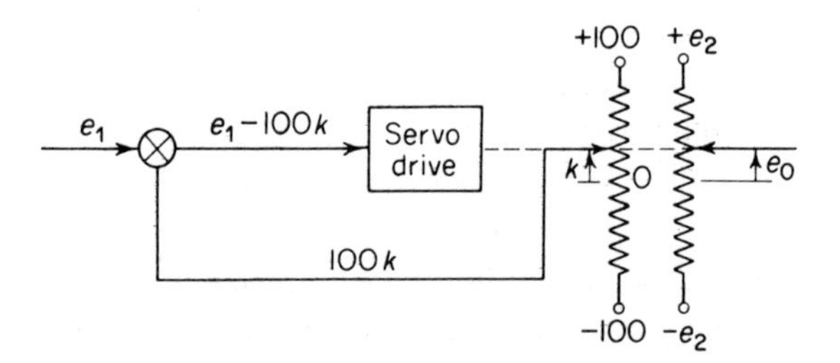

Figure 4.6-7. Servo-multiplier.

The function of the servo drive is to position the slider of the ganged potentiometer to zero the error between the output of the first potentiometer and the input voltage e_1. Since the output of each potentiometer is proportional to k we have

$$e_1 = 100k \quad \text{and} \quad e_0 = ke_2$$

Eliminating k, we obtain the equation

$$e_0 = \frac{e_1 e_2}{100} \tag{4.6-7}$$

(*g*) *Computer Circuit for the Single Degree of Freedom System.* The use of the analog computer for the solution of the single-degree-of-freedom linear system is demonstrated by the circuit of Fig. 4.6-8. The equation represented is

$$m\ddot{x} + c\dot{x} + kx = F(t)$$

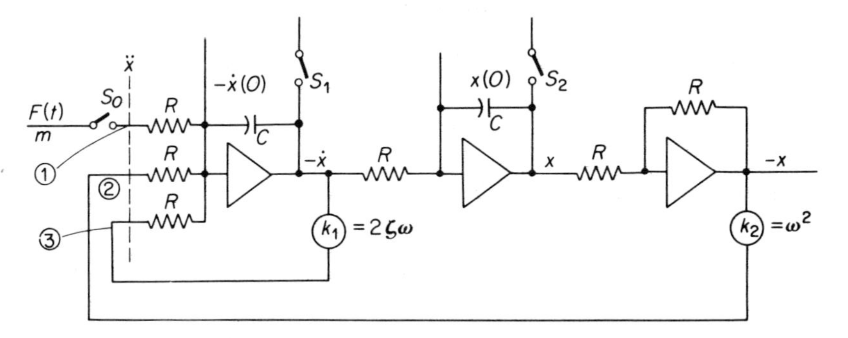

Figure 4.6-8. Analog circuit for the single degree of freedom system.

which is rearranged to

$$\ddot{x} = -2\zeta\omega_n\dot{x} - \omega_n^2 x + \frac{1}{m}F(t)$$

Assuming the input of the first amplifier to be $\ddot{x}$, its output is $-\dot{x}$, etc. The voltages in the three terminals ①, ②, and ③, which are equal to $\ddot{x}$ are those of the right side of the above equation. Note that the sign changes across each amplifier and that the potentiometer settings are for $2\zeta\omega_n$ and ω_n^2. The initial conditions $x(0)$ and $\dot{x}(0)$ are the voltages on the two capacitors at time $t = 0$. When the computer is set into operation by closing switch S_0, the switches S_1 and S_2 open simultaneously.

(*h*) *Scale Change.* If a linear transformation is made in the variables of the problem, the characteristics of the system remain unchanged, and we merely expand or contract the scale of the variables. In solving a problem on a computer, it is often necessary to make such a scale change because of the limitations of the computer. A change in the time scale may be necessary because the actual vibration may be too high in frequency for the computer and the recorder to follow. On the other hand, if the actual vibration is too low in frequency, the computer time may become too long, thereby introducing errors from drifts. Amplitude scaling is also necessary in order to operate the amplifiers within the limits of ± 100 volts. For maximum accuracy the peak response should be close to the ± 100-volt limits.

Assume that the equation for the actual system is

$$\ddot{x}(t) + 2\zeta\omega_n\dot{x}(t) + \omega_n^2 x(t) = \frac{1}{m}F(t) \tag{4.6-8}$$

with the initial conditions

$$\dot{x}(0) \quad \text{and} \quad x(0)$$

If we wish to change the time scale of the problem by α, we let

$$\tau = \alpha t \tag{4.6-9}$$

The derivatives are then related by the equations

$$\alpha \frac{d}{d\tau} = \frac{d}{dt}, \qquad \alpha^2 \frac{d^2}{d\tau^2} = \frac{d^2}{dt^2} \tag{4.6-10}$$

and the original differential equation with its initial conditions becomes

$$\begin{gathered} \alpha^2 \frac{d^2x(\tau)}{d\tau^2} + 2\zeta\omega_n\alpha\frac{dx(\tau)}{d\tau} + \omega_n^2 x(\tau) = \frac{1}{m}F(\tau) \\ \frac{dx(0)}{d\tau} = \frac{1}{\alpha}\frac{dx(0)}{dt}, \qquad x_\tau(0) = x_t(0) \end{gathered} \tag{4.6-11}$$

Dividing through by α^2, this equation takes the form

$$\frac{d^2x(\tau)}{d\tau^2} + 2\zeta\left(\frac{\omega_n}{\alpha}\right)\frac{dx(\tau)}{d\tau} + \left(\frac{\omega_n}{\alpha}\right)^2 x(\tau) = \frac{1}{\alpha^2}\frac{F(\tau)}{m} \tag{4.6-12}$$

which shows that the natural frequency of the system has been changed from ω_n to $\Omega = \omega_n/\alpha$. The damping factor ζ, however, has not been changed since the critical damping $c_{cr} = 2m\omega_n$ for the original equation has been changed to $c_{cr} = 2m\Omega$ for the new equation.

Theoretically, it is possible to devise a computer circuit to solve this new equation and interpret its results in terms of the original variables. However, there are problems concerning the orders of magnitudes of the voltages of the computer circuit which need further attention, and which can best be discussed in terms of the following example.

Example 4.6-1

The equation for a certain mechanical system excited by a step load $f(t) = 2000$ lb is given as

$$0.10\ddot{x} + 5\dot{x} + 4000x = 2000 \text{ lb}$$

with initial conditions

$$\dot{x}(0) = -20 \text{ in/sec} \quad \text{and} \quad x(0) = 0.25 \text{ in.}$$

Rewrite the equation for the computer and establish a workable circuit for its computation.

Solution: Writing the equation in the form

$$\dot{x} + 50\dot{x} + 40{,}000x = 20{,}000$$

the natural frequency of the system is found to be

$$\omega_n = \sqrt{k/m} = \sqrt{40{,}000} = 200 \text{ rad/sec}$$

This frequency is too high for the computer, so we arbitrarily choose $\alpha = 100$ to slow down the time scale by a factor of 100. The new equation with τ as the independent variable is then [see Eq. (4.6-12)]

$$\frac{d^2x(\tau)}{d\tau^2} + 0.50\frac{dx(\tau)}{d\tau} + 4.0x(\tau) = 2.0$$

with the initial conditions

$$\frac{dx(0)}{d\tau} = \frac{-20}{100} = -0.20, \qquad x_\tau(0) = 0.25$$

and we observe that the natural frequency has been reduced to $\Omega = 2$ rad/sec.

If we ignore orders of magnitudes, the following circuit of Fig. 4.6-9 would satisfy these equations. However, for the computer to give reliable results, the output voltages of the amplifiers must not exceed ± 100 volts, nor be too small. For these considerations, it is necessary to make an estimate of the peak displacement, velocity, and acceleration to be encountered in the revised differential equation, and establish proper scale factors which will give peak output voltages of amplifiers near their allowable maximum of ± 100 volts.

From the revised equation, we recognize that the damping will eventually eliminate all oscillations and the final displacement attained will be

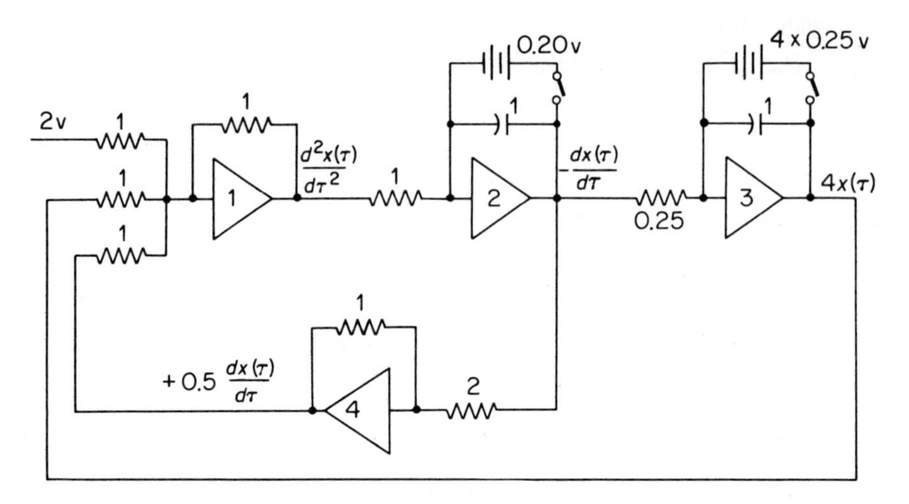

Figure 4.6-9. Analog circuit for single degree of freedom system.

equal to

$$x(\tau)_{\tau=\infty} = \frac{2.0}{4.0} = 0.5 \text{ in.}$$

We also note that without damping, the peak amplitude under a step function excitation is twice the value above, or

$$\underset{\max}{x(\tau)} = 1.0 \text{ in.}$$

The peak velocity and acceleration to be encountered can be estimated on the basis of harmonic motion to be

$$\left[\frac{dx(\tau)}{d\tau}\right]_{\max} = \Omega \underset{\max}{x(\tau)} = 2 \text{ in./sec}$$

$$\left[\frac{d^2x(\tau)}{d\tau^2}\right]_{\max} = \Omega^2 \underset{\max}{x(\tau)} = 4 \text{ in./sec}^2$$

If we now examine the circuit of Fig. 4.6-9 with the above maximum values, amplifiers 1, 2, and 3 are found to have peak output voltages of only 4, 2, and 4 volts, and no accuracy can be obtained from such a circuit.

To overcome this difficulty, we make a scale change so that with $\underset{\max}{x(\tau)} = 1$ in., the output will be more nearly equal to the allowable peak of 100 volts.

To avoid exceeding the 100-volt limit, we will let this voltage be 80 volts. By multiplying the revised differential equation by 20, we get

$$20\frac{d^2x(\tau)}{d\tau} = -10\frac{dx(\tau)}{d\tau} - 80x(\tau) + 40$$

The circuit for this scaled-up equation might now take the form shown in Fig. 4.6-10, where the initial voltages are scaled according to the output of the amplifiers, i.e., with $dx(0)/d\tau = -0.20$ in/sec, and the output of amplifier 2 equal to $-10(dx(\tau)/d\tau)$, its value should be $-10(-0.20) = +2$ volts. Similarly, the initial voltage of amplifier 3 should be $80 \times 0.25 = 20$ volts.

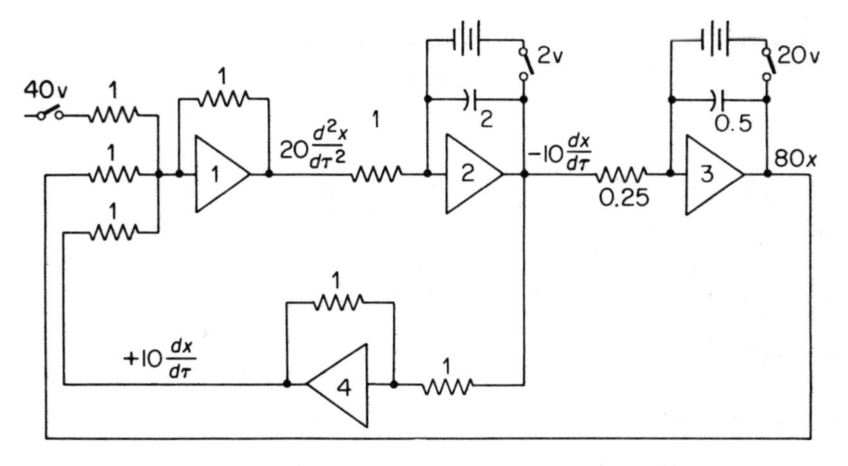

Figure 4.6-10.

We now have an output of 80 volts for amplifier 3, corresponding to the maximum expected amplitude of 1 in. However, since the maximum expected velocity is 2 in./sec, the peak output of amplifier 2 is only $10(dx/d\tau) = 20$ volts. It is preferable then to change the gain of amplifier 2 by a factor of 4 and reduce the gain of amplifier 3 accordingly, as shown in Fig. 4.6-11. The maximum expected value of the acceleration being 4 in./sec, the gain of amplifier 1 need not be changed. The gain of the amplifier is determined by the quantity RC which is unity for $R = 1$ megohm and $C = 1$ microfarad. Note also that these changes require a further change in the initial voltage of amplifier 2 from 2 volts to 8 volts.

Finally, it should be apparent that these computer circuits are not unique, and the same equation can be solved by different circuits. Thus the equation for this problem can also be solved by the circuit of Fig. 4.6-12.

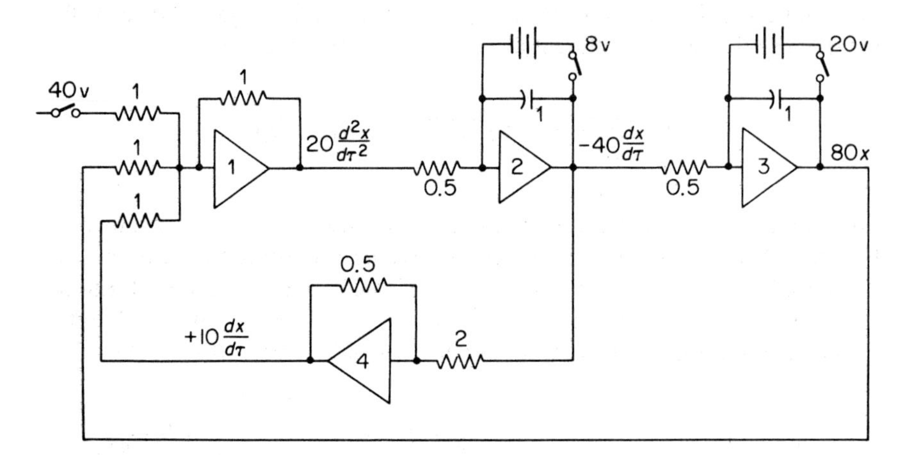

Figure 4.6-11.

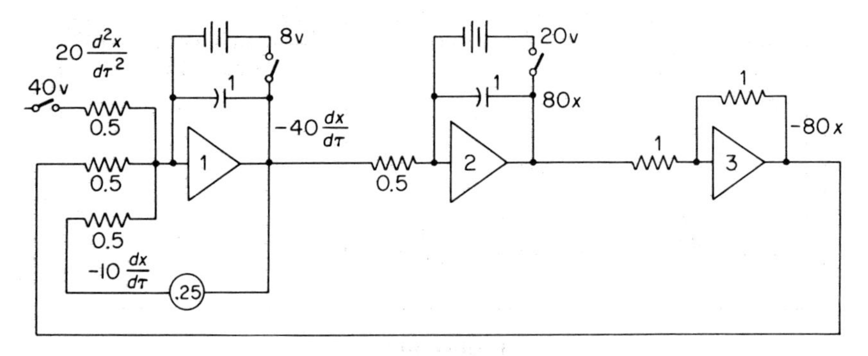

Figure 4-6.12.

4.7 FINITE DIFFERENCE DIGITAL COMPUTATION

When the differential equation cannot be integrated in closed form, numerical methods must be employed. This may well be the case when a spring-mass system is excited by a force that cannot be expressed by simple analytic functions, or when such a force is given only graphically or numerically.

Numerical integration is a procedure by which the differential equation of motion is solved progressively in time increments, starting from some time when the displacement and velocity are known. The solution is approximate, but as the time interval is reduced, the result approaches the true solution. Although there are a number of different numerical methods available, in this chapter we will consider only two methods, chosen for their simplicity. Merits of the various methods are associated with the errors and convergence of the procedure; these are discussed in many texts on numerical analysis.*,†

The basis of the numerical solution is essentially that of obtaining the numerical values of the unknown integral at pivotal points along the time axis. For this, the derivatives in the differential equation are approximated from a certain number of terms in the Taylor expansion. The Taylor expansion for x_{i+1} and x_{i-1} in terms of the pivotal point i are

$$\left.\begin{aligned} x_{i+1} &= x_i + \dot{x}_i\,\Delta t + \ddot{x}_i\frac{\Delta t^2}{2} + \dddot{x}_i\frac{\Delta t^3}{6} + \cdots \\ x_{i-1} &= x_i - \dot{x}_i\,\Delta t + \ddot{x}_i\frac{\Delta t^2}{2} - \dddot{x}_i\frac{\Delta t^3}{6} + \cdots \end{aligned}\right\} \quad i \geq 2 \qquad (4.7\text{-}1)$$

Subtracting and ignoring higher order terms, we obtain

$$\dot{x}_i\,\Delta t = \frac{1}{2}(x_{i+1} - x_{i-1}) \qquad i \geq 2 \qquad (4.7\text{-}2)$$

Adding, we find

$$\ddot{x}_i\,\Delta t^2 = x_{i-1} - 2x_i + x_{i+1} \qquad i \geq 2 \qquad (4.7\text{-}3)$$

These recurrence equations, together with the differential equations of motion, are sufficient for the numerical solution; however, some additional considerations are necessary to start the computational procedure.

When the initial acceleration (or force) is not zero, the simplest procedure is to assume it to remain constant during the first interval. Since sub-zero is not available in the digital computer, we use sub 1 for the initial values. We then have

$$\begin{aligned} \ddot{x}_2 &= \ddot{x}_1 \\ \dot{x}_2 &= \ddot{x}_1\,\Delta t \\ x_2 &= \frac{1}{2}\ddot{x}_1\,\Delta t^2 \end{aligned} \qquad (4.7\text{-}4)$$

The acceleration $\ddot{x}_1$ to be used in the above equation is determined from the

*A. Ralston and H. S. Wilf, *Mathematical Methods for Digital Computers Vols. I & II* (New York: John Wiley & Sons, 1968).

†M. G. Salvadori and M. L. Baron, *Numerical Methods in Engineering* (Prentice-Hall, Inc., 1952).

differential equation of motion and its initial conditions x_1 and $\dot{x}_1$

$$\ddot{x}_1 = \frac{1}{m}F(t_1) - \frac{c}{m}\dot{x}_1 - \frac{k}{m}x_1$$

The displacement x_2 and velocity $\dot{x}_2$ are then determined from Eq. (4.7-4) and substituted into the differential equation for $\ddot{x}_2$. Both x_2 and $\ddot{x}_2$ are then substituted into Eq. (4.7-3) to determine x_3 and the procedure is repeated. Usually, a time interval $\Delta t \leq \frac{1}{10}\tau$, where τ is the natural period, is sufficiently small to result in a satisfactory solution.

A flow diagram for the digital calculation is shown in Fig. 4.7-1. From

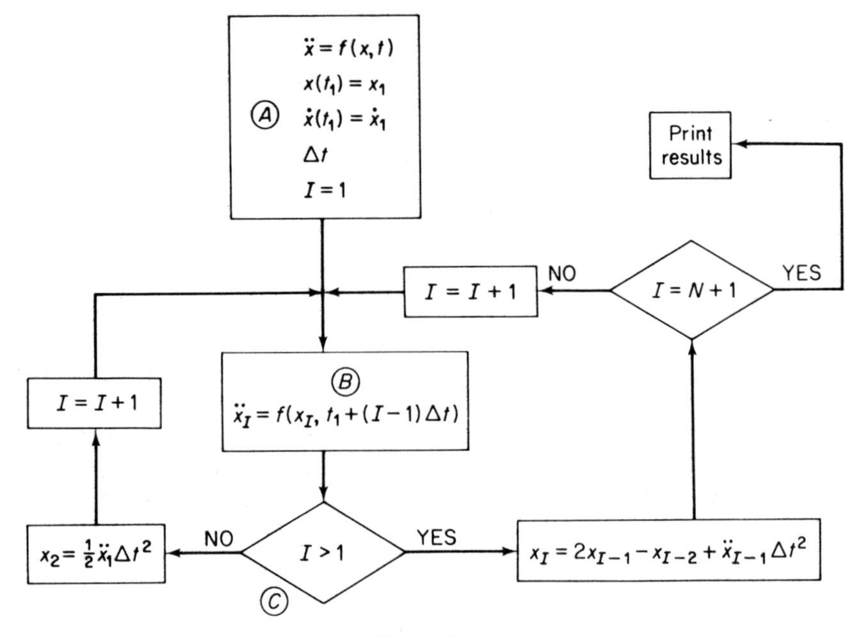

Figure 4.7-1.

the given data in block Ⓐ we proceed to block Ⓑ which is the differential equation. Going to Ⓒ for the first time, I is not greater than 1, and hence we proceed to the left where x_2 is calculated. Increasing I by 1 we complete the left loop to Ⓑ and Ⓒ where I is now equal to 2, so we proceed to the right to calculate x_I. Assuming N intervals of Δt, the path is to the NO direction and the right loop is repeated N times until $I = N + 1$, at which time the results are printed out.

The Fortran program for the computation is given after the flow diagram.

```
FORTRAN PROGRAM
        I = 1
20      ẍ_I = f(x_I, t_1 + (I − 1) Δt)
        IF (I.GT.1)  GO TO 30
        x_2 = ½ẍ_1 Δt² + ẋ_1 Δt + x_1
        I = I + 1   GO TO 20

30      x_I = 2x_{I−1} − x_{I−2} + ẍ_{I−1} Δt²
        IF (I = N + 1)  GO TO 40
        I = I + 1
        GO TO 20
40      PRINT
        STOP
        END
```

If the initial acceleration $\ddot{x}_1$ is zero, Eq. (4.7-4) results in $x_2 = 0$, and the calculation procedure cannot be started. This condition can be rectified by developing a new equation based on the assumption that during the first interval the acceleration varies linearly from $\ddot{x}_1$ to $\ddot{x}_2$ as follows

$$\ddot{x}_2 = \ddot{x}_1 + \alpha\,\Delta t$$

$$\dot{x}_2 = \ddot{x}_1\,\Delta t + \frac{1}{2}\alpha\,\Delta t^2$$

$$x_2 = \frac{1}{2}\ddot{x}_1\,\Delta t^2 + \frac{1}{6}\alpha\,\Delta t^3$$

Eliminating α in the last equation by the first equation we obtain

$$\begin{aligned} x_2 &= \frac{1}{2}\ddot{x}_1\,\Delta t^2 + \frac{1}{6}\left(\frac{\ddot{x}_2 - \ddot{x}_1}{\Delta t}\right)\Delta t^3 \\ &= \frac{1}{6}(2\ddot{x}_1 + \ddot{x}_2)\,\Delta t^2 \end{aligned} \tag{4.7-5}$$

With $\ddot{x}_1 = 0$, Eq. (4.7-5) and the differential equation must be solved by trial and error, i.e.

$$\begin{aligned} x_2 &= \tfrac{1}{6}\ddot{x}_2\,\Delta t^2 \\ \ddot{x}_2 &= \phi(x_2, F_2) \end{aligned} \tag{4.7-6}$$

In Sec. 4.8 the Runge-Kutta method, which is self-starting, is discussed.

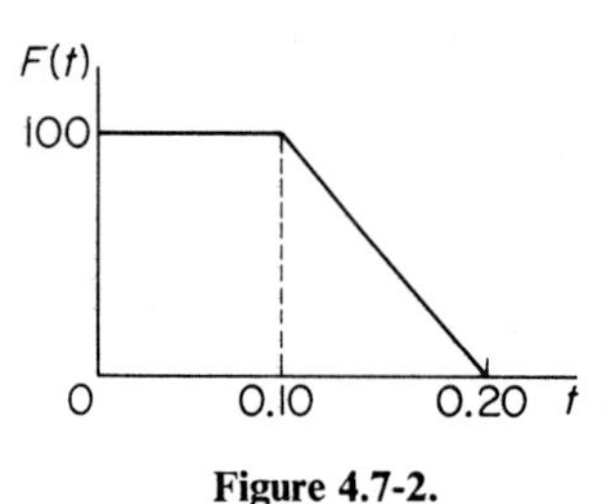

Figure 4.7-2.

Example 4.7-1

Solve numerically the differential equation $4\ddot{x} + 2000x = F(t)$ with initial conditions $x_1 = \dot{x}_1 = 0$ and $F(t)$ as given in Fig. 4.7-2.

Solution: The natural period of the system is first found as

$$\omega = \frac{2\pi}{\tau} = \sqrt{\frac{2000}{4}} = 22.4 \text{ rad/sec}$$

$$\tau = \frac{2\pi}{22.4} = 0.281 \text{ sec}$$

For this problem we will carry out the computation by slide rule, and in order to keep the number of computations to a minimum, we choose $t = 0.030$, which is approximately $\frac{1}{10}\tau$.

From the differential equation, the acceleration is

$$\ddot{x} = \tfrac{1}{4}F(t) - 500x$$

and with $x_1 = 0$ we obtain

$$\ddot{x}_1 = \tfrac{1}{4} \times 100 - 0 = 25$$

Eq. (4.7-4) then gives

$$x_2 = \tfrac{1}{2}(25)(.030)^2 = 0.0113$$

Using $x_2 = 0.0113$ we return to the differential equation to determine $\ddot{x}_2$

$$\ddot{x}_2 = \tfrac{1}{4}(100) - 500(0.0113)$$
$$= 25 - 5.65 = 19.35.$$

The quantities x_2 and $\ddot{x}_2$ are now substituted into Eq. (4.7-3) with $i = 2$

$$19.35(.030)^2 = 0 - 2(0.0113) + x_3$$
$$x_3 = 0.0400$$

We now have x_2, and x_3; $\ddot{x}_3$ can be computed from the differential equation. Equation (4.7-3) then gives x_4, and the procedure can be repeated between the differential equation and Eq. (4.7-3). The following table illustrates the sequence of calculation, the results of which are plotted in Fig. 4.7-3.

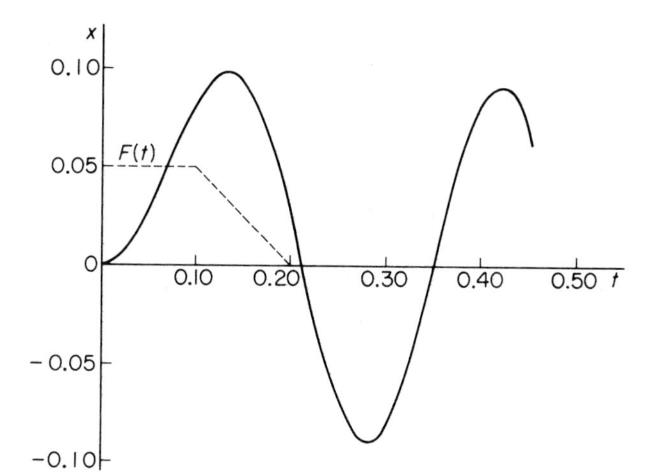

Figure 4.7-3.

i	t	$\frac{1}{4}F(t)$	$500x$	$\ddot{x}$	$\ddot{x}\Delta t^2$	x
1	0	25.0	0	25	.0225	0
2	.03	25.0	5.65	19.35	.0174	.0133*
3	.06	25.0	20.00	5.00	.0045	.0400†
4	.09	25.0	36.60	−11.60	−.0104	.0732
5	.120	20.0	48.0	−28.0	−.0252	.0960
6	.150	12.5	46.80	−34.3	−.0309	.0936
7	.180	5.0	30.15	−25.15	−.0226	.0603
8	.210	0	2.20	− 2.20	−.0020	.0044
9	.240	0	−26.75	26.75	.0241	−.0535
10	.270	0	−43.65	43.65	.0393	−.0873
11	.300	0	−40.90	40.90	.0368	−.0818
12	.330	0	−19.75	19.75	.0178	−.0395
13	.360	0	10.30	−10.30	−.0093	.0206
14	.390	0	35.70	−35.70	−.0321	.0714
15	.420	0	45.05	−45.05	−.0405	.0901
16	.450	0				.0683

* From Eq. (4.7-4) $x_2 = \frac{1}{2}\ddot{x}_1\Delta t^2 = \frac{1}{2}(.0225) = .0113$
† From Eq. (4.7-3) $x_3 = -x_1 + 2x_2 + \ddot{x}_2\Delta t^2 = -0 + 2(.0113) + .0174 = 0.040$

Example 4.7-2

Solve by the digital computer the problem of a spring-mass system excited by a triangular pulse. The differential equation of motion and the initial conditions are given as

$$0.5\ddot{x} + 8\pi^2 x = F(t)$$

$$x_1 = \dot{x}_1 = 0$$

The triangular force is defined in Fig. 4.7-4.

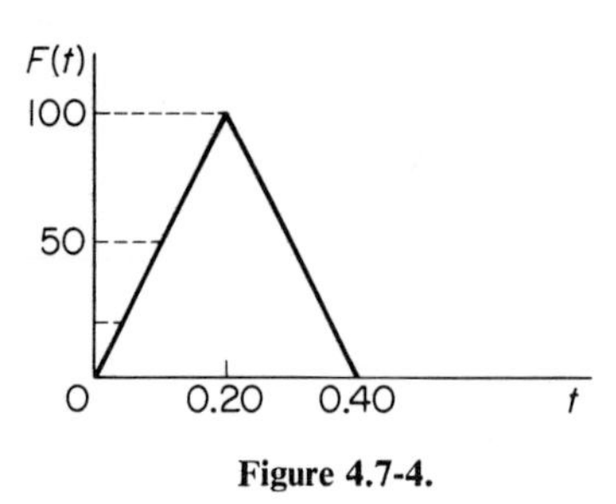

Figure 4.7-4.

Solution: The natural period of the system is

$$\tau = \frac{2\pi}{\omega} = \frac{2\pi}{4\pi} = 0.50$$

The time increment will be chosen as $t = .05$, and the differential equation is reorganized as

$$\ddot{x} = 2F(t) - 16\pi^2 x$$

This equation is to be solved together with the recurrence equation (Eq. 4.7-3)

$$x_{i+1} = \ddot{x}_i\,\Delta t^2 - x_{i-1} + 2x_i$$

Since the force and the acceleration are zero at $t = 0$, it is necessary to start the computational process with Eqs. (4.7-6), which are

$$x_2 = \tfrac{1}{6}\ddot{x}_2(.05)^2 = .000417\ddot{x}_2$$
$$\ddot{x}_2 = 2F(.05) - 16\pi^2 x_2 = 50 - 158x_2$$

Their simultaneous solution leads to

$$x_2 = \frac{(.05)^2 F(.05)}{3 + 8\pi^2(.05)^2} = .0195$$
$$\ddot{x}_2 = 46.91$$

The flow diagram for the computation is shown in Fig. 4.7.-5. With $\Delta t = 0.05$, the time duration for the force must be divided into regions $I = 1$ to 5, $I = 6$ to 9 and $I > 9$. The index I controls the computation path on the diagram.

The Fortran program can be written in many ways, one of which is shown in Fig. 4.7-6, and the results, Fig. 4.7-7, can also be plotted by the computer, as presented in Fig. 4.7-8. A smaller Δt would have resulted in a smoother plot.

The response x vs. t indicates a maximum at $x \simeq 1.97$ in. Since $k = 8\pi^2 = 79.0$, and $F_0 = 100$, a point on the response spectrum of Prob. 4-23 is verified as

$$\left(\frac{xk}{F_0}\right)_{max} = \frac{1.97 \times 79}{100} = 1.54$$

$$\frac{t_1}{\tau} = \frac{0.4}{0.5} = 0.80$$

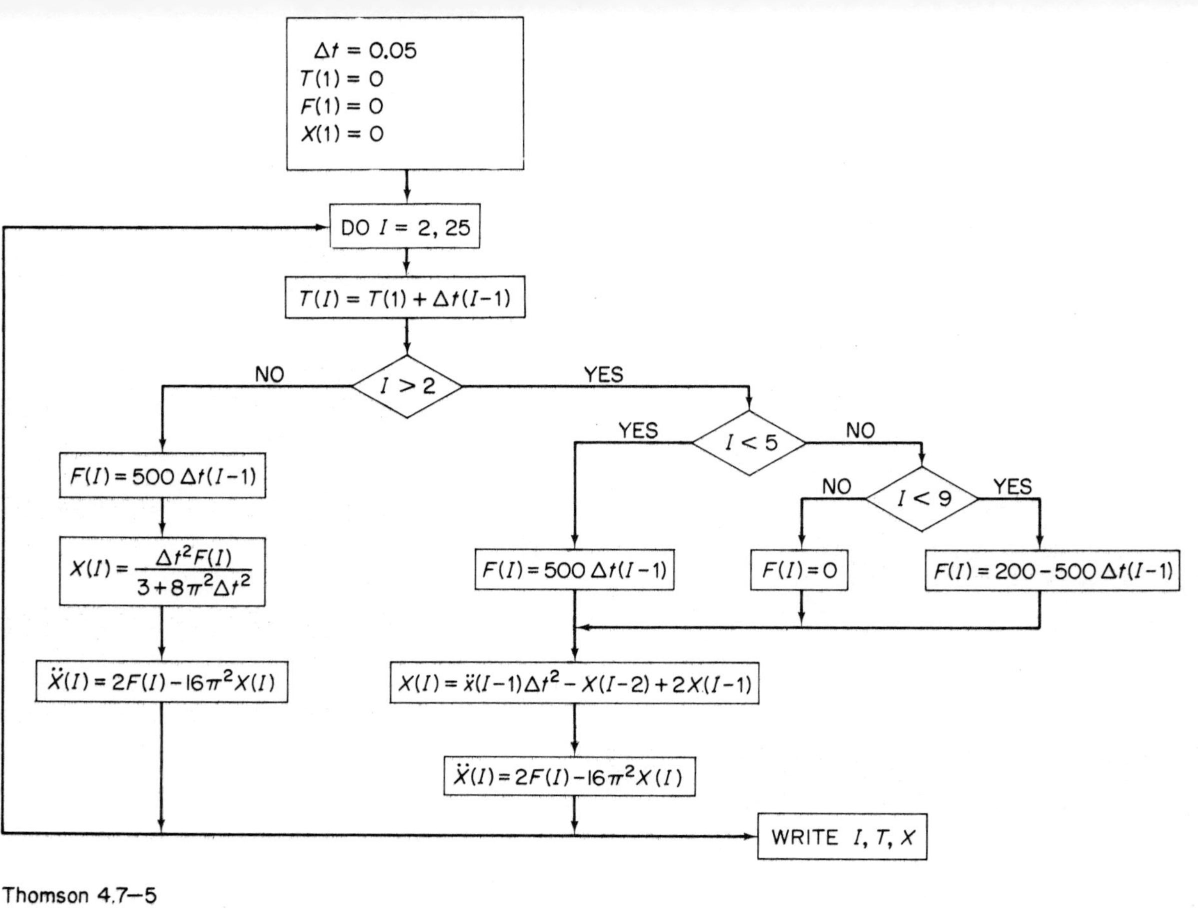

Δt = 0.05
T(1) = 0
F(1) = 0
X(1) = 0
DO I = 2, 25
T(I) = T(1) + Δt(I−1)
I > 2
NO
YES
F(I) = 500 Δt(I−1)
X(I) = Δt²F(I) / (3+8π²Δt²)
Ẍ(I) = 2F(I) − 16π²X(I)
I < 5
YES
NO
I < 9
NO
YES
F(I) = 500 Δt(I−1)
F(I) = 0
F(I) = 200 − 500 Δt(I−1)
X(I) = ẍ(I−1)Δt² − X(I−2) + 2X(I−1)
Ẍ(I) = 2F(I) − 16π²X(I)
WRITE I, T, X
Thomson 4.7–5

Figure 4.7-5.

```
                 C
                 C     VIBRATION PROBLEM
                 C     .
ISN 0002               DIMENSION X(25),DX2(25),F(25),T(25),J(25),VAR(25)
ISN 0003               PI2=3.1415**2
ISN 0004               DT=0.05
ISN 0005               DT2=DT**2
ISN 0006               X(1)=0.0
ISN 0007               DX2(1)=0.0
ISN 0008               F(1)=0.0
ISN 0009               T(1)=0.0
ISN 0010               J(1)=1
ISN 0011               DO 1 I=2,25
ISN 0012               J(I)=I
ISN 0013               T(I)=DT*(I-1)
ISN 0014               IF (I .GT. 2) GO TO 2
ISN 0016               F(I)=500*DT*(I-1)
ISN 0017               X(I)=(DT2*F(1) )/(3+8*PI2*DT2)
ISN 0018               DX2(I)=2*F(I)-16*PI2*X(I)
ISN 0019               GO TO 1
ISN 0020             2 IF(I .LE. 5) F(I)=500*DT*(I-1)
ISN 0022               IF (I .GT. 5 .AND. I .LT. 9) F(I)=200-500*DT*(I-1)
ISN 0024               IF (I .GE. 9) F(I)=0.0
ISN 0026               X(I)=DX2(I-1)*DT2-X(I-2)+2*X(I-1)
ISN 0027               DX2(I)=2*F(I)-16*PI2*X(I)
ISN 0028             1 CONTINUE
ISN 0029               WRITE(6,3)
ISN 0030             3 FORMAT(41H1   J   TIME    DISPL    ACCLRTN    FORCE)
ISN 0031               WRITE(6,4) (J(I),T(I),X(I),DX2(I),F(I),I=1,25)
ISN 0032             4 FORMAT(3X,I2,2X,F6.4,3X,F6.3,3X,F7.2,4X,F7.2)
                 C
                 C     PLOTTING
                 C
ISN 0033               DO 5 I=1,25
ISN 0034             5 VAR(I)=X(I)*10
ISN 0035               CALL PLOT1(VAR,25)
ISN 0036               STOP
ISN 0037               END
```

Figure 4.7-6.

J	TIME	DISPL	ACCLRTN	FORCE
1	0.0	0.0	0.0	0.0
2	0.0500	0.020	46.91	25.00
3	0.1000	0.156	75.31	50.00
4	0.1500	0.481	73.97	75.00
5	0.2000	0.992	43.44	100.00
6	0.2500	1.610	-104.25	75.00
7	0.3000	1.968	-210.78	50.00
8	0.3500	1.799	-234.10	25.00
9	0.4000	1.045	-165.01	0.00
10	0.4500	-0.122	19.22	0.0
11	0.5000	-1.240	195.86	0.0
12	0.5500	-1.869	295.19	0.0
13	0.6000	-1.760	277.98	0.0
14	0.6500	-0.957	151.04	0.0
15	0.7000	0.225	-35.52	0.0
16	0.7500	1.318	-208.06	0.0
17	0.8000	1.890	-298.47	0.0
18	0.8500	1.717	-271.05	0.0
19	0.9000	0.865	-136.64	0.0
20	0.9500	-0.328	51.72	0.0
21	1.0000	-1.391	219.66	0.0
22	1.0500	-1.906	300.89	0.0
23	1.1000	-1.668	263.33	0.0
24	1.1500	-0.772	121.83	0.0
25	1.2000	0.429	-67.77	0.0

Figure 4.7-7.

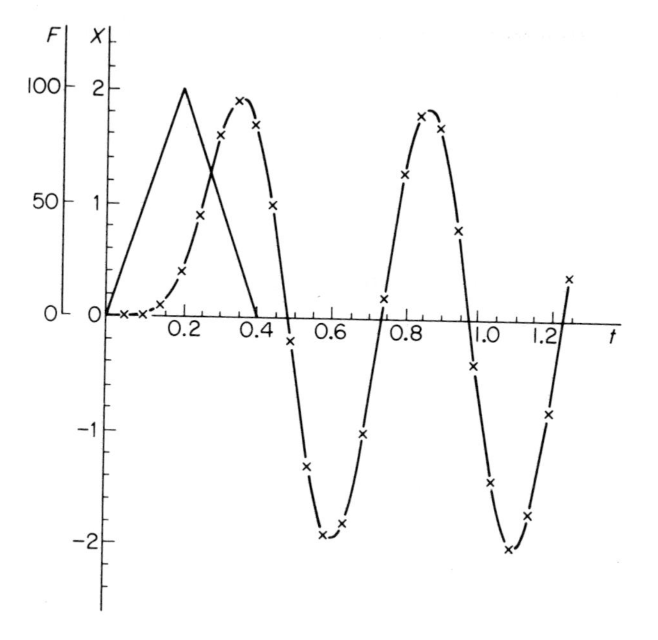

Figure 4.7-8.

4.8 THE RUNGE-KUTTA COMPUTATION

The Runge-Kutta computation procedure is quite popular since it is self-starting and results in good accuracy. A brief discussion of its basis is presented here.

Consider the differential equation for the single degree of freedom system.

$$m\frac{d^2x}{dt^2} + c\frac{dx}{dt} + kx = F(t)$$

Letting $y = dx/dt$ the above equation can be written as

$$\frac{dy}{dt} = f(t, x, y) \tag{4.8-1}$$

Both x and y in the neighborhood of x_i and y_i can be expressed in terms of the Taylor series. Letting the time increment be $h = \Delta t$

$$x = x_i + \left(\frac{dx}{dt}\right)_i h + \left(\frac{d^2x}{dt^2}\right)_i \frac{h^2}{2} + \cdots$$

$$y = y_i + \left(\frac{dy}{dt}\right)_i h + \left(\frac{d^2y}{dt^2}\right)_i \frac{h^2}{2} + \cdots$$

Instead of using these expressions, it is possible to replace the first derivative by an average slope and ignore higher order derivatives

$$x = x_i + \left(\frac{dx}{dt}\right)_{i\,av} h \tag{4.8-2}$$

$$y = y_i + \left(\frac{dy}{dt}\right)_{i\,av} h \tag{4.8-3}$$

If we used Simpson's rule, the average slope in the interval h becomes, i.e.

$$\left(\frac{dy}{dt}\right)_{i\,av} = \frac{1}{6}\left[\left(\frac{dy}{dt}\right)_{t_i} + 4\left(\frac{dy}{dt}\right)_{t_i+h/2} + \left(\frac{dy}{dt}\right)_{t_i+h}\right] \tag{4.8-4}$$

The Runge-Kutta method is very similar to the above computations, except that the center term of Eq. (4.8-4) is split into two terms and four values of t, x, y, and f are computed for each point i as follows

t	x	$y = \dot{x}$	$f = \dot{y} = \ddot{x}$
$T_1 = t_i$	$X_1 = x_i$	$Y_1 = y_i$	$F_1 = f(T_1, X_1, Y_1)$
$T_2 = t_i + \frac{h}{2}$	$X_2 = x_i + Y_1\frac{h}{2}$	$Y_2 = y_i + F_1\frac{h}{2}$	$F_2 = f(T_2, X_2, Y_2)$
$T_3 = t_i + \frac{h}{2}$	$X_3 = x_i + Y_2\frac{h}{2}$	$Y_3 = y_i + F_2\frac{h}{2}$	$F_3 = f(T_3, X_3, Y_3)$
$T_4 = t_i + h$	$X_4 = x_i + Y_3 h$	$Y_4 = y_i + F_3 h$	$F_4 = f(T_4, X_4, Y_4)$

These quantities are then used in the following recurrance formula

$$x_{i+1} = x_i + \frac{h}{6}[Y_1 + 2Y_2 + 2Y_3 + Y_4] \tag{4.8-5}$$

$$y_{i+1} = y_i + \frac{h}{6}[F_1 + 2F_2 + 2F_3 + F_4] \tag{4.8-6}$$

where it is recognized that the four values of Y divided by 6 represent an average slope dx/dt and the four values of F divided by 6 results in an average of dy/dt as defined by Eqs. (4.8-2) and (4.8-3).

PROBLEMS

4-1 Show that the time t_p corresponding to the peak response for the impulsively excited spring-mass system is given by the equation

$$\tan \sqrt{1 - \zeta^2}\, \omega_n t_p = \sqrt{1 - \zeta^2}/\zeta$$

4-2 Determine the peak displacement for the impulsively excited spring-mass system, and show that it can be expressed in the form

$$\frac{x_{\text{peak}}\sqrt{km}}{\hat{F}} = \exp\left(-\frac{\zeta}{\sqrt{1-\zeta^2}}\tan^{-1}\frac{\sqrt{1-\zeta^2}}{\zeta}\right)$$

Plot this result as a function of ζ.

4-3 Show that the time t_p corresponding to the peak response of the damped spring-mass system excited by a step force F_0 is $\omega_n t_p = \pi/\sqrt{1-\zeta^2}$.

4-4 For the system of Prob. 3, show that the peak response is equal to

$$\left(\frac{xk}{F_0}\right)_{\max} = 1 + \exp\left(-\frac{\zeta\pi}{\sqrt{1-\zeta^2}}\right)$$

4-5 A rectangular pulse of height F_0 and duration t_0 is applied to an undamped spring-mass system. Considering the pulse to be the sum of two step pulses, as shown in Fig. P4-5, determine its response for $t > t_0$ by the superposition of the undamped solutions.

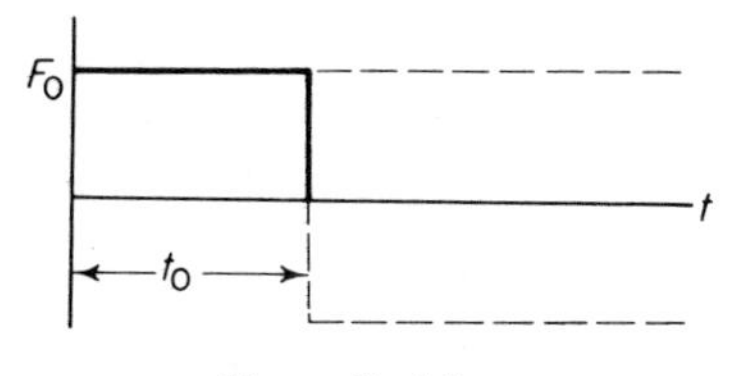

Figure P. 4-5.

4-6 If an arbitrary force $f(t)$ is applied to an undamped oscillator which has initial conditions other than zero, show that the solutuon must be in the form

$$x(t) = x_0 \cos \omega_n t + \frac{v_0}{\omega_n}\sin \omega_n t + \frac{1}{m\omega_n}\int_0^t f(\xi)\sin \omega_n(t-\xi)\,d\xi$$

4-7 Show that the response to a unit step function, designated by $h(t)$, is related to the impulsive response $g(t)$ by the equation $g(t) = \dot{h}(t)$.

4-8 Show that the convolution integral can also be written in terms of $h(t)$ as

$$x(t) = f(0)h(t) + \int_0^t \dot{f}(\xi)h(t-\xi)\,d\xi$$

where $h(t)$ is the response to a unit step function.

4-9 In Sec. 4.4 the subsidiary equation for the viscously damped spring-mass system was given by Eq. (4.4-1). Evaluate the second term due to initial conditions by the inverse transforms.

4-10 An undamped spring-mass system is given a base excitation of $\dot{y}(t) = 20(1 - 5t)$. If the natural frequency of the system is $\omega_n = 10\ \text{sec}^{-1}$, determine the maximum relative displacement.

4-11 A sinusoidal pulse is considered to be the superposition of two sine waves as shown in Fig. P4-11. Show that its solution is

$$\left(\frac{xk}{F_0}\right) = \frac{1}{(\tau/2t_1 - 2t_1/\tau)}\left(\sin\frac{2\pi t}{\tau} - \frac{2t_1}{\tau}\sin\frac{\pi t}{t_1}\right) \qquad t < t_1$$

$$\left(\frac{xk}{F_0}\right) = \frac{1}{(\tau/2t_1 - 2t_1/\tau)}\left[\left(\sin\frac{2\pi t}{\tau} - \frac{2t_1}{\tau}\sin\frac{\pi t}{t_1}\right)\right.$$
$$\left. + \left(\sin 2\pi\frac{t - t_1}{\tau} - \frac{2t_1}{\tau}\sin\pi\frac{t - t_1}{t_1}\right)\right], \qquad t > t_1$$

where $\tau = 2\pi/\omega$.

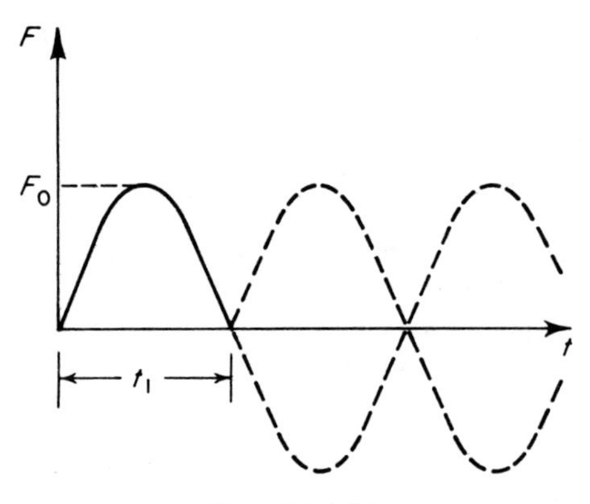

Figure P. 4-11.

4-12 For the triangular pulse shown in Fig. P4-12 show that the response is

$$x = \frac{2F_0}{k}\left(\frac{t}{t_1} - \frac{\tau}{2\pi t_1}\sin 2\pi\frac{t}{\tau}\right), \qquad 0 < t < \tfrac{1}{2}t_1$$

$$x = \frac{2F_0}{k}\left\{1 - \frac{t}{t_1} + \frac{\tau}{2\pi t_1}\left[2\sin\frac{2\pi}{\tau}\left(t - \frac{1}{2}t_1\right) - \sin 2\pi\frac{t}{\tau}\right]\right\}, \qquad \tfrac{1}{2}t_1 < t < t_1$$

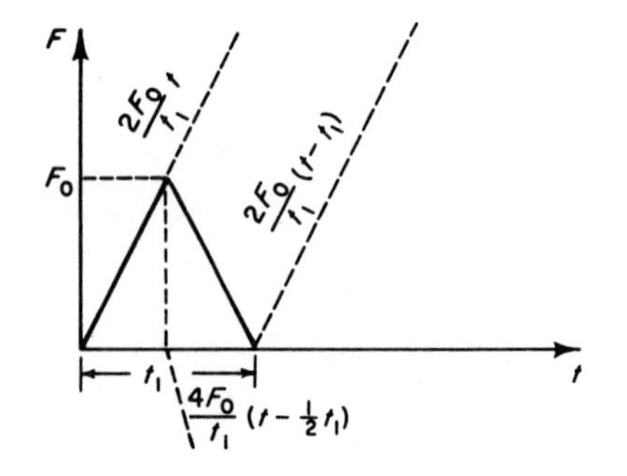

Figure P. 4-12.

$$x = \frac{2F_0}{k}\left\{\frac{\tau}{2\pi t_1}\left[2\sin\frac{2\pi}{\tau}\left(t - \frac{1}{2}t_1\right) - \sin\frac{2\pi}{\tau}(t - t_1) - \sin 2\pi\frac{t}{\tau}\right]\right\}, \quad t > t_1$$

4-13 A spring-mass system slides down a smooth 30° inclined plane as shown in Fig. P4-13. Determine the time elapsed from first contact of the spring until it breaks contact again.

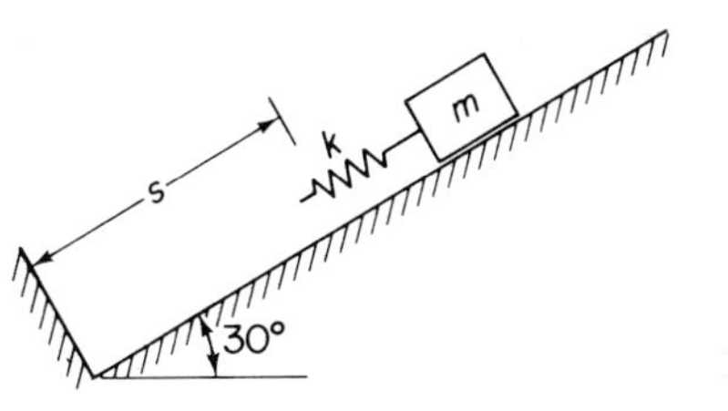

Figure P. 4-13.

4-14 Referring to Example 4.4-2, determine the rattle space required if the suspended system has a natural frequency of 10 cps and the box is to be dropped through a height of 1.0 in.

4-15 A 38.6 lb weight is supported on several springs whose combined stiffness is 6.40 lb/in. If the system is lifted so that the bottom of the springs are just free and released, determine the maximum displacement of m, and the time for maximum compression.

4-16 A delicate instrument is supported in a box by springs, as shown in Fig. P4-16, where its natural frequency is 10 cps. In the supported position, the clearance space between the instrument and the box walls is 1.0 inch. If the box is accidently dropped through a height of 20 inches, will the instrument hit the walls?

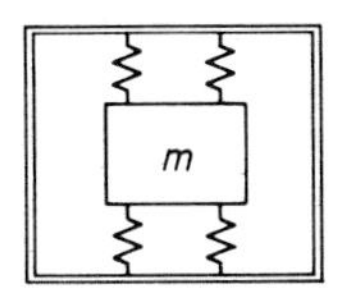

Figure P. 4-16.

4-17 A spring-mass system of Fig. P4-17 has a Coulomb damper which exerts a constant friction force f. For a base excitation, show that the solution is

$$\frac{\omega_n z}{v_0} = \frac{1}{\omega_n t_1}\left(1 + \frac{f t_1}{m v_0}\right)(1 - \cos\omega_n t) - \sin\omega_n t$$

where the base velocity of Prob. 4-26 is assumed.

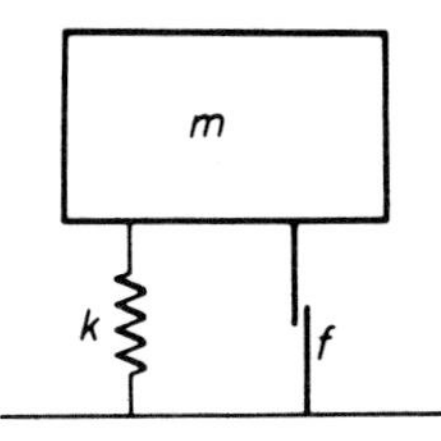

Figure P. 4-17.

4-18 Show that the peak response for Prob. 4-17 is

$$\frac{\omega_n z_{\max}}{v_0} = \frac{1}{\omega_n t_1}\left(1 + \frac{f t_1}{m v_0}\right)\left\{1 - \frac{\frac{1}{\omega_n t_1}\left(1 + \frac{f t_1}{m v_0}\right)}{\sqrt{1 + \left[\frac{1}{\omega_n t_1}\left(1 + \frac{f t_1}{m v_0}\right)\right]^2}}\right\}$$

$$- \frac{1}{\sqrt{1 + \left[\frac{1}{\omega_n t_1}\left(1 + \frac{f t_1}{m v_0}\right)\right]^2}}$$

By dividing by $\omega_n t_1$, the quantity $z_{\max}/v_0 t_1$ can be plotted as a function of $\omega_n t_1$ with $f t_1/m v_0$ as parameter.

4-19 In Prob. 4-18 the maximum force transmitted to m is

$$F_{\max} = f + |k z_{\max}|$$

To plot this quantity in nondimensional form, multiply by $t_1/m v_0$ to obtain

$$\frac{F_{\max} t_1}{m v_0} = \frac{f t_1}{m v_0} + (\omega_n t_1)^2\left(\frac{z_{\max}}{v_0 t_1}\right)$$

which again can be plotted as a function of ωt_1 with parameter $f t_1/m v_0$. Plot $|\omega_n z_{\max}/v_0|$ and $|z_{\max}/v_0 t_1|$ as function of $\omega_n t_1$ for $f t_1/m v_0$ equal to 0, 0.20, and 1.0.

4-20 Show that the response spectrum for the rectangular pulse of time duration t_0 shown in Fig. P4-20 is given by

$$\left(\frac{xk}{F_0}\right)_{\max} = 2 \sin \frac{\pi t_0}{\tau}, \qquad \frac{t_0}{\tau} < 0.50$$

$$= 2 \qquad \frac{t_0}{\tau} > 0.50$$

where

$$\tau = \frac{2\pi}{\omega_n}$$

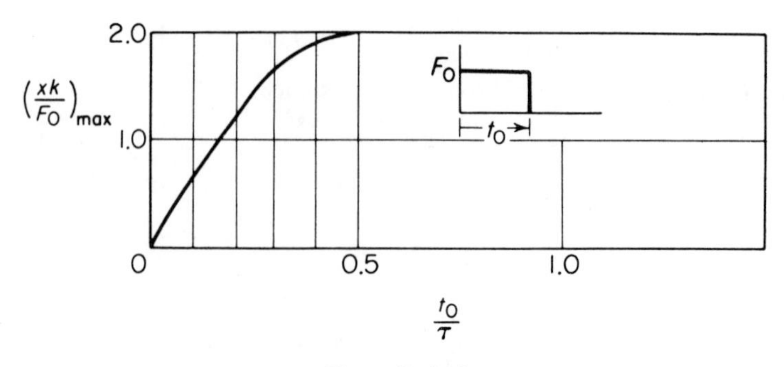

Figure P. 4-20.

4-21 Shown in Fig. P4-21 is the response spectrum for the sine pulse. Show that for small values of t_1/τ the peak response occurs in the region $t > t_1$. Show that when $t_1/\tau = \frac{1}{2}$ the peak response occurs at $t = t_1$.

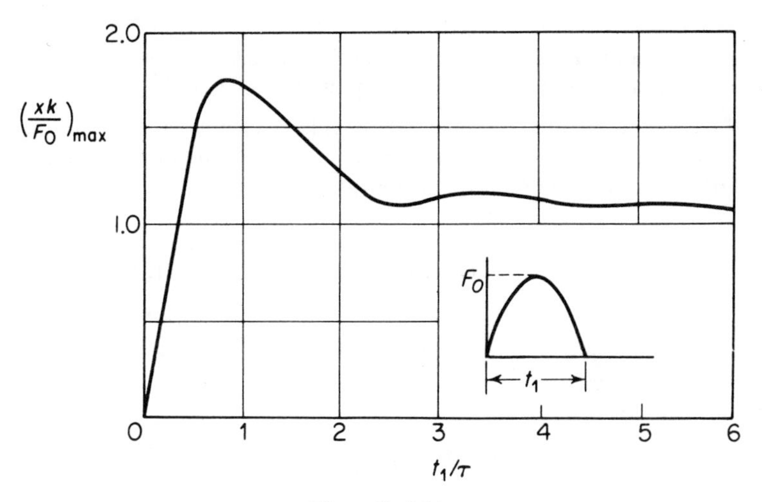

Figure P. 4-21.

4-22 An undamped spring-mass system with $w = 16.1$ lb has a natural period of 0.5 seconds. It is subjected to an impulse of 20 lb inches which has a triangular shape with time duration of 0.40 sec. Determine the maximum displacement of the mass.

4-23 For a triangular pulse of duration t_1, show that when $t_1/\tau = \frac{1}{2}$, the peak

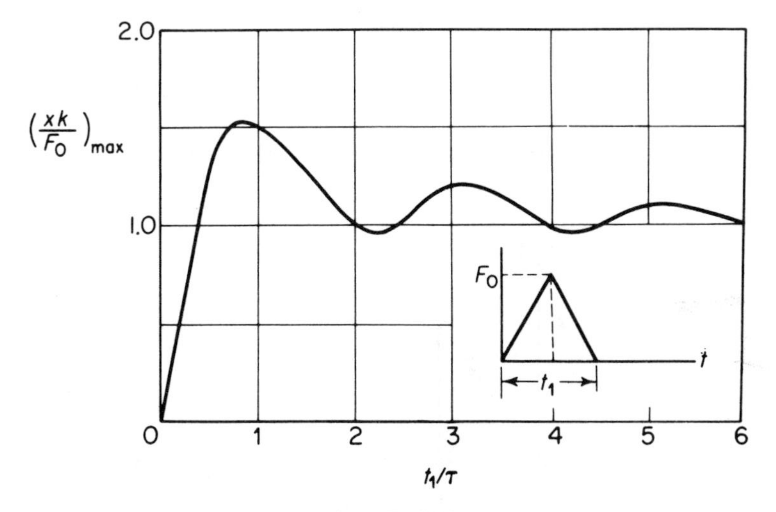

Figure P. 4-23.

response occurs at $t = t_1$, which can be established from the equation

$$2\cos\frac{2\pi t_1}{\tau}\left(\frac{t_p}{t_1} - 0.5\right) - \cos 2\pi\frac{t_1}{\tau}\left(\frac{t_p}{t_1} - 1\right) - \cos\frac{2\pi t_1}{\tau}\frac{t_p}{t_1} = 0$$

found by differentiating the equation for the displacement for $t > t_1$. The response spectrum for the triangular pulse is shown. in Fig. P4-23.

4-24 If the natural period τ of the oscillator is large compared to that of the pulse duration t_1, the maximum peak response will occur in the region $t > t_1$. For the undamped oscillator, the integrals written as

$$x = \frac{\omega_n}{k}\left\{\sin\omega_n t\int_0^t f(\xi)\cos\omega_n\xi\, d\xi - \cos\omega_n t\int_0^t f(\xi)\sin\omega_n\xi\, d\xi\right\}$$

will not change for $t > t_1$, since in this region $f(t) = 0$. Thus, by making the substitution

$$A\cos\phi = \omega_n\int_0^{t_1} f(\xi)\cos\omega_n\xi\, d\xi$$

$$A\sin\phi = \omega_n\int_0^{t_1} f(\xi)\sin\omega_n\xi\, d\xi$$

the response for $t > t_1$, is a simple harmonic motion with amplitude A. Discuss the nature of the response spectrum for this case.

4-25 An undamped spring-mass system, m, k, is given a force excitation $F(t)$ as shown in Fig. P4-25. Show that for $t < t_0$

$$\frac{kx(t)}{F_0} = \frac{1}{\omega_n t_0}(\omega_n t - \sin\omega_n t)$$

and for $t > t_0$

$$\frac{kx(t)}{F_0} = \frac{1}{\omega_n t_0}[\sin\omega_n(t - t_0) - \sin\omega_n t] + \cos\omega_n(t - t_0)$$

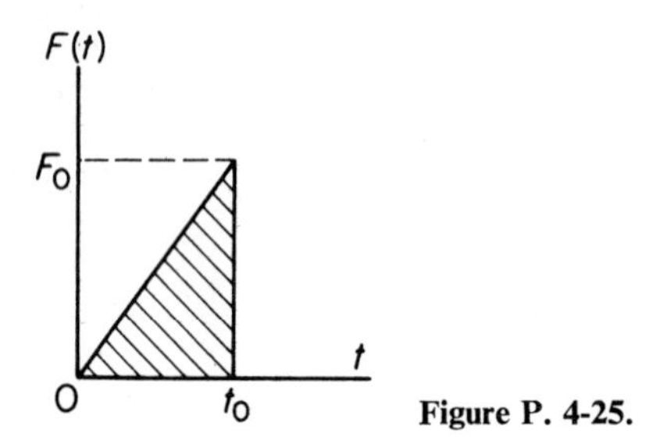

Figure P. 4-25.

4-26 The base of an undamped spring-mass system, m, k, is given a velocity pulse as shown in Fig. P4-26. Show that if the peak occurs at $t < t_1$, the response spectrum is given by the equation

$$\frac{\omega_n z_{\max}}{v_0} = \frac{1}{\omega_n t_1} - \frac{1}{\omega_n t_1\sqrt{1 + (\omega_n t_1)^2}} - \frac{\omega_n t_1}{\sqrt{1 + (\omega_n t_1)^2}}$$

Plot this result.

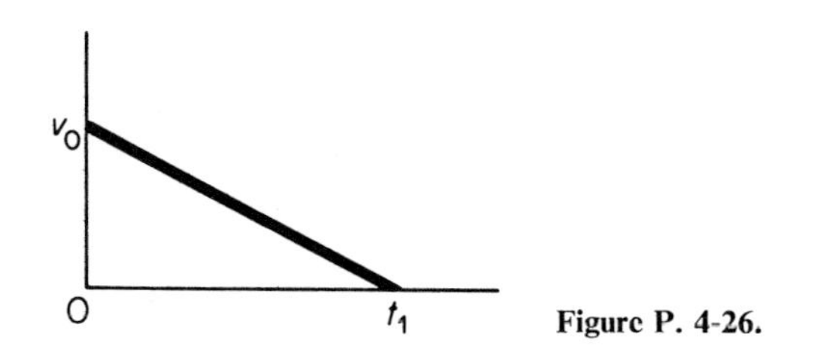

Figure P. 4-26.

4-27 In Prob. 4-26 if $t > t_1$, show that the solution is

$$\frac{\omega_n z}{v_0} = -\sin \omega_n t + \frac{1}{\omega_n t_1}[\cos \omega_n(t - t_1) - \cos \omega_n t]$$

4-28 Determine the time response for Prob. 4-10 using numerical integration.

4-29 Determine the time response for Prob. 4-22 using numerical integration.

4-30 Set up an analog computer circuit to solve the base-excited undamped system

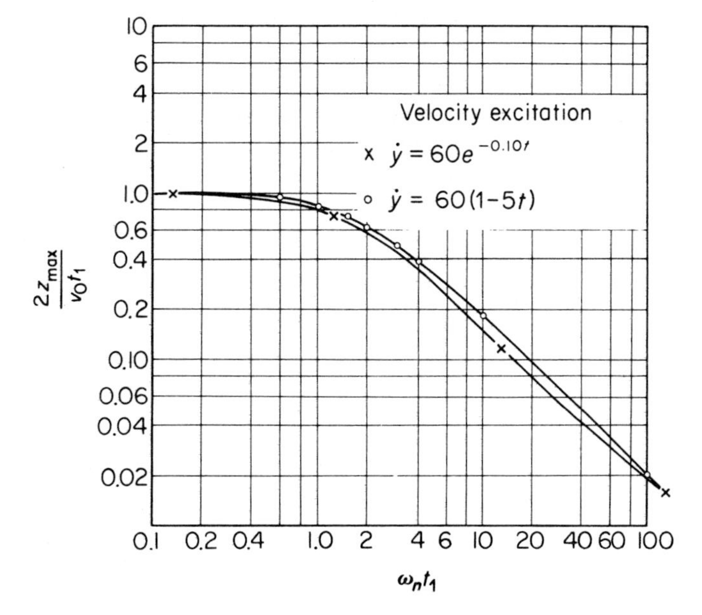

Figure P. 4-30.

of Prob. 4-26. Verify the response spectrum when the excitations are

$$\dot{y}(t) = 60e^{-t/10}, \qquad \dot{y}(t) = 60(1 - 5t)$$

The response spectra for the above excitations are shown to scale in Fig. P4-30.

4-31 A spring-mass system has the equation

$$\ddot{x} + 2\dot{x} + 100x = 0$$

with initial conditions $x(0) = 1.0$ in. and $\dot{x}(0) = 3$ in./sec. Slow down the computer equation by a factor 10, and determine the circuit diagram and scale factors for its efficient computation.

4-32 For a certain single degree of freedom system the following values are given: $m = 0.122$ lb sec^2/in., $k = 6100$ lb/in., $c = 0.10c_c$. Choose a computer time which is 500 times greater than the actual time and write the equation for the computer for an arbitrary excitation $F(\tau)$. Develop the circuit diagram with appropriate scale.

4-33 Write the equation of motion for a damped system with base excitation $y(t)$. Draw the analog computer circuit and show how the quantities $z = (x - y)$ and x can be measured.

4-34 A spring-mass system with viscous damping is initially at rest with zero displacement. If the system is activated by a harmonic force of frequency $\omega = \omega_n = \sqrt{k/m}$, determine the equation for its motion.

4-35 In Prob. 4-34 show that with small damping the amplitude will build up to a value $(1 - e^{-1})$ times the steady state value in the time $t = 1/f_n\delta$. (δ = logarithmic decrement).

4-36 Assume that a lightly damped system is driven by a force $F_0 \sin \omega_n t$ where ω_n is the natural frequency of the system. Determine the equation if the force is suddenly removed. Show that the amplitude decays to a value e^{-1} times the initial value in the time $t = 1/f_n\delta$.

5

Two Degrees of Freedom Systems

5.1 INTRODUCTION

When a system requires two coordinates to describe its motion, it is said to have two degrees of freedom. Such a system offers a simple introduction to the behavior of systems with several degrees of freedom.

A two degrees of freedom system will have two natural frequencies. When free vibration takes place at one of these natural frequencies, a definite relationship exists between the amplitudes of the two coordinates, and the configuration is referred to as the *normal mode*. The two degrees of freedom system will then have two normal mode vibrations corresponding to the two natural frequencies. Free vibration initiated under any condition will in general be the superposition of the two normal mode vibrations. However, forced harmonic vibration will take place at the frequency of the excitation, and the amplitude of the two coordinates will tend to a maximum at the two natural frequencies.

5.2 NORMAL MODE VIBRATION

Consider the undamped system of Fig. 5.2-1. Using coordinates x_1 and x_2 measured from inertial reference, the differential equations of motion for the system become

$$\begin{aligned} m\ddot{x}_1 &= -k(x_1 - x_2) - kx_1 \\ 2m\ddot{x}_2 &= k(x_1 - x_2) - kx_2 \end{aligned} \tag{5.2-1}$$

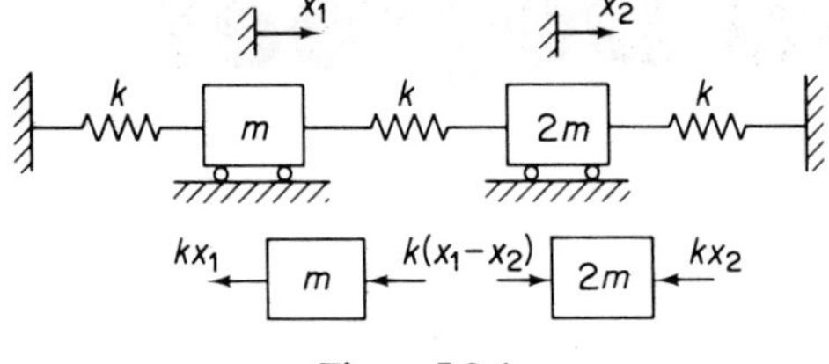

Figure 5.2-1.

We now define a normal mode oscillation as one in which each mass undergoes harmonic motion of the same frequency, passing simultaneously through the equilibrium position. For such motion we can let

$$\begin{aligned} x_1 &= A_1 e^{i\omega t} \\ x_2 &= A_2 e^{i\omega t} \end{aligned} \tag{5.2-2}$$

Substituting these into the differential equations gives

$$\begin{aligned} (2k - \omega^2 m)A_1 - kA_2 &= 0 \\ -kA_1 + (2k - 2\omega^2 m)A_2 &= 0 \end{aligned} \tag{5.2-3}$$

which are satisfied for any A_1 and A_2 if the following determinant is zero

$$\begin{vmatrix} (2k - \omega^2 m) & -k \\ -k & (2k - 2\omega^2 m) \end{vmatrix} = 0 \tag{5.2-4}$$

Letting $\omega^2 = \lambda$, the above determinant leads to the *characteristic equation*

$$\lambda^2 - \left(3\frac{k}{m}\right)\lambda + \frac{3}{2}\left(\frac{k}{m}\right)^2 = 0 \tag{5.2-5}$$

The two roots of this equation are

$$\lambda_1 = \left(\frac{3}{2} - \frac{1}{2}\sqrt{3}\right)\frac{k}{m} = 0.634\frac{k}{m}$$

and

$$\lambda_2 = \left(\frac{3}{2} + \frac{1}{2}\sqrt{3}\right)\frac{k}{m} = 2.366\frac{k}{m}$$

and the *natural frequencies* of the system are found to be

$$\omega_1 = \lambda_1^{1/2} = \sqrt{0.634\frac{k}{m}}$$

and (5.2-6)

$$\omega_2 = \lambda_2^{1/2} = \sqrt{2.366\frac{k}{m}}$$

Substitution of these natural frequencies into Eq. (5.2-3) enables one to find the ratio of the amplitudes. For $\omega_1^2 = 0.634k/m$, we obtain

$$\left(\frac{A_1}{A_2}\right)^{(1)} = \frac{k}{2k - \omega_1^2 m} = \frac{1}{2 - 0.634} = 0.731 \tag{5.2-7}$$

which is the amplitude ratio or *mode shape* corresponding to the first normal mode.

Similarly, using $\omega_2^2 = 2.366k/m$, we obtain

$$\left(\frac{A_1}{A_2}\right)^{(2)} = \frac{k}{2k - \omega_2^2 m} = \frac{1}{2 - 2.366} = -2.73 \tag{5.2-7}$$

for the mode shape corresponding to the second normal mode. We can display the two normal modes graphically as in Fig. 5.2-2. In the first normal mode, the two masses move in phase; in the second normal mode the masses move in opposition, or out of phase, with each other.

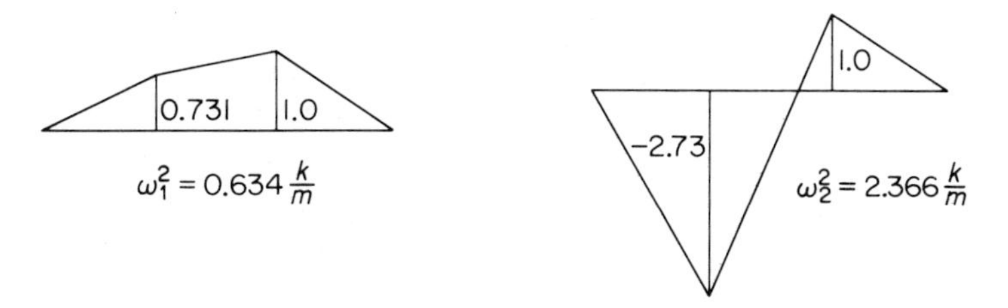

Figure 5.2-2. Normal modes of the system shown in Figure 5.2-1.

Example 5.2-1

For the system of Fig. 5.2-1, let the coupling spring at the center equal *nk* and compute the natural frequencies and mode shapes.

Solution: Let $k/m = \omega_{11}^2$, in which case the characteristic equation becomes

$$\omega^4 - \tfrac{3}{2}\omega_{11}^2(1+n)\omega^2 + \tfrac{1}{2}\omega_{11}^4(1+2n) = 0$$

The two normal mode frequencies in terms of ω_{11}^2 then become

$$\omega_{1,2}^2/\omega_{11}^2 = \frac{3}{4}(1+n) \pm \sqrt{\frac{9}{16}(1+n)^2 - \frac{1}{2}(1+2n)}$$

On varying the value of n, the following numerical values for $(\omega_1/\omega_{11})^2$ and $(\omega_2/\omega_{11})^2$ are found and plotted in Fig. 5.2-3. Note that $(\omega_1/\omega_{11})^2$ remains nearly constant.

Normal Mode Frequencies as Function of n

n	$(\omega_1/\omega_{11})^2$	$(\omega_2/\omega_{11})^2$
0	0.50	1.0
0.5	0.611	1.641
1.0	0.634	2.366
2.0	0.650	3.850
4.0	0.660	6.840
10.0	0.666	15.83
100.0	0.666	150.8
∞	0.666	∞

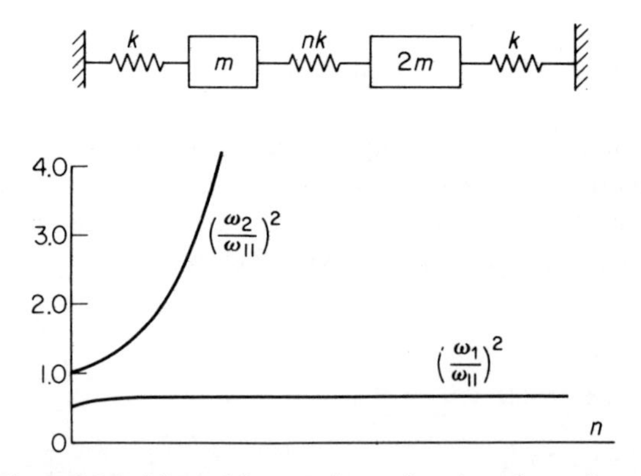

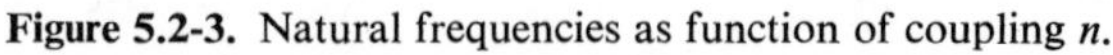

Figure 5.2-3. Natural frequencies as function of coupling n.

The mode shapes can now be found for any n as

$$\left(\frac{A_1}{A_2}\right)^{(1)} = \frac{1 + n - 2(\omega_1/\omega_{11})^2}{n}$$

$$\left(\frac{A_1}{A_2}\right)^{(2)} = \frac{1 + n - 2(\omega_2/\omega_{11})^2}{n}$$

For example, if $n = 4$, the two natural modes are as shown in Fig. 5.2-4.

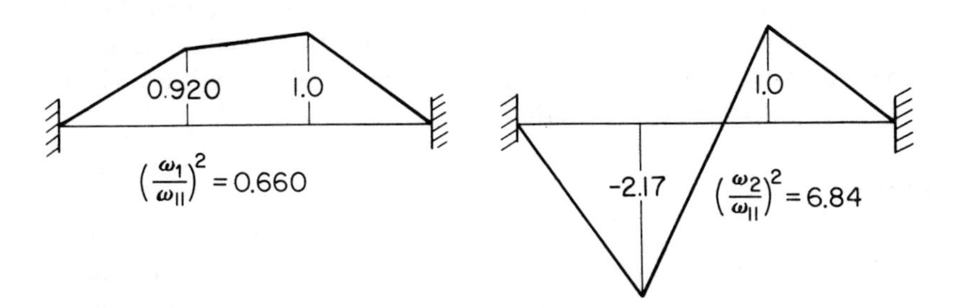

Figure 5.2-4. Normal modes of system shown in Figure 5.2-3 for $n = 4$.

Example 5.2-2

In Fig. 5.2-5 the two pendulums are coupled by means of a weak spring k, which is unstrained when the two pendulum rods are in the vertical position. Determine the normal mode vibrations.

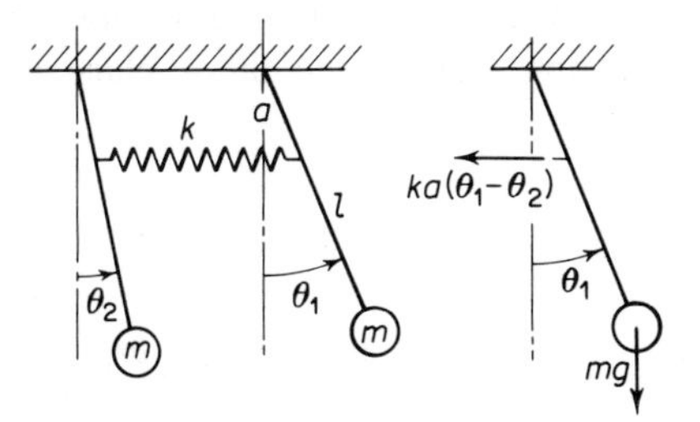

Figure 5.2-5. Coupled pendulum.

Solution: Assuming the counterclockwise angular displacements to be positive, and taking moments about the points of suspension, we obtain the following equations of motion for small oscillations

$$ml^2\ddot{\theta}_1 = -mgl\theta_1 - ka^2(\theta_1 - \theta_2)$$
$$ml^2\ddot{\theta}_2 = -mgl\theta_2 + ka^2(\theta_1 - \theta_2)$$

Assuming the normal mode solutions as

$$\theta_1 = A_1 \cos \omega t$$
$$\theta_2 = A_2 \cos \omega t$$

the natural frequencies and mode shapes are found to be

$$\omega_1 = \sqrt{\frac{g}{l}} \qquad \omega_2 = \sqrt{\frac{g}{l} + 2\frac{k}{m}\frac{a^2}{l^2}}$$

$$\left(\frac{A_1}{A_2}\right)^{(1)} = 1.0 \qquad \left(\frac{A_1}{A_2}\right)^{(2)} = -1.0$$

Thus in the first mode the two pendulums move in phase and the spring remains unstretched. In the second mode the two pendulums move in opposition and the coupling spring is actively involved with a node at its midpoint. Consequently, the natural frequency is higher.

Example 5.2-3

If the coupled pendulum of Example 5.2-2 is set into motion with initial conditions differing from those of the normal modes, the oscillations will contain both normal modes simultaneously. For example, if the initial conditions are $\theta_1(0) = A$ and $\theta_2(0) = 0$, the equations of motion will be

$$\theta_1(t) = \tfrac{1}{2}A \cos \omega_1 t + \tfrac{1}{2}A \cos \omega_2 t$$
$$\theta_2(t) = \tfrac{1}{2}A \cos \omega_1 t - \tfrac{1}{2}A \cos \omega_2 t$$

Consider the case where the coupling spring is very weak, and show that a beating phenomena takes place between the two pendulums.

Solution: The equations above can be rewritten as follows

$$\theta_1(t) = A \cos \left(\frac{\omega_1 - \omega_2}{2}\right)t \quad \cos \left(\frac{\omega_1 + \omega_2}{2}\right)t$$

$$\theta_2(t) = -A \sin \left(\frac{\omega_1 - \omega_2}{2}\right)t \quad \sin \left(\frac{\omega_1 + \omega_2}{2}\right)t$$

Since $(\omega_1 - \omega_2)$ is very small, $\theta_1(t)$ and $\theta_2(t)$ will behave like $\cos (\omega_1 + \omega_2)t/2$ and $\sin (\omega_1 + \omega_2 t)/2$ with slowly varying amplitudes as shown in Fig. 5.2-6. Since the system is conservative, energy is transferred from one pendulum to the other.

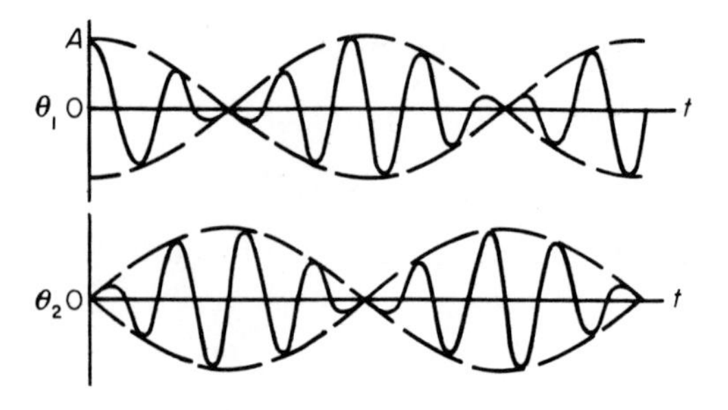

Figure 5.2-6. Exchange of energy between pendulums.

Example 5.2-4

If the masses and springs of the system shown in Fig. 5.2-7 are made equal to m and k as shown, the normal modes become

$$\omega_1^2 = \frac{k}{m} \qquad \omega_2^2 = \frac{3k}{m}$$

$$\frac{A_1}{A_2} = 1 \qquad \frac{A_1}{A_2} = -1$$

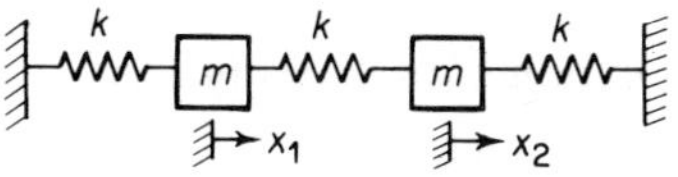

Figure 5.2-7.

Determine the free vibration of the system when the initial conditions are

$$x_1(0) = 5 \qquad x_2(0) = 0$$
$$\dot{x}_1(0) = 0 \qquad \dot{x}_2(0) = 0$$

Solution: Any free vibration can be considered to be the superposition of its normal modes. Thus the two displacements can be written as

$$\begin{aligned} x_1 &= A \sin(\omega_1 t + \psi_1) - B \sin(\omega_2 t + \psi_2) \\ x_2 &= A \sin(\omega_1 t + \psi_1) + B \sin(\omega_2 t + \psi_2) \end{aligned} \tag{a}$$

It should be noted here that the first terms on the right correspond to the first normal mode at the natural frequency ω_1. Its amplitude ratio is also $A_1/A_2 = A/A = 1$, which is the first normal mode shape. The second terms oscillate at frequency ω_2 with amplitude ratio $B_1/B_2 = -B/B = -1$, in conformity with the second normal mode vibration. The phase ψ_1 and ψ_2 simply allows the freedom of shifting the time origin and does not alter the character of the normal modes. The constants A, B, ψ_1 and ψ_2 are sufficient to satisfy the four initial conditions, which may be arbitrarily chosen

Letting $t = 0$ and $x_1(0) = 5$, $x_2(0) = 0$, we obtain

$$5 = A \sin \psi_1 - B \sin \psi_2$$
$$0 = A \sin \psi_1 + B \sin \psi_2$$

Thus by adding and subtracting we find

$$A \sin \psi_1 = 2.5$$
$$B \sin \psi_2 = -2.5$$

Differentiating Eq. (a) for the velocity and letting $t = 0$ we obtain two other equations

$$0 = \omega_1 A \cos \psi_1 - \omega_2 B \cos \psi_2$$
$$0 = \omega_1 A \cos \psi_1 + \omega_2 B \cos \psi_2$$

from which we find

$$\cos \psi_1 = 0 \quad \text{or} \quad \psi_1 = 90^\circ$$
$$\cos \psi_2 = 0 \quad \text{or} \quad \psi_2 = 90^\circ$$

The solution is then easily seen to be

$$x_1 = 2.5 \cos \sqrt{\frac{k}{m}}t + 2.5 \cos \sqrt{\frac{3k}{m}}t$$

$$x_2 = 2.5 \cos \sqrt{\frac{k}{m}}t - 2.5 \cos \sqrt{\frac{3k}{m}}t$$

which may be written in the matrix form

$$\begin{Bmatrix} x_1 \\ x_2 \end{Bmatrix} = 2.5\begin{Bmatrix} 1 \\ 1 \end{Bmatrix} \cos \sqrt{\frac{k}{m}}t - 2.5\begin{Bmatrix} -1 \\ 1 \end{Bmatrix} \cos \sqrt{\frac{3k}{m}}t$$

5.3 COORDINATE COUPLING

The differential equation of motion for the two degrees of freedom system are in general *coupled*, in that both coordinates appear in each equation. In the most general case the two equations for the undamped system have the form

$$m_{11}\ddot{x}_1 + m_{12}\ddot{x}_2 + k_{11}x_1 + k_{12}x_2 = 0$$
$$m_{21}\ddot{x}_1 + m_{22}\ddot{x}_2 + k_{21}x_1 + k_{22}x_2 = 0$$

These equations can be expressed in matrix form (See Appendix C) as

$$\begin{bmatrix} m_{11} & m_{12} \\ m_{21} & m_{22} \end{bmatrix}\begin{Bmatrix} \ddot{x}_1 \\ \ddot{x}_2 \end{Bmatrix} + \begin{bmatrix} k_{11} & k_{12} \\ k_{21} & k_{22} \end{bmatrix}\begin{Bmatrix} x_1 \\ x_2 \end{Bmatrix} = \begin{Bmatrix} 0 \\ 0 \end{Bmatrix}$$

which immediately reveals the type of coupling present. Mass or *dynamical coupling* exists if the mass matrix is nondiagonal, whereas stiffness or *static coupling* exists if the stiffness matrix is nondiagonal.

It is also possible to establish the type of coupling from the expressions for the kinetic and potential energies. Cross products of coordinates in either expression denote coupling, dynamic or static, depending on whether they are found in T or U. The choice of coordinates establishes the type of coupling, and both dynamic and static coupling may be present.

It is possible to find a coordinate system which has neither form of coupling. The two equations are then decoupled and each equation may be solved independently of the other. Such coordinates are called *principal coordinates* (also called *normal coordinates*).

Although it is always possible to decouple the equations of motion for the undamped system, this is not always the case for a damped system. The following matrix equations show a system which has zero dynamic and static coupling, but the coordinates are coupled by the damping matrix.

$$\begin{bmatrix} m_{11} & 0 \\ 0 & m_{22} \end{bmatrix} \begin{Bmatrix} \ddot{x}_1 \\ \ddot{x}_2 \end{Bmatrix} + \begin{bmatrix} c_{11} & c_{12} \\ c_{21} & c_{22} \end{bmatrix} \begin{Bmatrix} \dot{x}_1 \\ \dot{x}_2 \end{Bmatrix} + \begin{bmatrix} k_{11} & 0 \\ 0 & k_{22} \end{bmatrix} \begin{Bmatrix} x_1 \\ x_2 \end{Bmatrix} = \begin{Bmatrix} 0 \\ \end{Bmatrix}$$

If in the above equation $c_{12} = c_{21} = 0$, then the damping is said to be *proportional* (proportional to the stiffness or mass matrix), and the system equations become uncoupled.

Example 5.3-1

Figure 5.3-1 shows a rigid bar with its center of mass not coinciding with its geometric center, i.e., $l_1 \neq l_2$, and supported by two springs, k_1, k_2. It represents a two degree of freedom system since two coordinates are necessary to describe its motion. The choice of the coordinates will define the type of coupling which can be immediately determined from the mass and stiffness matrices. Mass or *dynamical coupling* exists if the mass matrix is nondiagonal, whereas stiffness or *static coupling* exists if the stiffness matrix is nondiagonal. It is also possible to have both forms of coupling.

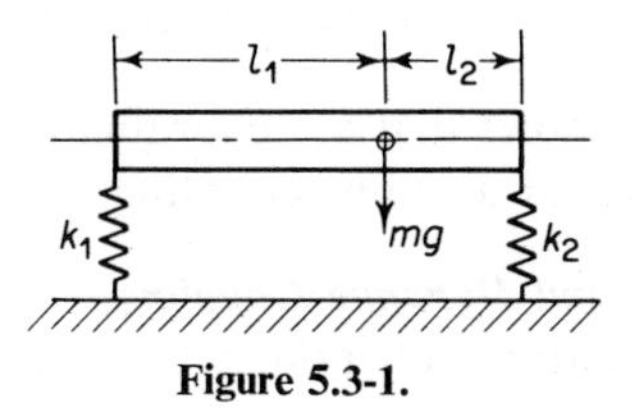

Figure 5.3-1.

Static Coupling. Choosing coordinates x and θ, where x is the linear displacement of the center of mass, the system will have static coupling as shown by the matrix equation

$$\begin{bmatrix} m & 0 \\ 0 & J \end{bmatrix} \begin{Bmatrix} \ddot{x} \\ \ddot{\theta} \end{Bmatrix} + \begin{bmatrix} (k_1 + k_2) & (k_2 l_2 - k_1 l_1) \\ (k_2 l_2 - k_1 l_1) & (k_1 l_1^2 + k_2 l_2^2) \end{bmatrix} \begin{Bmatrix} x \\ \theta \end{Bmatrix} = \begin{Bmatrix} 0 \\ \end{Bmatrix}$$

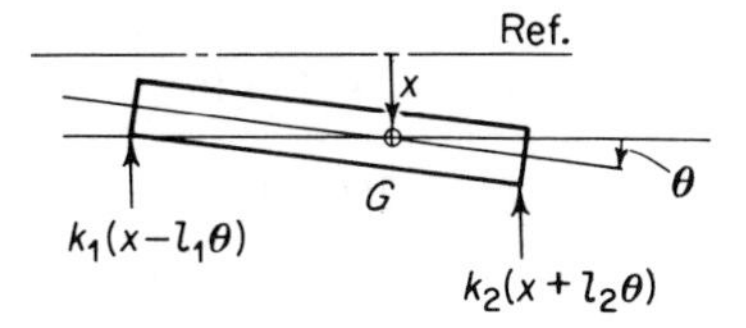

Figure 5.3-2. Coordinates leading to static coupling.

If $k_1 l_1 = k_2 l_2$, the coupling disappears, and we obtain uncoupled x and θ vibrations.

Dynamic Coupling. There is some point C along the bar where a force applied normal to the bar produces pure translation; i.e., $k_1 l_3 = k_2 l_4$. The equations of motion in terms of x_C and θ can be shown to be

$$\begin{bmatrix} m & me \\ me & J_C \end{bmatrix} \begin{Bmatrix} \ddot{x}_C \\ \ddot{\theta} \end{Bmatrix} + \begin{bmatrix} (k_1 + k_2) & 0 \\ 0 & (k_1 l_3^2 + k_2 l_4^2) \end{bmatrix} \begin{Bmatrix} x_C \\ \theta \end{Bmatrix} = \begin{Bmatrix} 0 \end{Bmatrix}$$

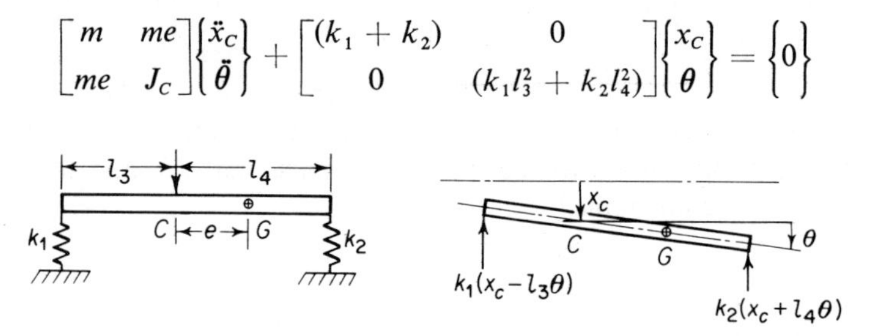

Figure 5.3-3. Coordinates leading to dynamic coupling.

which shows that the coordinates chosen eliminated the static coupling and introduced dynamic coupling.

Static and Dynamic Coupling. If we choose $x = x_1$ at the end of the bar, the equations of motion become

$$\begin{bmatrix} m & ml_1 \\ ml_1 & J_1 \end{bmatrix} \begin{Bmatrix} \ddot{x}_1 \\ \ddot{\theta} \end{Bmatrix} + \begin{bmatrix} (k_1 + k_2) & k_2 l \\ k_2 l & k_2 l^2 \end{bmatrix} \begin{Bmatrix} x_1 \\ \theta \end{Bmatrix} = \begin{Bmatrix} 0 \end{Bmatrix}$$

and both static and dynamic coupling are now present.

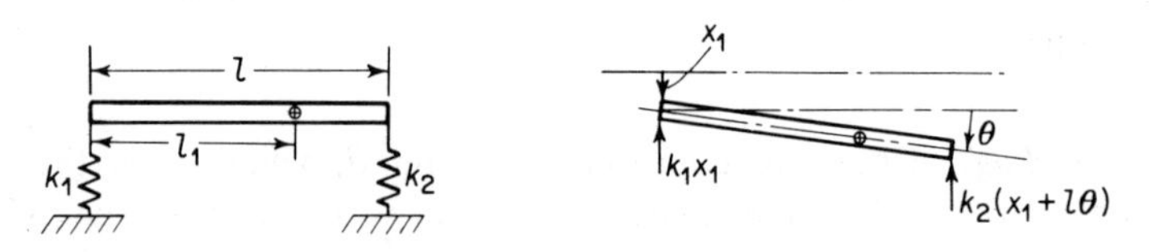

Figure 5.3-4. Coordinates leading to static and dynamic coupling.

Example 5.3-2

Determine the normal modes of vibration of an automobile simulated by the simplified two degrees of freedom system with the following numerical values

$$W = 3220 \text{ lb} \qquad l_1 = 4.5 \text{ ft} \qquad k_1 = 2400 \text{ lb/ft}$$

$$J_C = \frac{W}{g} r^2 \qquad l_2 = 5.5 \text{ ft} \qquad k_2 = 2600 \text{ lb/ft}$$

$$r = 4 \text{ ft} \qquad l = 10 \text{ ft}$$

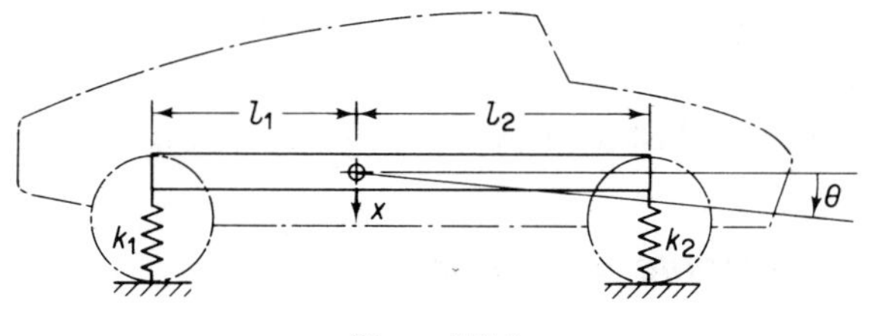

Figure 5.3-5.

The equations of motion indicate static coupling.

$$m\ddot{x} + k_1(x - l_1\theta) + k_2(x + l_2\theta) = 0$$
$$J_C\ddot{\theta} - k_1(x - l_1\theta)l_1 + k_2(x + l_2\theta)l_2 = 0$$

Assuming harmonic motion, we have

$$\begin{bmatrix} (k_1 + k_2 - \omega^2 m) & -(k_1 l_1 - k_2 l_2) \\ -(k_1 l_1 - k_2 l_2) & (k_1 l_1^2 + k_2 l_2^2 - \omega^2 J_C) \end{bmatrix} \begin{Bmatrix} x \\ \theta \end{Bmatrix} = \begin{Bmatrix} 0 \end{Bmatrix}$$

From the determinant of the matrix equation, the two natural frequencies are

$$\omega_1 = 6.90 \text{ rad/sec} = 1.10 \text{ cps}$$
$$\omega_2 = 9.06 \text{ rad/sec} = 1.44 \text{ cps}$$

The amplitude ratios for the two frequencies are

$$\left(\frac{x}{\theta}\right)_{\omega_1} = -14.6 \text{ ft/rad} = -3.06 \text{ in./deg}$$

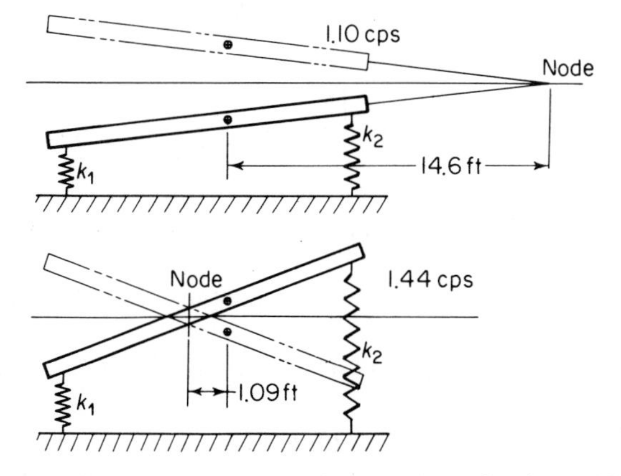

Figure 5.3-6. Normal modes of system shown in Figure 5.3-5.

$$\left(\frac{X}{\theta}\right)_{\omega_2} = 1.09 \text{ ft/rad} = 0.288 \text{ in./deg}$$

The mode shapes are illustrated by the diagrams of Fig. 5.3-5.

5.4 FORCED HARMONIC VIBRATION

We consider here a system excited by a harmonic force $F_1 \sin \omega t$. Assuming the motion to be represented by the matrix equation

$$\begin{bmatrix} m_1 & 0 \\ 0 & m_2 \end{bmatrix}\begin{Bmatrix} \ddot{x}_1 \\ \ddot{x}_2 \end{Bmatrix} + \begin{bmatrix} k_{11} & k_{12} \\ k_{12} & k_{22} \end{bmatrix}\begin{Bmatrix} x_1 \\ x_2 \end{Bmatrix} = \begin{Bmatrix} F_1 \\ 0 \end{Bmatrix} \sin \omega t \tag{5.4-1}$$

we assume the solution to be

$$\begin{Bmatrix} x_1 \\ x_2 \end{Bmatrix} = \begin{Bmatrix} X_1 \\ X_2 \end{Bmatrix} \sin \omega t$$

Substituting this solution into the first equation, we obtain

$$\begin{bmatrix} (k_{11} - m_1\omega^2) & k_{12} \\ k_{12} & (k_{22} - m_2\omega^2) \end{bmatrix}\begin{Bmatrix} X_1 \\ X_2 \end{Bmatrix} = \begin{Bmatrix} F_1 \\ 0 \end{Bmatrix} \tag{5.4-2}$$

or, in simpler notation

$$[Z(\omega)]\begin{Bmatrix} X_1 \\ X_2 \end{Bmatrix} = \begin{Bmatrix} F_1 \\ 0 \end{Bmatrix}$$

Premultiplying by $[Z(\omega)]^{-1}$ we obtain (See Appendix C)

$$\begin{Bmatrix} X_1 \\ X_2 \end{Bmatrix} = [Z(\omega)]^{-1}\begin{Bmatrix} F_1 \\ 0 \end{Bmatrix} = \frac{adj[Z(\omega)]\begin{Bmatrix} F_1 \\ 0 \end{Bmatrix}}{|Z(\omega)|} \tag{5.4-3}$$

Referring to Eq. (5.4-2), the determinant $|Z(\omega)|$ can be expressed as

$$|Z(\omega)| = m_1 m_2(\omega_1^2 - \omega^2)(\omega_2^2 - \omega^2) \tag{5.4-4}$$

where ω_1 and ω_2 are the normal mode frequencies. Thus Eq. (5.4-3) becomes

$$\begin{Bmatrix} X_1 \\ X_2 \end{Bmatrix} = \frac{1}{|Z(\omega)|}\begin{bmatrix} (k_{22} - m_2\omega^2) & -k_{12} \\ -k_{12} & (k_{11} - m_1\omega^2) \end{bmatrix}\begin{Bmatrix} F \\ 0 \end{Bmatrix} \tag{5.4-5}$$

or

$$X_1 = \frac{(k_{22} - m_2\omega^2)F}{m_1m_2(\omega_1^2 - \omega^2)(\omega_2^2 - \omega^2)}$$
$$X_2 = \frac{-k_{12}F}{m_1m_2(\omega_1^2 - \omega^2)(\omega_2^2 - \omega^2)} \tag{5.4-6}$$

Example 5.4-1

Apply Eqs. (5.4-6) of Sec. 5.4 to the system shown in Fig. 5.4-1 when m_1 is excited by the force $F_1 \sin \omega t$. Plot its frequency response curve.

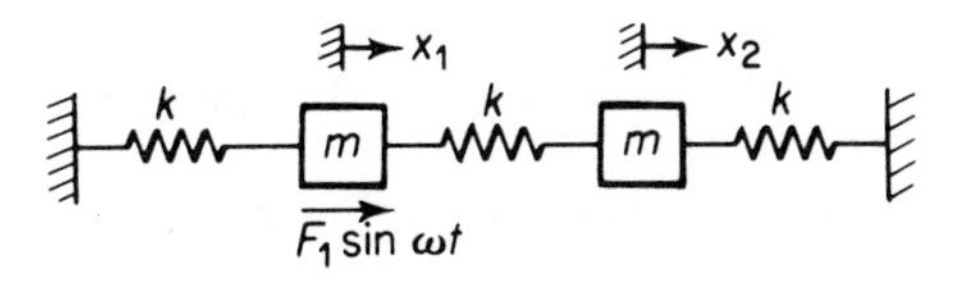

Figure 5.4-1

Solution: The equation of motion in matrix form is

$$\begin{bmatrix} m & 0 \\ 0 & m \end{bmatrix}\begin{Bmatrix} \ddot{x}_1 \\ \ddot{x}_2 \end{Bmatrix} + \begin{bmatrix} 2k & -k \\ -k & 2k \end{bmatrix}\begin{Bmatrix} x_1 \\ x_2 \end{Bmatrix} = \begin{Bmatrix} F_1 \\ 0 \end{Bmatrix}\sin \omega t$$

Thus we have $k_{11} = k_{22} = 2k$; $k_{12} = k_{21} = -k$; $\omega_1^2 = k/m$; and $\omega_2^2 = 3k/m$. Eqs. (5.4-6) of Sec. 5.4 therefore become

$$X_1 = \frac{(2k - m\omega^2)F_1}{m^2(\omega_1^2 - \omega^2)(\omega_2^2 - \omega^2)}$$
$$X_2 = \frac{kF_1}{m^2(\omega_1^2 - \omega^2)(\omega_2^2 - \omega^2)}$$

It is convenient here to expand each of the above equations in partial fractions. For X_1 we obtain

$$\frac{(2k - m\omega^2)F_1}{m^2(\omega_1^2 - \omega^2)(\omega_2^2 - \omega^2)} = \frac{C_1}{(\omega_1^2 - \omega^2)} + \frac{C_2}{(\omega_2^2 - \omega^2)}$$

To solve for C_1, multiply by $(\omega_1^2 - \omega^2)$ and let $\omega = \omega_1$

$$C_1 = \frac{(2k - m\omega_1^2)F_1}{m^2(\omega_2^2 - \omega_1^2)} = \frac{F_1}{2m}$$

Similarly, C_2 is evaluated by multiplying by $(\omega_2^2 - \omega^2)$ and letting $\omega = \omega_2$

$$C_2 = \frac{(2k - m\omega_2^2)F_1}{m^2(\omega_1^2 - \omega_2^2)} = \frac{F_1}{2m}$$

An alternative form of X_1 is then

$$X_1 = \frac{F_1}{2m}\left[\frac{1}{\omega_1^2 - \omega^2} + \frac{1}{\omega_2^2 - \omega^2}\right]$$
$$= \frac{F_1}{2k}\left[\frac{1}{1 - (\omega/\omega_1)^2} + \frac{1}{3 - (\omega/\omega_1)^2}\right]$$

Treating X_2 in the same manner, its equation is

$$X_2 = \frac{F_1}{2k}\left[\frac{1}{1 - (\omega/\omega_1)^2} - \frac{1}{3 - (\omega/\omega_1)^2}\right]$$

The frequency response curve is shown in Fig. 5.4-2.

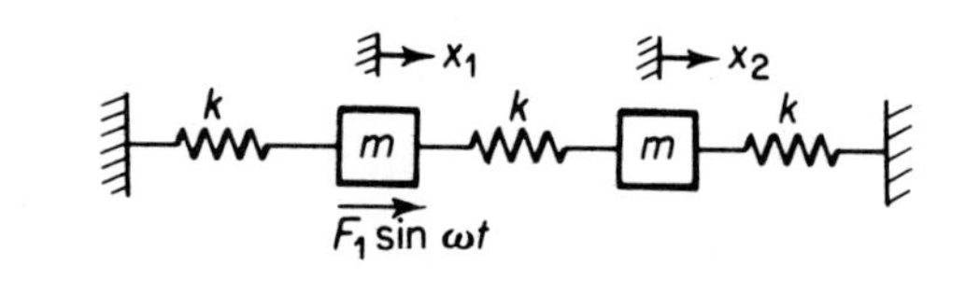

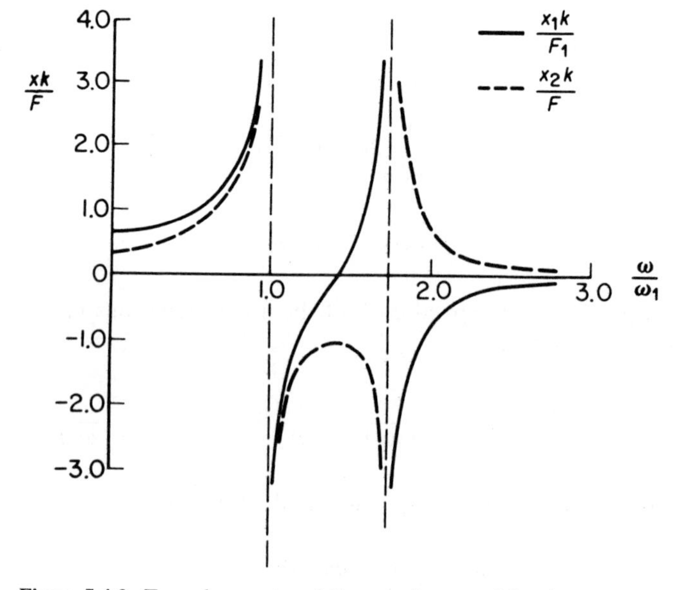

Figure 5.4-2. Forced response of the two degrees of freedom system.

5.5 VIBRATION ABSORBER

A spring-mass system k_2, m_2, tuned to the frequency of the exciting force such that $\omega^2 = k_2/m_2$, will act as a vibration absorber and reduce the motion of the main mass m_1 to zero. Making the substitution

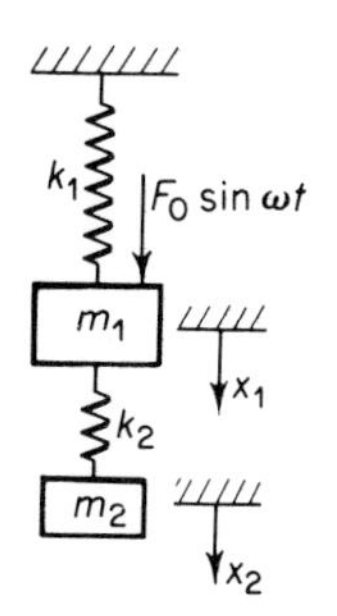

Figure 5.5-1. Vibration absorber.

$$\omega_{11}^2 = \frac{k_1}{m_1} \qquad \omega_{22}^2 = \frac{k_2}{m_2}$$

and assuming the motion to be harmonic, the equation for the amplitude X_1 can be shown to be equal to

$$\frac{X_1 k_1}{F_0} = \frac{\left[1 - \left(\frac{\omega}{\omega_{22}}\right)^2\right]}{\left[1 + \frac{k_2}{k_1} - \left(\frac{\omega}{\omega_{11}}\right)^2\right]\left[1 - \left(\frac{\omega}{\omega_{22}}\right)^2\right] - \frac{k_2}{k_1}} \tag{5.5-1}$$

Figure 5.5-2 shows a plot of this equation with $\mu = m_2/m_1$ as parameter. Note that $k_2/k_1 = \mu(\omega_{22}/\omega_{11})^2$. Since the system is one of two degrees of freedom, two natural frequencies exist. These are shown against μ in Fig. 5.5-3.

So far nothing has been said about the size of the absorber mass. At $\omega = \omega_{22}$, the amplitude $X_1 = 0$, but the absorber mass undergoes an

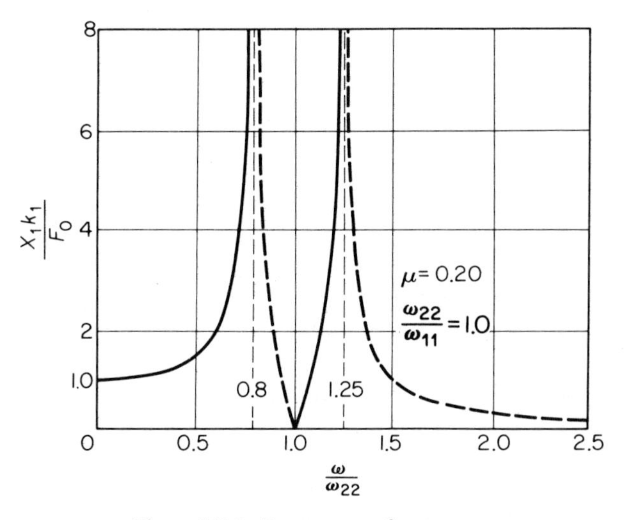

Figure 5.5-2. Response vs. frequency.

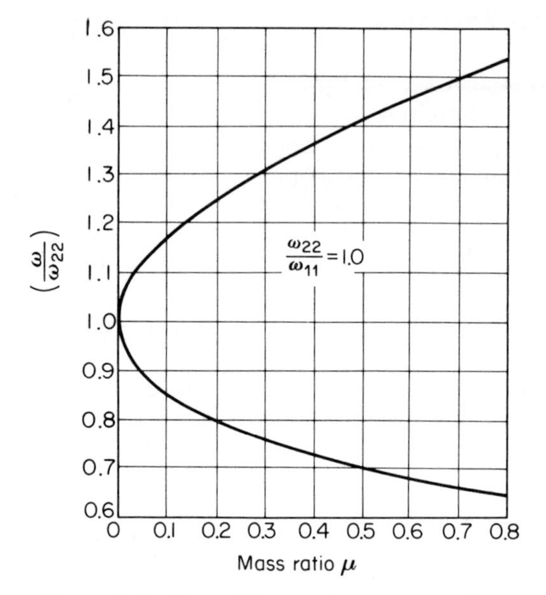

Figure 5.5-3. Natural frequencies vs. $\mu = m_2/m_1$.

amplitude equal to

$$X_2 = -\frac{F_0}{k_2} \tag{5.5-2}$$

Since the force acting on m_2 is

$$k_2 X_2 = \omega^2 m_2 X_2 = -F_0$$

the absorber system k_2, m_2 exerts a force equal and opposite to the disturbing force. Thus the size of k_2 and m_2 depends on the allowable value of X_2.

5.6 THE CENTRIFUGAL PENDULUM VIBRATION ABSORBER

The vibration absorber of Sec. 5.5 is only effective at one frequency, $\omega = \omega_{22}$. Also, with resonant frequencies on each side of ω_{22}, the usefulness of the spring-mass absorber is narrowly limited.

For a rotating system such as the automobile engine, the exciting torques are proportional to the rotational speed n, which may vary over a wide range. Thus for the absorber to be effective, its natural frequency must also be proportional to the speed. The characteristics of the centrifugal pendulum are ideally suited for this purpose.

Figure 5.6-1 shows the essentials of the centrifugal pendulum. It is a

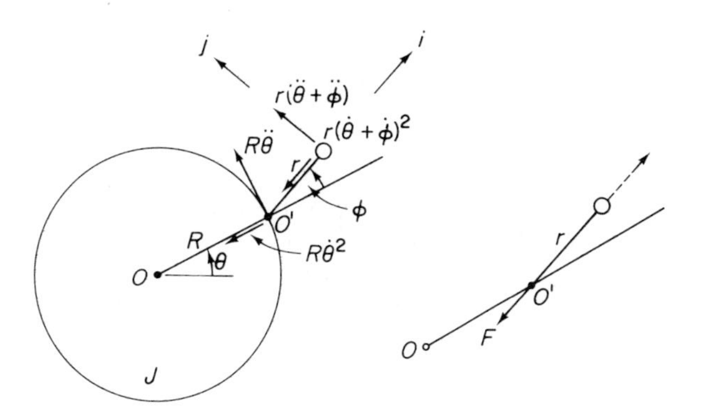

Figure 5.6-1. Centrifugal pendulum.

two degree of freedom nonlinear system; however, we will limit the oscillations to small angles, thereby reducing its complexity.

Placing the coordinates through point O′ parallel and normal to r, the line r rotates with angular velocity $(\dot{\theta} + \dot{\phi})$. The acceleration of m is equal to the vector sum of the acceleration of O′ and the acceleration of m relative to O′.

$$a_m = [R\ddot{\theta} \sin \phi - R\dot{\theta}^2 \cos \phi - r(\dot{\theta} + \dot{\phi})^2]i + [R\ddot{\theta} \cos \phi + R\dot{\theta}^2 \sin \phi + r(\ddot{\theta} + \ddot{\phi})]j \tag{5.6-1}$$

Since the moment about O′ is zero, we have

$$m[R\ddot{\theta} \cos \phi + R\dot{\theta}^2 \sin \phi + r(\ddot{\theta} + \ddot{\phi})]r = 0 \tag{5.6-2}$$

Assuming ϕ to be small, we let $\cos \phi = 1$ and $\sin \phi = \phi$, and arrive at the equation for the pendulum

$$\ddot{\phi} + \left(\frac{R}{r}\dot{\theta}^2\right)\phi = -\left(\frac{R + r}{r}\right)\ddot{\theta} \tag{5.6-3}$$

The above equation contains both θ and ϕ and hence a second equation is required, namely the torque equation for the wheel. Acting on the wheel are the exciting torque T and the pendulum torque which is found from the cross product of the wheel radius R and the pendulum force ma_m. It is evident from Eq. (5.6-1) and (5.6-3) that even for the small angle approximation we will end up with two nonlinear differential equations and that a simple solution is not possible without further approximation.

If we assume the motion of the wheel to be a steady rotation n plus a small sinusoidal oscillation, we can write

$$\begin{aligned} \theta &= nt + \theta_0 \sin \omega t \\ \dot{\theta} &= n + \omega\theta_0 \cos \omega t \cong n \\ \ddot{\theta} &= -\omega^2\theta_0 \sin \omega t \end{aligned} \tag{5.6-4}$$

Then Eq. (5.6-3) becomes

$$\ddot{\phi} + \left(\frac{R}{r} n^2\right)\phi = \left(\frac{R+r}{r}\right)\omega^2\theta_0 \sin \omega t \tag{5.6-3$'$}$$

and we recognize the natural frequency of the pendulum to be

$$\omega_n = n\sqrt{\frac{R}{r}} \tag{5.6-5}$$

Assuming a steady state solution $\phi = \phi_0 \sin(\omega t - \alpha)$, the amplitude ratio becomes

$$\frac{\theta_0}{\phi_0} = \frac{\dfrac{n^2R}{r} - \omega^2}{\left(\dfrac{R+r}{r}\right)\omega^2} \tag{5.6-6}$$

which clearly indicates that the oscillation θ_0 of the wheel becomes zero when $\omega = n\sqrt{R/r}$.

Also by recognizing that the largest term in Eq. (5.6-1) is due to $\dot{\theta}^2 = n^2$, the pendulum torque opposing the disturbing torque T becomes

$$M \cong m(R + r)n^2R\phi \tag{5.6-7}$$

5.7 THE VIBRATION DAMPER

In contrast to the vibration absorber, where the exciting force is opposed by the absorber, energy is dissipated by the vibration damper. Figure 5.7-1 represents a friction type of vibration damper, commonly known as the Lanchester damper, which has found practical use in torsional systems such as gas and diesel engines in limiting the amplitudes of vibration at critical speeds. The damper consists of two flywheels a free to rotate on the shaft and driven only by means of the friction rings b when the normal pressure is maintained by the spring-loaded bolts c.

When properly adjusted, the flywheels rotate with the shaft for small oscillations. However, when the torsional oscillations of the shaft tend to become large, the flywheels will not follow the shaft because of their large

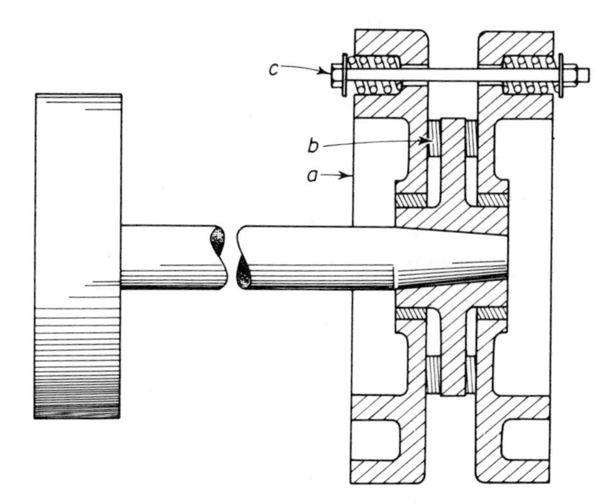

Figure 5.7-1. Torsional vibration damper.

inertia, and energy is dissipated by friction due to the relative motion. The dissipation of energy thus limits the amplitude of oscillation, thereby preventing high torsional stresses in the shaft.

In spite of the simplicity of the torsional damper the mathematical analysis for its behavior is rather complicated. For instance, the flywheels may slip continuously, for part of the cycle, or not at all, depending on the pressure exerted by the spring bolts. If the pressure on the friction ring is either too great for slipping or zero, no energy is dissipated, and the damper becomes ineffective. Obviously, maximum energy dissipation takes place at some intermediate pressure, resulting in optimum damper effectiveness.

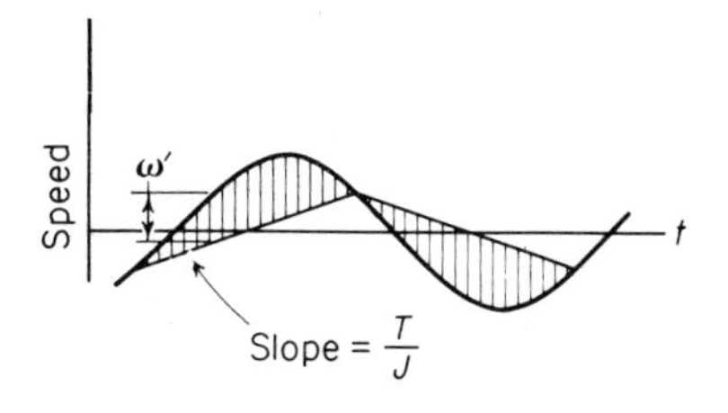

Figure 5.7-2. Torsional damper under continuous slip.

To obtain an insight to the problem, we consider briefly the case where the flywheels slip continuously. Assuming the shaft hub to be oscillating about its mean angular speed, as shown in Fig. 5.7-2, the flywheels will be acted upon by a constant frictional torque T while slipping. The acceleration of the flywheel, represented by the slope of the velocity curve, will hence be constant and equal to T/J, where J is the moment of inertia of the flywheels, and its velocity will be represented by a series of straight lines. The velocity of the flywheels will be increasing while the shaft speed is greater than that of the flywheels and decreasing when the shaft speed drops below that of the flywheels, as shown in the diagram.

The work done by the damper,

$$W = \int T d\theta = T \int \omega' \, dt \tag{5.7-1}$$

where ω' is the relative velocity, is equal to the product of the torque T and the shaded area of Fig. 5.7-2. Since this shaded area is small for large T and large for small T, the maximum energy is dissipated for some intermediate value of T.*

Obviously, the damper should be placed in a position where the amplitude of oscillation is the greatest. This position generally is found on the side of the shaft away from the main flywheel, since the node is usually near the largest mass.

The Untuned Viscous Vibration Damper. In a rotating system such as an automobile engine, the disturbing frequencies for torsional oscillations are proportional to the rotational speed. However, there is generally more than one such frequency, and the centrifugal pendulum has the disadvantage that several pendulums tuned to the order number of the disturbance must be used. In contrast to the centrifugal pendulum, the untuned viscous torsional damper is effective over a wide operating range. It consists of a free rotational mass within a cylindrical cavity filled with viscous fluid, as shown in Fig. 5.7-3. Such a system is generally incorporated into the end pulley of a crankshaft which drives the fan belt, and is often referred to as the Houdaille damper.

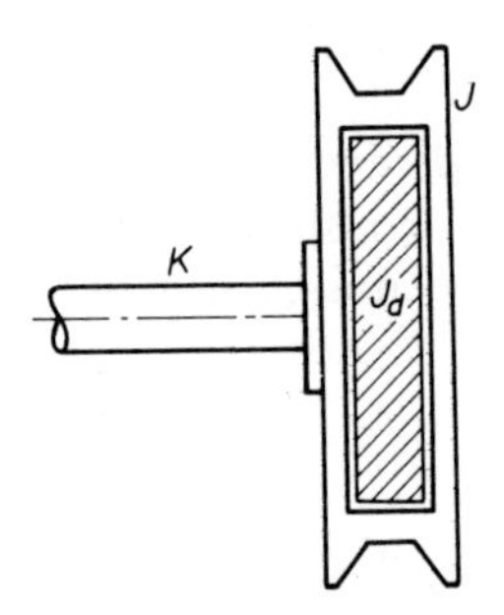

Figure 5.7-3. Untuned viscous damper.

We can examine the untuned viscous damper as a two degrees of freedom system by considering the crankshaft, to which it is attached, as being fixed at one end with the damper at the other end. With the torsional stiffness of the shaft equal to K in. lb/rad the damper can be considered to be excited by a harmonic torque $M_0 e^{i\omega t}$. The damper torque results from the viscosity of the fluid within the pulley cavity, and we will assume it to be proportional to the relative rotational speed between the pulley and the free mass. Thus the two equations of motion for the pulley and the free mass are

*J. P. Den Hartog and J. Ormondroyd, "Torsional-Vibration Dampers," *Trans. ASME* APM-52-13 (September–December, 1930), pp. 133–152.

$$J\ddot{\theta} + K\theta + c(\dot{\theta} - \dot{\varphi}) = M_0 e^{i\omega t}$$
$$J_d\ddot{\varphi} - c(\dot{\theta} - \dot{\varphi}) = 0 \tag{5.7-2}$$

Assuming the solution to be in the form

$$\theta = \theta_0 e^{i\omega t}$$
$$\varphi = \varphi_0 e^{i\omega t} \tag{5.7-3}$$

where θ_0 and φ_0 are complex amplitudes, their substitution into the differential equations results in

$$\left[\left(\frac{K}{J} - \omega^2\right) + i\frac{c\omega}{J}\right]\theta_0 - \frac{ic\omega}{J}\varphi_0 = \frac{M_0}{J}$$

and

$$\left(-\omega^2 + i\frac{c\omega}{J_d}\right)\varphi_0 = \frac{ic\omega}{J_d}\theta_0 \tag{5.7-4}$$

Eliminating φ_0 between the two equations, the expression for the amplitude θ_0 of the pulley becomes

$$\frac{\theta_0}{M_0} = \frac{(\omega^2 J_d - ic\omega)}{[\omega^2 J_d(K - J\omega^2)] + ic\omega[\omega^2 J_d - (K - J\omega^2)]} \tag{5.7-5}$$

Letting $\omega_n^2 = K/J$ and $\mu = J_d/J$, the critical damping is

$$c_c = 2J\omega_n, \quad c = \frac{c}{c_c}2J\omega_n = 2\zeta J\omega_n$$

The amplitude equation then becomes

$$\left|\frac{K\theta_0}{M_0}\right| = \sqrt{\frac{\mu^2(\omega/\omega_n)^2 + 4\zeta^2}{\mu^2(\omega/\omega_n)^2(1 - \omega^2/\omega_n^2)^2 + 4\zeta^2[\mu(\omega/\omega_n)^2 - (1 - \omega^2/\omega_n^2)]^2}} \tag{5.7-6}$$

which indicates that $|K\theta_0/M_0|$ is a function of three parameters, ζ, μ, and (ω/ω_n).

If μ is held constant and $|K\theta_0/M_0|$ plotted as a function of (ω/ω_n), the curve for any ζ will appear somewhat similar to that of a single degree of freedom system with a single peak. Of interest are the two extreme values of $\zeta = 0$ and $\zeta = \infty$. When $\zeta = 0$, we have an undamped system with resonant frequency $\omega_n = \sqrt{K/J}$, and the amplitude will be infinite at this frequency. If $\zeta = \infty$, the damper mass and the wheel will move together as a single mass, and again we have an undamped system but with natural frequency of $\sqrt{k/(J + J_d)}$.

Thus, like the Lanchester damper of the previous section, there is an optimum damping ζ_0 for which the peak amplitude is a minimum as shown in Fig. 5.7-4. The result can be presented as a plot of the peak values as a function of ζ for any given μ, as shown in Fig. 5.7-5.

It can be shown that the optimum damping is equal to

$$\zeta_0 = \frac{1}{\sqrt{2(1+\mu)(2+\mu)}} \tag{5.7-7}$$

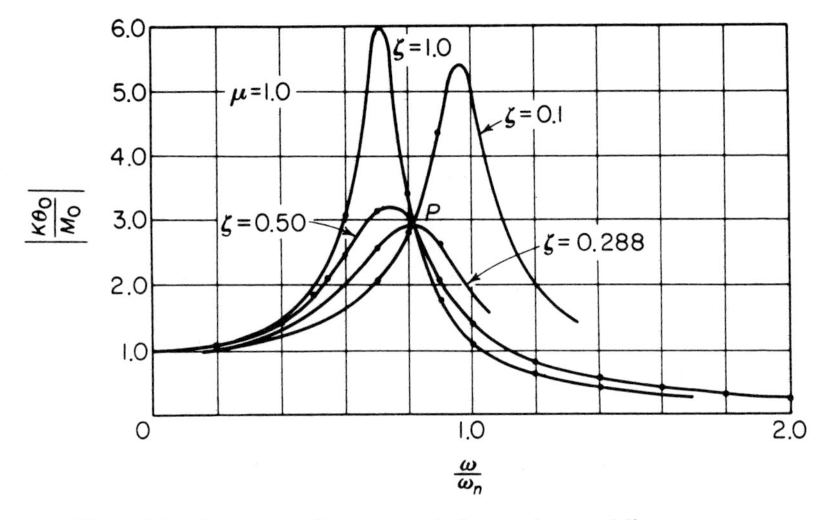

Figure 5.7-4. Response of an untuned viscous damper (all curves pass through *P*).

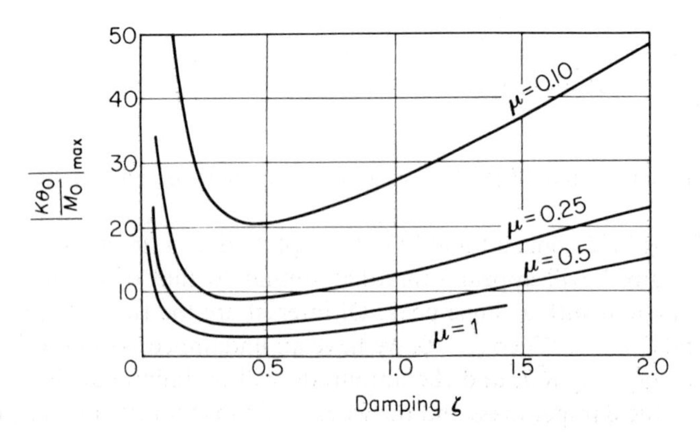

Figure 5.7-5. Plot of peak values vs. ζ.

and that the peak amplitude for optimum damping is found at a frequency equal to

$$\frac{\omega}{\omega_n} = \sqrt{2/(2 + \mu)} \tag{5.7-8}$$

These conclusions can be arrived at by observing that the curves of Fig. 5.7-4 all pass through a common point P, regardless of the numerical values of ζ. Thus, by equating the equation for $|K\theta_0/M|$ for $\zeta = 0$ and $\zeta = \infty$, Eq. 5.7-8 is found. The curve for optimum damping then must pass through P with a zero slope, so that if we substitute $(\omega/\omega_n)^2 = 2/(2 + \mu)$ into Eq. 5.7-6 and equate it to the amplitude as found in the undamped curve for the same frequency, the expression for ζ_0 is found. It is evident that these conclusions apply also to the linear spring-mass system of Fig. 5.7-6, which is a special case of the damped vibration absorber with the damper spring equal to zero.

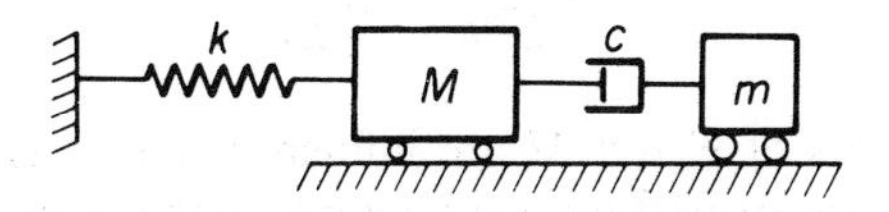

Figure 5.7-6. Untuned viscous damper.

5.8 GYROSCOPIC EFFECT ON ROTATING SHAFTS

A rotating wheel and shaft with angular momentum H can, under certain conditions, introduce a gyroscopic moment, thereby coupling the deflection and slope to produce a two degrees of freedom problem. We will illustrate this effect in terms of a wheel rotating on an overhanging shaft, as shown in Fig. 5.8-1.

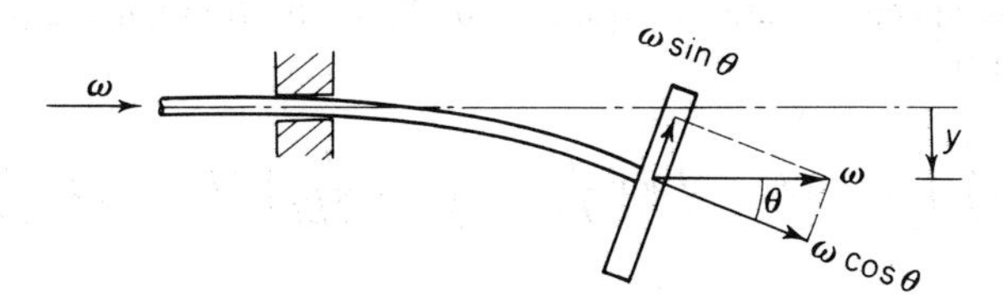

Figure 5.8-1. Gyroscopic effect.

If the rotational speed of the shaft is ω, its components parallel and normal to the face of the wheel are $\omega \sin\theta$ and $\omega \cos\theta$. Thus the angular momentum in these directions are $J_d\omega \sin\theta$ and $J_p\omega \cos\theta$, where J_p and J_d are the moments of inertia of the wheel along the polar axis and its diameter.

Resolving these vectors along the direction of ω and perpendicular to it, the component normal to ω is

$$H_n = J_p\omega \cos\theta \sin\theta - J_d\omega \sin\theta \cos\theta = (J_p - J_d)\omega \cos\theta \sin\theta \quad (5.8\text{-}1)$$

If the deflection plane of the shaft whirls with angular speed ω_1, a moment on the wheel equal to $H_n\omega_1$ is necessary, the component of H parallel to ω undergoing no change. Acting on the shaft is then an opposite moment

$$M = -H_n\omega_1 = -(J_p - J_d)\omega\omega_1 \cos\theta \sin\theta \cong -(J_p - J_d)\omega\omega_1\theta \quad (5.8\text{-}2)$$

To take account of this moment on the deflection shape of the shaft, we can write the equations for the deflection and slope at the end of the shaft

$$\begin{aligned} y &= F\frac{l^3}{3EI} + M\frac{l^2}{2EI} \\ \theta &= F\frac{l^2}{2EI} + M\frac{l}{EI} \end{aligned} \quad (5.8\text{-}3)$$

where the coefficients of F and M are influence functions of deflection and slope due to unit force or unit moment acting on the shaft end, and F and M are the force and moment on the end of the shaft. The force F is simply $m\omega_1^2 y$, and M is the gyroscopic couple $-(J_p - J_d)\omega\omega_1\theta$ so that Eqs. 5.8-3 become

$$\begin{aligned} y &= \left(m\omega_1^2\frac{l^3}{3EI}\right)y - \left[(J_p - J_d)\omega\omega_1\frac{l^2}{2EI}\right]\theta \\ \theta &= \left(m\omega_1^2\frac{l^2}{2EI}\right)y - \left[(J_p - J_d)\omega\omega_1\frac{l}{EI}\right]\theta \end{aligned} \quad (5.8\text{-}4)$$

For a wheel or disk with an unbalance, we have found in Sec. 3.4 that the whirling speed ω_1 can be equal to ω. Thus the frequency equation may take the form

$$\left(1 - \frac{m\omega^2 l^3}{3EI}\right)\left[1 + (J_p - J_d)\omega^2\frac{l}{EI}\right] + \left(m\omega^2\frac{l^2}{2EI}\right)\left[(J_p - J_d)\omega^2\frac{l^2}{2EI}\right] = 0$$

$$(5.8\text{-}5)$$

For a wheel approaching a thin disk, $J_p = 2J_d$, the frequency equation reduces to

$$\omega^4 + \frac{12EI}{mJ_d l^3}\left(\frac{ml^2}{3} - J_d\right)\omega^2 - \frac{12}{mJ_d}\left(\frac{EI}{l^2}\right)^2 = 0 \quad (5.8\text{-}6)$$

Since, in the absence of the gyroscopic couple, the natural frequency of the system is $\omega_y = \sqrt{3EI/ml^3}$, we can rewrite the frequency equation as

$$\omega^4 + 4\omega_y^2\left(\frac{1}{\alpha} - 1\right)\omega^2 - \frac{4}{\alpha}\omega_y^4 = 0 \tag{5.8-7}$$

where $\alpha = 3J_d/ml^2$ can be viewed as a coupling term. The relationship between $(\omega/\omega_y)^2$ and α is shown in Fig. 5.8-2. For very large values of α, the ratio θ/y approaches zero and the natural frequency of the system tends to the value $\omega = \sqrt{12EI/kl^3}$.

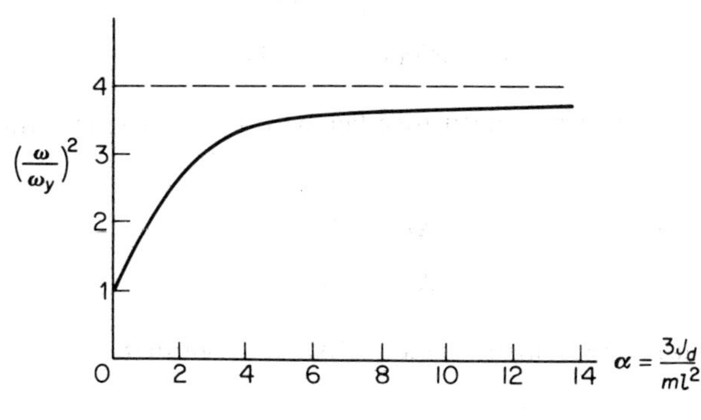

Figure 5.8-2. Solution of Eq. 5.8-7.

5.9 DIGITAL COMPUTATION

The finite difference method of Sec. 4.7 can easily be extended to the solution of systems with two degrees of freedom. The procedure is illustrated by the following problem which is programmed and solved by the digital computer.

The system to be solved is shown in Fig. 5.9-1. To avoid confusion with subscripts, we let the displacements be x and y.

$k_1 = 200$ lb/in.
$k_2 = 100$ lb/in.
$m_1 = 0.50$ lb sec^2/in.
$m_2 = 0.20$ lb sec^2/in.
$F = 100$ lb (step function)

$$\begin{cases} F = 0 \text{ for } t < 0 \\ \quad = F \text{ for } t > 0 \end{cases}$$

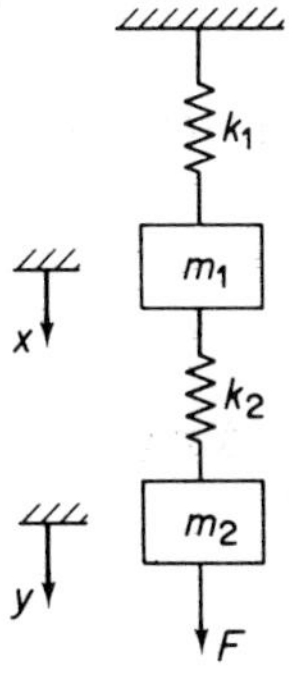

Figure 5.9-1.

Initial conditions:

$$x = \dot{x} = y = \dot{y} = F = 0$$

The subscripts for x and y then indicate the time sequence of the computation.

The equations of motion are

$$0.50\ddot{x} = -200x + 100(y - x)$$
$$0.20\ddot{y} = -100(y - x) + 100$$

which can be rearranged to

$$\ddot{x} = -600x + 200y$$
$$\ddot{y} = 500(x - y + 1)$$

These equations are to be solved together with the recurrence equations of Sec. 4.7.

$$x_{i+1} = \ddot{x}_i \Delta t^2 + 2x_i - x_{i-1}$$
$$y_{i+1} = \ddot{y}_i \Delta t^2 + 2y_i - y_{i-1}$$

To establish a reasonable value for Δt, we note that

$$\sqrt{\frac{k_1}{m_1}} = 20 \qquad \tau_1 = 0.314$$
$$\sqrt{\frac{k_2}{m_2}} = 22.4 \qquad \tau_2 = 0.280$$

We therefore arbitrarily choose a value $\Delta t = 0.020$ sec. It is also noted that the initial accelerations are $\ddot{x}_1 = 0$ and $\ddot{y}_1 = 500$, which requires us to use Eq. 4.7-4 for y and Eq. 4.7-6 for x. Using $\ddot{y}_1 = 500$ we have

$$y_2 = \tfrac{1}{2}(500)(.02)^2 = 0.10$$

The quantities x_2 and $\ddot{x}_2$ must be solved simultaneously from the equations

$$x_2 = \tfrac{1}{6}\ddot{x}_2 \Delta t^2$$
$$\ddot{x}_2 = -600x_2 + 200y_2$$

Eliminating $\ddot{x}_2$ the equation for x_2 becomes

$$x_2 = \frac{33.33y_2\Delta t^2}{1 + 100\Delta t^2}$$

The flow diagram for the computation is shown in Fig. 5.9-2 and the Fortran program is presented in Fig. 5.9-3. The computed results and the plot for x and y are shown in Fig. 5.9-4.

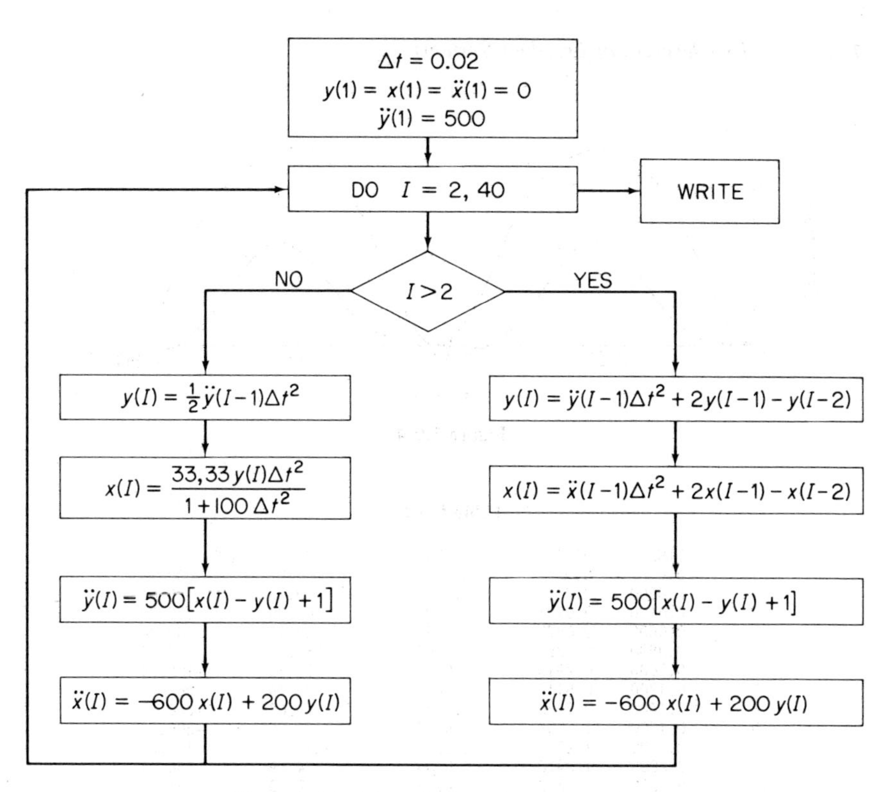

Figure 5.9-2. Flow diagram for computation.

```
ISN 0002          DIMENSION X(40),Y(40),DX2(40),DY2(40),T(40),J(40)
ISN 0003          J(1)=1
ISN 0004          DT=0.02
ISN 0005          DT2=DT**2
ISN 0006          DX2(1)=0.0
ISN 0007          DY2(1)=500
ISN 0008          X(1)=0.0
ISN 0009          Y(1)=0.0
ISN 0010          T(1)=0.0
ISN 0011          DO 100 I=2,40
ISN 0012          J(I)=I
ISN 0013          T(I)=DT*(I-1)
ISN 0014          IF(1.GT.2) GO TO 200
ISN 0016          Y(1)=DY2(I-1)*DT2/2
ISN 0017          X(I)=33.33*Y(I)*DT2/(1+100*DT2)
ISN 0018          DY2(I)=500*(X(I)-Y(I)+1)
ISN 0019          DX2(I)=-600*X(I)+200*Y(I)
ISN 0020          GO TO 100
ISN 0021   200     Y(I)=DY2(I-1)*DT2+2*Y(I-1)-Y(I-2)
ISN 0022           X(I)=DX2(I-1)*DT2+2*X(I-1)-X(I-2)
ISN 0023           DY2(I)=500*(X(I)-Y(I)+1)
ISN 0024           DX2(I)=-600*(I)+200*Y(I)
ISN 0025   100     CONTINUE
ISN 0026           WRITE (6,300)
ISN 0027   300     FORMAT (50 H1  J   TIME    DISP X    DISP-Y   ACC-X    ACC-Y)
ISN 0028           WRITE (6,400) (J(I),T(I),X(I),Y(I),DX2(I),DY2(I),I=1,40)
ISN 0029   400     FORMAT (IX, I2,3X,F7.4,3X,F6.4,3X,F7.4,3X,F9.2 ,3X,F9.2)
ISN 0030           STOP
ISN 0031           END
```

Figure 5.9-3. Fortran program for computation.

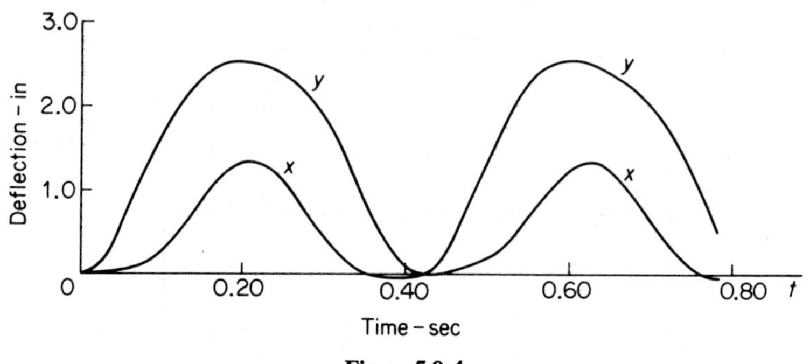

Figure 5.9-4.

Table 5.9-1.

J	TIME	DISP_X	DISP-Y	ACC-X	ACC-Y
1	0.0	0.0	0.0	0.0	500.00
2	0.0200	0.0013	0.1000	19.23	450.64
3	0.0400	0.0103	0.3803	69.90	315.00
4	0.0600	0.0472	0.7865	128.99	130.34
5	0.0800	0.1357	1.2449	167.55	-54.59
6	0.1000	0.2913	1.6815	161.53	-195.09
7	0.1200	0.5114	2.0400	101.14	-264.27
8	0.1400	0.7720	2.2928	-4.67	-260.37
9	0.1600	1.0308	2.4414	-130.19	-205.32
10	0.1800	1.2375	2.5080	-240.88	-135.25
11	0.2000	1.3478	2.5204	-304.59	-86.31
12	0.2200	1.3363	2.4983	-302.10	-81.02
13	0.2400	1.2039	2.4438	-233.58	-119.95
14	0.2600	0.9781	2.3413	-118.61	-181.60
15	0.2800	0.7049	2.1662	10.31	-230.65
16	0.3000	0.4358	1.8988	118.30	-231.52
17	0.3200	0.2140	1.5388	179.37	-162.42
18	0.3400	0.0639	1.1139	184.41	-24.96
19	0.3600	-.0123	0.6789	143.18	154.37
20	0.3800	-.0313	0.3057	79.95	351.47
21	0.4000	-.0184	0.0651	24.04	458.26
22	0.4200	0.0042	0.0078	-0.97	498.20
23	0.4400	0.0264	0.1498	14.10	438.32
24	0.4600	0.0543	0.4671	60.86	293.59
25	0.4800	0.1065	0.9018	116.49	102.31
26	0.5000	0.2052	1.3775	152.36	-86.13
27	0.5200	0.3650	1.8187	144.76	-226.87
28	0.5400	0.5826	2.1692	84.28	-293.29
29	0.5600	0.8339	2.4023	-19.90	-284.19
30	0.5800	1.0773	2.5218	-142.03	-222.24
31	0.6000	1.2639	2.5524	-247.86	-144.24
32	0.6200	1.3513	2.5252	-305.73	-86.97
33	0.6400	1.3164	2.4633	-297.19	-73.45
34	0.6600	1.1627	2.3720	-223.20	-104.68
35	0.6800	0.9196	2.2389	-104.01	-159.61
36	0.7000	0.6350	2.0419	27.37	-203.43
37	0.7200	0.3613	1.7635	135.90	-201.08
38	0.7400	0.1420	1.4047	195.74	-131.34
39	0.7600	0.0010	0.9933	198.09	3.82
40	0.7800	-.0608	0.5835	153.20	177.83

PROBLEMS

5-1 Write the equations of motion for the system shown in Fig. P5-1 and determine its natural frequencies and mode shapes.

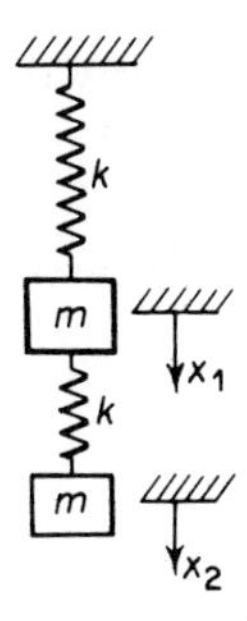

Figure P. 5-1.

5-2 Determine the normal modes and frequencies of the system shown in Fig. P5-2 when $n = 1$.

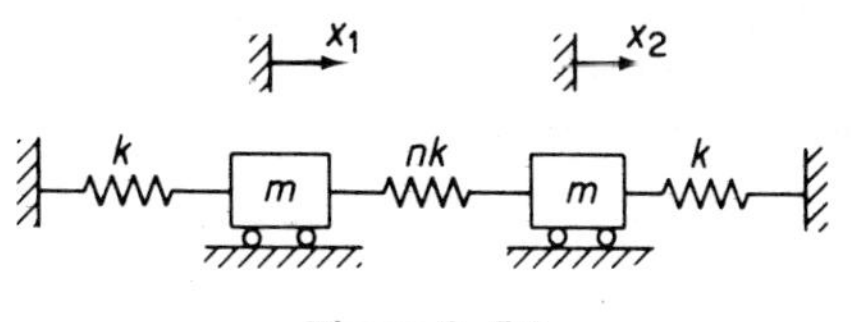

Figure P. 5-2.

5-3 For the system of Problem 5-2, determine the natural frequencies as a function of n.

5-4 Determine the natural frequencies and mode shapes of the system shown in Fig. P5-4.

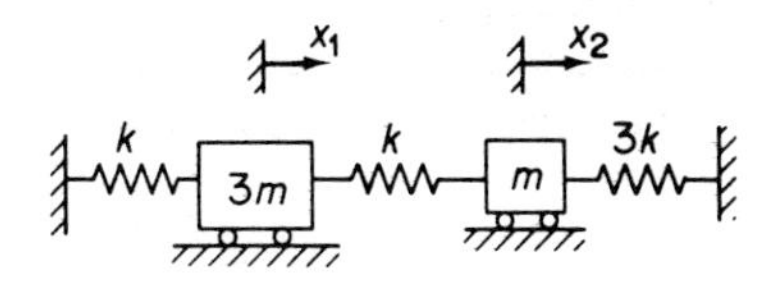

Figure P. 5-4.

5-5 Determine the normal modes of the torsional system shown in Fig. P5-5 for $K_1 = K_2$ and $J_1 = 2J_2$.

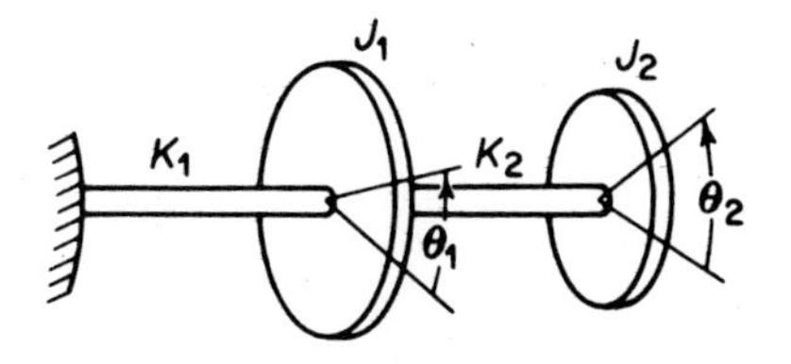

Figure P. 5-5.

5-6 If $K_1 = 0$ in the torsional system of Problem 5-5, the system becomes a degenerate two degrees of freedom system with only one natural frequency. Discuss the normal modes of this system as well as a linear spring-mass system equivalent to it. Show that the system can be treated as one of a single degree of freedom by using the coordinate $\phi = (\theta_1 - \theta_2)$.

5-7 Determine the natural frequency of the torsional system shown in Fig. P5-7, and draw the normal mode curve.

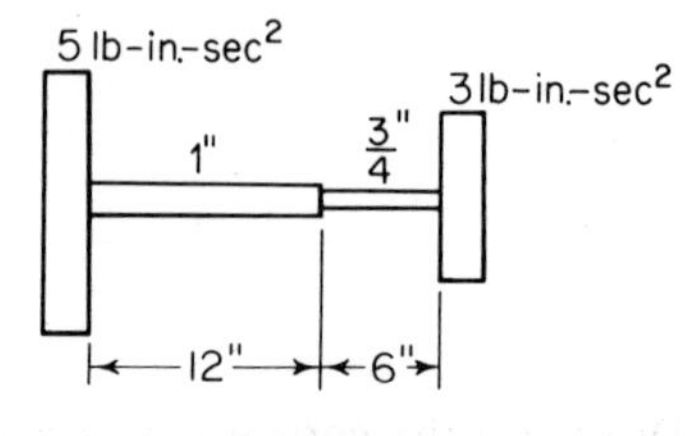

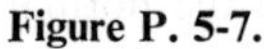
Figure P. 5-7.

5-8 An electric train made up of two cars of weight 50,000 lb each is connected by couplings of stiffness equal to 16,000 lb/in. as shown in Fig. P5-8. Determine the natural frequency of the system.

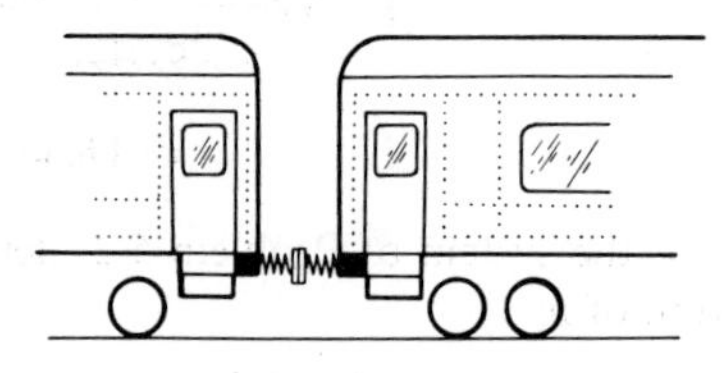

Figure P. 5-8.

5-9 Assuming small amplitudes, set up the differential equation of motion for the double pendulum using the coordinates shown in Fig. P5-9. Show that the natural frequencies of the system are given by the equation

$$\omega = \sqrt{\frac{g}{l}(2 \pm \sqrt{2})}$$

Determine the ratio of amplitudes x_1/x_2 and locate the nodes for the two modes of vibration.

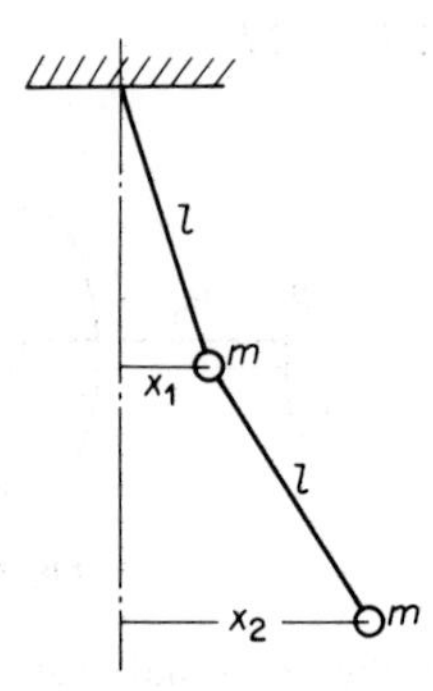

Figure P. 5-9.

5-10 Set up the equations of motion of the double pendulum in terms of angles θ_1 and θ_2 measured from the vertical.

5-11 Two masses m_1 and m_2 are attached to a light string with tension T, as shown in Fig. P5-11. Assuming that T remains unchanged when the masses are displaced normal to the string, write the equations of motion expressed in matrix form.

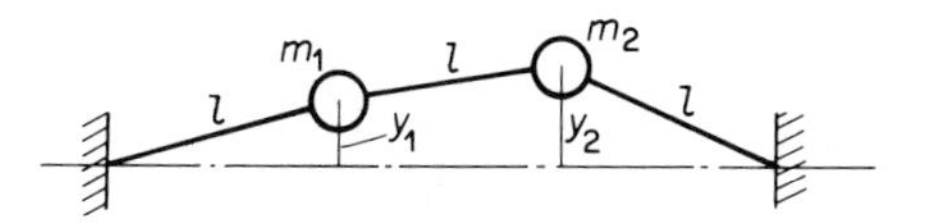

Figure P. 5-11.

5-12 In Problem 5-11 if the two masses are made equal, show that normal mode frequencies are $\omega = \sqrt{T/ml}$ and $\omega_2 = \sqrt{3T/ml}$. Establish the configuration for these normal modes.

5-13 In Problem 5-11 if $m_1 = 2m$, and $m_2 = m$, determine the normal mode frequencies and mode shapes.

5-14 A torsional system shown in Fig. P5-14 is composed of a shaft of stiffness K_1, a hub of radius r and moment of inertia J_1, four leaf springs of stiffness k_2, and an outer wheel of radius R and moment of inertia J_2. Set up the differential equations for torsional oscillation, assuming one end of the shaft to be fixed. Show that the frequency equation reduces to

$$\omega^4 - \left(\omega_{11}^2 + \omega_{22}^2 + \frac{J_2}{J_1}\omega_{22}^2\right)\omega^2 + \omega_{11}^2\omega_{22}^2 = 0$$

where ω_{11} and ω_{22} are uncoupled frequencies given by the expressions

$$\omega_{11}^2 = \frac{K_1}{J_1}, \qquad \omega_{22}^2 = \frac{4k_2R^2}{J_2}$$

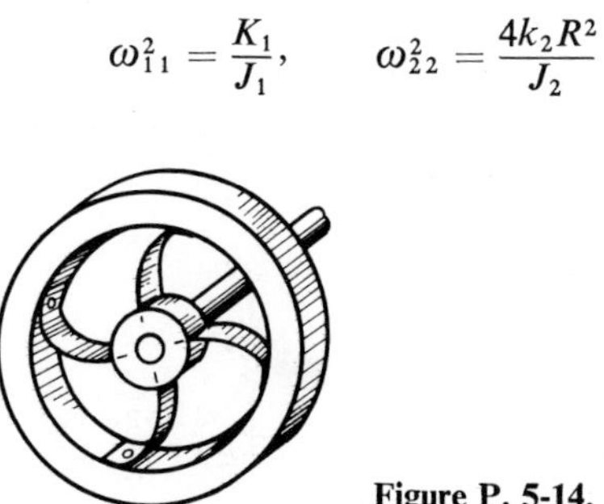

Figure P. 5-14.

5-15 Two equal pendulums free to rotate about the x-x axis are coupled together by a rubber hose of torsional stiffness k lb in./rad, as shown in Fig. P5-15. Determine the natural frequencies for the normal modes of vibration, and describe how these motions may be started.

If $l = 19.3$ in., mg $= 3.86$ lb, and $k = 2.0$ lb in./rad, determine the beat period for a motion started with $\theta_1 = 0$ and $\theta_2 = \theta_0$. Examine carefully the phase of the motion as the amplitude approaches zero.

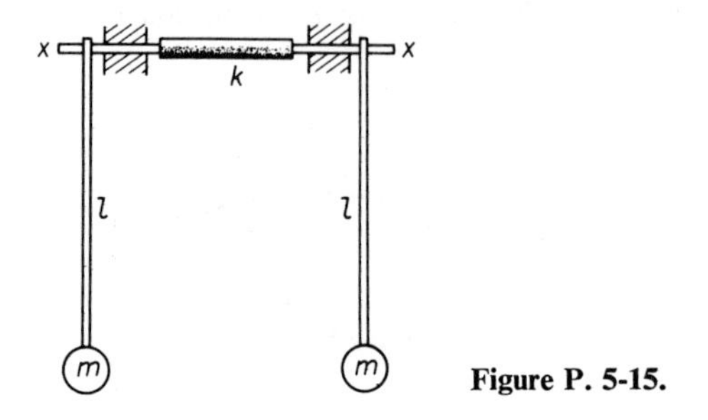

Figure P. 5-15.

5-16 Determine the equations of motion for the system of Problem 5-4 when the initial conditions are $x_1(0) = A$, $\dot{x}_1(0) = x_2(0) = \dot{x}_2(0) = 0$.

5-17 The double pendulum of Problem 5-9 is started with the following initial conditions: $x_1(0) = x_2(0) = X$, $\dot{x}_1(0) = \dot{x}_2(0) = 0$. Determine the equations of motion.

5-18 The lower mass of Problem 5-1 is given a sharp blow, imparting to it an initial velocity $\dot{x}_2(0) = V$. Determine the equation.

5-19 If the system of Problem 5-1 is started with initial conditions $x_1(0) = 0$, $x_2(0) = 1.0$, $\dot{x}_1(0) = \dot{x}_2(0) = 0$, show that the equations of motion are

$$x_1(t) = 0.447 \cos \omega_1 t - 0.447 \cos \omega_2 t$$
$$x_2(t) = 0.722 \cos \omega_1 t + 0.278 \cos \omega_2 t$$
$$\omega_1 = \sqrt{.382\ \mathrm{k/m}}, \qquad \omega_2 = \sqrt{2.618\ k/m}$$

5-20 Choose coordinates x for the displacement of c and θ clockwise for the rotation of the uniform bar shown in Fig. P5-20, and determine the natural frequencies and mode shapes.

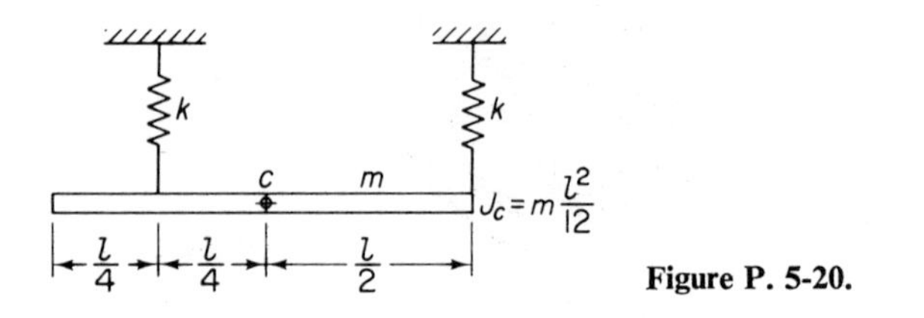

Figure P. 5-20.

5-21 Set up the matrix equation of motion for the system shown in Fig. P5-21, using coordinates x_1 and x_2 at m and $2m$. Determine the equation for the normal mode frequencies and describe the mode shapes.

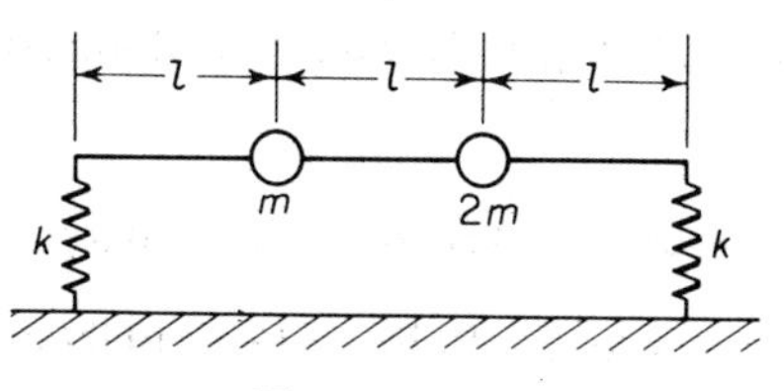

Figure P. 5-21.

5-22 In Problem 5-21, if the coordinates x at m and θ are used, what form of coupling will they result in?

5-23 Compare Problems 5-9 and 5-10 in matrix form and indicate the type of coupling present in each coordinate system.

5-24 The following information is given for a certain automobile shown in Fig. P5-24.

$$W = 3500 \text{ lb} \qquad k_1 = 2000 \text{ lb/ft}$$
$$l_1 = 4.4 \text{ ft} \qquad k_2 = 2400 \text{ lb/ft}$$
$$l_2 = 5.6 \text{ ft} \qquad r = 4 \text{ ft} = \text{radius of gyration about } c$$

Determine the normal modes of vibration and locate the node for each mode.

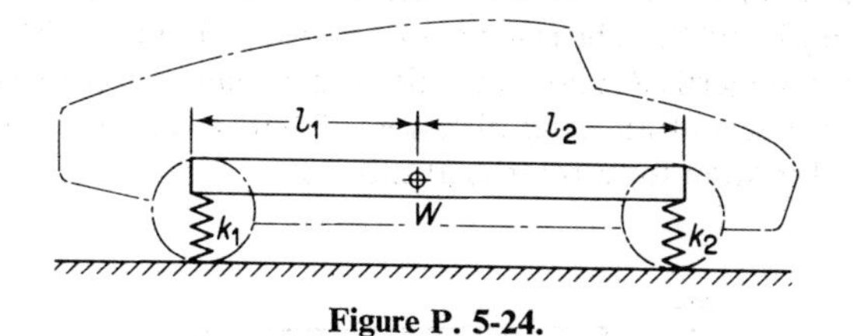

Figure P. 5-24.

5-25 An airfoil section to be tested in a wind tunnel is supported by a linear spring k and a torsional spring K, as shown in Fig. P5-25. If the center of gravity of the section is a distance e ahead of the point of support, determine the differential equations of motion of the system.

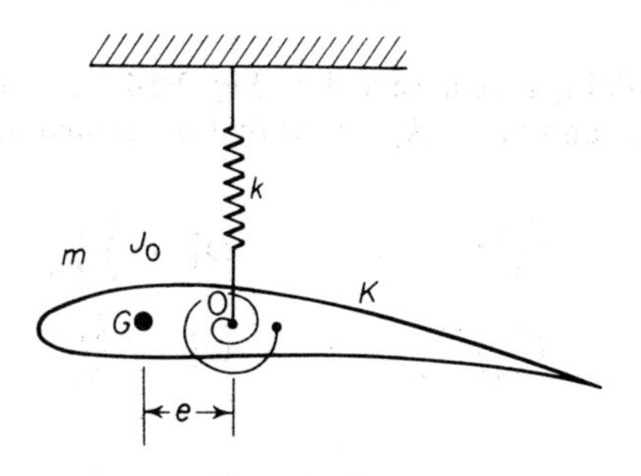

Figure P. 5-25.

5-26 Determine the natural frequencies and normal modes of the system shown in Fig. P5-26 when

$$gm_1 = 3.86 \text{ lb} \qquad k_1 = 20 \text{ lb/in.}$$

$$gm_2 = 1.93 \text{ lb} \qquad k_2 = 10 \text{ lb/in.}$$

When forced by $F_1 = F_0 \sin \omega t$, determine the equations for the amplitudes and plot them against ω/ω_{11}.

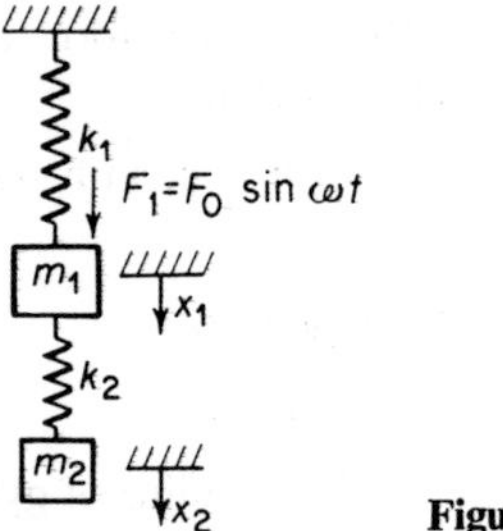

Figure P. 5-26.

5-27 A rotor is mounted in bearings which are free to move in a single plane as shown in Fig. P5-27. The rotor is symmetrical about O with total mass M and moment of inertia J_0 about an axis perpendicular to the shaft. If a small unbalance mr acts at an axial distance b from its center O, determine the equations of motion for a rotational speed ω.

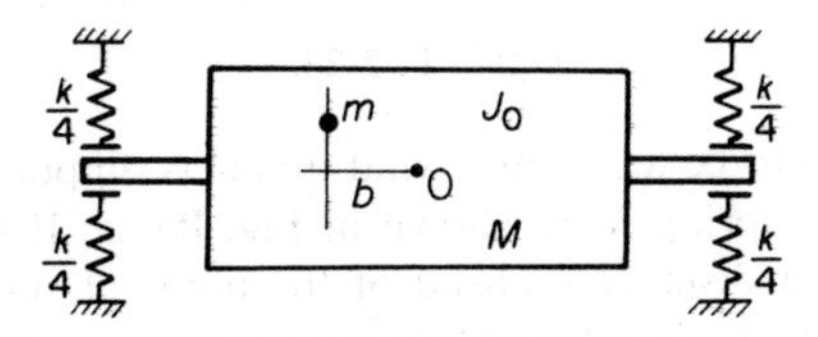

Figure P. 5-27.

5-28 A two-story building is represented in Fig. P5-28 by a lumped mass system where $m_1 = \frac{1}{2}m_2$ and $k_1 = \frac{1}{2}k_2$. Show that its normal modes are

$$\left(\frac{x_1}{x_2}\right)^{(1)} = 2, \qquad \omega_1^2 = \frac{1}{2}\frac{k_1}{m_1}$$

$$\left(\frac{x_1}{x_2}\right)^{(2)} = -1, \qquad \omega_2^2 = 2\frac{k_1}{m_1}$$

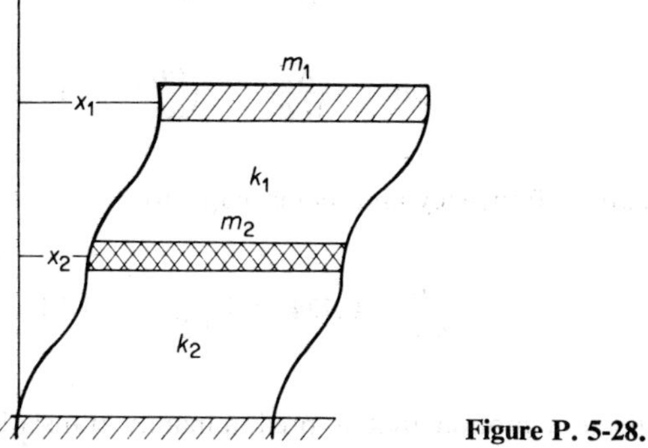

Figure P. 5-28.

5-29 In Problem 5-28, if a force is applied to m_1 to deflect it by unity, and the system is released from this position, determine the equation of motion of each mass by using the normal mode summation method.

5-30 In Problem 5-29, determine the ratio of the maximum shear in the first and second stories.

5-31 Repeat Problem 5-29 if the load is applied to m_2, displacing it by unity.

5-32 Assume in Problem 5-28 that an earthquake causes the ground to oscillate in the horizontal direction according to the equation $x_g = X_g \sin \omega t$. Determine the response of the building and plot it against ω/ω_1.

5-33 To simulate the effect of an earthquake on a rigid building, the base is assumed to be connected to the ground through two springs; K_h for the translational stiffness, and K_r for the rotational stiffness. If the ground is now given a harmonic motion $Y_g = Y_G \sin \omega t$, set up the equations of motion in terms of the coordinates shown in Fig. P5-33.

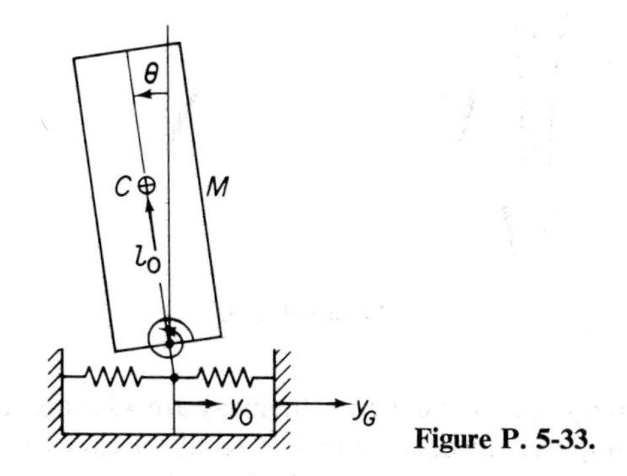

Figure P. 5-33.

5-34 Solve the equations of Problem 5-33 by letting

$$\omega_h^2 = \frac{K_h}{M}, \qquad \left(\frac{\rho_c}{l_0}\right)^2 = \frac{1}{3}$$

$$\omega_r^2 = \frac{K_r}{M\rho_c^2} \qquad \left(\frac{\omega_r}{\omega_h}\right)^2 = 4$$

The first natural frequency and mode shape are

$$\frac{\omega_1}{\omega_h} = 0.734 \text{ and } \frac{Y_0}{l_0\theta} = -1.14$$

which indicate a motion that is predominantly translational. Establish the second natural frequency and its mode ($Y_1 = Y_0 - 2l_0\theta_0 =$ displacement of top).

5-35 The response and mode configuration for Problems 5-33 and 5-34 are shown in Fig. P5-35. Verify the mode shapes for several values of the frequency ratio.

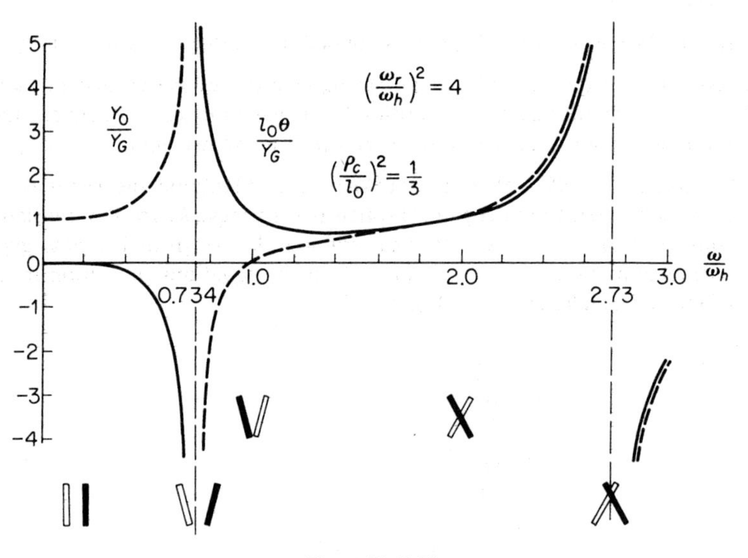

Figure P. 5-35.

5-36 The expansion joints of a concrete highway are 45 ft apart. These joints cause a series of impulses at equal intervals to affect cars traveling at a constant speed. Determine the speeds at which pitching motion and up-and-down motion are most apt to arise for the automobile of Problem 5-24.

5-37 For the system shown in Fig. P5-37, $W_1 = 200$ lb and the absorber weight $W_2 = 50$ lb. If W_1 is excited by a 2 lb in. unbalance rotating at 1800 rpm, determine the proper value of the absorber spring k_2. What will be the amplitude of W_2?

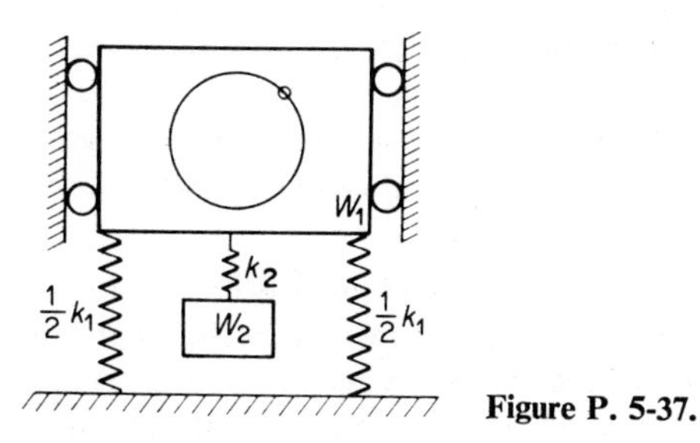

Figure P. 5-37.

5-38 In Problem 5-37, if a dashpot c is introduced between W_1 and W_2, determine the amplitude equations by the complex algebra method.

5-39 A flywheel of moment of inertia I has a torsional absorber of moment of inertia I_d free to rotate on the shaft and connected to the flywheel by four springs of stiffness k lb/in. as shown in Fig. P5-39. Set up the differential equations of motion for the system, and discuss the response of the system to an oscillatory torque.

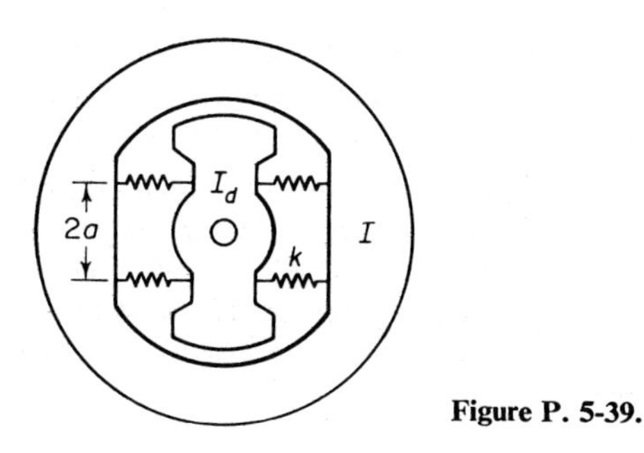

Figure P. 5-39.

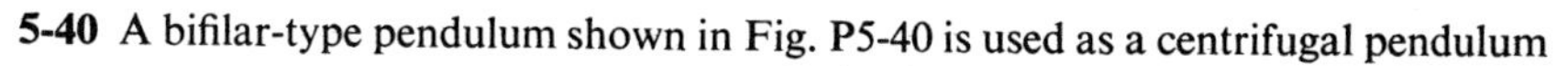

5-40 A bifilar-type pendulum shown in Fig. P5-40 is used as a centrifugal pendulum

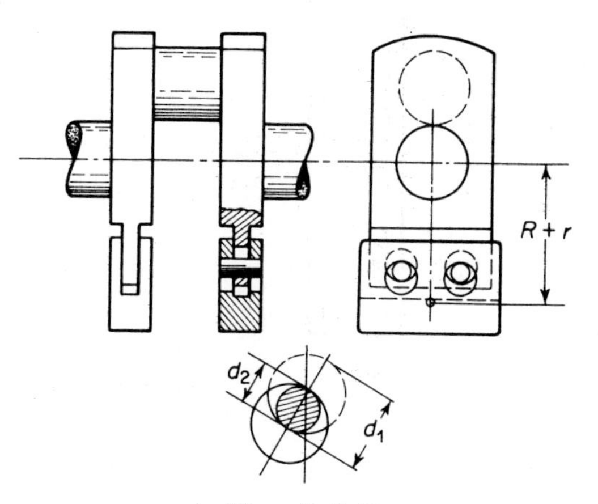

Figure P. 5-40.

to eliminate torsional oscillations. The U-shaped weight fits loosely and rolls on two pins of diameter d_2 within two larger holes of equal diameters d_1. With respect to the crank, the counterweight has a motion of curvilinear translation with each point moving in a circular path of radius $r = d_1 - d_2$. Prove that the U-shaped weight does indeed move in a circular path of $r = d_1 - d_2$.

5-41 A bifilar-type centrifugal pendulum is proposed to eliminate a torsional disturbance of frequency equal to four times the rotational speed. If the distance R to the center of gravity of the pendulum mass is made equal to 4.0 in. and $d_1 = \frac{3}{4}$ in., what must be the diameter d_2 of the pins?

5-42 A jig used to size coal contains a screen that reciprocates with a frequency of 600 cpm. The jig weighs 500 lb and has a fundamental frequency of 400 cpm. If an absorber weighing 125 lb is to be installed to eliminate the vibration of the jig frame, determine the absorber spring stiffness. What will be the resulting two natural frequencies of the system?

5-43 In a certain refrigeration plant a section of pipe carrying the refrigerant vibrated violently at a compressor speed of 232 rpm. To eliminate this difficulty it was proposed to clamp a spring-mass system to the pipe to act as an absorber. For a trial test a 2.0-lb absorber tuned to 232 cpm resulted in two natural frequencies of 198 and 272 cpm. If the absorber system is to be designed so that the natural frequencies lie outside the region 160 to 320 cpm, what must be the weight and spring stiffness?

5-44 A type of damper frequently used on automobile crankshafts is shown in Fig. P5-44. J represents a solid disk free to spin on the shaft, and the space between the disk and case is filled with a silicone oil of coefficient of viscosity μ. The damping action results from any relative motion between the two. Derive an equation for the damping torque exerted by the disk on the case due to a relative velocity of ω.

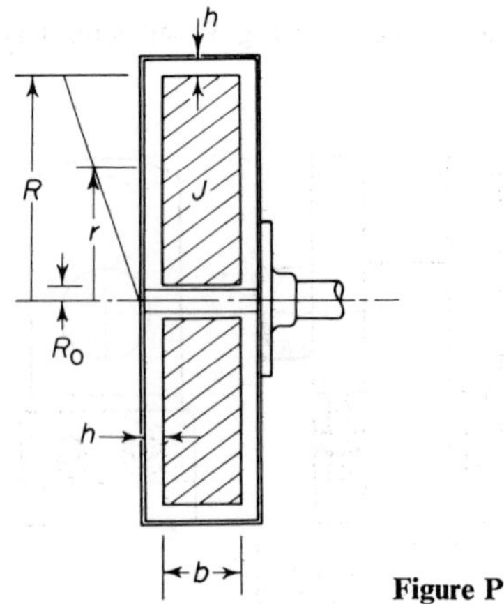

Figure P. 5-44.

5-45 For the Houdaille viscous damper with mass ratio $\mu = 0.25$, determine the optimum damping ζ_0 and the frequency at which the damper is most effective.

5-46 If the damping for the viscous damper of Problem 5-45 is equal to $\zeta = 0.10$, determine the peak amplitude as compared to the optimum.

5-47 Establish the relationships given by Eqs. (5.7-7) and (5.7-8) of Sec. 5.7.

5-48 A simply supported shaft of length l and stiffness EI has a thin but rigid disk keyed to it at the point $l/3$ as shown in Fig. P5-48. Establish the equations of motion for y and θ and plot $(\omega/\omega_y)^2$ vs. J_d/ml^2.

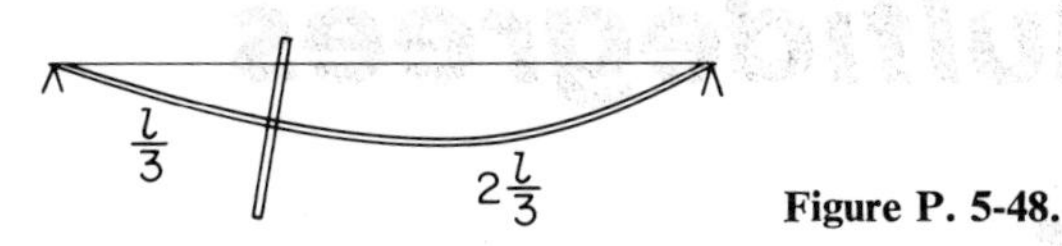

Figure P. 5-48.

5-49 Draw the flow diagram and develop the Fortran program for the computation of the response of the system shown in Prob. 5-4 when the mass $3m$ is excited by a rectangular pulse of magnitude 100 lb and duration $6\pi\sqrt{m/k}$ sec.

5-50 In Prob. 5-28 assume the following data, $k_1 = 4 \times 10^3$ lb/in, $k_2 = 6 \times 10^3$ lb/in, $m_1 = m_2 = 100$. Develop the flow diagram and the Fortran program for the case where the ground is given a displacement $y = 10'' \sin \pi t$ for 4 seconds.

6

Multidegrees of Freedom Systems

6.1 INTRODUCTION

The analysis of dynamical systems of several degrees of freedom is complicated by a large number of equations and many detailed computations. It is therefore desirable to approach the problem with a concise treatment that will clearly lead to the desired results without becoming entangled in intermediate details. In this respect matrix methods are ideally suited for our purpose in that large arrays of equations can be manipulated in short-hand notation. The large volume of computation that is generally required can be assigned to the digital computer, without which these problems become impractical.

In this chapter we will discuss the various matrix techniques applicable to the vibration of dynamical systems of multidegrees of freedom. Fundamental concepts essential in the formulation of the equations are discussed first and various concepts underlying vibration theory are developed in matrix notation. These concepts form the basis for the treatment and understanding of the behavior of large systems.

6.2 THE FLEXIBILITY AND STIFFNESS MATRIX

The flexibility influence coefficient a_{ij} is defined as the displacement at i due to a unit force applied at j. With forces f_1, f_2, and f_3 acting at stations 1, 2, and 3, the principle of superposition can be applied to determine the displacements in terms of the flexibility influence coefficients

$$\begin{aligned} x_1 &= a_{11}f_1 + a_{12}f_2 + a_{13}f_3 \\ x_2 &= a_{21}f_1 + a_{22}f_2 + a_{23}f_3 \\ x_3 &= a_{31}f_1 + a_{32}f_2 + a_{33}f_3 \end{aligned} \tag{6.2-1}$$

In matrix notation, the equation is

$$\{x\} = [a]\{f\} \tag{6.2-2}$$

where

$$[a] = \begin{bmatrix} a_{11} & a_{12} & a_{13} \\ a_{21} & a_{22} & a_{23} \\ a_{31} & a_{32} & a_{33} \end{bmatrix} \tag{6.2-3}$$

is the *flexibility matrix.*

If Eq. (6.2-2) is premultiplied by the inverse of the flexibility matrix $[a]^{-1}$, we obtain the equation

$$[a]^{-1}\{x\} = \{f\} = [k]\{x\} \tag{6.2-4}$$

We thus find that the inverse of the flexibility matrix is the stiffness matrix $[k]$

$$[a]^{-1} = [k]$$

or

$$[a] = [k]^{-1} \tag{6.2-5}$$

Writing out the terms of Eq. (6.2-4) we have

$$\begin{Bmatrix} f_1 \\ f_2 \\ f_3 \end{Bmatrix} = \begin{bmatrix} k_{11} & k_{12} & k_{13} \\ k_{21} & k_{22} & k_{23} \\ k_{31} & k_{32} & k_{33} \end{bmatrix} \begin{Bmatrix} x_1 \\ x_2 \\ x_3 \end{Bmatrix} \tag{6.2-6}$$

The various elements of the stiffness matrix have the following interpretation. If $x_1 = 1.0$ and $x_2 = x_3 = 0$, the forces at 1, 2 and 3 that are required to maintain this displacement according to Eq. (6.2-6) are k_{11}, k_{21} and k_{31}. Similarly, the forces f_1, f_2 and f_3 required to maintain the displacement configuration $x_1 = 0, x_2 = 1.0$ and $x_3 = 0$ are k_{12}, k_{22} and k_{32}. Thus, the general rule for establishing the stiffness elements of any column is to set the displacement corresponding to that column to unity with all other displacements equal to zero and measure the forces required at each station.

Example 6.2-1

Determine the influence coefficients for the points (1), (2) and (3) of the uniform cantilever beam shown in Fig. 6.2-1.

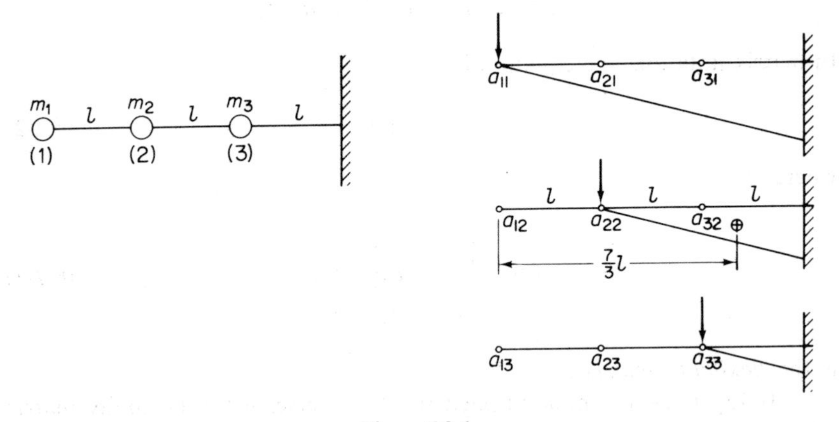

Figure 6.2-1.

Solution: The influence coefficients may be determined by placing unit loads at (1), (2) and (3) as shown, and calculating the deflections at these points. Using the area moment method,* the deflection at the various points is equal to the moment of the M/EI curve about the point in question. For example, the value of $a_{21} = a_{12}$ is found from Fig. 6.2-1 (b) as follows

$$a_{12} = \frac{1}{EI}\left[\frac{1}{2}(2l)^2 \times \frac{7}{3}l\right] = \frac{14}{3}\frac{l^3}{EI}$$

The other values (determined as above) are

$$a_{11} = \frac{27}{3}\frac{l^3}{EI} \qquad a_{21} = a_{12} = \frac{14}{3}\frac{l^3}{EI}$$

*Egor P. Popov, *Introduction to Mechanics of Solids* (Englewood Cliffs, N.J.: Prentice-Hall, Inc., 1968), p. 411.

$$a_{22} = \frac{8}{3}\frac{l^3}{EI} \qquad a_{23} = a_{32} = \frac{2.5}{3}\frac{l^3}{EI}$$

$$a_{33} = \frac{1}{3}\frac{l^3}{EI} \qquad a_{13} = a_{31} = \frac{4}{3}\frac{l^3}{EI}$$

The flexibility matrix can now be written as

$$a = \frac{l^3}{3EI}\begin{bmatrix} 27 & 14 & 4 \\ 14 & 8 & 2.5 \\ 4 & 2.5 & 1 \end{bmatrix}$$

and the symmetry about the diagonal should be noted.

Example 6.2-2

Fig. 6.2-2 shows a three degrees of freedom system. Determine the stiffness matrix.

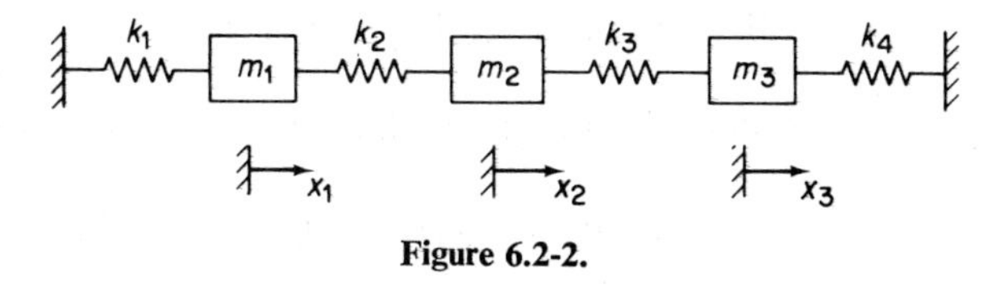

Figure 6.2-2.

Solution: Let $x_1 = 1.0$ and $x_2 = x_3 = 0$. The forces required at 1, 2 and 3, considering forces to the right as positive, are

$$f_1 = k_1 + k_2 = k_{11}$$
$$f_2 = -k_2 = k_{21}$$
$$f_3 = 0 = k_{31}$$

Repeat with $x_2 = 1$, and $x_1 = x_3 = 0$. The forces are now

$$f_1 = -k_2 = k_{12}$$
$$f_2 = k_2 + k_3 = k_{22}$$
$$f_3 = -k_3 = k_{32}$$

For the last column of k's, let $x_3 = 1$ and $x_1 = x_2 = 0$. The forces are

$$f_1 = 0 = k_{13}$$
$$f_2 = -k_3 = k_{23}$$
$$f_3 = k_3 + k_4 = k_{33}$$

The stiffness matrix can now be written as

$$K = \begin{bmatrix} (k_1 + k_2) & -k_2 & 0 \\ -k_2 & (k_2 + k_3) & -k_3 \\ 0 & -k_3 & (k_3 + k_4) \end{bmatrix}$$

6.3 RECIPROCITY THEOREM

The reciprocity theorem states that in a linear system $a_{ij} = a_{ji}$. For the proof of this theorem, we consider the work done by forces f_i and f_j where the order of loading is i followed by j and then by its reverse. Reciprocity results when we recognize that the work done is independent of the order of loading.

Applying f_i, the work done is $\frac{1}{2}f_i^2 a_{ii}$. Applying f_j, the work done by f_j is $\frac{1}{2}f_j^2 a_{jj}$. However, i undergoes further displacement $a_{ij}f_j$, and the additional work done by f_i becomes $a_{ij}f_j f_i$. Thus the total work done is

$$W = \tfrac{1}{2}f_i^2 a_{ii} + \tfrac{1}{2}f_j^2 a_{jj} + a_{ij}f_j f_i \tag{6.3-1}$$

We now reverse the order of loading, in which case the total work done is

$$W = \tfrac{1}{2}f_j^2 a_{jj} + \tfrac{1}{2}f_i^2 a_{ii} + a_{ji}f_i f_j \tag{6.3-2}$$

Since the work done in the two cases must be equal, we find that

$$a_{ij} = a_{ji} \tag{6.3-3}$$

6.4 EIGENVALUES AND EIGENVECTORS

For the undamped system of several degrees of freedom, the equations of motion expressed in matrix form become

$$[M]\{\ddot{x}\} + [k]\{x\} = \{0\} \tag{6.4-1}$$

where

$$M = \begin{bmatrix} m_{11} & m_{12}\cdots & \\ \vdots & & \\ m_{n1} & m_{n2}\cdots & m_{nn} \end{bmatrix} = \text{mass matrix (a square matrix)}$$

$$K = \begin{bmatrix} k_{11} & k_{12}\cdots & \\ \vdots & & \\ k_{n1} & k_{n2}\cdots & k_{nn} \end{bmatrix} = \text{stiffness matrix (a square matrix)}$$

$$X = \begin{Bmatrix} x_1 \\ x_2 \\ \cdot \\ \cdot \\ \cdot \\ x_n \end{Bmatrix} = \text{displacement vector (a column matrix)}$$

When there is no ambiguity, we will dispense with the brackets and braces and use capital letters and simply write the matrix equation as

$$M\ddot{X} + KX = 0$$

If we premultiply the above equation by M^{-1}, we obtain the following terms

$$M^{-1}M = I \quad \text{(a unit matrix)}$$
$$M^{-1}K = A \quad \text{(a dynamic matrix)}$$

and

$$I\ddot{X} + AX = 0 \tag{6.4-2}$$

Assuming harmonic motion $\ddot{X} = -\lambda X$, where $\lambda = \omega^2$, Eq. (6.4-2) becomes

$$[A - \lambda I]\{X\} = 0 \tag{6.4-3}$$

From Eq. (6.4-3) we form the determinant

$$|A - \lambda I| = 0 \tag{6.4-4}$$

which is the *characteristic equation* of the system. The roots λ_i of the characteristic equation are called *eigenvalues* and the natural frequencies of the system are determined from them by the relationship

$$\lambda_i = \omega_i^2 \tag{6.4-5}$$

By substituting λ_i into the matrix equation (6.4-3), we obtain the corresponding mode shape X_i which is called the *eigenvector*. Thus for an n-degrees of freedom system, there will be n eigenvalues and n eigenvectors.

It is also possible to find the eigenvectors from the adjoint matrix (See Appendix C) of the system. If, for conciseness, we make the abbreviation $B = A - \lambda I$ and start with the definition of the inverse

$$B^{-1} = \frac{1}{|B|} \operatorname{adj} B \tag{6.4-6}$$

we can premultiply by $|B|\,B$ to obtain

$$|B|\,I = B \operatorname{adj} B$$

or in terms of the original expression for B

$$|A - \lambda I|\,I = [A - \lambda I] \operatorname{adj} [A - \lambda I] \tag{6.4-7}$$

If now we let $\lambda = \lambda_i$, an eigenvalue, then the determinant on the left side of the equation is zero and we obtain

$$[0] = [A - \lambda_i I] \operatorname{adj} [A - \lambda_i I] \tag{6.4-8}$$

The above equation is valid for all λ_i and represents n equations for the n-degrees of freedom system. Comparing this equation with Eq. (6.4-3) for the ith mode

$$[A - \lambda_i I]\{X\}_i = 0$$

we recognize that the adjoint matrix, adj $[A - \lambda_i I]$, must consist of columns, each of which is the eigenvector X_i (multiplied by an arbitrary constant).

Example 6.4-1

Consider the system of Fig. 6.4-1.

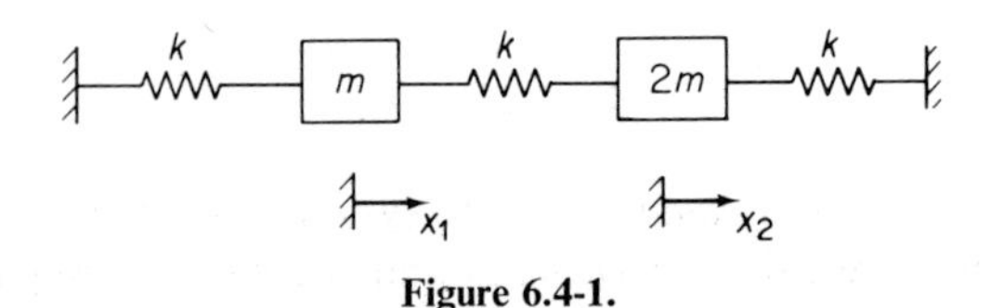

Figure 6.4-1.

The equations of motion can be expressed in matrix notation as

$$\begin{bmatrix} m & 0 \\ 0 & 2m \end{bmatrix} \begin{Bmatrix} \ddot{x}_1 \\ \ddot{x}_2 \end{Bmatrix} + \begin{bmatrix} 2k & -k \\ -k & 2k \end{bmatrix} \begin{Bmatrix} x_1 \\ x_2 \end{Bmatrix} = \begin{Bmatrix} 0 \\ 0 \end{Bmatrix} \tag{a}$$

Premultiplying by the inverse of the mass matrix

$$M^{-1} = \begin{bmatrix} \dfrac{1}{m} & 0 \\ 0 & \dfrac{1}{2m} \end{bmatrix}$$

and letting $\lambda = \omega^2$, Eq. (a) becomes

$$\begin{bmatrix} \left(2\frac{k}{m} - \lambda\right) & -\frac{k}{m} \\ -\frac{1}{2}\frac{k}{m} & \left(\frac{k}{m} - \lambda\right) \end{bmatrix} \begin{Bmatrix} x_1 \\ x_2 \end{Bmatrix} = 0 \tag{b}$$

The characteristic equation from the determinant of the above matrix is

$$\lambda^2 - 3\frac{k}{m}\lambda + \frac{3}{2}\left(\frac{k}{m}\right)^2 = 0 \tag{c}$$

from which the eigenvalues are found to be

$$\begin{aligned} \lambda_1 &= 0.634\frac{k}{m} \\ \lambda_2 &= 2.366\frac{k}{m} \end{aligned} \tag{d}$$

The eigenvectors can be found from Eq. (b) by substituting the above values of λ. We will, however, illustrate the use of the adjoint matrix in their evaluation.

The adjoint matrix from Eq. (b) is

$$\operatorname{adj}[A - \lambda_i I] = \begin{bmatrix} \left(\frac{k}{m} - \lambda_i\right) & \frac{k}{m} \\ \frac{k}{2m} & \left(\frac{2k}{m} - \lambda_i\right) \end{bmatrix} \tag{e}$$

Substituting $\lambda_1 = 0.634k/m$, we obtain from Eq. (e)

$$\begin{bmatrix} .366 & 1.000 \\ .500 & 1.366 \end{bmatrix} \frac{k}{m}$$

and each column normalized to unity results in the first eigenvector

$$\begin{bmatrix} .732 & .732 \\ 1.000 & 1.000 \end{bmatrix} \quad \text{or } X_1 = \begin{Bmatrix} .732 \\ 1.000 \end{Bmatrix}$$

Similarly, when $\lambda_2 = 2.366k/m$ is used, the second eigenvector is obtained from either column of Eq. (e) to be

$$X_2 = \begin{Bmatrix} -2.73 \\ 1.00 \end{Bmatrix}$$

The two normal nodes are shown in Fig. 6.4-2.

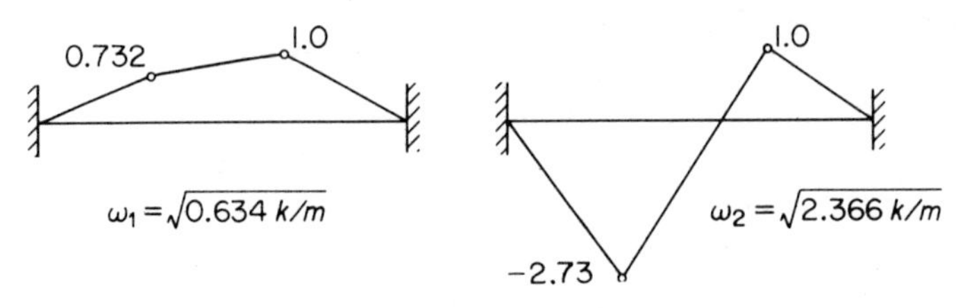

Figure 6.4-2.

6.5 ORTHOGONAL PROPERTIES OF THE EIGENVECTORS

The normal modes, or the eigenvectors of the system, can be shown to be *orthogonal* with respect to the mass and stiffness matrices as follows. Let the equation for the ith mode be

$$KX_i = \lambda_i MX_i \tag{6.5-1}$$

Premultiply by the transpose of mode j

$$X'_j KX_i = \lambda_i (X'_j MX_i) \tag{6.5-2}$$

Next, start with the equation for the jth mode and premultiply by X'_i to obtain

$$X'_i KX_j = \lambda_j (X'_i MX_j) \tag{6.5-3}$$

Since K and M are symmetric matrices, the following relationships hold*

$$\begin{aligned} X'_j MX_i &= X'_i MX_j \\ X'_j KX_i &= X'_i KX_j \end{aligned} \tag{6.5-4}$$

Thus, subtracting Eq. (6.5-3) from Eq. (6.5-2), we obtain

$$0 = (\lambda_i - \lambda_j) X'_i MX_j \tag{6.5-5}$$

If $\lambda_i \neq \lambda_j$, the above equation requires that

$$X'_i MX_j = 0 \tag{6.5-6}$$

It is also evident from Eq. (6.5-2) or Eq. (6.5-3) that as a consequence of Eq. (6.5-6)

$$X'_i KX_j = 0 \tag{6.5-7}$$

*See Appendix C.

Equations (6.5-6) and (6.5-7) define the *orthogonal* character of the normal modes.

Finally, if $i = j$, Eq. (6.5-5) is satisfied for any finite value of the products given by Eqs. (6.5-6) or (6.5-7). We therefore let

$$\begin{aligned} X_i'MX_i &= M_i \\ X_i'KX_i &= K_i \end{aligned} \tag{6.5-8}$$

These are called the *generalized mass* and the *generalized stiffness* respectively.

6.6 REPEATED ROOTS

When repeated roots are found in the characteristic equation, the corresponding eigenvectors are not unique, and a linear combination of such eigenvectors may also satisfy the equation of motion. To illustrate this point, let X_1 and X_2 be eigenvectors belonging to a common eigenvalue λ_0, and X_3 be a third eigenvector belonging to λ_3 that is different from λ_0. We can then write

$$\begin{aligned} AX_1 &= \lambda_0 X_1 \\ AX_2 &= \lambda_0 X_2 \\ AX_3 &= \lambda_3 X_3 \end{aligned} \tag{6.6-1}$$

By multiplying the second equation by a constant b and adding it to the first, we obtain another equation

$$A(X_1 + bX_2) = \lambda_0(X_1 + bX_2) \tag{6.6-2}$$

Thus a new eigenvector, $X_{12} = X_1 + bX_2$, which is a linear combination of the first two, also satisfies the basic equation

$$AX_{12} = \lambda_0 X_{12} \tag{6.6-3}$$

and hence no unique mode exists for λ_0.

Any of the modes corresponding to λ_0 must be orthogonal to X_3 if they are to be a normal mode. If all three modes are orthogonal, they are linearly independent and may be combined to describe the free vibration resulting from any initial condition.

Example 6.6-1

Consider the system shown in Fig. 6.6-1 where the connecting bar is rigid and negligible in weight.

The two normal modes of vibration are shown to be

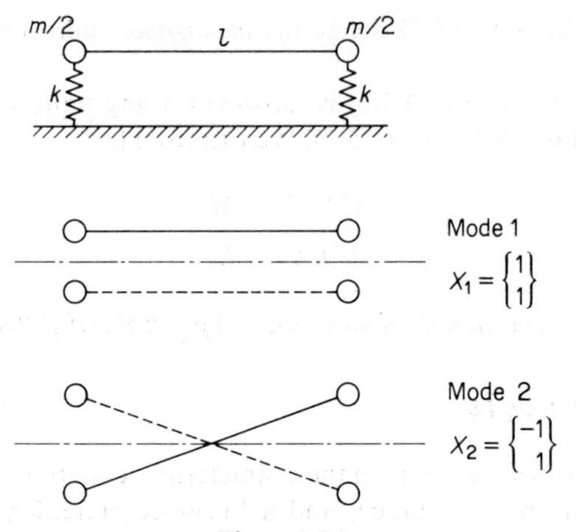

Figure 6.6-1.

translation and rotation, which are orthogonal. The natural frequencies for the two modes, however, are equal and can be calculated to be

$$\omega_n = \sqrt{\frac{2k}{m}}$$

The example illustrates that different eigenvectors may have equal eigenvalues.

Example 6.6-2

Determine the eigenvalues and eigenvectors when

$$A = \begin{bmatrix} 0 & -1 & 1 \\ -1 & 0 & 1 \\ 1 & 1 & 0 \end{bmatrix}$$

Solution: The characteristic equation $|A - \lambda I| = 0$ yields

$$(\lambda - 1)^2(\lambda + 2) = 0$$

so the eigenvalues are $\lambda_1 = 1$, $\lambda_2 = 1$, and $\lambda_3 = -2$.

Forming the adjoint matrix

$$\operatorname{adj}[A - \lambda I] = \begin{bmatrix} (\lambda^2 - 1) & -(\lambda - 1) & (\lambda - 1) \\ -(\lambda - 1) & (\lambda^2 - 1) & (\lambda - 1) \\ (\lambda - 1) & (\lambda - 1) & (\lambda^2 - 1) \end{bmatrix}$$

the eigenvector corresponding to $\lambda_3 = -2$ is found from any column of the above matrix

$$\begin{bmatrix} 3 & 3 & -3 \\ 3 & 3 & -3 \\ -3 & -3 & 3 \end{bmatrix} \text{ or } X_3 = \begin{Bmatrix} -1 \\ -1 \\ 1 \end{Bmatrix}$$

Substitution of $\lambda_1 = \lambda_2 = 1$ into the adjoint matrix leads to all zeros, so we return to the original matrix equation $[A - \lambda I]X = 0$ with $\lambda = 1$

$$\begin{aligned} -x_1 \quad -x_2 \quad +x_3 &= 0 \\ -x_1 \quad -x_2 \quad +x_3 &= 0 \\ x_1 \quad +x_2 \quad -x_3 &= 0 \end{aligned}$$

All three of these equations are of the form

$$x_1 = x_3 - x_2$$

and hence for the eigenvalue X_1 corresponding to $\lambda_1 = \lambda_2 = 1$ we can write

$$X_1 = \begin{Bmatrix} x_3 - x_2 \\ x_2 \\ x_3 \end{Bmatrix}$$

which is found to be orthogonal to X_3 for all values of x_2 and x_3, i.e.

$$(X_3)\{X_1\} = 0$$

Thus for $x_2 = x_3 = 1$, one could obtain

$$X_1 = \begin{Bmatrix} 0 \\ 1 \\ 1 \end{Bmatrix}$$

and for $x_3 = 1$ and $x_2 = -1$ the second eigenvector could be

$$X_2 = \begin{Bmatrix} 2 \\ -1 \\ 1 \end{Bmatrix}$$

As shown previously by Eq. (6.6-2), X_1 and X_2 are not unique, and any linear combination of X_1 and X_2 will also satisfy the original matrix equation.

6.7 THE MODAL MATRIX P

In Chapter 5 we found that static or dynamic coupling results from the choice of coordinates, and that for an undamped system, there exists a set of principal coordinates that will express the equations of motion in the uncoupled form. Such uncoupled coordinates are desirable since each equation can be solved independently of the others.

For a lumped mass multidegrees of freedom system, coordinates chosen at each mass point will result in a mass matrix that is diagonal, but the stiffness matrix will contain off-diagonal terms indicating static coupling. Coordinates chosen in another way may result in dynamic coupling or both dynamic and static coupling.

It is possible to uncouple the equations of motion of an n-degrees of freedom system, provided we know beforehand the normal modes of the system. When the n normal modes (or eigenvalues) are assembled into a square matrix with each normal mode represented by a column, we call it the *modal matrix P*. Thus the modal matrix for a three degrees of freedom system may appear as

$$P = \begin{bmatrix} \begin{Bmatrix} x_1 \\ x_2 \\ x_3 \end{Bmatrix}_1 & \begin{Bmatrix} x_1 \\ x_2 \\ x_3 \end{Bmatrix}_2 & \begin{Bmatrix} x_1 \\ x_2 \\ x_3 \end{Bmatrix}_3 \end{bmatrix} = [X_1 \quad X_2 \quad X_3] \tag{6.7-1}$$

The modal matrix makes it possible to include all the orthogonality relations of Sec. 6.5 into one equation. For this operation we need also the transpose of P, which is

$$P' = \begin{bmatrix} (x_1 & x_2 & x_3)_1 \\ (x_1 & x_2 & x_3)_2 \\ (x_1 & x_2 & x_3)_3 \end{bmatrix} = [X_1 \quad X_2 \quad X_3]' \tag{6.7-2}$$

with each row corresponding to a mode. If we now form the product $P'MP$ or $P'KP$, the result will be a diagonal matrix since the off-diagonal terms simply express the orthogonality relations which are zero.

As an example consider a two degrees of freedom system. Performing the indicated operation with the modal matrix, we have

$$\begin{aligned} P'MP &= [X_1 \quad X_2]'[M][X_1 \quad X_2] \\ &= \begin{bmatrix} X_1'MX_1 & X_1'MX_2 \\ X_2'MX_1 & X_2'MX_2 \end{bmatrix} \\ &= \begin{bmatrix} M_1 & 0 \\ 0 & M_2 \end{bmatrix} \end{aligned} \tag{6.7-3}$$

In the above equation, the off-diagonal terms are zero because of orthogonality, and the diagonal terms are the generalized mass M_i.

It is evident that a similar formulation applies also to the stiffness matrix K that results in the following equation

$$P'KP = \begin{bmatrix} K_1 & 0 \\ 0 & K_2 \end{bmatrix} \tag{6.7-4}$$

The diagonal terms here are the generalized stiffness K_i.

If each of the columns of the modal matrix P is divided by the square root of the generalized mass M_i, the new matrix is called the *weighted modal matrix* and designated as $\tilde{P}$. It is easily seen that the diagonalization of the mass matrix by the weighted modal matrix results in the unit matrix

$$\tilde{P}'M\tilde{P} = I \tag{6.7-5}$$

Since $K_i/M_i = \lambda_i$, the stiffness matrix treated similarly by the weighted modal matrix becomes a diagonal matrix of the eigenvalues

$$\tilde{P}'K\tilde{P} = \begin{bmatrix} \lambda_1 & & & & 0 \\ & \lambda_2 & & & \\ & & \cdot & & \\ & & & \cdot & \\ 0 & & & \cdot & \\ & & & & \lambda_n \end{bmatrix} = \Lambda \tag{6.7-6}$$

Example 6.7-1

Consider the symmetrical two degrees of freedom system shown in Fig. 6.7-1. The equation of motion in matrix form is

$$\begin{bmatrix} m & 0 \\ 0 & m \end{bmatrix} \begin{Bmatrix} \ddot{x}_1 \\ \ddot{x}_2 \end{Bmatrix} + \begin{bmatrix} 2k & -k \\ -k & 2k \end{bmatrix} \begin{Bmatrix} x_1 \\ x_2 \end{Bmatrix} = 0 \tag{a}$$

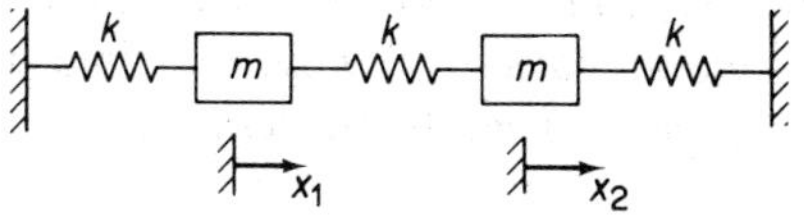

Figure 6.7-1.

and the eigenvalues and eigenvectors can be shown to equal

$$\lambda_1 = \omega_1^2 = \frac{k}{m} \qquad \lambda_2 = \omega_2^2 = 3\frac{k}{m}$$

$$\begin{Bmatrix} x_1 \\ x_2 \end{Bmatrix}_{\lambda_1} = \begin{Bmatrix} 1 \\ 1 \end{Bmatrix}, \quad \begin{Bmatrix} x_1 \\ x_2 \end{Bmatrix}_{\lambda_2} = \begin{Bmatrix} -1 \\ 1 \end{Bmatrix} \tag{b}$$

The generalized mass for both modes is $2m$, and the modal matrix and the weighted modal matrix are

$$P = \begin{bmatrix} 1 & -1 \\ 1 & 1 \end{bmatrix} \qquad \tilde{P} = \frac{1}{\sqrt{2m}} \begin{bmatrix} 1 & -1 \\ 1 & 1 \end{bmatrix}$$

To decouple the original equation we will use $\tilde{P}$ in the transformation

$$\begin{Bmatrix} x_1 \\ x_2 \end{Bmatrix} = \frac{1}{\sqrt{2m}} \begin{bmatrix} 1 & -1 \\ 1 & 1 \end{bmatrix} \begin{Bmatrix} y_1 \\ y_2 \end{Bmatrix} \tag{d}$$

and premultiply by $\tilde{P}'$ to obtain

$$\tilde{P}'M\tilde{P}\ddot{Y} + \tilde{P}'K\tilde{P}Y = 0$$

or

$$\begin{bmatrix} 1 & 0 \\ 0 & 1 \end{bmatrix} \begin{Bmatrix} \ddot{y}_1 \\ \ddot{y}_2 \end{Bmatrix} + \frac{k}{m} \begin{bmatrix} 1 & 0 \\ 0 & 3 \end{bmatrix} \begin{Bmatrix} y_1 \\ y_2 \end{Bmatrix} = 0 \tag{e}$$

Thus Eq. (a) has been transformed to the uncoupled Eq. (e) by the coordinate transformation of Eq. (d). The coordinates y_1 and y_2 are referred to as *principal* or *normal coordinates.*

6.8 FORCED VIBRATION AND COORDINATE DECOUPLING

The equations of motion of an n-degrees of freedom system with viscous damping and arbitrary excitation $F(t)$ can be presented in the matrix form

$$M\ddot{X} + C\dot{X} + KX = F \tag{6.8-1}$$

If the damping matrix C is proportional to either the mass or the stiffness matrix or to a linear combination of these matrices, then the damping is called *proportional damping* and can be expressed as

$$C = \alpha M + \beta K \tag{6.8-2}^*$$

where α and β are constants.

For the case of proportional damping, the equations of motion represented by Eq. (6.8-1) can be uncoupled by either the modal matrix P or by the weighted modal matrix $\tilde{P}$ of the corresponding free vibrating system.

*It can be shown that $C = \alpha M^n + \beta K^m$ can also be diagonalized (See Problems 6-12 and 6-13).

Using $\tilde{P}$, let

$$X = \tilde{P}Y \tag{6.8-3}$$

where Y is another column matrix. Equation (6.8-3) represents a transformation of coordinates from X to Y. Substituting Eq. (6.8-3) into (6.8-1) and premultiplying by $\tilde{P}'$, we obtain

$$\tilde{P}'M\tilde{P}\ddot{Y} + \tilde{P}'C\tilde{P}\dot{Y} + \tilde{P}'K\tilde{P}Y = \tilde{P}'F \tag{6.8-4}$$

With C equal to Eq. (6.8-2) and recognizing Eqs. (6.7-5) and (6.7-6), the above equation becomes

$$I\ddot{Y} + (\alpha I + \beta\Lambda)\dot{Y} + \Lambda Y = \tilde{P}'F \tag{6.8-5}$$

Since all the coefficients on the left side of this equation are diagonal matrices, Eq. (6.8-5) represents a set of uncoupled second order equations of the form

$$\ddot{y}_i + (\alpha + \beta\omega_i^2)\dot{y}_i + \omega_i^2 y_i = Q_i(t) \tag{6.8-6}$$

The solution to the above equations may be conveniently carried out by Laplace transformation.

If the damping matrix is not proportional, the equations of motion will be coupled by the damping matrix, and the set of equations must be solved simultaneously or by the state-space method of Sec. 6.10.

6.9 FORCED NORMAL MODES OF DAMPED SYSTEMS

In normal mode vibration, every point of the system undergoes harmonic motion and passes through the equilibrium position simultaneously. We have found that such motion is possible for free undamped vibration.

Normal mode type of vibration is also possible in a damped system if it is excited by a number of harmonic forces equal to the number of degrees of freedom of the system. To show this we consider a viscously damped system of n-degrees of freedom, excited by harmonic forces of frequency ω. Its equation of motion is

$$[m]\{\ddot{x}\} + [c]\{\dot{x}\} + [k]\{x\} = \{F\} \sin \omega t \tag{6.9-1}$$

The problem of concern here is the conditions that will produce a solution of the form

$$\{x\} = \{X\} \sin (\omega t - \theta) \tag{6.9-2}$$

Such a problem has been examined by several investigators.* Their

*B. M. Fraejis de Veubeke, "Déphasages Characteristiques et Vibrations Forcés d'un Systéme Amorti," Académie Royale de Belgique, *Bulletin de la Classe des Sciences*, Series 5, Vol. XXXIV (1948), pp. 626.

conclusions are that for a given frequency of excitation ω, n solutions of the type described by Eq. (6.9-2) exist, where each of these modes is associated with a definite phase θ_i and a distribution of the force vector $\{F\}_i$ which is required for its excitation. The response under these conditions is termed *forced normal modes* of the damped system, in that every point in the system moves in phase and passes through its equilibrium position simultaneously with respect to the other points. As in the case of the free undamped vibration, orthogonality relations exist between the modes.

If Eq. (6.9-2) is substituted into Eq. (6.9-1) and the coefficients of like terms are equated, we obtain the two equations

$$[([k] - [m]\omega^2)\sin\theta - [c]\omega\cos\theta]\{X\} = \{0\} \tag{6.9-3}$$

$$[([k] - [m]\omega^2)\cos\theta + [c]\omega\sin\theta]\{X\} = \{F\} \tag{6.9-4}$$

Rearranging Eq. (6.9-3) to the form

$$[\lceil 1 \rfloor \tan\theta - ([k] - [m]\omega^2)^{-1}[c]\omega]\{X\} = \{0\} \tag{6.9-5}$$

it is evident that there are n values of $\tan\theta_i$ corresponding to the n eigenvalues of Eq. (6.9-5), and for each $\tan\theta_i$ there is a corresponding eigenvector $\{X\}_i$. The required forcing functions $\{F\}_i$ are then obtained by substitution of θ_i and $\{X\}_i$ into Eq. (6.9-4).

The orthogonality relations are obtained by rewriting Eq. (6.9-3) for the ith eigenvalue and eigenvector, premultiplying by the transpose of the jth eigenvector, and repeating the procedure with i and j interchanged.

$$\tan\theta_i\{X\}'_j([k] - [m]\omega^2)\{X\}_i - \omega\{X\}'_j[c]\{X\}_i = 0$$

$$\tan\theta_j\{X\}'_i([k] - [m]\omega^2)\{X\}_j - \omega\{X\}'_i[c]\{X\}_j = 0$$

Due to the symmetry of $[m]$, $[k]$, and $[c]$, we then obtain, by subtracting, the following relations for $\tan\theta_i \neq \tan\theta_j$

$$\{X\}'_j([k] - [m]\omega^2)\{X\}_i = 0 \tag{6.9-6}$$

$$\{X\}'_j[c]\{X\}_i = 0 \tag{6.9-7}$$

Similarly, from Eq. (6.9-4) we obtain the third relation

$$\{X\}'_j\{F\}_i = 0 \tag{6.9-8}$$

Transformation of Coordinates. Considerable simplification results from the transformation of coordinates, utilizing either the modal matrix $[P]$ of the undamped system or the weighted normalized modal matrix $[\tilde{P}]$, which is the modal matrix $[P]$ with the ith columns divided by the square root of its generalized mass $(\{X\}'_i[m]\{X\}_i)^{1/2}$ (See Sec. 6.7). If the transformation

$$\{x\} = [\tilde{P}]\{y\} \tag{6.9-9}$$

is made in Eq. (6.9-1), its premultiplication by the transpose $[\tilde{P}]'$ will result in the equation

$$\lfloor 1 \rceil\{\ddot{y}\} + [C]\{\dot{y}\} + \lfloor \lambda_i \rceil\{y\} = [\tilde{P}]'\{F\} \sin \omega t \tag{6.9-10}$$

where

$$\begin{aligned} \lfloor \lambda_i \rceil = [\tilde{P}]'[k][\tilde{P}] &= \text{squares of the undamped natural frequencies} \\ [C] = [\tilde{P}]'[c][\tilde{P}] &= \text{symmetric damping matrix} \\ \lfloor 1 \rceil = [\tilde{P}]'[m][\tilde{P}] &= \text{unity matrix} \end{aligned}$$

Assuming a solution of the form

$$\{y\} = \{Y\} \sin (\omega t - \phi) \tag{6.9-11}$$

the results of the previous section are replaced by the following

$$\left[\lfloor 1 \rceil \tan \phi - \omega \left\lceil \frac{1}{\lambda_i - \omega^2} \right\rfloor [C]\right]\{Y\} = 0 \tag{6.9-12}$$

$$[(\lfloor \lambda_i \rceil - \omega^2 \lfloor 1 \rceil) \cos \phi + \omega[C] \sin \phi]\{Y\} = [\tilde{P}]'\{F\} \tag{6.9-13}$$

$$\{Y\}_j'(\lfloor \lambda_i \rceil - \lfloor 1 \rceil \omega^2)\{Y\}_i = 0 \tag{6.9-14}$$

$$\{Y\}_j'[C]\{Y\}_i = 0 \tag{6.9-15}$$

$$\{Y\}_j'[\tilde{P}]'\{F\}_i = 0 \tag{6.9-16}$$

If the modal matrix $[P]$ is used instead of the weighted normalized matrix $[\tilde{P}]$, both the mass and the stiffness matrices will be diagonalized, but the mass matrix will not be a unit matrix.

Numerical Example

Figure 6.9-1 shows a two degree of freedom system whose equation of motion is

$$m\begin{bmatrix} 1 & 0 \\ 0 & 1 \end{bmatrix}\begin{Bmatrix} \ddot{x}_1 \\ \ddot{x}_2 \end{Bmatrix} + c\begin{bmatrix} 2 & -1 \\ -1 & 1 \end{bmatrix}\begin{Bmatrix} \dot{x}_1 \\ \dot{x}_2 \end{Bmatrix} + k\begin{bmatrix} 2 & -1 \\ -1 & 2 \end{bmatrix}\begin{Bmatrix} x_1 \\ x_2 \end{Bmatrix} = \begin{Bmatrix} F_1 \\ F_2 \end{Bmatrix} \sin \omega t$$

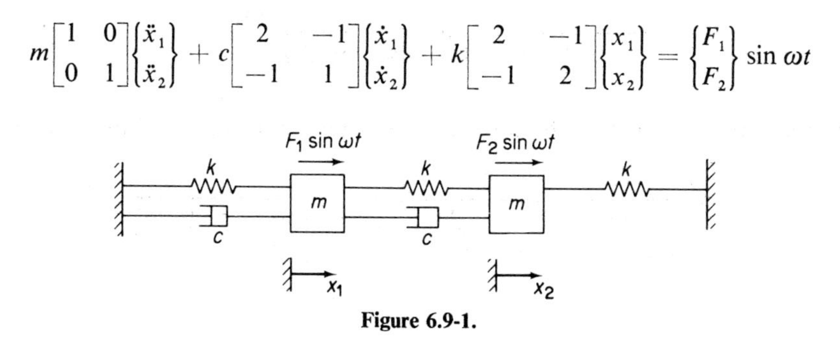

Figure 6.9-1.

Its normal modes are defined by

$$\lambda_1 = \frac{k}{m} \qquad \left(\frac{x_1}{x_2}\right)_1 = 1$$

$$\lambda_2 = 3\frac{k}{m} \qquad \left(\frac{x_1}{x_2}\right)_2 = -1$$

Since the generalized mass is $2m$ for both modes, the modal matrix and the weighted normalized matrix are

$$[P] = \begin{bmatrix} 1 & -1 \\ 1 & 1 \end{bmatrix} \qquad [\tilde{P}] = \frac{1}{\sqrt{2m}}\begin{bmatrix} 1 & -1 \\ 1 & 1 \end{bmatrix}$$

Using $[\tilde{P}]$, we have

$$\lfloor \lambda_i \rceil = \frac{k}{m}\begin{bmatrix} 1 & 0 \\ 0 & 3 \end{bmatrix}$$

$$[C] = \frac{c}{2m}\begin{bmatrix} 1 & -1 \\ -1 & 5 \end{bmatrix}$$

and Eq. (6.9-12) becomes

$$\left[\lfloor 1 \rceil \tan\phi - \frac{\omega c}{2k}\begin{bmatrix} \dfrac{1}{1 - \dfrac{\omega^2 m}{k}} & 0 \\ 0 & \dfrac{1}{3 - \dfrac{\omega^2 m}{k}} \end{bmatrix}\begin{bmatrix} 1 & -1 \\ -1 & 5 \end{bmatrix} \right]\begin{Bmatrix} Y_1 \\ Y_2 \end{Bmatrix} = 0$$

Letting $\mu = (2k/\omega c)\tan\phi$, the above equation reduces to

$$\begin{bmatrix} \left(\mu - \dfrac{1}{1 - \dfrac{\omega^2 m}{k}}\right) & \left(\dfrac{1}{1 - \dfrac{\omega^2 m}{k}}\right) \\ \left(\dfrac{1}{3 - \dfrac{\omega^2 m}{k}}\right) & \left(\mu - \dfrac{5}{3 - \dfrac{\omega^2 m}{k}}\right) \end{bmatrix}\begin{Bmatrix} Y_1 \\ Y_2 \end{Bmatrix} = 0$$

The eigenvalue μ is then found from the determinant of the above matrix

$$\mu^2 - \mu\left[\frac{1}{1 - \dfrac{\omega^2 m}{k}} + \frac{5}{3 - \dfrac{\omega^2 m}{k}}\right] + \frac{4}{\left(1 - \dfrac{\omega^2 m}{k}\right)\left(3 - \dfrac{\omega^2 m}{k}\right)} = 0$$

and the eigenvector is determined from the ratio

$$\frac{Y_1}{Y_2} = \frac{\dfrac{1}{1 - \dfrac{\omega^2 m}{k}}}{\dfrac{1}{1 - \dfrac{\omega^2 m}{k}} - \mu}$$

To obtain numerical values, it is necessary to specify the excitation frequency ω or $\omega^2 m/k$. For $\omega^2 m/k = 0.50$, we obtain

$$\mu_1 = 1.105 \qquad \mu_2 = 2.895$$

$$\{Y\}_{\mu_1} = \begin{matrix} 2.24 \\ 1.00 \end{matrix} \qquad \{Y\}_{\mu_2} = \begin{matrix} -2.24 \\ 1.00 \end{matrix}$$

These quantities satisfy the orthogonality relations given by Eqs. (6.9-14) and (6.9-15). By Eq. (6.9-16), we can find the ratio of the forces. The equation

$$\{Y\}'_{\mu_2}[\tilde{P}]'\{F\}_{\mu_1} = 0$$

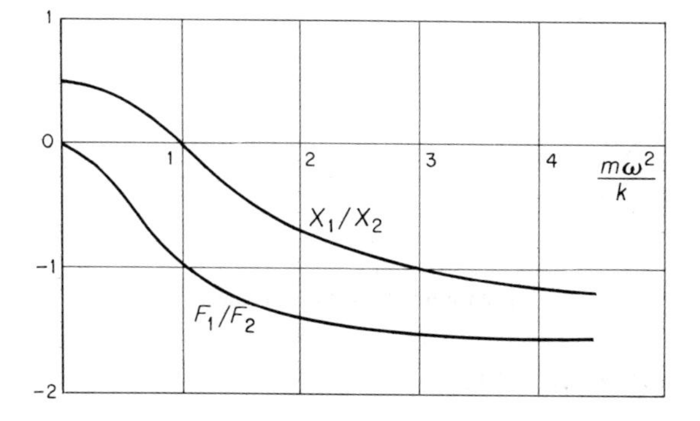

Figure 6.9-2. First mode.

results in $(F_1/F_2)_{\mu_1} = -0.382$. Similarly, $\{Y\}_{\mu_1}[\tilde{P}]'\{F\}_{\mu_2} = 0$ yields

$$\left(\frac{F_1}{F_2}\right)_{\mu_2} = -2.61$$

Finally, the real amplitudes $\{X\}$ are found from the transformation $\{X\} = [\tilde{P}]\{Y\}$ to be

$$\{X\}_{\mu_1} = \begin{Bmatrix} .382 \\ 1.000 \end{Bmatrix} \qquad \{X\}_{\mu_2} = \begin{Bmatrix} 2.61 \\ 1.00 \end{Bmatrix}$$

To complete the problem, other frequencies $\omega^2 m/k$ must be chosen and the above computation repeated. Plots of the amplitude and force ratios are shown for the two modes in Figs. 6.9-2 and 6.9-3.

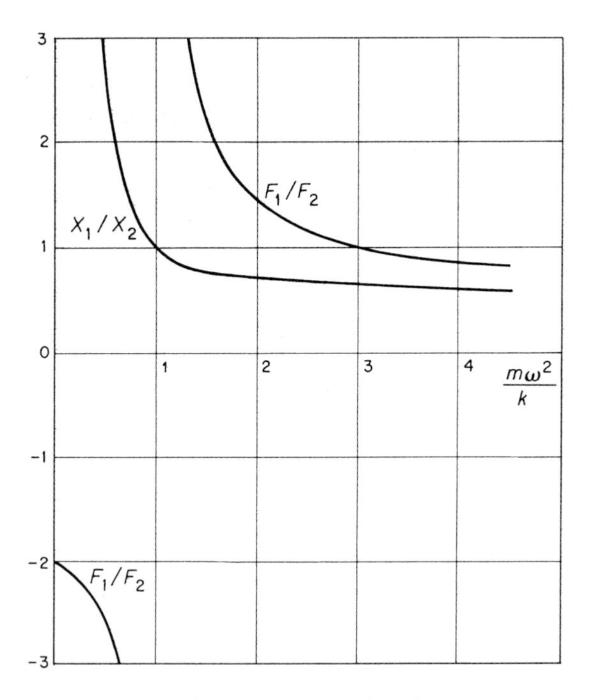

Figure 6.9-3. Second mode.

6.10 STATE SPACE METHOD

In the more general case where proportional damping does not exist, the second order system of equations of the form of Eq. (6.8-4) cannot be decoupled unless they are reformulated in terms of first order equations. For this reformulation, each of the original variables and their derivatives are assigned new variables, referred to as *state variables*, and hence second derivatives become first derivatives of the new state variables. Although the order of the derivatives is reduced by this procedure, the number of variables is now doubled, thereby adding to the burden of computation. The use of the digital computer is therefore essential for numerical computation.

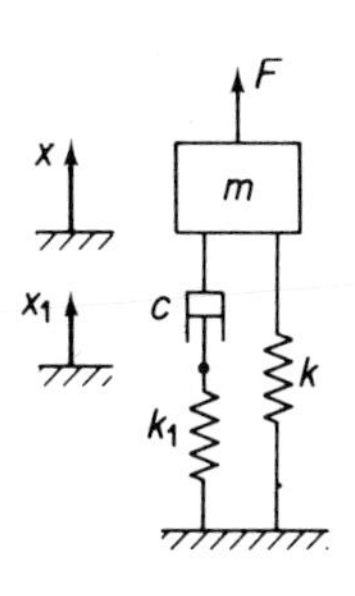

Figure 6.10-1.

Example 6.10-1

Consider the viscoelastically damped system of Fig. 6.10-1. The system differs from the viscously damped system by the addition of the spring k_1 which introduces one more coordinate x_1 to the system. The equations of motion for the system in inertial coordinates x and x_1 are

$$\begin{aligned} m\ddot{x} &= -kx - c(\dot{x} - \dot{x}_1) + F \\ 0 &= c(\dot{x} - \dot{x}_1) - k_1 x_1 \end{aligned} \tag{6.10-1}$$

Making the following substitutions

$$\omega_0^2 = \frac{k}{m}, \qquad \beta = \frac{k_1}{m}, \qquad \alpha = \frac{k_1}{c}$$

the equations are rewritten as

$$\begin{aligned} \ddot{x} &= -\omega_0^2 x - \beta x_1 + \frac{F}{m} \\ \dot{x}_1 &= \dot{x} - \alpha x_1 \end{aligned} \tag{6.10-2}$$

We now choose *state variables* z as follows

$$\begin{aligned} x_1 &= z_1 \\ \dot{x}_1 &= \dot{z}_1 \\ x &= z_2 \\ \dot{x} &= z_3 = \dot{z}_2 \end{aligned} \tag{6.10-3}$$

and rewrite the equations of motion in terms of z

$$\begin{Bmatrix} \dot{z}_1 \\ \dot{z}_2 \\ \dot{z}_3 \end{Bmatrix} = \begin{bmatrix} -\alpha & 0 & 1 \\ 0 & 0 & 1 \\ -\beta & -\omega_0^2 & 0 \end{bmatrix} \begin{Bmatrix} z_1 \\ z_2 \\ z_3 \end{Bmatrix} + \begin{Bmatrix} 0 \\ 0 \\ \frac{F}{m} \end{Bmatrix} \tag{6.10-4}$$

In abbreviated notation the above equation is

$$\dot{z} = Az + u \tag{6.10-5}$$

where

$$z = \begin{Bmatrix} z_1 \\ z_2 \\ z_3 \end{Bmatrix} \quad u = \begin{Bmatrix} 0 \\ 0 \\ \frac{F}{m} \end{Bmatrix} \quad \text{and} \quad A = \begin{bmatrix} -\alpha & 0 & 1 \\ 0 & 0 & 1 \\ -\beta & -\omega_0^2 & 0 \end{bmatrix}$$

The solution of this first order equation is well known and can be written in closed form as

$$z = e^{At}z_0 + e^{At}\int_0^t e^{-A\tau}u(\tau)\,d\tau \tag{6.10-6}$$

This apparently simple equation is actually not at all simple and requires considerable numerical treatment that is presented in the following section.

Solution 1: Consider the homogeneous equation

$$\dot{z} = Az \tag{6.10-7}$$

with its solution

$$z = e^{At}z_0 \tag{6.10-8}$$

In this equation, the term e^{At} requires interpretation. One procedure would be to consider the Laplace transform solution of the homogeneous equation and compare results. Letting $\bar{z}(s)$ be the Laplace transform of the column vector z, we obtain

$$s\bar{z}(s) - z_0 = A\bar{z}(s)$$

or (6.10-9)

$$[sI - A]\bar{z}(s) = z_0$$

Thus

$$z = \mathcal{L}^{-1}[sI - A]^{-1}z_0 \tag{6.10-10}$$

from which we conclude that

$$\begin{aligned} e^{At} &= \mathcal{L}^{-1}[sI - A]^{-1} \\ &= \mathcal{L}^{-1}\frac{\text{adj}[sI - A]}{|sI - A|} \end{aligned} \tag{6.10-11}$$

It is evident here that the roots s_i of the characteristic equation $|sI - A| = 0$ must be evaluated and that the right side of the equation after the inversion process will be a square matrix whose elements are $e^{s_i t}$ multiplied by constants. If repeated roots are present, terms such as $te^{s_i t}$ will also appear in the matrix.

Solution 2: We can also examine the solution to the state space equation as an eigenvalue, eigenvector problem. The eigenvalues are obtained from the characteristic equation

$$|A - \lambda I| = 0 \tag{6.10-12}$$

and the eigenvectors from one of the columns of the adjoint matrix *adj* $[\lambda I - A]$ with λ_i for the ith mode.

Before preceding with this problem, we introduce here a *diagonalization technique* which will be essential to the solution. The homogeneous equation from Eq. (6.10-5) is first written for the ith mode as

$$[A - \lambda_i]Z_i = 0 \tag{6.10-13}$$

where the eigenvalues are assumed to be distinct.

There exists n (for this problem, $n = 3$) such equations which we will rearrange as

$$AZ_i = \lambda_i Z_i \tag{6.10-14}$$

These n matrix equations can be assembled into a single matrix equation in terms of the modal matrix P and a diagonal matrix of the eigenvalues defined as

$$\Lambda = \begin{bmatrix} \lambda_1 & & & & 0 \\ & \lambda_2 & & & \\ & & \cdot & & \\ & & & \cdot & \\ & & & & \cdot \\ 0 & & & & \lambda_n \end{bmatrix} \tag{6.10-15}$$

The assembled equation then becomes

$$AP = P\Lambda \tag{6.10-16}$$

which can be easily verified as n equations of the type designated as Eq. (6.10-14). If the above equation is premultiplied by P^{-1} we obtain

$$P^{-1}AP = \Lambda \tag{6.10-17}$$

and the matrix A is diagonalized in terms of the eigenvalues of the system.

Returning now to the solution of Eq. (6.10-5), introduce the coordinate transformation

$$Z = Py \tag{6.10-18}$$

to obtain

$$P\dot{y} = PAy + u \tag{6.10-19}$$

Premultiplying by P^{-1} we have

$$\begin{aligned} \dot{y} &= P^{-1}APy + P^{-1}u \\ &= \Lambda y + P^{-1}u \end{aligned} \tag{6.10-20}$$

Equation (6.10-20) is now decoupled, and the solution for y_i is simply obtained by Laplace transformation.

Considering only the homogeneous equation

$$\dot{y} = \Lambda y \tag{6.10-21}$$

the solution is

$$y = e^{\Lambda t} y_0 \tag{6.10-22}$$

or

$$\begin{Bmatrix} y_1 \\ y_2 \\ y_3 \end{Bmatrix} = \begin{bmatrix} e^{\lambda_1 t} & 0 & 0 \\ 0 & e^{\lambda_2 t} & 0 \\ 0 & 0 & e^{\lambda_3 t} \end{bmatrix} \begin{Bmatrix} y_1 \\ y_2 \\ y_3 \end{Bmatrix}_0 \tag{6.10-23}$$

To transform to the original coordinates, we note that $y = P^{-1}z$. Substituting into Eq. (6.10-22) and premultiplying by P, we obtain

$$z = Pe^{\Lambda t}P^{-1}z_0 \qquad (6.10\text{-}24)$$

Comparing with Eq. (6.10-8) we conclude that

$$e^{At} = Pe^{\Lambda t}P^{-1} \qquad (6.10\text{-}25)$$

which is again a square nxn matrix.

PROBLEMS

6-1 Determine the stiffness matrix for the system shown in Fig. P6-1 and establish the flexibility matrix by its inverse.

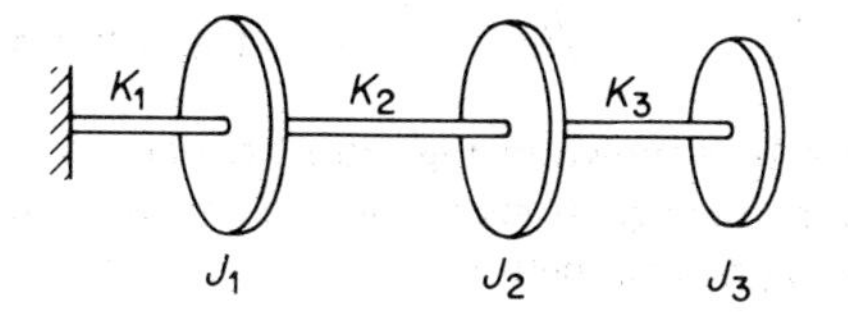

Figure P. 6-1.

6-2 Establish the stiffness and flexibility matrices for Prob. 6-9.

6-3 A simply supported uniform beam of length l is loaded with weights at positions $0.25l$ and $0.6l$. Determine the flexibility influence coefficients for these positions.

6-4 Determine the flexibility matrix for the cantilever beam shown in Fig. P6-4 and calculate the stiffness matrix from its inverse.

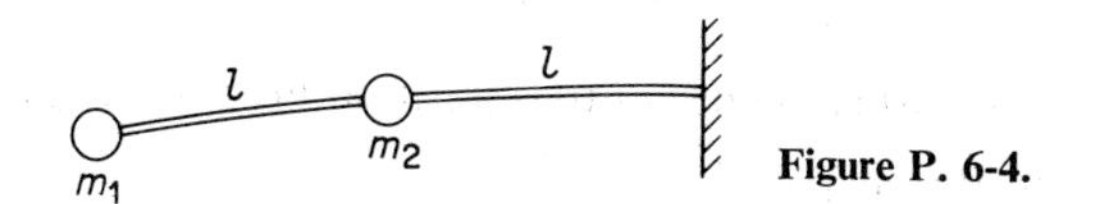

Figure P. 6-4.

6-5 Consider a system with n springs in series as presented in Fig. P6-5 and show that the stiffness matrix is a band matrix along the diagonal.

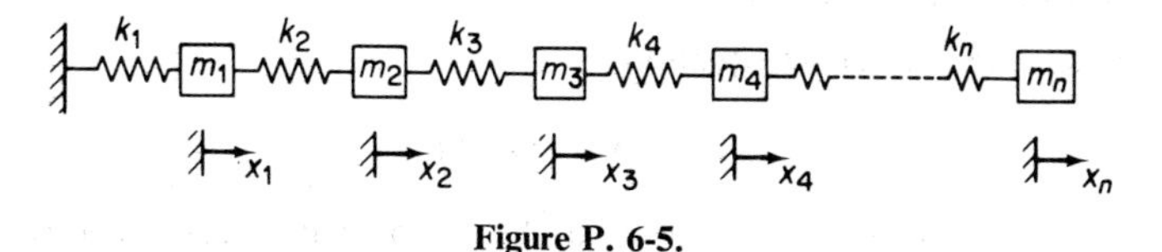

Figure P. 6-5.

6-6 Determine the flexibility matrix for the system of Prob. 6-5.

6-7 Using the adjoint matrix, determine the normal modes of the spring-mass system shown in Fig. P6-7.

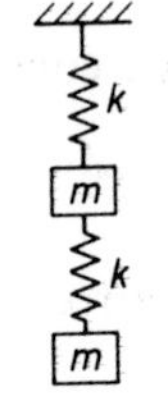

Figure P. 6-7.

6-8 For the system shown in Fig. P6-8, write the equations of motion in matrix form and determine the normal modes from the adjoint matrix.

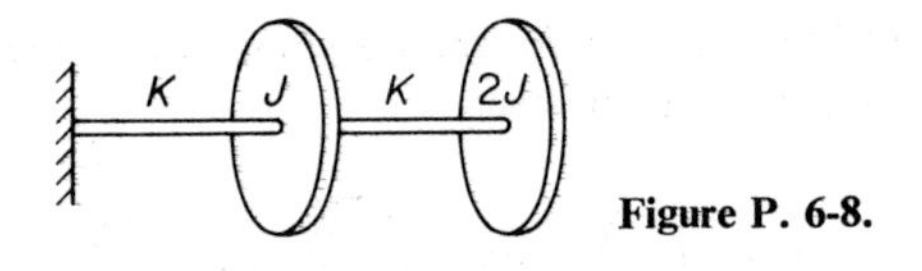

Figure P. 6-8.

6-9 The two uniform bars shown in Fig. P6-9 are of equal length but of different masses. Determine the equations of motion and the natural frequencies and mode shapes using matrix methods.

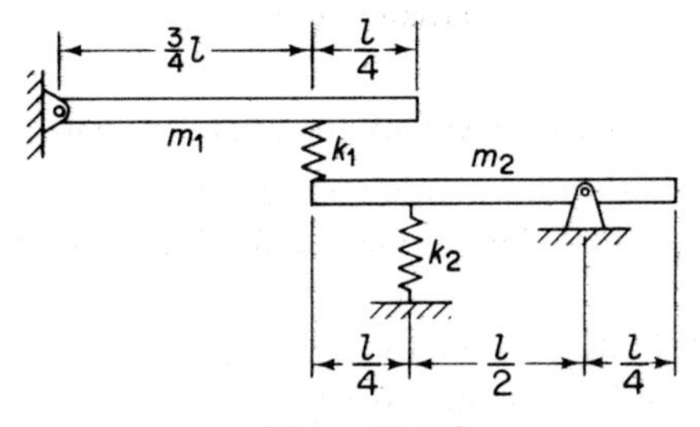

Figure P. 6-9.

6-10 Show that the normal modes of the system of Prob. 6-8 are orthogonal.

6-11 Verify the relationship of Eq. (6.5-7)

$$X_i' K X_j = 0$$

by applying it to Prob. 6-8.

6-12 Starting with the matrix equation

$$K\phi_s = \omega_s M\phi_s$$

premultiply first by KM^{-1} and, using the orthogonality relation $\phi_r' M\phi_s = 0$, show that

$$\phi_r' K M^{-1} K \phi_s = 0$$

Repeat to show that

$$\phi_r' [K M^{-1}]^h K \phi_s = 0$$

for $h = 1, 2, \ldots n$, where n = number of degrees of freedom of the system.

6-13 In a manner similar to Prob. 6-12, show that

$$\phi_r' [M K^{-1}]^h M \phi_s = 0, \qquad h = 1, 2, \ldots$$

6-14 Determine the modal matrix P and the weighted modal matrix $\tilde{P}$ for the system shown in Fig. P6-14. Show that P or $\tilde{P}$ will diagonalize the stiffness matrix.

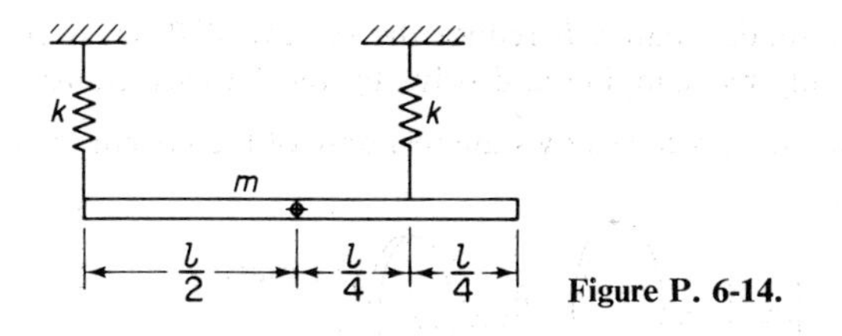

Figure P. 6-14.

6-15 For the system shown in Fig. P6-15, choose coordinates x_1 and x_2 at the ends of the bar and determine the type of coupling this introduces.

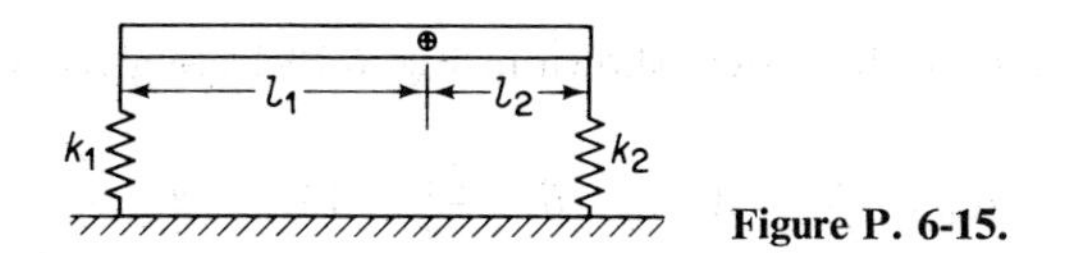

Figure P. 6-15.

6-16 Write the expressions for the kinetic and potential energies of the system shown in Fig. P6-15 for other sets of coordinates and note that coupling exists if cross products of coordinates appear in either T or U.

6-17 Determine the modal matrix P and the weighted modal matrix $\tilde{P}$ for the system shown in Fig. P6-17 and diagonalize the stiffness matrix thereby decoupling the equations.

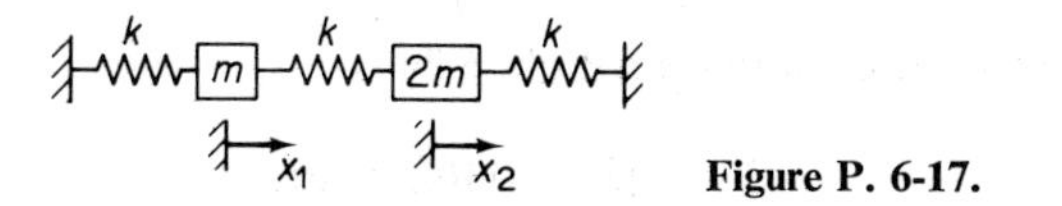

Figure P. 6-17.

6-18 Determine P for the system of Prob. 6-14 and decouple the equations.

6-19 Determine $\tilde{P}$ for the double pendulum with coordinates θ_1 and θ_2. Show that $\tilde{P}$ decouples the equations of motion.

6-20 Determine by the method of Laplace transformation the solution to the forced vibration problem shown in Fig. P6-20.

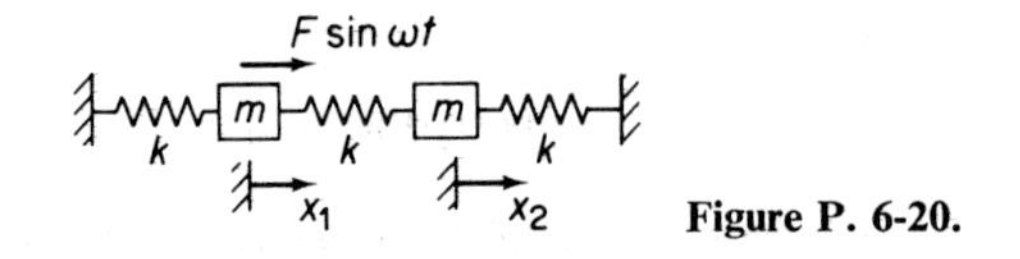

Figure P. 6-20.

6-21 Determine the damping matrix for the system presented in Fig. P6-21 and show that it is not proportional.

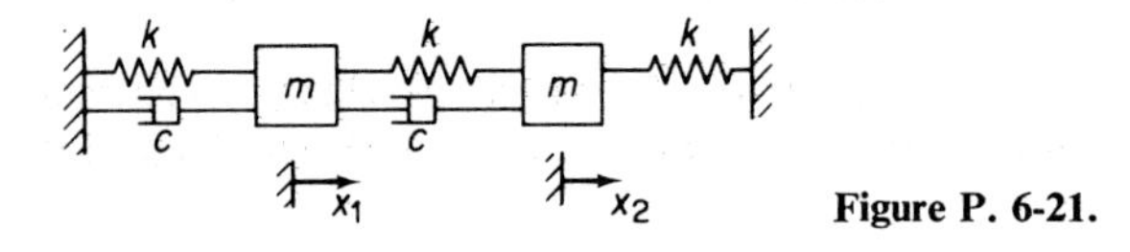

Figure P. 6-21.

6-22 Using the modal matrix $\tilde{P}$ reduce the system of Prob. 6-21 to one which is coupled only by damping and solve by the Laplace transform method.

6-23 Determine the forced steady state response of the system shown in Fig. P6-23.

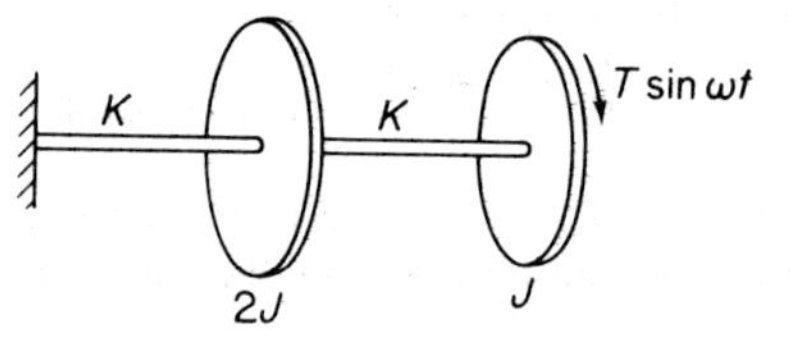

Figure P. 6-23.

6-24 The constants of the system shown in Fig. 6.10-1 are given as

$$\omega_0^2 = \frac{k}{m} = 100, \qquad \beta = \frac{k_1}{m} = 10, \qquad \alpha = \frac{k_1}{c} = 4$$

Determine the system matrix $[A - \lambda I]$.

6-25 For Prob. 6-24, show that the *adj* $[A - \lambda I]$ is

$$\begin{bmatrix} (\lambda^2 + 100) & -100 & \lambda \\ -10 & \lambda(4 + \lambda) + 10 & (4 + \lambda) \\ -10\lambda & -100(4 + \lambda) & \lambda(4 + \lambda) \end{bmatrix}$$

6-26 The characteristic equation for Prob. 6-24 is

$$\lambda^3 + 4\lambda^2 + 110\lambda + 400 = 0$$

which has roots

$$\lambda_1 = -3.676, \qquad \lambda_{2,3} = -0.1619 \pm i10.43$$

Using λ_1, show that each column of the adjoint matrix is proportional to the eigenvector of mode 1, which can be reduced to

$$\begin{Bmatrix} z_1 \\ z_2 \\ z_3 \end{Bmatrix} = \begin{Bmatrix} -11.35 \\ 1.0 \\ -\ 3.676 \end{Bmatrix}$$

6-27 If $\lambda_2 = -0.1619 + i10.43$ is substituted into *adj* $[A - \lambda I]$ of Prob. 6-24, show that each column reduces to the second eigenvector, which is

$$\begin{Bmatrix} z_1 \\ z_2 \\ z_3 \end{Bmatrix} = \begin{Bmatrix} 0.876 + i.338 \\ 1.00 \\ -0.1619 + i10.43 \end{Bmatrix}$$

6-28 Show that the modal matrix of Prob. 6-24 is

$$P = \begin{bmatrix} -11.35 & 0.876 + i.338 & 0.876 - i.338 \\ 1.00 & 1.00 & 1.00 \\ -3.676 & -0.1619 + i10.43 & -0.1619 - i10.43 \end{bmatrix}$$

6-29 Show, by comparing the viscoelastic system of Fig. 6.10-1 to the viscously damped system, that the equivalent viscous damping and the equivalent stiffness are

$$c_{eq} = \frac{c}{1 + \left(\frac{\omega c}{k_1}\right)^2}$$

$$k_{eq} = \frac{k + (k_1 + k)\left(\frac{\omega c}{k_1}\right)^2}{1 + \left(\frac{\omega c}{k_1}\right)^2}$$

6-30 Consider a single degree of freedom viscously damped system

$$m\ddot{x} + c\dot{x} + kx = F \sin \omega t$$

and express this in the state space matrix equation.

6-31 Solve the state space equation of Prob. 6-30 and compare with the solution of the second order equation.

7

Lumped Parameter Systems

7.1 INTRODUCTION

When the degrees of freedom of a system become large, the problem of obtaining numerical results becomes difficult. It is necessary to rely on the high-speed electronic computer for the solution. Although the problem of finding the eigenvalues and eigenvectors of a matrix equation is routinely handled by the electronic computer, there are approximate and other alternative procedures which are often useful. In particular, the concept of breaking up a complicated system into subsystems with simple elastic and dynamic properties is a useful one in rendering manageable systems whose solutions appear obscured in complexity.

In this chapter the basic ideas underlying some of these concepts will be discussed and illustrated by simple examples.

7.2 CHARACTERISTIC EQUATION

Two alternate procedures are available to the eigenvalue, eigenvector problem. The matrix equation

$$M\ddot{X} + KX = 0 \tag{7.2-1}$$

can first be premultipled by M^{-1}, and with the assumption of harmonic motion $\ddot{X} = -\lambda X$, where $\lambda = \omega^2$, we obtain the equation

$$[A - \lambda I]X = 0 \tag{7.2-2}$$

The matrix $A = M^{-1}K$ in the above equation is frequently referred to as the *system matrix* since all of the dynamical properties of the system are defined by this matrix.

As an alternativc to the above equation, we can premultiply Eq. (7.2-1) by K^{-1} to obtain the equation

$$[A^{-1} - \lambda^{-1}I]X = 0 \tag{7.2-3}$$

where $A^{-1} = K^{-1}M$ and K^{-1} is the *flexibility matrix* $[a]$. Thus the characteristic equation from which the eigenvalues are found may be written in either of the following forms

$$|A - \lambda I| = 0 \tag{7.2-4}$$

or

$$|A^{-1} - \lambda^{-1}I| = 0 \tag{7.2-5}$$

Equation (7.2-4) is based on the stiffness formulation, whereas Eq. (7.2-5) is the formulation based on flexibility. The first equation leads to an nth degree algebraic equation in $\lambda = \omega^2$, whereas the second equation results in an nth degree algebraic equation in $\lambda^{-1} = \omega^{-2}$.

7.3 METHOD OF INFLUENCE COEFFICIENTS

Equation (7.2-3) is an abbreviated form of the equations of motion formulated on the basis of the flexibility influence coefficients. We now wish to bring out the details of this equation for further examination.

Starting with Eq. (6.2-1), which is repeated below

$$\begin{aligned} x_1 &= a_{11}f_1 + a_{12}f_2 + a_{13}f_3 \\ x_2 &= a_{21}f_1 + a_{22}f_2 + a_{23}f_3 \\ x_3 &= a_{31}f_1 + a_{32}f_2 + a_{33}f_3 \end{aligned} \tag{6.2-1}$$

we assume harmonic motion and replace the forces f_i by the inertia forces $-m_i\ddot{x}_i = \omega^2 m_i x_i$. Equation (6.2-1) then becomes

$$\begin{Bmatrix} x_1 \\ x_2 \\ x_3 \end{Bmatrix} = \omega^2 \begin{bmatrix} a_{11}m_1 & a_{12}m_2 & a_{13}m_3 \\ a_{21}m_1 & a_{22}m_2 & a_{23}m_3 \\ a_{31}m_1 & a_{32}m_2 & a_{33}m_3 \end{bmatrix} \begin{Bmatrix} x_1 \\ x_2 \\ x_3 \end{Bmatrix} \tag{7.3-1}$$

or

$$\left[A^{-1} - \left(\frac{1}{\omega^2}\right)I\right]\{X\} = 0$$

where A^{-1} is equal to the square matrix on the right side of Eq. (7.3-1).

The characteristic equation is represented by the determinant

$$\begin{vmatrix} \left(a_{11}m_1 - \frac{1}{\omega^2}\right) & (a_{12}m_2) & (a_{13}m_3) \\ (a_{21}m_1) & \left(a_{22}m_2 - \frac{1}{\omega^2}\right) & (a_{23}m_3) \\ (a_{31}m_1) & (a_{32}m_2) & \left(a_{33}m_3 - \frac{1}{\omega^2}\right) \end{vmatrix} = 0 \tag{7.3-3}$$

which on expansion leads to a third degree equation in $(1/\omega^2)$

$$\left(\frac{1}{\omega^2}\right)^3 - (a_{11}m_1 + a_{22}m_2 + a_{33}m_3)\left(\frac{1}{\omega^2}\right)^2 + \cdots = 0 \tag{7.3-3}$$

If the roots of this equation are $1/\omega_1^2$, $1/\omega_2^2$ and $1/\omega_3^2$, the above equation can be factored into the following form

$$\left(\frac{1}{\omega^2} - \frac{1}{\omega_1^2}\right)\left(\frac{1}{\omega^2} - \frac{1}{\omega_2^2}\right)\left(\frac{1}{\omega^2} - \frac{1}{\omega_3^2}\right) = 0$$

or

$$\left(\frac{1}{\omega^2}\right)^3 - \left(\frac{1}{\omega_1^2} + \frac{1}{\omega_2^2} + \frac{1}{\omega_3^2}\right)\left(\frac{1}{\omega^2}\right)^2 + \cdots = 0 \tag{7.3-4}$$

It is evident here that the coefficient of $(1/\omega^2)^2$ is equal to the sum of the roots of the characteristic equation and is also equal to the sum of the diagonal terms of A^{-1}, which is called the *trace* of the matrix (See Appendix C)

$$\text{trace } A^{-1} = \sum_{i=1}^{3} \left(\frac{1}{\omega_i^2}\right) \tag{7.3-5}$$

Example 7.3-1

Using the influence coefficients of Example (6.2-1), determine the matrix equation for the normal modes of the system shown in Fig. 6.2-1.

Solution: The inverse of the system matrix A is

$$A^{-1} = K^{-1}M = [a]M$$
$$= \frac{l^3}{3EI}\begin{bmatrix} 27 & 14 & 4 \\ 14 & 8 & 2.5 \\ 4 & 2.5 & 1 \end{bmatrix}\begin{bmatrix} m_1 & 0 & 0 \\ 0 & m_2 & 0 \\ 0 & 0 & m_3 \end{bmatrix}$$

Equation (7.2-3) then becomes

$$\left[[a][m] - \frac{1}{\omega^2}\begin{bmatrix} 1 & 0 & 0 \\ 0 & 1 & 0 \\ 0 & 0 & 1 \end{bmatrix}\right]\{X\} = 0$$

or

$$\begin{Bmatrix} x_1 \\ x_2 \\ x_3 \end{Bmatrix} = \frac{\omega^2 l^3}{3EI}\begin{bmatrix} 27 & 14 & 4 \\ 14 & 8 & 2.5 \\ 4 & 2.5 & 1 \end{bmatrix}\begin{bmatrix} m_1 & 0 & 0 \\ 0 & m_2 & 0 \\ 0 & 0 & m_3 \end{bmatrix}\begin{Bmatrix} x_1 \\ x_2 \\ x_3 \end{Bmatrix}$$

Example 7.3-2

Given the equation

$$\begin{Bmatrix} x_1 \\ x_2 \\ x_3 \end{Bmatrix} = \begin{bmatrix} a_{11} & a_{12} & a_{13} \\ a_{21} & a_{22} & a_{23} \\ a_{31} & a_{32} & a_{33} \end{bmatrix}\begin{Bmatrix} f_1 \\ f_2 \\ f_3 \end{Bmatrix}$$

determine the stiffness matrix from Cramer's rule.

Solution: From Cramer's rule, f_1 can be written as

$$f_1 = \frac{\begin{vmatrix} x_1 & a_{12} & a_{13} \\ x_2 & a_{22} & a_{23} \\ x_3 & a_{32} & a_{33} \end{vmatrix}}{|a|}$$

Letting $x_1 = 1$ and $x_2 = x_3 = 0$, k_{11} is found to be

$$k_{11} = \frac{\begin{vmatrix} a_{22} & a_{23} \\ a_{32} & a_{33} \end{vmatrix}}{|a|}$$

Letting $x_2 = 1$ and $x_1 = x_3 = 0$, k_{12} is found to be

$$k_{12} = \frac{-\begin{vmatrix} a_{12} & a_{13} \\ a_{32} & a_{33} \end{vmatrix}}{|a|} = k_{21}$$

Similarly, all other terms can be found. It should be noted that the above procedure is simply that of inverting the matrix $[a]$.

7.4 RAYLEIGH PRINCIPLE

For the single degree of freedom conservative system, the natural frequency can be found by equating the maximum kinetic energy to the maximum potential energy. Rayleigh* showed that this procedure can also be applied to systems of higher degrees of freedom provided a reasonable distribution of the deflection is assumed. The method can be conveniently discussed in terms of matrix notation as follows.

Let M and K be the mass and stiffness matrices and X the assumed displacement vector for the amplitude of vibration. Then for harmonic motion, the maximum kinetic and potential energies can be written as

$$T_{\max} = \tfrac{1}{2}\omega^2 X'MX \tag{7.4-1}$$

and

$$U_{\max} = \tfrac{1}{2}X'KX \tag{7.4-2}$$

Equating the two and solving for ω^2, we obtain the Rayleigh quotient

$$\omega^2 = \frac{X'KX}{X'MX} \tag{7.4-3}$$

This quotient approaches the lowest natural frequency (or fundamental frequency) from the high side, and its value is somewhat insensitive to the choice of the assumed amplitudes. To show these qualities, we will express

*John W. Strutt, Baron Rayleigh, *The Theory of Sound* (2nd rev. ed.) (New York: Dover Publications, 1937), Vol. 1, pp. 109–10.

the assumed displacement curve in terms of the normal modes X_i as follows

$$X = X_1 + C_2X_2 + C_3X_3 + \cdots \tag{7.4-4}$$

Then

$$X'KX = X'_1KX_1 + C_2^2X'_2KX_2 + C_3^2X'_3KX_3 + \cdots$$

and

$$X'MX = X'_1MX_1 + C_2^2X'_2MX_2 + C_3^2X'_3MX_3 + \cdots$$

where cross terms of the form X'_iKX_j and X'_iMX_j have been eliminated by the orthogonality conditions.

Noting that

$$X'_iKX_i = \omega_i^2X'_iMX_i \tag{7.4-5}$$

the Rayleigh quotient becomes

$$\omega^2 = \omega_1^2\left\{1 + C_2^2\left(\frac{\omega_2^2}{\omega_1^2} - 1\right)\frac{X'_2MX_2}{X'_1MX_1} + \cdots\right\} \tag{7.4-6}$$

If X'_iMX_i is normalized to the same number, the above equation reduces to

$$\omega^2 = \omega_1^2\left\{1 + C_2^2\left(\frac{\omega_2^2}{\omega_1^2} - 1\right) + \cdots\right\} \tag{7.4-7}$$

It is evident, then, that ω^2 is greater than ω_1^2 because $\omega_2^2/\omega_1^2 > 1$. Since C_2 represents the deviation of the assumed amplitudes from the exact amplitudes X_1, the error in the computed frequency is only proportional to the square of the deviation of the assumed amplitudes from their exact values.

This analysis shows that if the exact fundamental deflection (or mode) X_1 is assumed, the fundamental frequency found by this method will be the correct frequency, since C_2, C_3, etc., will then be zero. For any other curve, the frequency determined will be higher than the fundamental. This fact can be explained on the basis that any deviation from the natural curve requires additional constraints, a condition that implies greater stiffness and higher frequency. In general, the use of the static deflection curve of the elastic body results in a fairly accurate value of the fundamental frequency. If greater accuracy is desired, the approximate curve can be repeatedly improved.

Another form of Rayleigh's quotient that gives a better estimate of the fundamental frequency can be obtained by starting from the equation of motion based on the flexibility influence coefficient

$$\begin{aligned} X &= aM\ddot{X} \\ &= \omega^2 aMX \end{aligned} \tag{7.4-8}$$

Premultiplying by $X'M$ we obtain

$$X'MX = \omega^2 X'MaMX$$

and the Rayleigh quotient becomes

$$\omega^2 = \frac{X'MX}{X'MaMX} \tag{7.4-9}$$

When X is expressed in terms of the normal modes as before, the orthogonality conditions will again eliminate all terms for which $i \neq j$, and the estimate of the fundamental frequency becomes

$$\omega^2 = \omega_1^2 \left\{1 + C_2^2\left(1 - \frac{\omega_1^2}{\omega_2^2}\right) + \cdots \right. \tag{7.4-10}$$

Since $(1 - \omega_1^2/\omega_i^2)$ is smaller than $(\omega_i^2/\omega_1^2 - 1)$ where $\omega_i > \omega_1$, Eq. (7.4-9) results in a better estimate of the fundamental frequency.

In this section we wish to extend Rayleigh's method to beam vibrations. Letting m be the mass per unit length along the beam and y the amplitude of the assumed deflection curve, the kinetic energy is expressed by the equation

$$T_{\max} = \tfrac{1}{2}\int \dot{y}^2\, dm = \tfrac{1}{2}\omega^2 \int y^2\, dm$$

where ω is the fundamental frequency in radians per second.

The potential energy of the beam is determined by the work done on the beam which is stored as elastic energy. Letting M be the bending moment and θ the slope of the elastic curve, the work done is equal to

$$U = \tfrac{1}{2}\int M\, d\theta$$

Since the deflection in beams is generally small, the following geometric relations are assumed to hold (See Fig. 7.4-1).

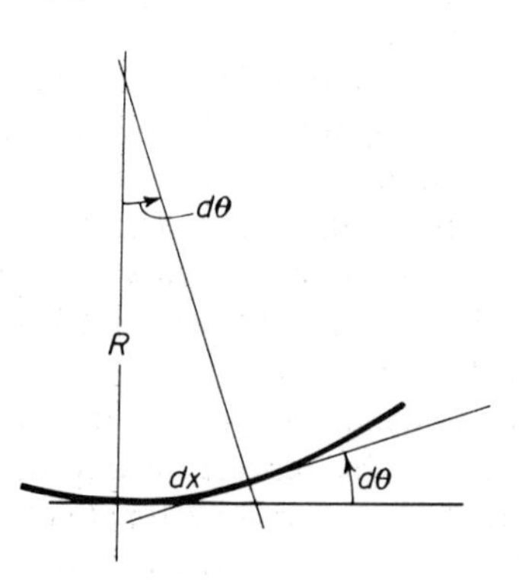

Figure 7.4-1.

$$\theta = \frac{dy}{dx}, \qquad \frac{1}{R} = \frac{d\theta}{dx} = \frac{d^2y}{dx^2}$$

In addition to these relations we have, from the theory of beams, the flexure formula

$$\frac{1}{R} = \frac{M}{EI}$$

where EI is the flexural rigidity of the beam and R is the radius of curvature. Substituting for $d\theta$ and $1/R$, U may be written as

$$U_{\max} = \frac{1}{2}\int \frac{M^2}{EI}\,dx = \frac{1}{2}\int EI\left(\frac{d^2y}{dx^2}\right)^2 dx$$

Equating the kinetic and potential energies, the fundamental frequency of the beam is determined from the equation

$$\omega^2 = \frac{\int EI(d^2y/dx^2)^2\,dx}{\int y^2\,dm} \tag{7.4-11}$$

Example 7.4-1

In applying this procedure to a simply supported beam of uniform cross section, shown in Fig. 7.4-2, we assume the deflection to be represented by a sine curve as follows

$$y = \left(y_0 \sin\frac{\pi x}{l}\right)\sin\omega t$$

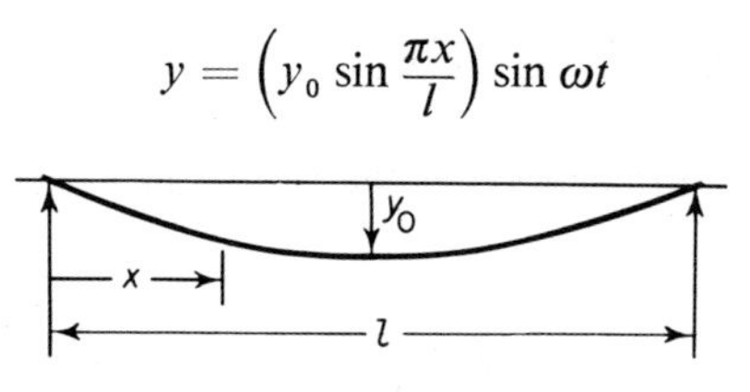

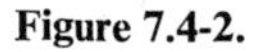
Figure 7.4-2.

where y_0 is the maximum deflection at mid-span. The second derivative then becomes

$$\frac{d^2y}{dx^2} = -\left(\frac{\pi}{l}\right)^2 y_0 \sin\frac{\pi x}{l}\sin\omega t$$

Substituting into Eq. (7.4-11) we obtain

$$\omega^2 = \frac{EI\left(\frac{\pi}{l}\right)^4 \int_0^l \sin^2\frac{\pi x}{l}\,dx}{\frac{w}{g}\int_0^l \sin^2\frac{\pi x}{l}\,dx} = \pi^4\frac{gEI}{wl^4}$$

The fundamental frequency is therefore found to be

$$\omega_1 = \pi^2\sqrt{gEI/wl^4}$$

In this case the assumed curve happened to be the correct curve, and the exact frequency is obtained by Rayleigh's method. Any other curve assumed for the case will result in a constant greater than π^2 in the frequency equation.

Example 7.4-2

If the distance between the ends of the beam of Fig. 7.4-2 is rigidly fixed, a tensile stress σ will be developed by the lateral deflection. Account for this additional strain energy in the frequency equation.

Solution: Due to the lateral deflection, the length dx of the beam is increased by an amount

$$[\sqrt{1 + (dy/dx)^2} - 1]\,dx \cong \frac{1}{2}\left(\frac{dy}{dx}\right)^2 dx$$

The additional strain energy in the element dx is

$$dU = \tfrac{1}{2}\sigma A\epsilon\,dx = \tfrac{1}{2}EA\epsilon^2\,dx$$

where A is the cross-sectional area, σ the stress due to tension, and $\epsilon = \frac{1}{2}(dy/dx)^2$ is the unit strain.

Equating the kinetic energy to the total strain energy of bending and tension, we obtain

$$\frac{1}{2}\omega^2 \int y^2\,dm = \frac{1}{2}\int EI\left(\frac{d^2y}{dx^2}\right)^2 dx + \frac{1}{2}\int \frac{EA}{4}\left(\frac{dy}{dx}\right)^4 dx$$

The above equation then leads to the frequency equation

$$\omega_1^2 = \frac{\displaystyle\int EI\left(\frac{d^2y}{dx^2}\right)^2 dx + \int \frac{EA}{4}\left(\frac{dy}{dx}\right)^4 dx}{\displaystyle\int y^2\,dm}$$

which contains an additional term due to the tension.

Example 7.4-3

Consider next the cantilever beam shown in Fig. 7.4-3. We will assume here that the amplitude of the beam at any point x is given with sufficient accuracy by the statical deflection curve of a massless cantilever beam with a concentrated load at the end. Writing this equation in the form

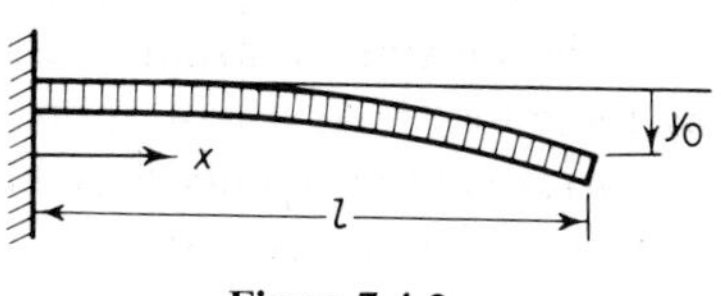

Figure 7.4-3.

$$y = \tfrac{1}{2}y_0\left[3\left(\frac{x}{l}\right)^2 - \left(\frac{x}{l}\right)^3\right]$$

where $y_0 = Pl^3/3EI$ is the amplitude of the free end, the stiffness at the free end becomes $k = P/y_0 = 3EI/l^3$. The potential energy which is equal to the work done, is then

$$U_{\max}{}^* = \tfrac{1}{2}ky_0^2 = \frac{3EI}{2l^3}\,y_0^2$$

The kinetic energy is next determined by integrating one half the product of the mass and the square of the velocity over the length of the beam

$$T_{\max} = \frac{w}{2g}\int_0^l (\omega y)^2\,dx = \frac{w}{2g}\left(\frac{\omega y_0}{2}\right)^2\int_0^l\left[3\left(\frac{x}{l}\right)^2 - \left(\frac{x}{l}\right)^3\right]^2 dx$$
$$= \tfrac{1}{2}\left(\frac{33wl}{140g}\right)\omega^2 y_0^2$$

The above equation indicates that for the assumed deflection curve, the continuous beam of w lb/ft is equivalent in vibration characteristics to that of a weightless beam with a concentrated weight ($\frac{33}{140}\,wl$) at the end.

Equating the two energies, the fundamental frequency of vibration in radians per second becomes

$$\omega_1 = \sqrt{\frac{(3EI/l^3)g}{\frac{33}{140}wl}} = 3.56\sqrt{gEI/wl^4}$$

The exact solution for this case is

$$\omega_1 = 3.515\sqrt{gEI/wl^4}$$

In general, the deflection curve assumed for the problem should satisfy the boundary conditions of deflection, slope, shear, and moment. These conditions are satisfied by the static deflection curve which generally results in a frequency of acceptable accuracy.

*This result can also be found from the equation $U_{\max} = \frac{1}{2}\int_0^l EI\left(\frac{dy^2}{dx^2}\right)^2 dx$.

If a beam is represented by a series of lumped weights $W_1, W_2, W_3, \ldots$, the maximum strain energy can be determined from the work done by these loads. As a first approximation, the static deflection $y_1, y_2, y_3, \ldots$ of corresponding points may be used, in which case the maximum kinetic and potential energies are

$$T_{\max} = \frac{1}{2}\frac{\omega^2}{g}[W_1 y_1^2 + W_2 y_2^2 + W_3 y_3^2 + \cdots] \tag{7.4-12}$$

$$U_{\max} = \tfrac{1}{2}[W_1 y_1 + W_2 y_2 + W_3 y_3 + \cdots] \tag{7.4-13}$$

By equating the two, the frequency equation is established as

$$\omega_1^2 = \frac{g \sum Wy}{\sum Wy^2} \tag{7.4-14}$$

Example 7.4-4

To illustrate the use of this equation, we will find the first approximation to the fundamental frequency of lateral vibration for the system shown in Fig. 7.4-4.

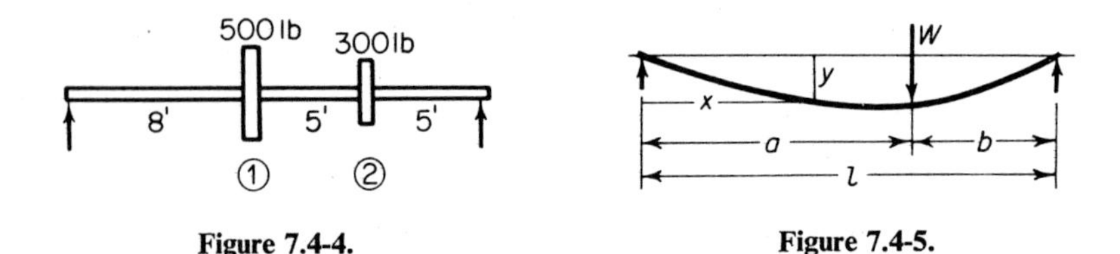

Figure 7.4-4. **Figure 7.4-5.**

Referring to Fig. 7.4-5, the deflection at any point x, due to a concentrated load W a distance a and b from the ends, can be determined from the equation

$$y_x = \frac{Wbx}{6EIl}(l^2 - x^2 - b^2)$$

which can be found in any standard text on strength of materials.* The deflections at the loads can be obtained from the super-position of the two loads, shown in Fig. 7.4-6

$$y_1' = \frac{300 \times 5 \times 8}{6 \times 18 \times EI}(18^2 - 8^2 - 5^2) \times 12^3 = \frac{45.2 \times 10^6}{EI} \text{ in.}$$

$$y_2' = \frac{300 \times 5 \times 13}{6 \times 18 \times EI}(18^2 - 13^2 - 5^2) \times 12^3 = \frac{40.7 \times 10^6}{EI} \text{ in.}$$

*Egor P. Popov, *Introduction to Mechanics of Solids* (Englewood Cliffs, N. J.: Prentice-Hall, Inc,. 1968).

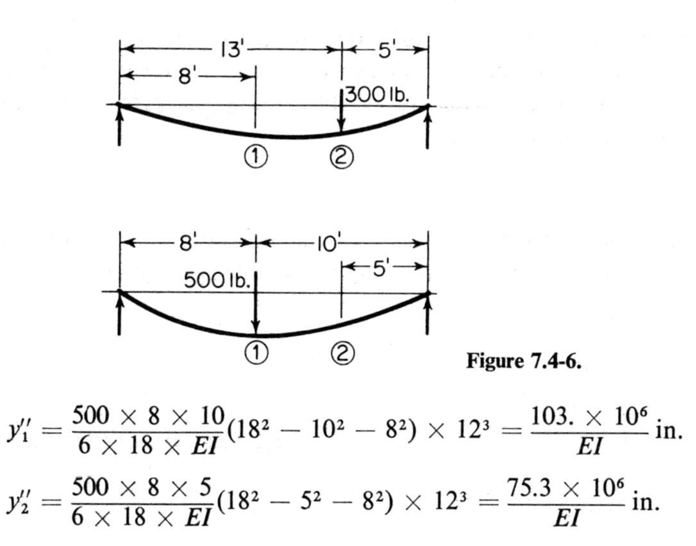

Figure 7.4-6.

$$y_1'' = \frac{500 \times 8 \times 10}{6 \times 18 \times EI}(18^2 - 10^2 - 8^2) \times 12^3 = \frac{103. \times 10^6}{EI} \text{ in.}$$

$$y_2'' = \frac{500 \times 8 \times 5}{6 \times 18 \times EI}(18^2 - 5^2 - 8^2) \times 12^3 = \frac{75.3 \times 10^6}{EI} \text{ in.}$$

Adding y' and y'', the deflections at 1 and 2 become

$$y_1 = 148 \times \frac{10^6}{EI}, \qquad y_2 = 116 \times \frac{10^6}{EI}$$

Substituting into Eq. (7.4-14), the first approximation for the fundamental frequency is

$$\omega_1 = \sqrt{\frac{g \sum Wy}{\sum Wy^2}} = \sqrt{\frac{386(500 \times 148 + 300 \times 116)EI}{(500 \times 148^2 + 300 \times 116^2)10^6}}$$
$$= 0.0017\sqrt{EI} \text{ rad/sec}$$

If further accuracy is desired, a better approximation to the dynamic curve can be made by using dynamic loads in place of the static weights. Since the dynamic load is $m\omega^2 y$, which is proportional to the deflection, we can recalculate the deflection with the modified weights W_1 and $W_2(y_2/y_1)$.

The concept of dynamic loads can also be used, starting with a much simpler curve than the static curve. Assuming such a curve to be $y(x)$, the dynamic loading per unit length is $\omega^2 m(x)y(x)$ which must equal the change in shear along the beam

$$dV = \omega^2 m(x)y(x)\, dx$$

as shown in Fig. 7.4-7. Since $dM = V\,dx$, the moment M can be found by integrating and substituting into the equation

$$U_{\max} = \frac{1}{2}\int \frac{M^2}{EI}\, dx$$

which then is proportional to ω^4. Actually the equation for T_{max} is not so sensitive to the inaccuracies of the assumed curve, whereas the strain energy which depends on the curvature, could be very much in error and hence must be computed with care.

Example 7.4-5

Determine the fundamental frequency of the uniform cantilever beam shown in Fig. 7.4-8, using the simple curve $y = cx^2$.

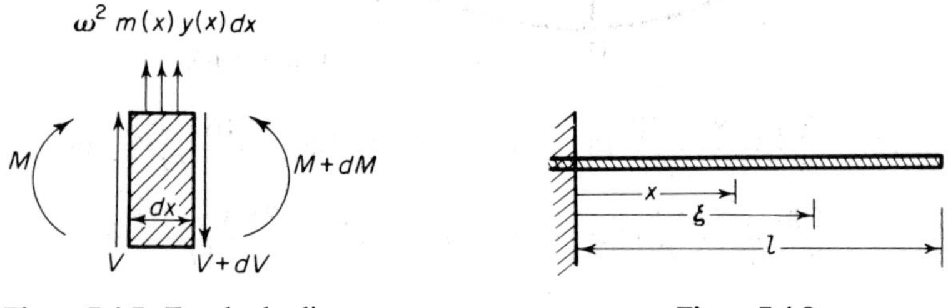

Figure 7.4-7. Free-body diagram of beam element.

Figure 7.4-8.

Solution: If we use Eq. (7.4-11), we would find the result to be very much in error since the above curve does not satisfy the boundary conditions at the free end. By using Eq. (7.4-11) we obtain

$$\omega = 4.47\sqrt{EI/ml^4}$$

whereas the exact value is

$$\omega_1 = 3.52\sqrt{EI/ml^4}$$

Acceptable results using the given curve can be found by the procedure outlined in the previous section.

$$V(\xi) = \omega^2 \int_\xi^l mc\xi^2 \, d\xi = \frac{\omega^2 mc}{3}(l^3 - \xi^3)$$

and the bending moment becomes

$$M(x) = \int_x^l V(\xi)\, d\xi = \frac{\omega^2 mc}{3}\int_x^l (l^3 - \xi^3)\, d\xi$$

$$= \frac{\omega^2 mc}{12}(3l^4 - 4l^3 x + x^4)$$

The maximum strain energy is found by substituting $M(x)$ into U_{max}.

$$U_{\max} = \frac{1}{2EI}\left(\frac{\omega^2 mc}{12}\right)^2 \int_0^l (3l^4 - 4l^3 x + x^4)^2\, dx$$

$$= \frac{\omega^4}{2EI}\frac{m^2c^2}{144}\frac{312}{135}l^9$$

The maximum kinetic energy is

$$T_{\max} = \tfrac{1}{2}\int_0^l \dot{y}^2 m\, dx = \tfrac{1}{2}c^2\omega^2 m \int_0^l x^4\, dx = \tfrac{1}{2}c^2\omega^2 m \frac{l^5}{5}$$

By equating these results, we obtain

$$\omega_1 = \sqrt{12.47\, EI/ml^4} = 3.53\sqrt{EI/ml^4}$$

which is very close to the exact result.

7.5 DUNKERLEY'S FORMULA*

Rayleigh's principle, which gives the upper bound to the fundamental frequency, can now be complemented by Dunkerley's formula, which results in a lower bound to the fundamental frequency. From Eqs. (7.3-3) and (7.3-4), it is easily seen that for an n-degree of freedom system, the relation is

$$\frac{1}{\omega_1^2} + \frac{1}{\omega_2^2} + \cdots + \frac{1}{\omega_n^2} = a_{11}m_1 + a_{22}m_2 + \cdots + a_{nn}m_n \qquad (7.5\text{-}1)$$

Since a_{ii} is the influence coefficient equal to the deflection at i resulting from a unit load at i, its reciprocal must be the stiffness coefficient k_{ii}, equal to the force per unit deflection at i. Also

$$\omega_{ii} = \sqrt{\frac{k_{ii}}{m_i}} \qquad (7.5\text{-}2)$$

is the natural frequency of the system when only m_i is present. We can therefore write Eq. (7.5-1) in the following equivalent forms

$$\frac{1}{\omega_1^2} + \frac{1}{\omega_2^2} + \cdots + \frac{1}{\omega_n^2} = a_{11}m_1 + a_{22}m_2 + \cdots + a_{nn}m_n$$

$$= \frac{m_1}{k_{11}} + \frac{m_2}{k_{22}} + \cdots + \frac{m_n}{k_{nn}} \qquad (7.5\text{-}3)$$

$$= \frac{1}{\omega_{11}^2} + \frac{1}{\omega_{22}^2} + \cdots + \frac{1}{\omega_{nn}^2}$$

*S. Dunkerley, "On the Whirling and Vibration of Shafts," *Phil. Trans. Roy. Soc.*, 185 (1895), 269–360.

The estimate to the fundamental frequency is made by recognizing that ω_2, ω_3, etc., are natural frequencies of higher modes and hence $1/\omega_2^2$, $1/\omega_3^2$, etc., can be neglected in the left side of Eq. (7.5-3). By neglecting these terms, $1/\omega_1^2$ is larger than its true value and therefore ω_1 is smaller than the exact value of the fundamental frequency.

Example 7.5-1

Dunkerley's equation is useful for estimating the fundamental frequency of a structure undergoing vibration testing. Natural frequencies of structures are often determined by attaching to the structure an eccentric mass exciter, and noting the frequencies corresponding to the maximum amplitude. The frequencies so measured represent those of the structure plus exciter and may deviate considerably from the natural frequencies of the structure itself when the mass of the exciter is a substantial percentage of the total mass. In such cases the fundamental frequency of the structure by itself may be determined by the following equation

$$\frac{1}{\omega_1^2} = \frac{1}{\omega_{11}^2} + \frac{1}{\omega_{22}^2} \qquad (a)$$

where ω_1 = fundamental frequency of structure plus exciter,
ω_{11} = fundamental frequency of the structure by itself,
ω_{22} = natural frequency of exciter mounted on the structure in the absence of other masses.

It is sometimes convenient to express this equation in another form, for instance

$$\frac{1}{\omega_1^2} = \frac{1}{\omega_{11}^2} + a_{22}m_2 \qquad (b)$$

where m_2 is the mass of the concentrated weight or exciter and a_{22} the influence coefficient of the structure at the point of attachment of the exciter.

Example 7.5-2

An airplane rudder tab showed a resonant frequency of 30 cps when vibrated by an eccentric mass shaker weighing 1.5 lb. By attaching an additional weight of 1.5 lb to the shaker, the resonant frequency was lowered to 24 cps. Determine the true natural frequency of the tab.

Solution: The measured resonant frequencies are those due to the total mass of the tab and shaker. Letting f_{11} be the true natural frequency of the tab, and substituting into Eq. (*b*) of Example 7.5-1, we obtain

$$\frac{1}{(2\pi \times 30)^2} = \frac{1}{(2\pi f_{11})^2} + \frac{1.5}{386} a_{22}$$

$$\frac{1}{(2\pi \times 24)^2} = \frac{1}{(2\pi f_{11})^2} + \frac{3.0}{386} a_{22}$$

Eliminating a_{22}, the true natural frequency is

$$f_{11} = 45.3 \text{ cps}$$

The rigidity of stiffness of the tab at the point of attachment of the shaker may be determined from $1/a_{22}$ which from the same equations is found to be

$$k_2 = \frac{1}{a_{22}} = \frac{1}{0.00407} = 246 \text{ lb/in.}$$

Example 7.5-3

Determine the fundamental frequency of a uniformly loaded cantilever beam with a concentrated mass M at the end, equal to the mass of the uniform beam (See Fig. 7.5-1).

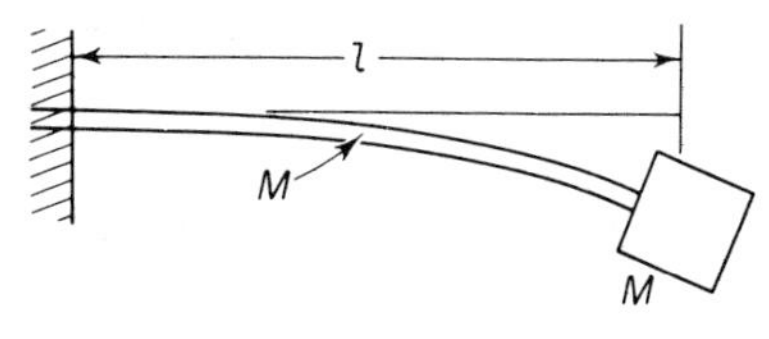

Figure 7.5-1.

Solution: The frequency equation for the uniformly loaded beam by itself is

$$\omega_{11}^2 = 3.515\left(\frac{EI}{Ml^3}\right)$$

For the concentrated mass by itself attached to a weightless cantilever beam, we have

$$\omega_{22}^2 = 3.00\left(\frac{EI}{Ml^3}\right)$$

Substituting into Dunkerley's formula rearranged in the following form, the natural frequency of the system is determined as

$$\omega_1^2 = \frac{\omega_{11}^2 \omega_{22}^2}{\omega_{11}^2 + \omega_{22}^2} = \frac{3.515 \times 3.0}{3.515 + 3.0}\left(\frac{EI}{Ml^3}\right) = 2.41\left(\frac{EI}{Ml^3}\right)$$

This result may be compared to the frequency equation obtained by

Rayleigh's method which is

$$\omega_1^2 = \frac{3EI}{(1 + \frac{33}{140})Ml^3} = 2.43\left(\frac{EI}{Ml^3}\right)$$

Example 7.5-4

The natural frequency of a given airplane wing in torsion is 1600 cpm. What will be the new torsional frequency if a 1000-lb fuel tank is hung at a position one-sixth of the semi-span from the center line of the airplane such that its moment of inertia about the torsional axis is 1800 lb in./sec² ? The torsional stiffness of the wing at this point is 60×10^6 lb in/rad.

Solution: The frequency of the tank attached to the weightless wing is

$$f_{22} = \frac{1}{2\pi}\sqrt{\frac{60 \times 10^6}{1800}} = 29.1 \text{ cps} = 1745 \text{ cpm}$$

The new torsional frequency with the tank, from Eq. (*a*) of Example 7.5-1 then becomes

$$\frac{1}{f_1^2} = \frac{1}{1600^2} + \frac{1}{1745^2}, \qquad f_1 = 1180 \text{ cpm}$$

Example 7.5-5

The fundamental frequency of a uniform beam of mass M, simply supported as in Fig. 7.5-2 is equal to $\pi^2\sqrt{EI/Ml^3}$. If a lumped mass m_0 is attached to the beam at $x = l/3$, determine the new fundamental frequency.

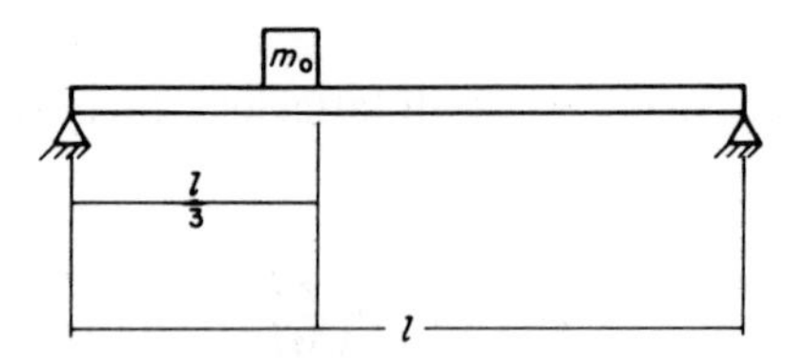

Figure 7.5-2.

Solution: Starting with Eq. (*b*) of Example 7.5-1, we let ω_{11} be the fundamental frequency of the uniform beam and ω_1 the new fundamental frequency with m_0 attached to the beam. Multiplying through Eq. (*b*) by ω_1^2, we have

$$1 = \left(\frac{\omega_1}{\omega_{11}}\right)^2 + a_{22}m_0\omega_{11}^2\left(\frac{\omega_1}{\omega_{11}}\right)^2$$

or

$$\left(\frac{\omega_1}{\omega_{11}}\right)^2 = \frac{1}{1 + a_{22} m_0 \omega_{11}^2}$$

The quantity a_{22} is the influence coefficient at $x = l/3$ due to a unit load applied at the same point. It can be found from the beam formula in Example 7.4-4 to be

$$a_{22} = \frac{8}{6 \times 81} \frac{l^3}{EI}$$

Substituting $\omega_{11}^2 = \pi^4 EI/Ml^3$ together with a_{22}, we obtain the convenient formula

$$\left(\frac{\omega_1}{\omega_{11}}\right)^2 = \frac{1}{1 + \frac{8\pi^4}{6 \times 81} \frac{m_0}{M}} = \frac{1}{1 + 1.6 \frac{m_0}{M}}$$

7.6 METHOD OF MATRIX ITERATION

The equations of motion, formulated either on the basis of the stiffness equation or the flexibility equation, are similar in form and appear as

$$\begin{Bmatrix} x_1 \\ x_2 \\ \cdot \\ \cdot \\ \cdot \\ x_n \end{Bmatrix} = \lambda \begin{bmatrix} a_{11} & a_{12} & \cdots & a_{1n} \\ a_{21} & a_{22} & \cdots & \\ \cdot & & & \\ \cdot & & & \\ \cdot & & & \\ a_{n1} & a_{n2} & \cdots & a_{nn} \end{bmatrix} \begin{Bmatrix} x_1 \\ x_2 \\ \cdot \\ \cdot \\ \cdot \\ x_n \end{Bmatrix} \qquad (7.6\text{-}1)$$

where λ is equal to $1/\omega^2$ for the stiffness formulation, and ω^2 for the flexibility formulation.

The iteration procedure is started by assuming a set of deflections for the right column of Eq. (7.6-1) and performing the indicated operation, which results in a column of numbers. This is then normalized by making one of the amplitudes equal to unity and dividing each term of the column by the particular amplitude which was normalized. The procedure is then repeated with the normalized column until the amplitudes stabilize to a definite pattern.

As will be shown in Sec. 7.7, the iteration process converges to the lowest value of λ so that for the equation formulated on the flexibility influence coefficients, the fundamental or the lowest mode of vibration is found. Likewise, for the equation formulated on the basis of the stiffness influence coefficients, the convergence is to the highest mode which corresponds to the lowest value of $\lambda = 1/\omega^2$.

Example 7.6-1

The uniform beam of Fig. 7.6-1, free to vibrate in the plane shown, has two concentrated weights $W_1 = 500$ lb and $W_2 = 100$ lb. Determine the fundamental frequency of the system.

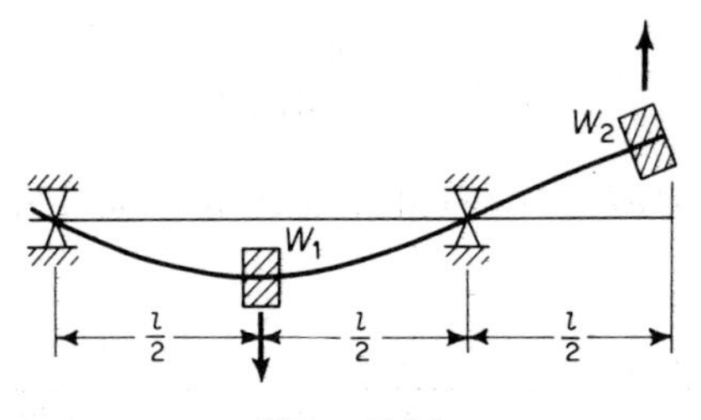

Figure 7.6-1.

Solution: The influence coefficients for this problem, determined from deflection equations of beams by placing a unit load at positions 1 and 2, are

$$a_{11} = \frac{l^3}{48EI} = \tfrac{1}{6}a_{22}, \qquad a_{12} = a_{21} = \frac{l^3}{32EI} = \tfrac{1}{4}a_{22}, \qquad a_{22} = \frac{l^3}{8EI}$$

Substituting into Eq. (7.3-1)

$$\begin{bmatrix} x_1 \\ x_2 \end{bmatrix} = \frac{\omega^2 l^3}{8EIg}\begin{bmatrix} \frac{500}{6} & \frac{100}{4} \\ \frac{500}{4} & 100 \end{bmatrix}\begin{bmatrix} x_1 \\ x_2 \end{bmatrix}$$

Starting with $x_1 = x_2 = 1.0$ for the right column, we obtain

$$\begin{bmatrix} x_1 \\ x_2 \end{bmatrix} = \frac{\omega^2 l^3}{8EIg}\begin{bmatrix} 108.3 \\ 225.0 \end{bmatrix} = \frac{108.3\omega^2 l^3}{8EIg}\begin{bmatrix} 1.00 \\ 2.08 \end{bmatrix}$$

If the procedure is repeated with $x_1 = 1.0$ and $x_2 = 2.08$, the second result is

$$\begin{bmatrix} x_1 \\ x_2 \end{bmatrix} = \frac{\omega^2 l^3}{8EIg}\begin{bmatrix} 135.3 \\ 333.0 \end{bmatrix} = \frac{135.3\omega^2 l^3}{8EIg}\begin{bmatrix} 1.00 \\ 2.46 \end{bmatrix}$$

By repeating the procedure a few more times the deflections will converge to

$$\begin{bmatrix} 1.00 \\ 2.60 \end{bmatrix} = \frac{148.3\omega^2 l^3}{8EIg}\begin{bmatrix} 1.00 \\ 2.60 \end{bmatrix}$$

The fundamental frequency from the above equation using $g = 386$ in./sec^2

is

$$\omega = \sqrt{8EIg/148.3l^3} = 4.56\sqrt{EI/l^3}$$

and the amplitude ratio is found to be

$$\frac{x_1}{x_2} = \frac{1.0}{2.60}$$

If only the fundamental frequency is of interest, sufficient accuracy can be obtained from the results of the first and second iterations. From the first iteration the inertia forces are $500\omega^2/g$ and $208\omega^2/g$. These forces produce deflections obtained in the second iteration, which are $x_1 = 135.3\omega^2 l^3/8EIg = 16.92\omega^2 l^3/EIg$ and $x_2 = 2.46x_1$. The work done by these forces is then

$$U = \frac{1}{2}(500 + 208 \times 2.46)\frac{\omega^2}{g}x_1 = \frac{1}{2} \times 1012 \times \frac{\omega^2 x_1}{g}$$

and the corresponding kinetic energy is

$$T = \frac{1}{2}(500 + 100 \times 2.46^2)\frac{\omega^2}{g}x_1^2 = \frac{1}{2} \times 1105 \times \frac{\omega^2 x_1^2}{g}$$

Equating the two, the fundamental frequency is found as

$$\omega = \sqrt{\frac{1012 \times 386}{1105 \times 16.92}\frac{EI}{l^3}} = 4.57\sqrt{\frac{EI}{l^3}}$$

7.7 CALCULATION OF HIGHER MODES

When the equations of motion are formulated in terms of the flexibility influence coefficients, the iteration procedure converges to the lowest mode present in the assumed deflection. It is evident that if the lowest mode is absent in the assumed deflection, the iteration technique will converge to the next lowest, or the second, mode.

Letting the assumed curve X be expressed by the sum of the normal modes X_i

$$X = C_1X_1 + C_2X_2 + C_3X_3 + \cdots \tag{7.7-1}$$

To distinguish between the assumed curve X and the normal modes X_i in the above equation, we will designate the normal modes as

$$X_i = \begin{Bmatrix} x_1 \\ x_2 \\ x_3 \end{Bmatrix}_i$$

and the assumed curve as

$$X = \begin{Bmatrix} \bar{x}_1 \\ \bar{x}_2 \\ \bar{x}_3 \end{Bmatrix}$$

We will now impose the condition $C_1 = 0$ to remove the first mode from the assumed deflection X. For this, we introduce the orthogonality relationship by premultiplying Eq. (7.7-1) by X'_1M, which eliminates all terms on the right side except the first term.

$$X'_1MX = C_1X'_1MX_1 \tag{7.7-2}$$

Equating the left side of the above equation to zero, C_1 becomes zero and the first mode is eliminated from Eq. (7.7-1).

$$\begin{aligned} X'_1MX &= (x_1 \quad x_2 \quad x_3)\begin{bmatrix} m_1 & 0 & 0 \\ 0 & m_2 & 0 \\ 0 & 0 & m_3 \end{bmatrix}\begin{Bmatrix} \bar{x}_1 \\ \bar{x}_2 \\ \bar{x}_3 \end{Bmatrix} = 0 \\ &= m_1x_1\bar{x}_1 + m_2x_2\bar{x}_2 + m_3x_3\bar{x}_3 = 0 \end{aligned} \tag{7.7-3}$$

From the above equation, we obtain

$$\begin{aligned} \bar{x}_1 &= -\frac{m_2}{m_1}\left(\frac{x_2}{x_1}\right)\bar{x}_2 - \frac{m_3}{m_1}\left(\frac{x_3}{x_1}\right)\bar{x}_3 \\ \bar{x}_2 &= \bar{x}_2 \\ \bar{x}_3 &= \bar{x}_3 \end{aligned} \tag{7.7-4}$$

where the last two equations in the above set are introduced merely as identities. Rewriting in matrix form, Eq. (7.7-4) becomes

$$\begin{aligned} \{X\} &= \begin{bmatrix} 0 & -\frac{m_2}{m_1}\left(\frac{x_2}{x_1}\right) & -\frac{m_3}{m_1}\left(\frac{x_3}{x_1}\right) \\ 0 & 1 & 0 \\ 0 & 0 & 1 \end{bmatrix}\{X\} \\ &= SX \end{aligned} \tag{7.7-5}$$

Since this equation is the result of $C_1 = 0$, the first mode has been swept out of the assumed deflection by the *sweeping matrix S*. When this equation is substituted into the original matrix equation

$$X = \omega^2 aMX \tag{7.7-6}$$

the result is

$$X = \omega^2 aMSX \tag{7.7-7}$$

The iteration procedure applied to Eq. (7.7-7) will converge to the second mode.

For the third and higher modes, the sweeping procedure is repeated, making $C_1 = C_2 = 0$, etc. This reduces the order of the matrix equation by one each time; however, the convergence for higher modes becomes more critical if impurities are introduced through the sweeping matrices. It is well to check the highest mode by the inversion of the original matrix equation, which should be equal to the equation formulated in terms of the stiffness influence coefficients.

Example 7.7-1

Write the matrix equation, based on flexibility influence coefficients, for the system shown in Fig. 7.7-1 and determine all the natural modes.

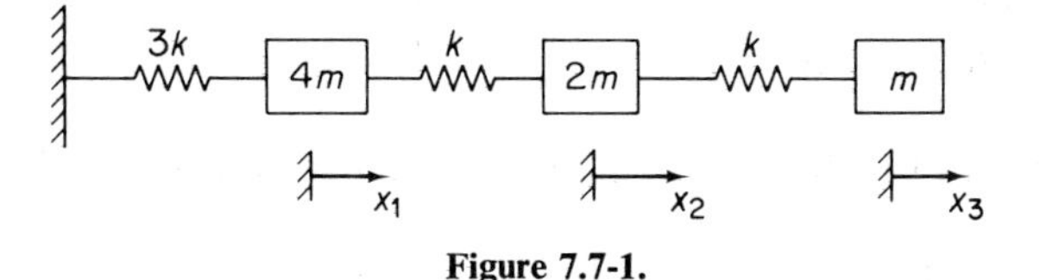

Figure 7.7-1.

Solution: The influence coefficients are found by applying a unit load, one at a time, to points 1, 2, and 3.

$$a_{11} = a_{21} = a_{12} = a_{31} = a_{13} = \frac{1}{3k}$$

$$a_{22} = a_{32} = a_{23} = \left(\frac{1}{3k} + \frac{1}{k}\right) = \frac{4}{3k}$$

$$a_{33} = \frac{1}{3k} + \frac{1}{k} + \frac{1}{k} = \frac{7}{3k}$$

The equations of motion in matrix form are then

$$\begin{Bmatrix} x_1 \\ x_2 \\ x_3 \end{Bmatrix} = \frac{\omega^2}{3k} \begin{bmatrix} 1 & 1 & 1 \\ 1 & 4 & 4 \\ 1 & 4 & 7 \end{bmatrix} \begin{bmatrix} 4m & 0 & 0 \\ 0 & 2m & 0 \\ 0 & 0 & m \end{bmatrix} \begin{Bmatrix} x_1 \\ x_2 \\ x_3 \end{Bmatrix}$$

$$\begin{Bmatrix} x_1 \\ x_2 \\ x_3 \end{Bmatrix} = \frac{\omega^2 m}{3k} \begin{bmatrix} 4 & 2 & 1 \\ 4 & 8 & 4 \\ 4 & 8 & 7 \end{bmatrix} \begin{Bmatrix} x_1 \\ x_2 \\ x_3 \end{Bmatrix}$$

Starting with arbitrary values of $x_1x_2x_3$, the above equation converges to the first mode, which is

$$\begin{Bmatrix} x_1 \\ x_2 \\ x_3 \end{Bmatrix} = \frac{\omega^2 m}{3k} 14.32 \begin{Bmatrix} 0.25 \\ 0.79 \\ 1.00 \end{Bmatrix}$$

The fundamental frequency is then found to be

$$\omega_1 = \sqrt{\frac{3k}{14.32m}} = 0.457\sqrt{\frac{k}{m}}$$

To determine the second mode, we form the sweeping matrix given by Eq. (7.7-5)

$$S = \begin{bmatrix} 0 & -\frac{1}{2}\left(\frac{0.79}{0.25}\right) & -\frac{1}{4}\left(\frac{1.00}{0.25}\right) \\ 0 & 1 & 0 \\ 0 & 0 & 1 \end{bmatrix} = \begin{bmatrix} 0 & -1.58 & -1 \\ 0 & 1 & 0 \\ 0 & 0 & 1 \end{bmatrix}$$

The new equation for the second mode iteration is from Eq. (7.7-6)

$$\begin{Bmatrix} x_1 \\ x_2 \\ x_3 \end{Bmatrix} = \frac{\omega^2 m}{3k} \begin{bmatrix} 4 & 2 & 1 \\ 4 & 8 & 4 \\ 4 & 8 & 7 \end{bmatrix} \begin{bmatrix} 0 & -1.58 & -1 \\ 0 & 1 & 0 \\ 0 & 0 & 1 \end{bmatrix} \begin{Bmatrix} x_1 \\ x_2 \\ x_3 \end{Bmatrix}$$

$$= \frac{\omega^2 m}{3k} \begin{bmatrix} 0 & -4.32 & -3.0 \\ 0 & 1.67 & 0 \\ 0 & 1.67 & 3.0 \end{bmatrix} \begin{Bmatrix} x_1 \\ x_2 \\ x_3 \end{Bmatrix}$$

Starting the iteration process with arbitrary amplitudes, the above equation converges to the second mode, which is

$$\begin{Bmatrix} x_1 \\ x_2 \\ x_3 \end{Bmatrix} = \frac{\omega^2 m}{3k} 3 \begin{Bmatrix} -1.0 \\ 0 \\ 1.0 \end{Bmatrix}$$

The natural frequency of the second mode is therefore found to be

$$\omega_2 = \sqrt{\frac{k}{m}}$$

For the determination of the third mode, we impose the conditions

$C_1 = C_2 = 0$ from the orthogonality equation (7.7-3)

$$C_1 = \sum_{i=1}^{3} m_i(x_i)_1 \bar{x}_i = 4(0.25)\bar{x}_1 + 2(0.79)\bar{x}_2 + 1(1.0)\bar{x}_3 = 0$$

$$C_2 = \sum_{i=1}^{3} m_i(x_i)_2 \bar{x}_i = 4(-1.0)\bar{x}_1 + 2(0)\bar{x}_2 + 1(1.0)\bar{x}_3 = 0$$

From these two equations we obtain

$$\bar{x}_1 = 0.25\bar{x}_3 \qquad \bar{x}_2 = -0.79\bar{x}_3$$

which can be expressed by the matrix equation

$$\begin{Bmatrix} x_1 \\ x_2 \\ x_3 \end{Bmatrix} = \begin{bmatrix} 0 & 0 & 0.25 \\ 0 & 0 & -0.79 \\ 0 & 0 & 1.00 \end{bmatrix} \begin{Bmatrix} \bar{x}_1 \\ \bar{x}_3 \\ \bar{x}_3 \end{Bmatrix}$$

This matrix is devoid of the first two modes and can be used as a sweeping matrix for the third mode. Applying this to the original equation, we obtain

$$\begin{Bmatrix} x_1 \\ x_2 \\ x_3 \end{Bmatrix} = \frac{\omega^2 m}{3k} \begin{bmatrix} 4 & 2 & 1 \\ 4 & 8 & 4 \\ 4 & 8 & 7 \end{bmatrix} \begin{bmatrix} 0 & 0 & 0.25 \\ 0 & 0 & -0.79 \\ 0 & 0 & 1.00 \end{bmatrix} \begin{Bmatrix} \bar{x}_1 \\ \bar{x}_2 \\ \bar{x}_3 \end{Bmatrix}$$

The above equation results immediately in the third mode, which is

$$\begin{Bmatrix} x_1 \\ x_2 \\ x_3 \end{Bmatrix} = \frac{\omega^2 m}{3k} 1.68 \begin{Bmatrix} 0.25 \\ -0.79 \\ 1.00 \end{Bmatrix}$$

The natural frequency of the third mode is then found to be

$$\omega_3 = \sqrt{\frac{3k}{1.68m}} = 1.34\sqrt{\frac{k}{m}}$$

7.8 TRANSFER MATRICES* (HOLZER-TYPE PROBLEMS)

In the method of transfer matrices, a large system is broken down into subsystems with simple elastic and dynamical properties. The formulation is in terms of the *state vector*, which is a column matrix of the displacements and internal forces; the *point matrix*, which contains the dynamical proper-

*E. C. Pestel and F. A. Leckie, *Matrix Methods in Elastomechanics* (New York: McGraw-Hill Book Co., 1963).

ties of the subsystem; and the *field matrix*, which describes the elastic properties of the subsystem. In terms of these quantities, the calculations are made to proceed from one end of the system to the other, the natural frequencies being established by satisfying the appropriate boundary conditions.

The Spring-Mass System. Figure 7.8-1 shows a part of a linear spring-mass system with one of the subsections isolated. The nth section consists of the mass m_n with displacement x_n and the spring of stiffness k_n, whose ends have displacements x_n and x_{n-1}. When necessary to do so, we designate quantities to the left and right of the element by superscripts L and R.

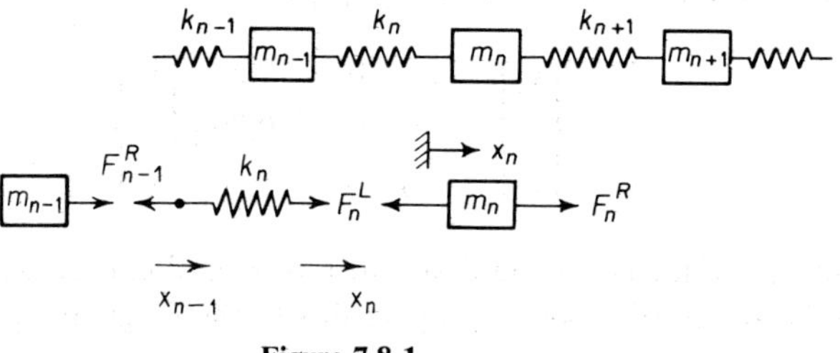

Figure 7.8-1.

For the mass m_n, Newton's second law is

$$m_n\ddot{x}_n = F^R_n - F^L_n$$

which for harmonic motion becomes

$$F^R_n = -\omega^2 m_n x_n + F^L_n \tag{7.8-1}$$

Since the displacement on either side of m_n is the same, we have the identity

$$x_n = x^R_n = x^L_n \tag{7.8-2}$$

Equations (7.8-1) and (7.8-2) can now be assembled into a single matrix equation

$$\begin{Bmatrix} x \\ F \end{Bmatrix}^R_n = \begin{bmatrix} 1 & 0 \\ -\omega^2 m & 1 \end{bmatrix}_n \begin{Bmatrix} x \\ F \end{Bmatrix}^L_n \tag{7.8-3}$$

where $\begin{Bmatrix} x \\ F \end{Bmatrix}$ is the *state vector* and the square matrix is the point matrix.

Next we examine the spring k_n whose end forces are equal

$$F^R_{n-1} = F^L_n \tag{7.8-4}$$

The spring force is related to the spring modulus k_n by the equation

$$x_n^L - x_{n-1}^R = \frac{F_{n-1}^R}{k_n} \tag{7.8-5}$$

Equations (7.8-4) and (7.8-5) are now assembled in matrix form

$$\begin{Bmatrix} x \\ F \end{Bmatrix}_n^L = \begin{bmatrix} 1 & \frac{1}{k} \\ 0 & 1 \end{bmatrix}_n \begin{Bmatrix} x \\ F \end{Bmatrix}_{n-1}^R \tag{7.8-6}$$

where the square matrix above is the field matrix.

We now relate the quantities at station n in terms of quantities at station $n - 1$ by substituting Eq. (7.8-6) into (7.8-3)

$$\begin{aligned} \begin{Bmatrix} x \\ F \end{Bmatrix}_n^R &= \begin{bmatrix} 1 & 0 \\ -\omega^2 m & 1 \end{bmatrix} \begin{bmatrix} 1 & \frac{1}{k} \\ 0 & 1 \end{bmatrix} \begin{Bmatrix} x \\ F \end{Bmatrix}_{n-1}^R \\ &\xrightarrow{=} \begin{bmatrix} 1 & \frac{1}{k} \\ -\omega^2 m & \left(1 - \frac{\omega^2 m}{k}\right) \end{bmatrix}_n \begin{Bmatrix} x \\ F \end{Bmatrix}_{n-1}^R \end{aligned} \tag{7.8-7}$$

Since the state vector at $n - 1$ is transferred to the state vector at n through the square matrix above, it is called the *transfer matrix* for section n. With known values of the state vector at station 1 and a chosen value of ω^2, it is possible to progressively compute the state vectors to the last station n. Depending on the boundary conditions, either x_n or F_n can be plotted as a function of ω^2; the natural frequencies of the system are established when the boundary conditions are satisfied. Problems of this type, where only one displacement is associated with each mass, are called *Holzer*-type problems. Holzer* developed a tabular method of this type which he applied to the multimass torsional problem.

7.9 TORSIONAL SYSTEM

Signs are often a source of confusion in rotating systems, and it is necessary to clearly define the sense of positive quantities. The coordinate along the rotational axis is considered positive towards the right. If a cut is made across the shaft, the face with the outward normal towards the positive coordinate direction is called the positive face. Positive torques and positive angular

*H. Holzer, *Die Berechnung der Drehschwingungen* (Berlin: Springer-Verlag, 1921).

displacements are indicated on the positive face by arrows pointing positively according to the right hand screw rule as shown in Fig. 7.9-1.

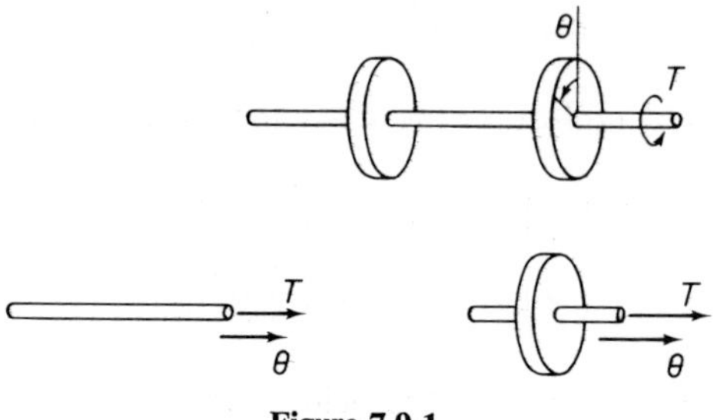

Figure 7.9-1.

With this definition, the development of the transfer matrix of the torsional system is identical to that of the linear spring-mass system with $\begin{Bmatrix} \theta \\ T \end{Bmatrix}$ as the state vector. We isolate the nth section as in Fig. 7.9-2 and write the dynamical equation for the point matrix and the elastic equation for the field matrix. They are

$$\begin{Bmatrix} \theta \\ T \end{Bmatrix}_n^R = \begin{bmatrix} 1 & 0 \\ -\omega^2 J & 1 \end{bmatrix}_n \begin{Bmatrix} \theta \\ T \end{Bmatrix}_n^L \tag{7.9-1}$$

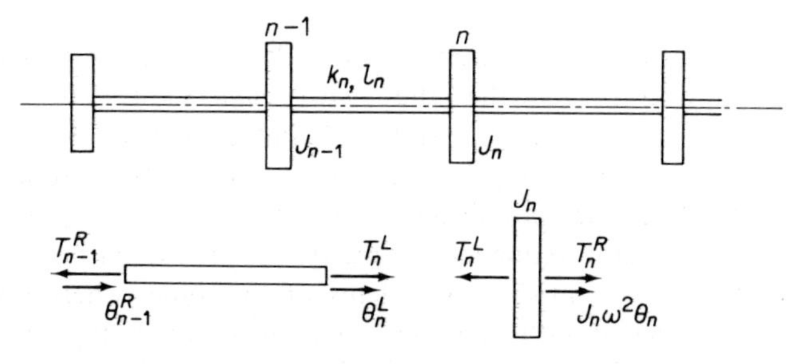

Figure 7.9-2.

and

$$\begin{Bmatrix} \theta \\ T \end{Bmatrix}_n^L = \begin{bmatrix} 1 & \dfrac{1}{K} \\ 0 & 1 \end{bmatrix}_n \begin{Bmatrix} \theta \\ T \end{Bmatrix}_{n-1}^R \tag{7.9-2}$$

which combine to

$$\begin{Bmatrix} \theta \\ T \end{Bmatrix}_n^R = \begin{bmatrix} 1 & \dfrac{1}{K} \\ -\omega^2 J & \left(1 - \dfrac{\omega^2 J}{K}\right) \end{bmatrix}_n \begin{Bmatrix} \theta \\ T \end{Bmatrix}_{n-1}^R \tag{7.9-3}$$

We thus note that each of Eqs. (7.9-1), (7.9-2) and (7.9-3) is identical to those of the linear spring-mass system.

In the development so far, the stations were numbered in increasing order from left to right with the transfer matrix also progressing to the right. The arrow under the equal sign in Eqs. (7.8-7) and (7.9-3) indicate this direction of progression. In some problems it is convenient to proceed with the transfer matrix in the opposite direction, in which case we need only to invert Eqs. (7.8-7) or (7.9-3). We then obtain the relationship

$$\begin{Bmatrix} \theta \\ T \end{Bmatrix}^R_{n-1} \underset{\leftarrow}{=} \begin{bmatrix} \left(1 - \dfrac{\omega^2 J}{K}\right) & -\dfrac{1}{K} \\ \omega^2 J & 1 \end{bmatrix} \begin{Bmatrix} \theta \\ T \end{Bmatrix}^R_n \tag{7.9-4}$$

The arrow now indicates that the transfer matrix progresses from right to left with the order of the station numbering unchanged. The student should verify this equation, starting with the free-body development.

Example 7.9-1

Determine the natural frequencies and the mode shapes of the system shown in Fig. 7.9-3

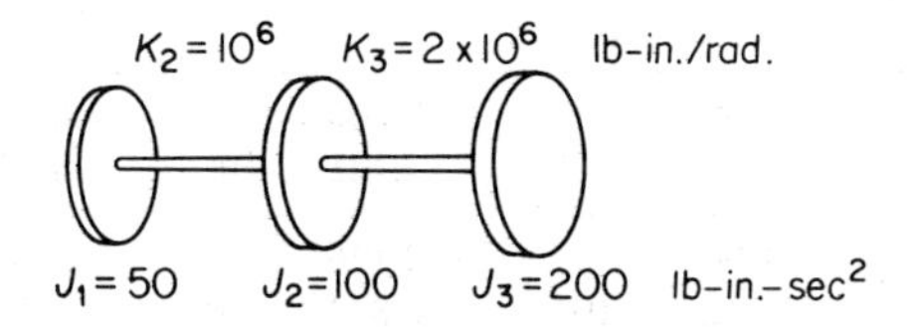

Figure 7.9-3.

Solution: Starting at the right of disk 1, with the state vector

$$\begin{bmatrix} \theta \\ T \end{bmatrix}^R_1 = \begin{bmatrix} 1.00 \\ -50\omega^2 \end{bmatrix}$$

the state vectors at station 2 and station 3 are

$$\begin{bmatrix} \theta \\ T \end{bmatrix}^R_2 = \begin{bmatrix} 1 & 10^{-6} \\ -100\omega^2 & \left(1 - \dfrac{100\omega^2}{10^6}\right) \end{bmatrix}_2 \begin{bmatrix} 1.0 \\ -50\omega^2 \end{bmatrix}_1$$

$$\begin{bmatrix} \theta \\ T \end{bmatrix}^R_3 = \begin{bmatrix} 1 & \frac{1}{2} \times 10^{-6} \\ -200\omega^2 & \left(1 - \dfrac{200\omega^2}{2 \times 10^6}\right) \end{bmatrix}_3 \begin{bmatrix} \theta \\ T \end{bmatrix}^R_2$$

Thus by assuming different values of ω, the quantities θ and T can be found, first for station 2 and then for station 3. Since the end 3 is free, the frequencies which result in $T_3^R = 0$ are the natural frequencies of the system. The following table gives the state vectors at each station for three values of ω, and Fig. 7.9-4 is a plot of T_3^R indicating that the natural frequencies of the system

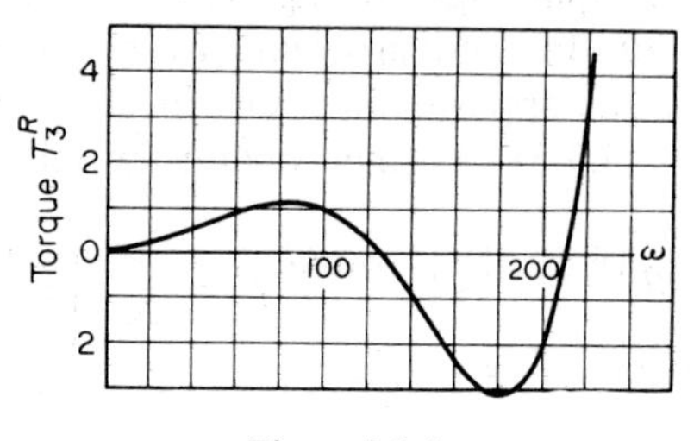

Figure 7.9-4.

are $\omega_1 = 126$ rad/sec and $\omega_2 = 210$ rad/sec. The torques T_3 for $\omega = 126$ and 210 are not zero, but are close enough to zero to approximate the condition $T_3 = 0$. The mode shapes for the two natural frequencies are also shown in Fig. 7.9-5

ω rad/sec	$\begin{bmatrix} \theta \\ T \end{bmatrix}_1^R$	$\begin{bmatrix} \theta \\ T \end{bmatrix}_2^R$	$\begin{bmatrix} \theta \\ T \end{bmatrix}_3^R$
126	1.00 0.794×10^6	0.206 1.121×10^6	-0.355 -0.009×10^6
150	1.00 1.126×10^6	-0.126 0.842×10^6	-0.547 -1.618×10^6
210	1.00 2.205×10^6	-1.205 -3.104×10^6	0.347 -0.044×10^6

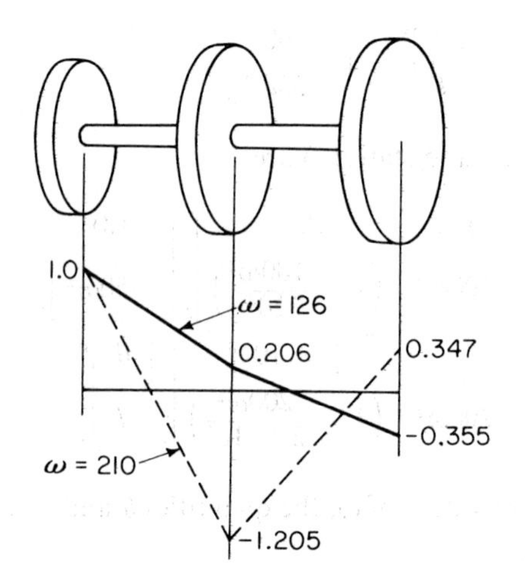

Figure 7.9-5.

Example 7.9-2. Digital Computation of Torsional Problem

The digital computer program for the Holzer-type problem is illustrated for the torsional system of Fig. 7.9-6. The program is written in such a manner that, by changing the data, it is applicable to any other torsional system.

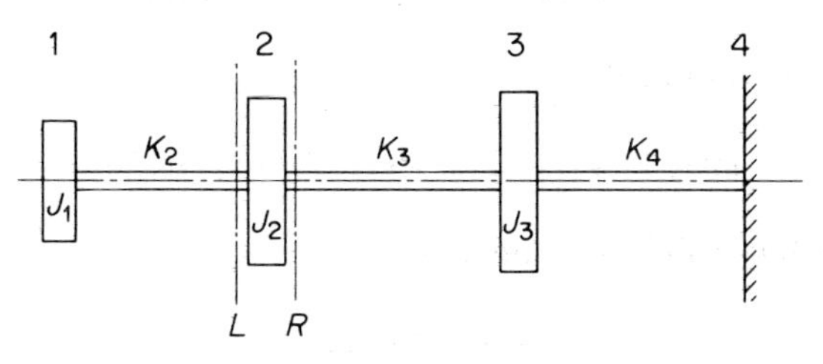

Figure 7.9-6.

Instead of dealing directly with Eq. (7.9-3) we will work with Eqs. (7.9-2) and (7.9-1), which are equivalent. They are

$$\begin{Bmatrix} \theta \\ T \end{Bmatrix}^L_{n+1} = \begin{bmatrix} 1 & \dfrac{1}{K} \\ 0 & 1 \end{bmatrix}_{n+1} \begin{Bmatrix} \theta \\ T \end{Bmatrix}^R_n \tag{7.9-2}$$

$$\begin{Bmatrix} \theta \\ T \end{Bmatrix}^R_{n+1} = \begin{bmatrix} 1 & 0 \\ -\omega^2 J & 1 \end{bmatrix}_{n+1} \begin{Bmatrix} \theta \\ T \end{Bmatrix}^L_{n+1} \tag{7.9-1}$$

When written out, these equations are

$$\theta(I, N+1) = \theta(I, N) + TR(I, N)/K(N+1) \tag{a}$$

$$TL(I, N+1) = TR(I, N) \tag{b}$$

$$TR(I, N+1) = TL(I, N+1) - \lambda(I)*J(N+1)*\theta(I, N+1) \tag{c}$$

where

$$\lambda = \omega^2$$

$$TL = \text{torque to the left of disk}$$

$$TR = \text{torque to the right of disk}$$

The index N defines the position along the structure and the index I defines the frequency to be used. For the computer program some notation changes are required to conform to the Fortran language. For example, the stiffness K and the moment of inertia of the disk J are designated as SK and SJ, and θ is written out.

The three equations above are to be solved for θ and TR at each point

N of the structure and for various values of λ. At the natural frequencies, θ must be zero at the fixed end.

The frequency range can be scanned by choosing an initial ω and an increment $\Delta\omega$. For example in this problem we choose the frequencies

$$\omega = 40, 60, 80, - - - - - - - 620.$$

which can be programmed as

$$\omega(I) = 40 + (I - 1)*20, \quad I = 1 \text{ to } 30$$

The corresponding $\lambda(I)$ is computed as

$$\lambda(I) = \omega(I)**2$$

Starting with the boundary conditions

$$\theta(I, 1) = 1$$
$$TR(I, 1) = -\lambda(I)*J(1)$$

the equations (a), (b), and (c) are computed for each N (position in structure), keeping I (or frequency) fixed. Having reached $N = 4$, I is advanced an integer to the next frequency and the process is repeated. These operations are clearly seen in the flow diagram of Fig. 7.9-7.

The computer results are shown in the following sections. Plotted in Fig. 7.9-8 is the angle θ_4 against ω. The natural frequencies of the system correspond to frequencies where θ_4 becomes zero, and it is seen from Fig. 7.9-8 that they are

$$\omega_1 = 160$$
$$\omega_2 = 356$$
$$\omega_3 = 552$$

The angle θ_i of each point is shown for $\omega_1 = 160$ in Fig. 7.9-9.

Systems with Damping. When damping is included, the form of the transfer matrix is not altered, but the mass and stiffness elements become complex quantities. This can be easily shown by writing the equations for the nth subsystem shown in Fig. 7.9-10. The torque equation for disk n is

$$-\omega^2 J_n \theta_n = T_n^R - T_n^L - i\omega c_n \theta_n$$

or

$$(i\omega c_n - \omega^2 J_n)\theta_n = T_n^R - T_n^L \tag{7.9-5}$$

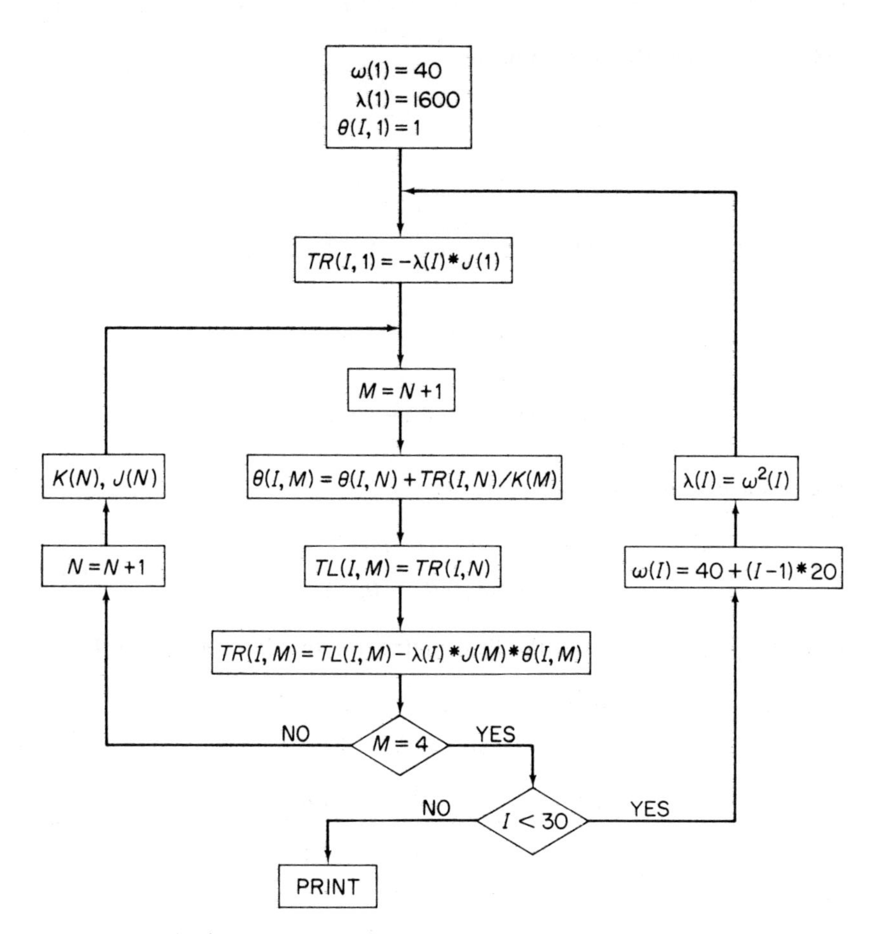

Figure 7.9-7. Flow diagram for torsional problem.

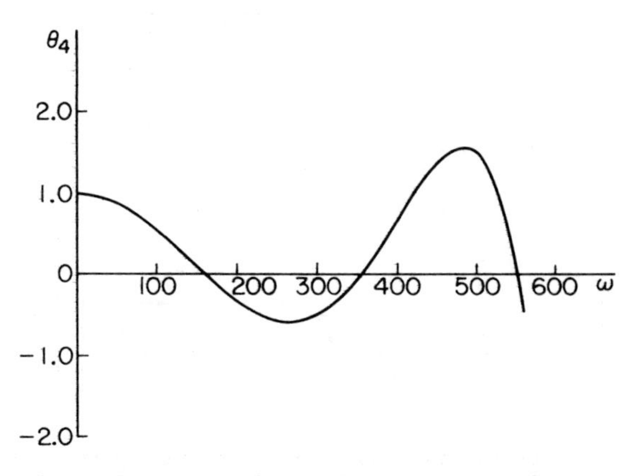

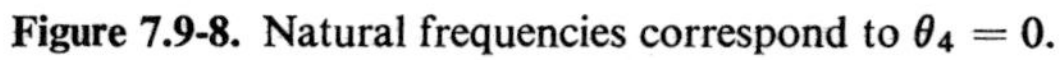

Figure 7.9-8. Natural frequencies correspond to $\theta_4 = 0$.

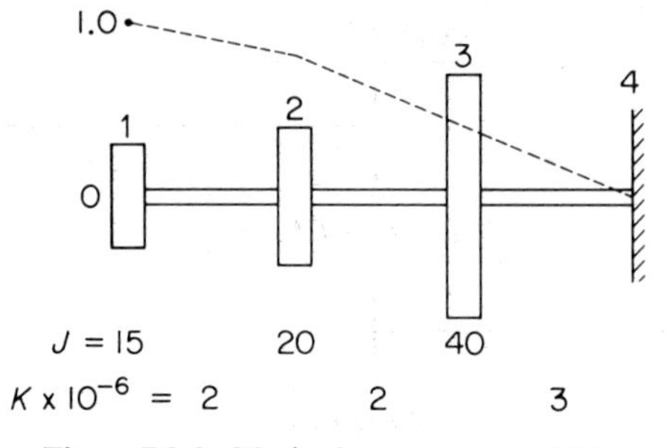

Figure 7.9-9. Mode shapes at $\omega_1 = 160$.

The elastic equation for the nth shaft is

$$\begin{aligned} T_n^L &= K_n(\theta_n - \theta_{n-1}) + i\omega g_n(\theta_n - \theta_{n-1}) \\ &= (K_n + i\omega g_n)(\theta_n - \theta_{n-1}) \end{aligned} \tag{7.9-6}$$

Thus the point matrix and the field matrix for the damped system become

$$\begin{Bmatrix} \theta \\ T \end{Bmatrix}_n^R = \begin{bmatrix} 1 & 0 \\ (i\omega c - \omega^2 J) & 1 \end{bmatrix}_n \begin{Bmatrix} \theta \\ T \end{Bmatrix}_n^L \tag{7.9-7}$$

$$\begin{Bmatrix} \theta \\ T \end{Bmatrix}_n^L = \begin{bmatrix} 1 & \dfrac{1}{(K + i\omega g)} \\ 0 & 1 \end{bmatrix}_n \begin{Bmatrix} \theta \\ T \end{Bmatrix}_{n-1}^R \tag{7.9-8}$$

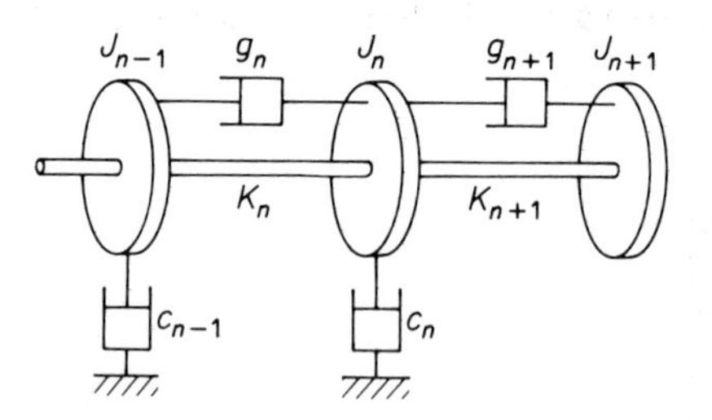

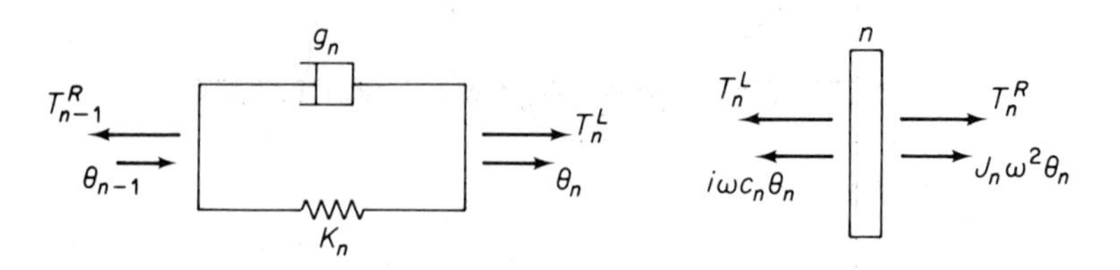

Figure 7.9-10. Torsional system with damping.

which are identical to the undamped case except for the mass and stiffnes elements; these elements are now complex.

Example 7.9-3

The torsional system of Fig. 7.9-11 is excited by a harmonic torque at a point to the right of disk 4. Determine the torque-frequency curve and establish the first natural frequency of the system.

$$J_1 = J_2 = 500 \text{ lb in.sec}^2$$
$$J_3 = J_4 = 1000 \text{ lb in.sec}^2$$
$$K_2 = K_3 = K_4 = 10^6 \text{ lb in./rad}$$
$$c_2 = 10^4 \text{ lb in. sec/rad}$$
$$g_4 = 2 \times 10^4 \text{ lb in. sec/rad}$$

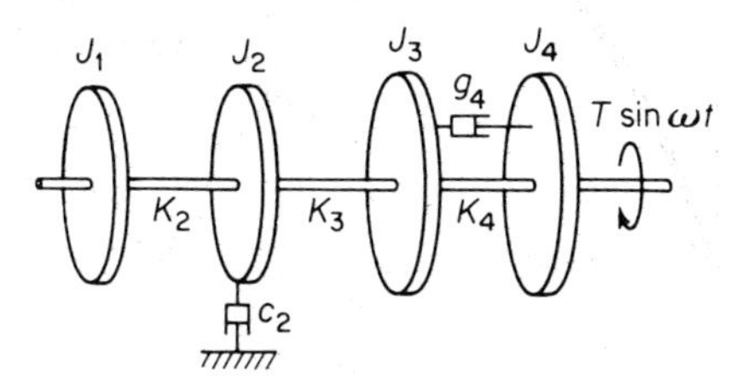

Figure 7.9-11.

Solution: The numerical computations for $\omega^2 = 1000$ are shown in the first accompanying table. The complex mass and stiffness terms are first tabulated for each station n. Substituting into the point and field matrices, i.e., Eqs. (7.9-7) and (7.9-8), the complex amplitude and torque for each station are found, as in the second table.

n	$(\omega^2 J_n - i\omega c_n)10^{-6}$	$(K_n + i\omega g_n)10^{-6}$
1	$0.50 + 0.0i$	
2	$0.50 - 0.316i$	$1.0 + 0.0i$
3	$1.0 + 0.0i$	$1.0 + 0.0i$
4	$1.0 + 0.0i$	$1.0 + 0.635i$

n	θ_n	T_n^R (for $\omega^2 = 1000$)
1	$1.0 + 0.0i$	$(-0.50 + 0.0i) \times 10^6$
2	$0.50 + 0.0i$	$(-0.750 + 0.158i) \times 10^6$
3	$-0.250 + 0.158i$	$(-0.50 + 0.0i) \times 10^6$
4	$-0.607 + 0.384i$	$(0.107 - 0.384i) \times 10^6$

The above computations are repeated for a sufficient number of frequencies to plot the torque-frequency curve of Fig. 7.9-12. The plot shows the

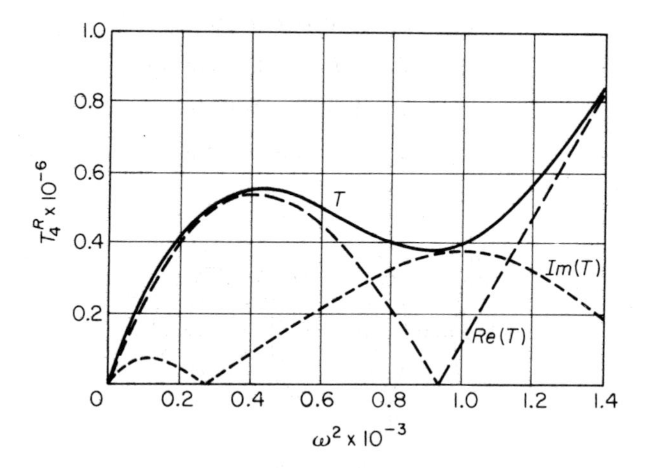

Figure 7.9-12. Torsional-frequency curve for damped torsional system of Figure 7.9-11.

real and imaginary parts of T_4^R as well as their resultant, which in this problem is the exciting torque. For example, the resultant torque at $\omega^2 = 1000$ is $10^6\sqrt{0.107^2 + 0.384^2} = 0.394 \times 10^6$ in. lb. The first natural frequency of the system from this diagram is found to be approximately $\omega = \sqrt{930} = 30.5$ rad/sec, where the natural frequency is defined as that frequency of the undamped system which requires no torque to sustain the motion.

Example 7.9-4

In Fig. 7.9-11 if $T = 2000$ in. lb and $\omega = 31.6$ rad/sec, determine the amplitude of the second disk.

Solution: The table above indicates that a torque of 394,000 in. lb will produce an amplitude of $\theta_2 = 0.50$ radian. Since amplitude is proportional to torque, the amplitude of the second disk for the specified torque is $0.50 \times \frac{2}{394} = 0.00254$ radian.

7.10 GEARED SYSTEM

Consider the geared torsional system of Fig. 7.10-1, where the speed ratio of shaft 2 to shaft 1 is n. The system can be reduced to an equivalent single shaft system as follows.

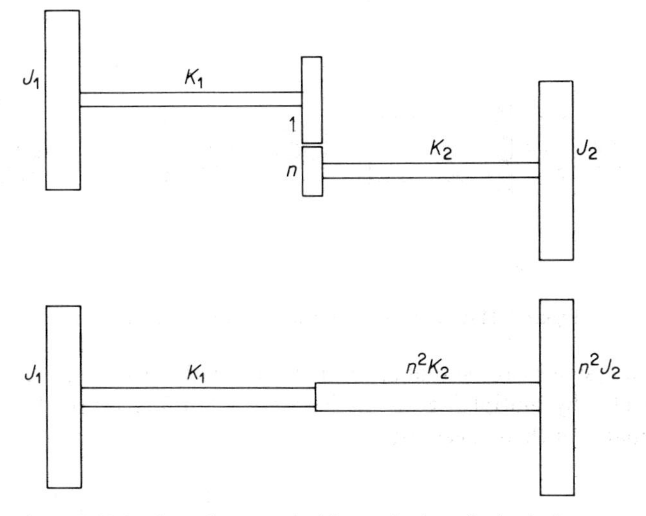

Figure 7.10-1. Geared system and its equivalent single shaft system.

With the speed of shaft 2 equal to $\dot{\theta}_2 = n\dot{\theta}_1$, the kinetic energy of the system is

$$T = \tfrac{1}{2}J_1\dot{\theta}_1^2 + \tfrac{1}{2}J_2n^2\dot{\theta}_1^2 \tag{7.10-1}$$

Thus the equivalent inertia of disk 2 referred to shaft 1 is n^2J_2.

To determine the equivalent stiffness of shaft 2 referred to shaft 1, clamp disks 1 and 2 and apply a torque to gear 1, rotating it through an angle θ_1. Gear 2 will then rotate through the angle $\theta_2 = n\theta_1$, which will also be the twist in shaft 2. The potential energy of the system is then

$$U = \tfrac{1}{2}K_1\theta_1^2 + \tfrac{1}{2}K_2n^2\theta_1^2 \tag{7.10-2}$$

and the equivalent stiffness of shaft 2 referred to shaft 1 is n^2K_2.

The rule for geared systems is thus quite simple: *multiply all stiffness and inertias of the geared shaft by n^2*, where n is the speed ratio of the geared shaft to the reference shaft.

7.11 BRANCHED SYSTEMS

Branched systems are frequently encountered; some common examples are the dual propeller system of a marine installation and the drive shaft and differential of an automobile, which are shown in Fig. 7.11-1.

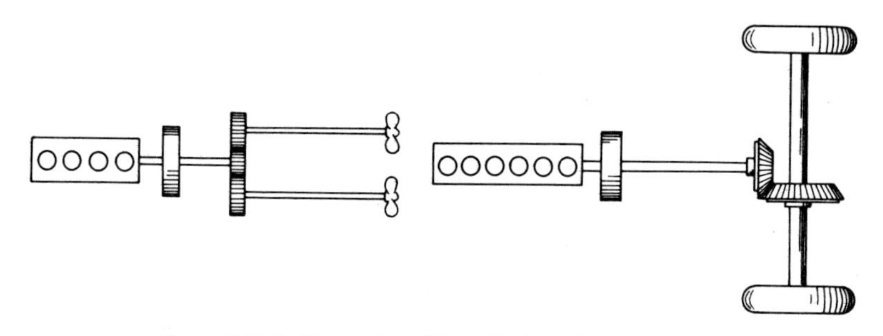

Figure 7.11-1. Examples of branched torsional systems.

Such systems can be reduced to the form with one-to-one gears shown in Fig. 7.11-2 by mutiplying all the inertias and stiffnesses of the branches by the squares of their speed ratios.

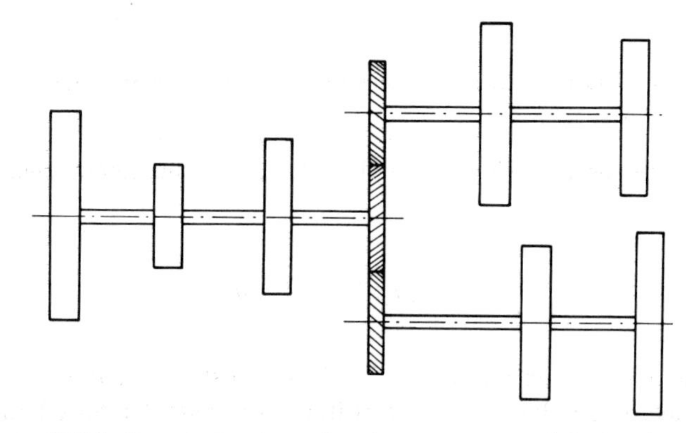

Figure 7.11-2. Branched system reduced to common speeds by 1 to 1 gears.

Example 7.11-1

Outline the matrix procedure for solving the torsional branched system of Fig. 7.11-3.

Solution: We first convert to a system having one-to-one gears by multiplying the stiffness and inertia of branch B by n^2, as shown in Fig. 7.11-3(b). We can then proceed from station 0 through to station 3, taking note of the fact that gear B introduces a torque T_{B1}^R on gear A.

Fig. 7.11-4 shows the free-body diagram of the two gears. With T_{B1}^R shown as positive torque, the torque exerted on gear A by gear B is negative

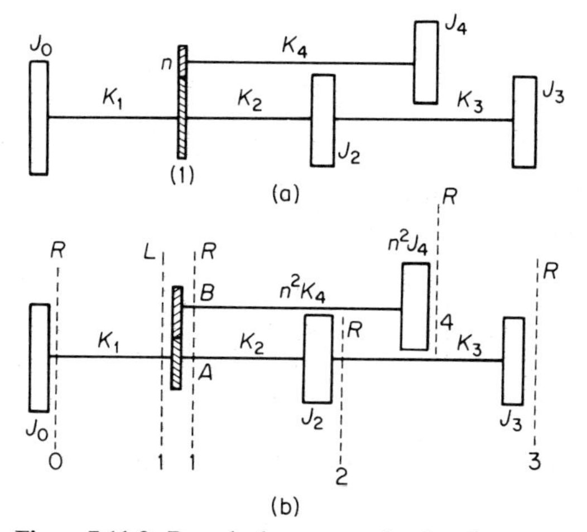

Figure 7.11-3. Branched system and reduced system.

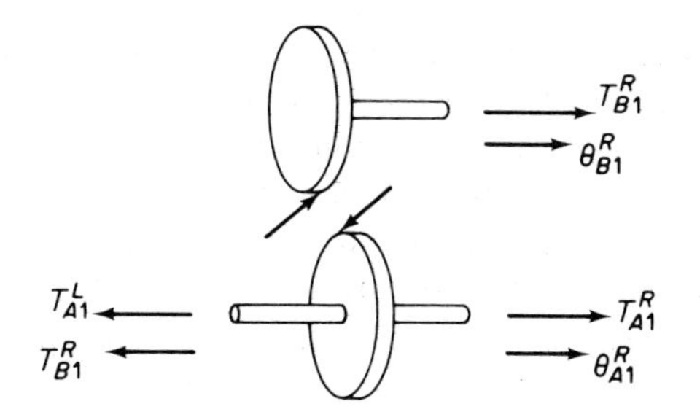

Figure 7.11-4.

as shown. The torque balance on gear A is then

$$T_{A1}^R = T_{A1}^L + T_{B1}^R \tag{7.11-1}$$

and we need now to express T_{B1}^R in terms of the angular displacement θ_1 of shaft A.

Using Eq. (7.9-4) and noting that $T_{B4}^R = 0$, we have for shaft B

$$\begin{Bmatrix} \theta_B \\ T_B \end{Bmatrix}_1^R = \begin{bmatrix} \left(1 - \dfrac{\omega^2 n^2 J_4}{n^2 K_4}\right) & -\dfrac{1}{n^2 K_4} \\ \omega^2 n^2 J_4 & 1 \end{bmatrix} \begin{Bmatrix} \theta_B \\ 0 \end{Bmatrix}_4^R \tag{7.11-2}$$

Since $\theta_{B1}^R = -\theta_{A1}^L = -\theta_{A1}^R$, we obtain

$$\theta_{B1}^R = \left(1 - \frac{\omega^2 J_4}{K_4}\right)\theta_{B4}^R = -\theta_{A1}^L \tag{7.11-3}$$

$$T_{B1}^R = \omega^2 n^2 J_4 \theta_{B4}^R \tag{7.11-4}$$

Eliminating θ_{B4}^R

$$T_{B1}^R = \frac{-\omega^2 n^2 J_4}{\left(1 - \dfrac{\omega^2 J_4}{K_4}\right)} \theta_{A1}^L \tag{7.11-5}$$

Substituting Eq. (7.11-5) into Eq. (7.11-1), the transfer function of shaft A across the gears becomes

$$\begin{Bmatrix} \theta_A \\ T_A \end{Bmatrix}_1^R \overrightarrow{=} \begin{bmatrix} 1 & 0 \\ \dfrac{-\omega^2 J_4}{\left(1 - \dfrac{\omega^2 J_4}{K_4}\right)} & 1 \end{bmatrix} \begin{Bmatrix} \theta_A \\ T_A \end{Bmatrix}_1^L \tag{7.11-6}$$

It is now possible to proceed along shaft A from $1R$ to $3R$ in the usual manner.

7.12 BEAMS

When a beam is replaced by lumped masses connected by massless beam sections, a method developed by N. O. Myklestad* can be used to compute progressively the deflection, slope, moment, and shear from one station to the next, in a manner similar to the Holzer method. Again it is advantageous, from the point of view of conciseness and efficiency of computation, to express these equations in matrix form.

(*a*) *Uncoupled Flexural Vibrations.* Figure 7.12-1 shows a typical section

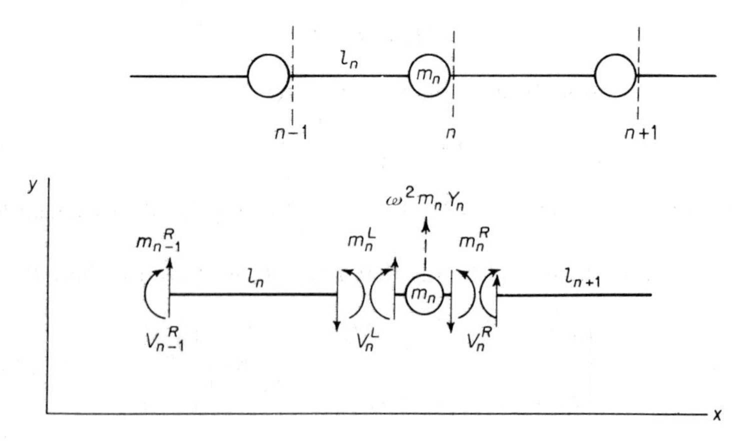

Figure 7.12-1. Idealized beam with lumped masses.

*N. O. Myklestad, "A New Method of Calculating Natural Modes of Uncoupled Bending Vibration of Airplane Wings and Other Types of Beams, *Jour. Aero. Sci.* (April 1944), pp. 153–62.

of an idealized beam with lumped masses. Examining the nth section, the forces and moments acting on the mass and beam section are indicated by the free-body diagram. From them the equations for the shear and moment are found to be

$$\begin{aligned} M_n^L &= M_n^R & V_{n-1}^R &= V_n^L \\ V_n^L &= V_n^R - \omega^2 m_n y_n & M_{n-1}^R &= M_n^L - V_n^L l_n \end{aligned} \tag{7.12-1}$$

For the elastic deformation of the nth beam section, we refer to Fig. 7.12-2. The deflection and slope at the ends are then related by the equations

$$\begin{aligned} y_n &= y_{n-1} + l_n\theta_{n-1} + M_n^L \frac{l_n^2}{2(EI)_n} - V_n^L \frac{l_n^3}{3(EI)_n} \\ \theta_n &= \theta_{n-1} + \frac{M_n^L l_n}{(EI)_n} - \frac{V_n^L l_n^2}{2(EI)_n} \end{aligned} \tag{7.12-2}$$

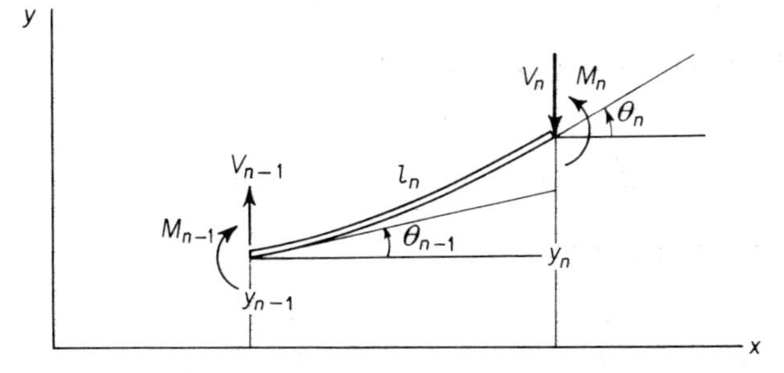

Figure 7.12-2. Elastic deformation of a beam section.

where the various influence coefficients used are based on uniform section, and are

(1) Slope of station n referred to tangent at $n - 1$

$$= \frac{l}{EI} \quad \text{due to a unit moment at } n$$

$$= \frac{l^2}{2EI} \quad \text{due to a unit shear at } n$$

(2) Deflection of station n measured from the tangent at $n - 1$

$$= \frac{l^2}{2EI} \quad \text{due to a unit moment at } n$$

$$= \frac{l^3}{3EI} \quad \text{due to a unit shear at } n$$

Expressing M_n and V_n in Eq. (7.12-2) in terms of M_{n-1} and V_{n-1} from Eq. (7.12-1), these equations can be rewritten as

$$y_n = y_{n-1} + l_n\theta_{n-1} + \frac{M^R_{n-1}l^2_n}{2(EI)_n} + \frac{V^R_{n-1}l^3_n}{6(EI)_n} \tag{7.12-3}$$

$$\theta_n = 0 + \theta_{n-1} + \frac{M^R_{n-1}l_n}{(EI)_n} + \frac{V^R_{n-1}l^2_n}{2(EI)_n}$$

$$M^L_n = 0 + 0 + M^R_{n-1} + V^R_{n-1}l_n$$

$$V^L_n = 0 + 0 + 0 + V^R_{n-1}$$

and expressed as a field matrix.

$$\begin{bmatrix} y \\ \theta \\ M \\ V \end{bmatrix}^L_n \underset{\rightarrow}{=} \begin{bmatrix} 1 & l & \frac{l^2}{2EI} & \frac{l^3}{6EI} \\ 0 & 1 & \frac{l}{EI} & \frac{l^2}{2EI} \\ 0 & 0 & 1 & l \\ 0 & 0 & 0 & 1 \end{bmatrix}_n \begin{bmatrix} y \\ \theta \\ M \\ V \end{bmatrix}^R_{n-1} \tag{7.12-4}$$

For the point mass m_n, we have

$$y^R_n = y^L_n \qquad M^R_n = M^L_n$$

$$\theta^R_n = \theta^L_n \qquad V^R_n = V^L_n + \omega^2 m_n y^L_n$$

which leads to the point matrix

$$\begin{bmatrix} y \\ \theta \\ M \\ V \end{bmatrix}^R_n = \begin{bmatrix} 1 & 0 & 0 & 0 \\ 0 & 1 & 0 & 0 \\ 0 & 0 & 1 & 0 \\ \omega^2 m & 0 & 0 & 1 \end{bmatrix}_n \begin{bmatrix} y \\ \theta \\ M \\ V \end{bmatrix}^L_n \tag{7.12-5}$$

Substituting Eq. (6.12-4) for the column on the right side of Eq. (7.12-5), the final equation relating the state vectors at n and $n - 1$ is found to be

$$\begin{bmatrix} y \\ \theta \\ M \\ V \end{bmatrix}^R_n = \begin{bmatrix} 1 & 0 & 0 & 0 \\ 0 & 1 & 0 & 0 \\ 0 & 0 & 1 & 0 \\ m\omega^2 & 0 & 0 & 1 \end{bmatrix}_n \begin{bmatrix} 1 & l & \frac{l^2}{2EI} & \frac{l^3}{6EI} \\ 0 & 1 & \frac{l}{EI} & \frac{l^2}{2EI} \\ 0 & 0 & 1 & l \\ 0 & 0 & 0 & 1 \end{bmatrix}_n \begin{bmatrix} y \\ \theta \\ M \\ V \end{bmatrix}^R_{n-1}$$

$$\underset{\rightarrow}{=} \begin{bmatrix} 1 & l & \frac{l^2}{2EI} & \frac{l^3}{6EI} \\ 0 & 1 & \frac{l}{EI} & \frac{l^2}{2EI} \\ 0 & 0 & 1 & l \\ \omega^2 m & \omega^2 ml & \frac{\omega^2 ml^2}{2EI} & 1 + \omega^2 m\frac{l^3}{6EI} \end{bmatrix}_n \begin{bmatrix} y \\ \theta \\ M \\ V \end{bmatrix}^R_{n-1} \tag{7.12-6}$$

For any frequency ω, Eq. (7.12-6) enables us to start at the left boundary 0 and proceed to the right boundary N, these quantities being linearly related by the equation

$$\begin{bmatrix} y \\ \theta \\ M \\ V \end{bmatrix}_N = \begin{bmatrix} u_{11} & u_{12} & u_{13} & u_{14} \\ u_{21} & u_{22} & u_{23} & u_{24} \\ u_{31} & u_{32} & u_{33} & u_{34} \\ u_{41} & u_{42} & u_{43} & u_{44} \end{bmatrix} \begin{bmatrix} y \\ \theta \\ M \\ V \end{bmatrix}_O \tag{7.12-7}$$

Generally, two of the boundary conditions at each end are known, so that the frequencies which satisfy these conditions are the natural frequencies of the beam.

Example 7.12-1

A cantilever beam, fixed at the left end, is represented by several lumped masses. Determine the boundary equations leading to the natural frequencies.

Solution: At station 0, $y_0 = \theta_0 = 0$, and from Eq. (7.12-7) we obtain

$$M_N = u_{33}M_0 + u_{34}V_0$$
$$V_N = u_{43}M_0 + u_{44}V_0$$

where M_0 and V_0 are unknown and M_N and V_N must be zero. The boundary conditions are then satisfied if the determinant of the equation is zero, or

$$\begin{vmatrix} u_{33} & u_{34} \\ u_{43} & u_{44} \end{vmatrix} = 0$$

so that this quantity may be plotted as a function of ω to establish the natural frequencies of the beam.

(*b*) *Rotating Beams.* In this section we will examine rotating beams, such as propellers and turbine blades, for vibration perpendicular to the plane of rotation. Due to centrifugal force, we will need to consider terms in addition to the beam analysis of the previous section.

The centrifugal force, which is normal to the axis of rotation, is shown in Fig. 7.12-3 and is equal to $m_n\Omega^2x_n$ for mass m_n. The additional quantity that must be introduced is then the axial force

$$F_n^L = F_n^R + m_n\Omega^2x_n \tag{7.12-8}$$

where

$$F_n^L = \sum_{i=n}^{N} m_i\Omega^2x_i \tag{7.12-9}$$

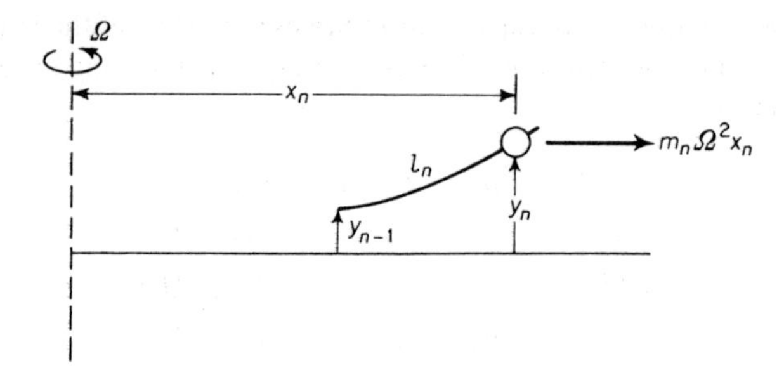

Figure 7.12-3. Centrifugal force on a rotating beam.

Because of this term, the moment equation is modified to

$$M_{n-1}^R = M_n^L - V_n^L l_n - F_n^L(y_n^L - y_{n-1}^R)$$

The deflection and slope are also influenced by F_n^L, and we account for it by considering only the component of F_n^L normal to the beam as a shear load

$$y_n^L = y_{n-1}^R + l_n\theta_{n-1}^R - \theta_n^L F_n^L \frac{l^3}{3(EI)_n} + M_n^L \frac{l_n^2}{2(EI)_n} - V_n^L \frac{l_n^3}{3(EI)_n}$$

$$\theta_n^L = \theta_{n-1}^R + \frac{M_n^L l_n}{(EI)_n} - \frac{V_n^L l_n^2}{2(EI)_n} - \frac{\theta_n^L F_n^L l^2}{2(EI)_n}$$

These equations can now be rearranged for calculation to proceed from right to left as follows

$$y_{n-1}^R = y_n^L - \theta_n^L\left[l_n + \frac{F_n^L l_n^3}{6(EI)_n}\right] + M_n^L \frac{l_n^2}{2(EI)_n} - V_n^L \frac{l^3}{6(EI)_n}$$

$$\theta_{n-1}^R = \theta_n^L\left[1 + \frac{F_n^L l^2}{2(EI)_n}\right] - M_n^L \frac{l_n}{(EI)_n} + V_n^L \frac{l^2}{2(EI)_n}$$

$$M_{n-1}^R = M_n^L\left[1 + \frac{F_n^L l_n^2}{2(EI)_n}\right] - V_n^L\left[l_n + \frac{F_n^L l_n^3}{6(EI)_n}\right] - \theta_n^L F_n^L\left[l_n + \frac{F_n^L l_n^3}{6(EI)_n}\right]$$

$$V_{n-1}^R = V_n^L$$

Arranged in matrix form, these equations appear as

$$\begin{bmatrix} y \\ \theta \\ M \\ V \end{bmatrix}_{n-1}^R \underset{\leftarrow}{=} \begin{bmatrix} 1 & -\left(l + \frac{Fl^3}{6EI}\right) & \frac{l^2}{2EI} & -\frac{l^3}{6EI} \\ 0 & \left(1 + \frac{Fl^2}{2EI}\right) & -\frac{l}{EI} & \frac{l^2}{2EI} \\ 0 & -\left(l + \frac{Fl^3}{6EI}\right) & \left(1 + \frac{Fl^2}{2EI}\right) & -\left(l + \frac{Fl^3}{6EI}\right) \\ 0 & 0 & 0 & 1 \end{bmatrix}_n \begin{bmatrix} y \\ \theta \\ M \\ V \end{bmatrix}_n^L \tag{7.12-10}$$

To complete the problem, we need the point matrix for mass m_n

$$\begin{bmatrix} y \\ \theta \\ M \\ V \end{bmatrix}_n^L = \begin{bmatrix} 1 & 0 & 0 & 0 \\ 0 & 1 & 0 & 0 \\ 0 & 0 & 1 & 0 \\ -m\omega^2 & 0 & 0 & 1 \end{bmatrix} \begin{bmatrix} y \\ \theta \\ M \\ V \end{bmatrix}_n^R \tag{7.12-11}$$

Substituting this for the right column of Eq. (7.9-10), the final result becomes

$$\begin{bmatrix} y \\ \theta \\ M \\ V \end{bmatrix}_{n-1}^R \overset{=}{\leftarrow} \begin{bmatrix} \left(1 + \dfrac{m\omega^2 l^3}{6EI}\right) & -\left(l + \dfrac{Fl^3}{6EI}\right) & \dfrac{l^2}{2EI} & -\dfrac{l^3}{6EI} \\ -\dfrac{m\omega^2 l^2}{2EI} & \left(1 + \dfrac{Fl^2}{2EI}\right) & -\dfrac{l}{EI} & \dfrac{l^2}{2EI} \\ m\omega^2\left(l + \dfrac{Fl^3}{6EI}\right) & -F\left(l + \dfrac{Fl^3}{6EI}\right) & \left(1 + \dfrac{Fl^2}{2EI}\right) & -\left(l + \dfrac{Fl^3}{6EI}\right) \\ -m\omega^2 & 0 & 0 & 1 \end{bmatrix}_n \begin{bmatrix} y \\ \theta \\ M \\ V \end{bmatrix}_n^R \tag{7.12-12}$$

(*c*) *Coupled Flexure-Torsion Vibration.* Natural modes of vibration of airplane wings and other beam structures are often coupled flexure-torsion modes which for higher modes differ considerably from those of uncoupled modes. To treat such problems it is necessary to introduce one additional influence coefficient, h_n, defined as the angle of twist of station n relative to station $n - 1$, due to a unit torque at n. Referring to the beam section of Fig. 7.12-4, the equations pertaining to the torque T are

$$\begin{aligned} T_n^R - T_n^L &= J_n\ddot{\varphi}_n + m_n c_n \ddot{y}_n \\ &= -J_n\omega^2\varphi_n - m_n c_n \omega^2 y_n \end{aligned}$$

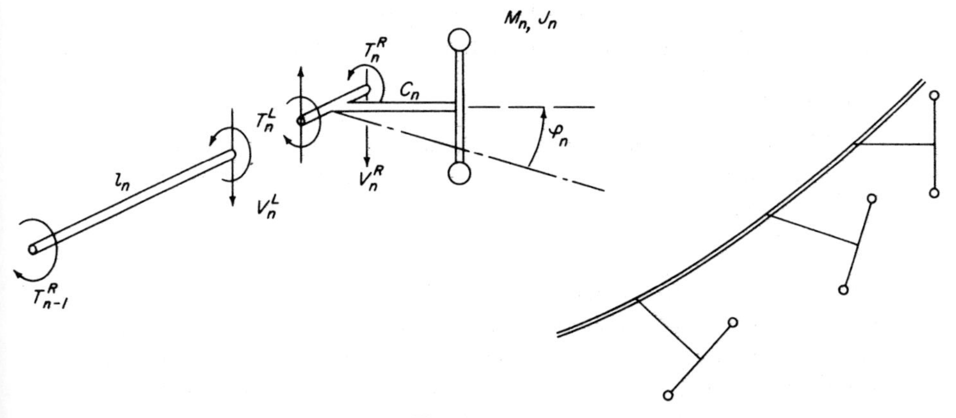

Figure 7.12-4.

$$\varphi_n^L - \varphi_{n-1}^R = T_n^L h_n$$
$$\varphi_n^R = \varphi_n^L$$

where $J_n = J_{ncg} + m_n c_n^2$ is the moment of inertia of the nth section about the elastic axis of the beam. The shear across the mass is

$$V_n^L - V_n^R = -m_n \omega^2 (y_n + c_n \varphi_n)$$

and the point matrix across m can be written as

$$\begin{bmatrix} y \\ \theta \\ M \\ V \\ \varphi \\ T \end{bmatrix}_n^R = \left[\begin{array}{cccc|cc} 1 & 0 & 0 & 0 & 0 & 0 \\ 0 & 1 & 0 & 0 & 0 & 0 \\ 0 & 0 & 1 & 0 & 0 & 0 \\ m\omega^2 & 0 & 0 & 1 & m\omega^2 c & 0 \\ \hline 0 & 0 & 0 & 0 & 1 & 0 \\ -mc\omega^2 & 0 & 0 & 0 & -J\omega^2 & 1 \end{array}\right] \begin{bmatrix} y \\ \theta \\ M \\ V \\ \varphi \\ T \end{bmatrix}_n^L \tag{7.12-13}$$

The field matrix between station $(n-1)R$ and $(n)L$ is the same as Eq. (7.12-4) with two additional equations

$$\begin{bmatrix} y \\ \theta \\ M \\ V \\ \varphi \\ T \end{bmatrix}_n^L = \left[\begin{array}{cccc|cc} 1 & l & \frac{l^2}{2EI} & \frac{l^3}{6EI} & 0 & 0 \\ 0 & 1 & \frac{l}{EI} & \frac{l^2}{2EI} & 0 & 0 \\ 0 & 0 & 1 & l & 0 & 0 \\ 0 & 0 & 0 & 1 & 0 & 0 \\ \hline 0 & 0 & 0 & 0 & 1 & h \\ 0 & 0 & 0 & 0 & 0 & 1 \end{array}\right]_n \begin{bmatrix} y \\ \theta \\ M \\ V \\ \varphi \\ T \end{bmatrix}_{n-1}^R \tag{7.12-14}$$

Thus by substituting Eq. (7.12-14) for the right column of Eq. (7.12-13), the state vectors at station $(n)R$ are related to the state vectors at station $(n-1)R$.

Example 7.12-2

Figure 7.12-5 shows the mass breakdown for a fighter-plane wing and fuselage for flexure-torsion vibration. Find the boundary equations for the determination of the symmetric flexure-torsion modes.

Solution: To use the matrix equation (7.12-14), we let the center line of the airplane be station 0 and let half of the mass and mass moment of inertia of the fuselage about the elastic axis be m_1 and J_1, with $l_1 = 0$. At the wing tip we put in station 7 with $m_7 = l_7 = 0$ to make use of the column matrix $[\ \ ]_7^L$ which now becomes equal to $[\ \ ]_6^R$.

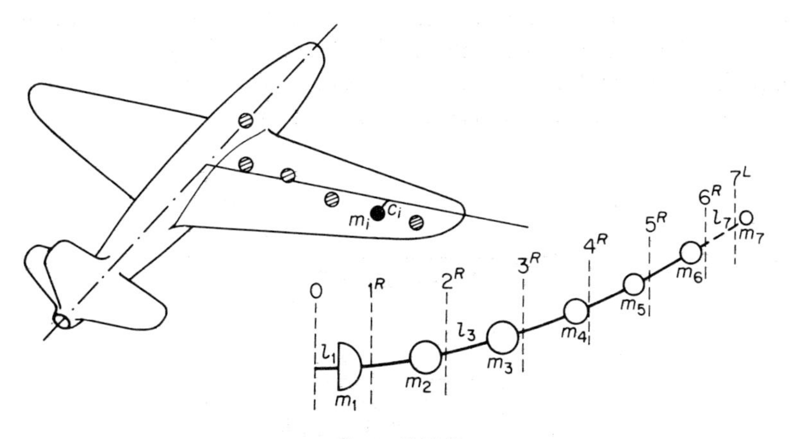

Figure 7.12-5.

For the symmetric modes, the bending slope θ_0^R, the shear V_0^R, and the twisting torque T_0^R are zero at the center line, whereas at the wing tip the moment, shear, and the torque are zero. The boundary equations from Eq. (7.12-14) then appear as

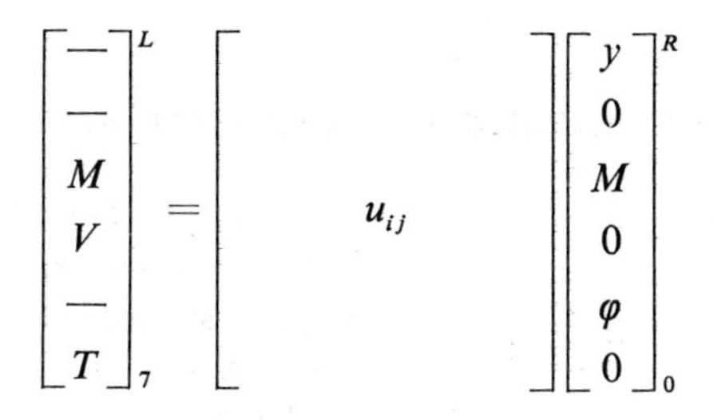

which may be rewritten as

$$\begin{bmatrix} M \\ V \\ T \end{bmatrix}_7^L = \begin{bmatrix} u_{31} & u_{33} & u_{35} \\ u_{41} & u_{43} & u_{45} \\ u_{61} & u_{63} & u_{65} \end{bmatrix} \begin{bmatrix} y \\ M \\ \varphi \end{bmatrix}_0^R$$

The quantities y_0, M_0, and φ_0 at the center line are unknown; however, the column matrix at the left for the moment, shear, and torque at the wing tip is zero. Thus the determinant of the u_{ij} which is a function of ω must be zero to satisfy the boundary conditions.

By plotting the quantity

$$D(\omega) = \begin{vmatrix} u_{31} & u_{33} & u_{35} \\ u_{41} & u_{43} & u_{45} \\ u_{61} & u_{63} & u_{65} \end{vmatrix}$$

against ω, the natural frequencies for the symmetric modes are established. The mode shapes are then established for the natural frequencies found, by computing y_n, θ_n, and φ_n. Figure 7.12-6 shows a typical contour plot for the second symmetric mode of a particular fighter plane.

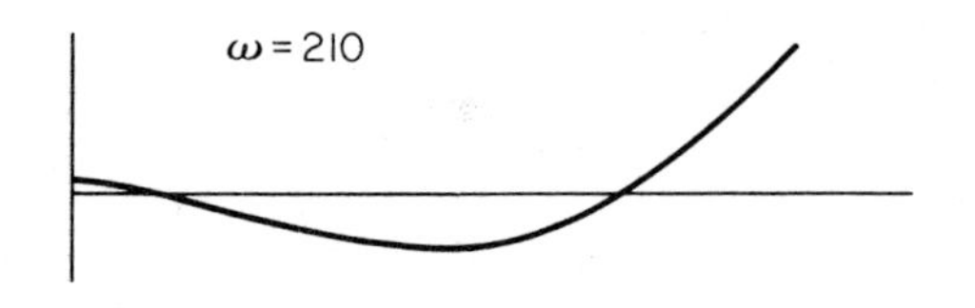

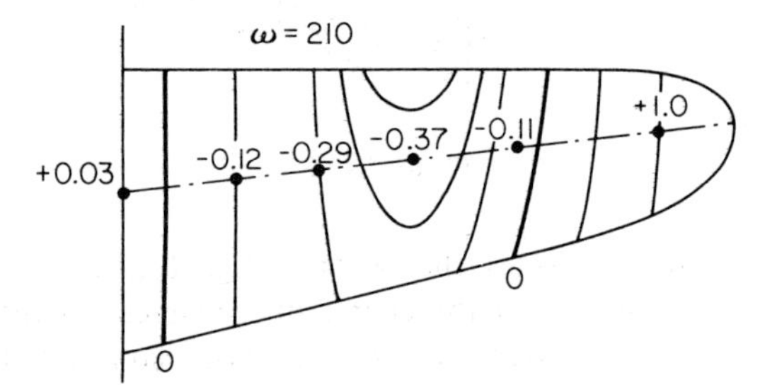

Figure 7.12-6.

7.13 REPEATED STRUCTURES AND TRANSFER MATRIX

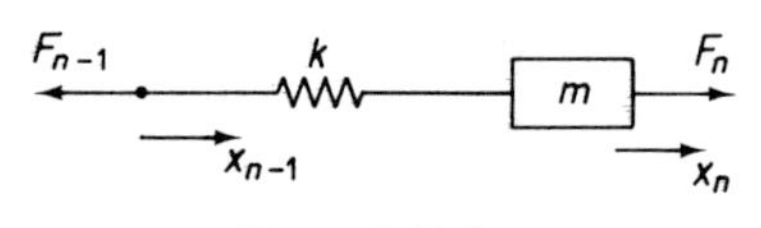

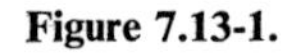

Figure 7.13-1.

The transfer matrix of the previous chapter, when applied to repeated identical sections, leads to some interesting results. It should be noted that the determinant of the transfer matrix is unity regardless of whether or not the system is damped. The following three cases are presented to verify the above statement. Case 1, with Fig. 7.13-1.

$$\begin{Bmatrix} F \\ x \end{Bmatrix}_n = \begin{bmatrix} \left(1 - \dfrac{\omega^2 m}{k}\right) & -\omega^2 m \\ \dfrac{1}{k} & 1 \end{bmatrix} \begin{Bmatrix} F \\ x \end{Bmatrix}_{n-1} \qquad (7.13\text{-}1)$$

This is the same equation as Eq. (7.8-7) with the state vector inverted. Case 2, with Fig. 7.13-2.

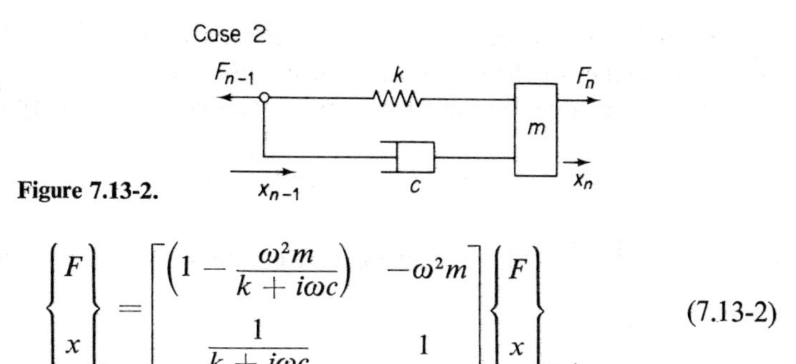

Figure 7.13-2.

$$\begin{Bmatrix} F \\ x \end{Bmatrix}_n = \begin{bmatrix} \left(1 - \dfrac{\omega^2 m}{k + i\omega c}\right) & -\omega^2 m \\ \dfrac{1}{k + i\omega c} & 1 \end{bmatrix} \begin{Bmatrix} F \\ x \end{Bmatrix}_{n-1} \tag{7.13-2}$$

Case 3, with Fig. 7.13-3.

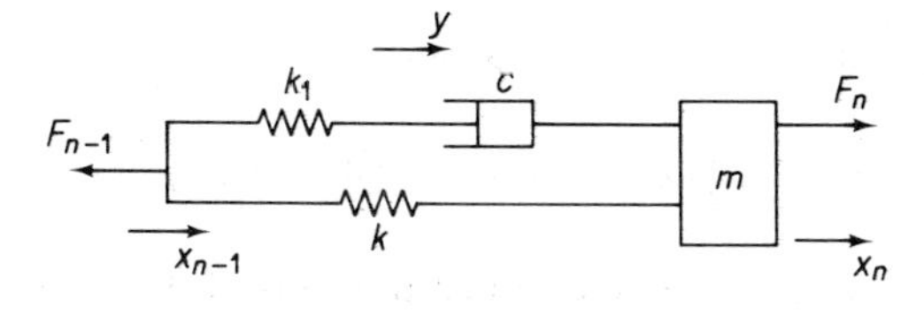

Figure 7.13-3.

$$\begin{Bmatrix} F \\ x \end{Bmatrix}_n = \begin{bmatrix} \left(1 - \dfrac{\omega^2 m[k_1 + \omega c]}{kk_1 + i\omega c(k_1 + k)}\right) & -\omega^2 m \\ \dfrac{k_1 + i\omega c}{kk_1 + i\omega c(k_1 + k)} & 1 \end{bmatrix} \begin{Bmatrix} F \\ x \end{Bmatrix}_{n-1} \tag{7.13-3}$$

The intermediate coordinate y has been eliminated in the above equation.

In each of the above cases, the transfer matrix is in the form

$$[T] = \begin{bmatrix} A & B \\ C & D \end{bmatrix} \tag{7.13-4}$$

and the determinant $AD - BC = 1.0$. Even for the transfer matrix of the beam section (i.e., Eq. 7.12-6), the determinant of the 4×4 matrix is unity, as one can easily show.

When the system has n identical sections, the transfer matrix procedure leads to the equation

$$\begin{Bmatrix} F \\ x \end{Bmatrix}_n = [T]^n \begin{Bmatrix} F \\ x \end{Bmatrix}_0 \tag{7.13-5}$$

and hence it is of interest to be able to calculate the nth power of the transfer matrix. This is done by first determining the eigenvalues μ and eigenvectors

ξ of the matrix $[T]$, which must not be confused with the natural frequencies and mode shapes of the system previously discussed.

The eigenvalues and eigenvectors of the matrix $[T]$ satisfy the equation

$$[T]\{\xi\} - \mu\{\xi\} = 0 \tag{7.13-6}$$

For $[T] = \begin{bmatrix} A & B \\ C & D \end{bmatrix}$, the eigenvalues are found from the characteristic equation

$$\begin{vmatrix} (A-\mu) & B \\ C & (D-\mu) \end{vmatrix} = 0 \tag{7.13-7}$$

which as a result of $AD - BC = 1$ gives

$$\mu_{1,2} = \tfrac{1}{2}(A + D) \pm \sqrt{\tfrac{1}{4}(A + D)^2 - 1} \tag{7.13-8}$$

The eigenvectors can only be determined in terms of its ratio

$$\begin{aligned} \left(\frac{\xi_1}{\xi_2}\right)_1 &= \frac{B}{\mu_1 - A} = r_1 \\ \left(\frac{\xi_1}{\xi_2}\right)_2 &= \frac{B}{\mu_2 - A} = r_2 \end{aligned} \tag{7.13-9}$$

We next form the modal matrix P of the eigenvector columns

$$[P] = \begin{bmatrix} r_1 & r_2 \\ 1 & 1 \end{bmatrix} \tag{7.13-10}$$

The two equations

$$[T]\{\xi\} = \mu\{\xi\}$$

for μ_1 and μ_2 may now be assembled as a single matrix equation (Also see Eq. (6.10-16).

$$[T][P] = [P][\Lambda]$$

where $[\Lambda] = \begin{bmatrix} \mu_1 & 0 \\ 0 & \mu_2 \end{bmatrix}$ = diagonal matrix of the eigenvalues.

By post-multiplying by $[P]^{-1}$ we obtain

$$[T] = [P][\Lambda][P]^{-1}$$

The square of the above equation is

$$[T]^2 = [T][T] = [P][\Lambda][P]^{-1}[P][\Lambda][P]^{-1} = [P][\Lambda]^2[P]^{-1}$$

where $[\Lambda]^2 = \begin{bmatrix} \mu_1^2 & 0 \\ 0 & \mu_2^2 \end{bmatrix}$

Repeated multiplication leads to the nth power

$$[T]^n = [P][\Lambda]^n[P]^{-1} \tag{7.13-12}$$

The boundary conditions can now be applied to the equation

$$\begin{Bmatrix} F \\ x \end{Bmatrix}_n = [T]^n \begin{Bmatrix} F \\ x \end{Bmatrix}_0 = \begin{bmatrix} t_{11} & t_{12} \\ t_{21} & t_{22} \end{bmatrix} \begin{Bmatrix} F \\ x \end{Bmatrix}_0$$

For example, if the end 0 is fixed and the end n is free, $x_0 = 0$ and $F_n = 0$, and we obtain

$$0 = t_{11}F_0$$

Since $F_0 \neq 0$, the natural frequencies are found from $t_{11} = 0$. In case damping is present, the elements of the transfer matrix are complex quantities. In this case, the end displacement x_n may be chosen as unity, and the force F_0 is found from

$$1 = t_{21}F_0$$

7.14 DIFFERENCE EQUATION

The difference equation offers another approach to the problem of repeated identical sections. As an example of repeating sections, consider the N-story building shown in Fig. 7.14-1, where the mass of each floor is m and the lateral or shear stiffness of each section between floors is k lb/in. The equation of motion for the nth mass is then

$$m\ddot{x}_n = k(x_{n+1} - x_n) - k(x_n - x_{n-1}) \tag{7.14-1}$$

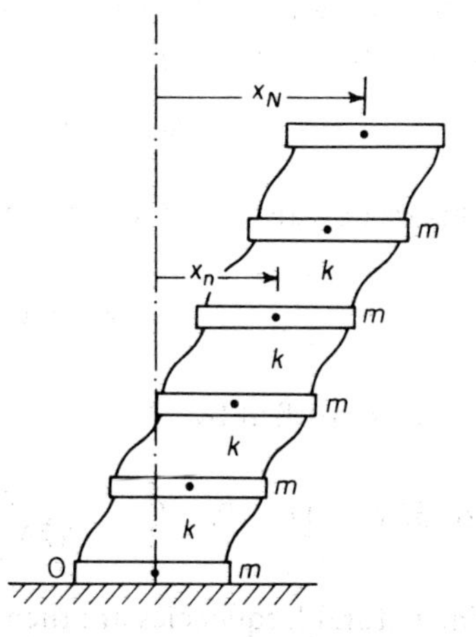

Figure 7.14-1. Repeated structure for difference equation analysis.

which for harmonic motion can be represented in terms of the amplitudes as

$$X_{n+1} - 2\left(1 - \frac{\omega^2 m}{2k}\right)X_n + X_{n-1} = 0 \tag{7.14-2}$$

The solution to this equation is found by substituting

$$X_n = e^{i\beta n} \tag{7.14-3}$$

which leads to the relationship

$$\begin{aligned}\left(1 - \frac{\omega^2 m}{2k}\right) &= \frac{e^{i\beta} + e^{-i\beta}}{2} = \cos \beta \\ \frac{\omega^2 m}{k} &= 2(1 - \cos \beta) = 4 \sin^2 \frac{\beta}{2}\end{aligned} \tag{7.14-4}$$

The general solution for X_n is

$$X_n = A \cos \beta n + B \sin \beta n \tag{7.14-5}$$

where A and B are evaluated from the boundary conditions. At the base, $n = 0$, the amplitude $X_0 = 0$, so that $A = 0$. At the top story, $n = N$, the equation of motion is

$$m\ddot{x}_N = -k(x_N - x_{N-1})$$

which, in terms of the amplitude, becomes

$$X_{N-1} = \left(1 - \frac{\omega^2 m}{k}\right)X_N \tag{7.14-6}$$

Substituting from the general solution, we obtain the following relationship for the evaluation of β

$$\sin \beta(N - 1) = [1 - 2(1 - \cos \beta)] \sin \beta N$$

This result can be reduced to the product form

$$2 \cos \beta(N + \tfrac{1}{2}) \sin \frac{\beta}{2} = 0 \tag{7.14-7}$$

which is satisfied by

$$\cos \beta(N + \tfrac{1}{2}) = 0, \quad \frac{\beta}{2} = \frac{\pi}{2(2N + 1)}, \frac{3\pi}{2(2N + 1)}, \frac{5\pi}{2(2N + 1)}, \cdots \tag{7.14-8}$$

The natural frequencies are then available from Eq. 7.14-4 as

$$\omega = 2\sqrt{k/m}\sin\frac{\beta}{2} \tag{7.14-9}$$

which lead to

$$\omega_1 = 2\sqrt{\frac{k}{m}}\sin\frac{\pi}{2(2N+1)}$$

$$\omega_2 = 2\sqrt{\frac{k}{m}}\sin\frac{3\pi}{2(2N+1)}$$

$$\vdots$$

$$\omega_N = 2\sqrt{\frac{k}{m}}\sin\frac{(2N-1)\pi}{2(2N+1)}$$

Figure 7.14-2 shows a graphical representation of these natural frequencies when $N = 4$.

The method of difference equation presented here is applicable to many other dynamical systems where repeating sections are present. The natural frequencies are always given by Eq. (7.14-9); however, the quantity β must be established for each problem from its boundary conditions.

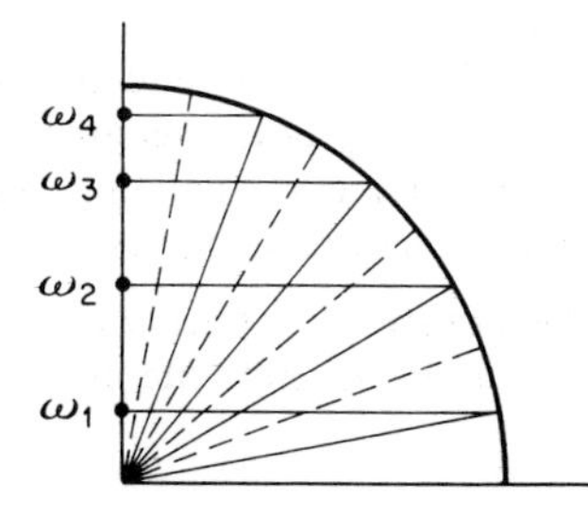

Figure 7.14-2. Natural frequencies of a repeated structure with $N = 4$.

PROBLEMS

7-1 Set up the matrix equation for the system shown in Fig. P7-1, in the form $\{\theta\} = \omega^2[a][J]\{\theta\}$.

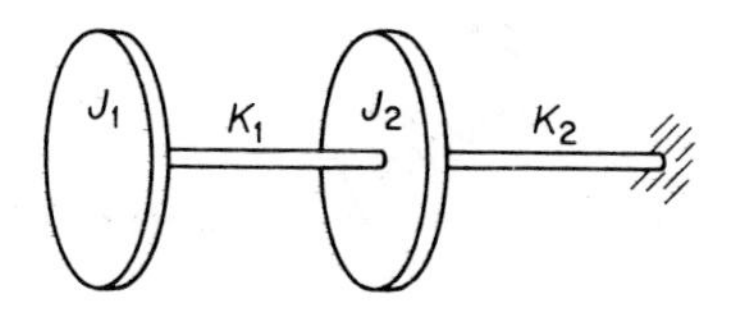

Figure P. 7-1.

7-2 Determine the influence coefficients for the spring-mass system shown in Fig. P7-2.

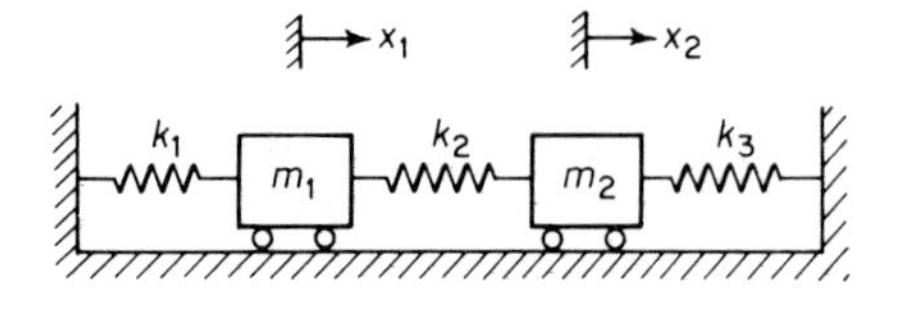

Figure P. 7-2.

7-3 Write the kinetic and potential energy expressions for the system of Prob. 7-2, when

$$k_1 = k, \qquad m_1 = m,$$
$$k_2 = 3k, \qquad m_2 = 2m$$
$$k_3 = 2k,$$

and determine the equation for ω^2 by equating the two energies. Letting $x_2/x_1 = n$, plot ω^2 versus n. Pick off the maximum and minimum values of ω^2 and the corresponding values of n, and show that they represent the two natural modes of the system.

7-4 Determine the influence coefficients for the two-mass cantilever beam shown in Fig. P7-4, and write its equation of motion in matrix form.

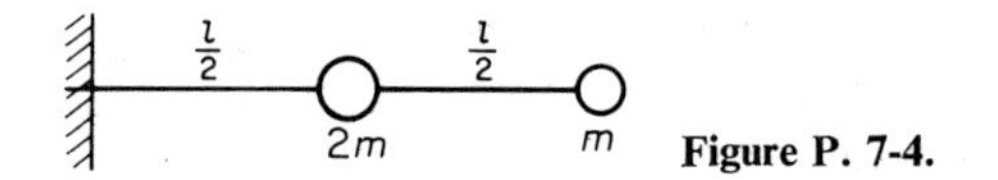

Figure P. 7-4.

7-5 Three equal springs of stiffness k lb/in. are joined at one end, the other ends being arranged symmetrically at 120° from each other, as shown in Fig. P7-5. Prove that the influence coefficients of the junction in a direction making an angle θ with any spring is independent of θ and equal to $1/1.5k$.

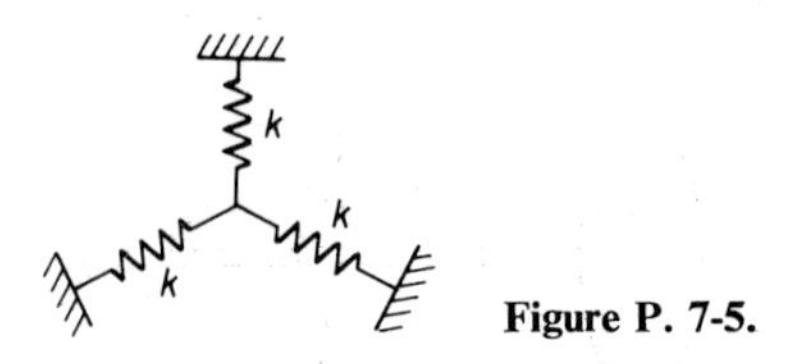

Figure P. 7-5.

7-6 Determine the influence coefficients for the triple pendulum as shown in Fig. P.7-6.

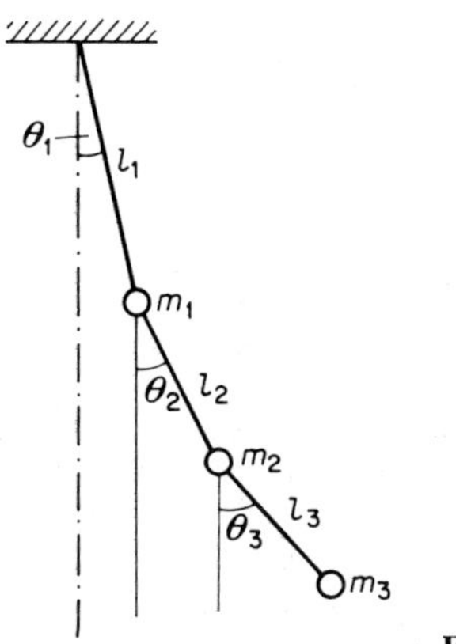

Figure P. 7-6.

7-7 Determine the influence coefficients for the spring-mass system of three degrees of freedom as shown in Fig. P7-7.

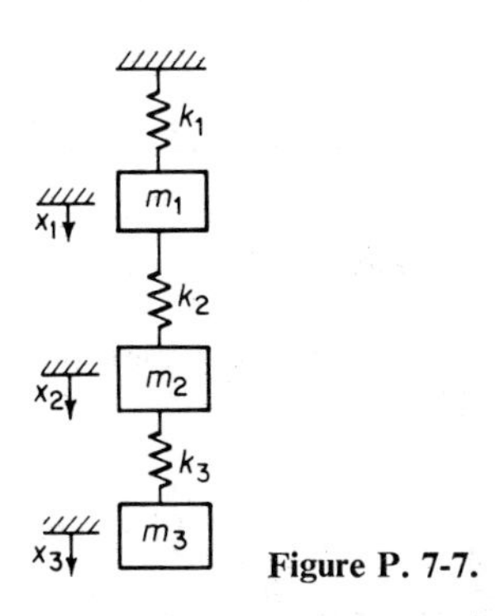

Figure P. 7-7.

7-8 Show that the frequency equation for a torsional system of three disks and two shafts shown in Fig. P7-8, is

$$\omega^4 - \left[\frac{K_1}{J_1} + \frac{K_2}{J_2}\left(1 + \frac{K_1}{K_2} + \frac{J_2}{J_3}\right)\right]\omega^2 + \frac{K_1}{J_1}\frac{K_2}{J_2}\left(\frac{J_1 + J_2 + J_3}{J_3}\right) = 0$$

Figure P. 7-8.

7-9 Derive the frequency equation for a linear spring-mass system containing three masses and two springs, and compare with the result of Prob. 7-8.

7-10 An equivalent torsional system of a propeller, radial engine, and superchanger is shown in Fig. P7-10. Determine the two natural frequencies by using matrix formulation.

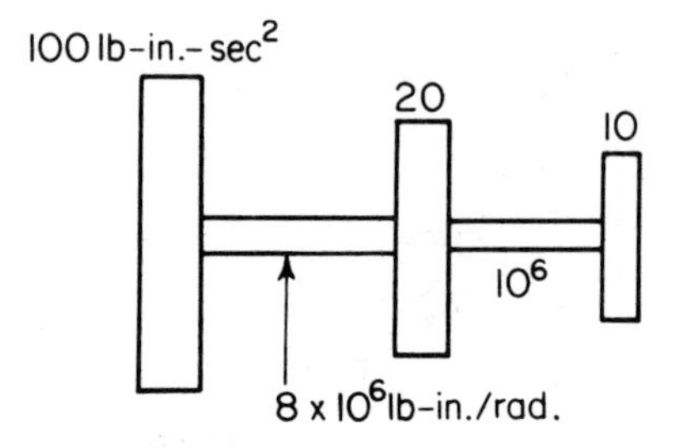

Figure P. 7-10.

7-11 Determine the natural modes of the simplified model of an airplane shown in Fig. P7-11 where $M/m = n$ and the beam of length l is uniform.

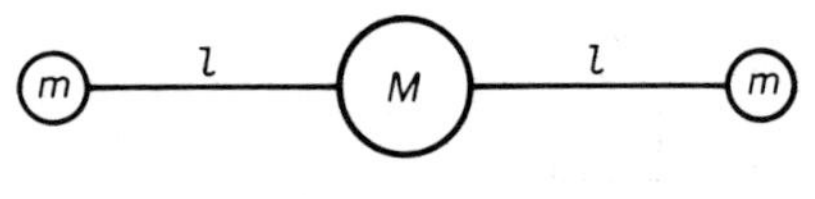

Figure P. 7-11.

7-12 Using Rayleigh's method, estimate the fundamental frequency of the lumped mass system shown in Fig. P7-12.

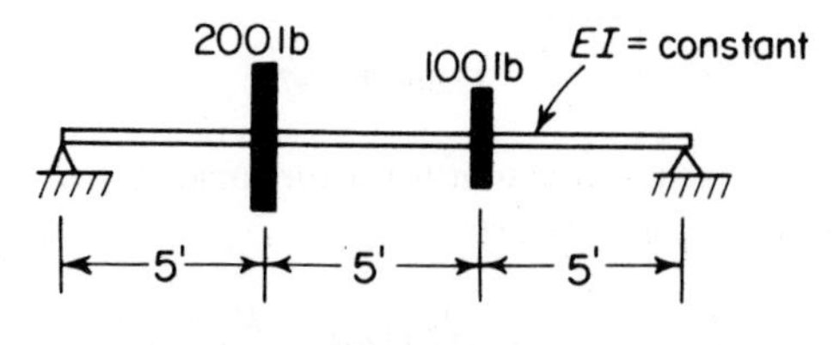

Figure P. 7-12.

7-13 Estimate the fundamental frequency of the lumped mass cantilever beam shown in Fig. P7-13.

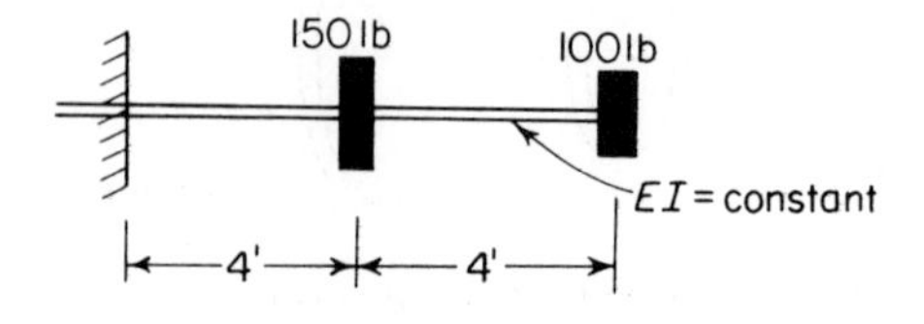

Figure P. 7-13.

7-14 A uniform beam of mass M and stiffness $K = EI/l^3$, shown in Fig. P7-14, is supported on equal springs with total vertical stiffness of k lb/in. Using Rayleigh's method with the deflection $y_{\max} = \sin(\pi x/l) + b$, show that the frequency equation becomes

$$\omega^2 = \frac{2k}{M}\left(\frac{\dfrac{K}{k}\dfrac{\pi^4}{4} + \dfrac{b^2}{2}}{\dfrac{1}{2} + \dfrac{4b}{\pi} + b^2}\right)$$

By $\partial\omega^2/\partial b = 0$, show that the lowest frequency results when

$$b = -\frac{\pi}{4}\left(\frac{1}{2} - \frac{K\pi^4}{2k}\right) \pm \sqrt{\left[\frac{\pi}{2}\left(\frac{1}{2} - \frac{K\pi^4}{2k}\right)\right]^2 + \frac{\pi^4 K}{2k}}$$

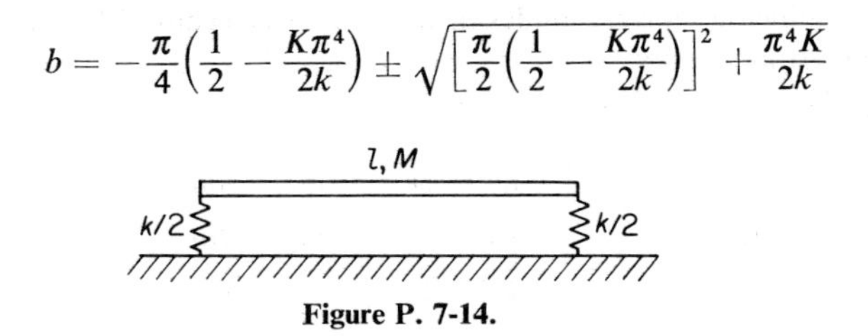

Figure P. 7-14.

7-15 Determine the fundamental frequency of the lumped mass beam of Prob. 7-12 using Eq. 7.4-3.

7-16 Determine the natural frequency for the system of Prob. 7-12 using Eq. 7.4-9.

7-17 Using Dunkerley's equation, determine the fundamental frequency of the three-mass cantilever beam shown in Fig. P7-17.

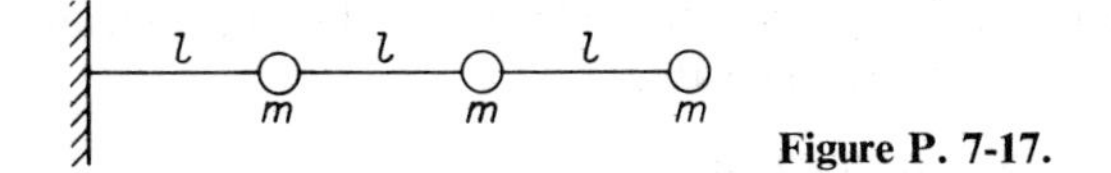

Figure P. 7-17.

7-18 Using Dunkerley's equation, determine the fundamental frequency of the beam shown in Fig. P7-18.

$$W_1 = W, \qquad W_2 = 4W, \qquad W_3 = 2W$$

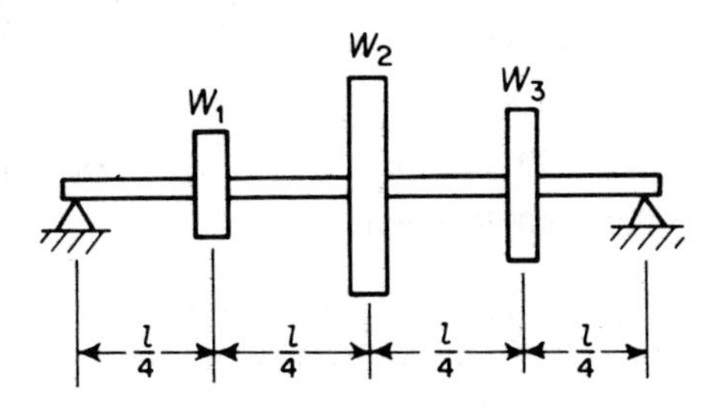

Figure P. 7-18.

7-19 A load of 100 lb at the wing tip of a fighter plane produced a corresponding deflection of 0.78 in. If the fundamental bending frequency of the same wing is 622 cpm, approximate the new bending frequency when a 320-lb fuel tank (including fuel) is attached to the wing tip.

7-20 A given beam was vibrated by an eccentric weight shaker weighing 12 lb at the mid-span, and resonance was found at 435 cps. With an additional weight of 10 lb, the resonant frequency was lowered to 398 cps. Determine the natural frequency of the beam.

7-21 Determine the two natural modes of the system of Prob. 7-10 and show that they are orthogonal.

7-22 Determine the normal modes of the cantilever beam of Prob. 7-13 and verify their orthogonality.

7-23 For the system of Prob. 7-7, let

$$k_1 = 3k, \qquad m_1 = 4m$$
$$k_2 = k, \qquad m_2 = 2m$$
$$k_3 = k, \qquad m_3 = m$$

Set up the matrix equation and determine the three principal modes by iteration. Check the orthogonality of the modes found.

7-24 Using Dunkerley's equation, calculate the fundamental frequency for Prob. 7-23 and compare with the results of matrix iteration.

7-25 Determine the three principal modes of the beam shown in Fig. P7-18 when $W_1 = W_2 = W_3$. Check the fundamental frequency by Dunkerley's equation.

7-26 Show that Dunkerley's equation always results in a fundamental frequency that is lower than the true value.

7-27 Using Holzer's method in matrix form, determine the first two natural frequencies and normal modes of the torsional system shown in Fig. P7-27 with the following values of J and K

$$J_1 = J_2 = J_3 = 10 \text{ lb in. sec}^2$$
$$J_4 = 20 \text{ lb in. sec}^2$$
$$K_1 = K_2 = 1.5 \times 10^6 \text{ lb in./rad}$$
$$K_3 = 2.0 \times 10^6 \text{ lb in./rad}$$

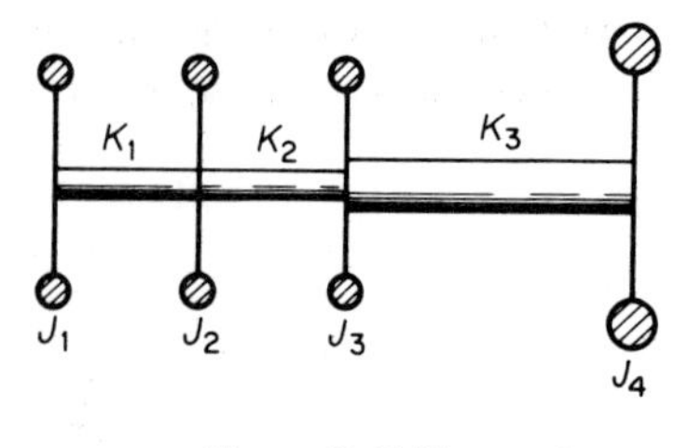

Figure P. 7-27.

7-28 A fighter-plane wing is reduced to a series of disks and shafts for Holzer's analysis as shown in Fig. P7-28. Determine the first two natural frequencies for symmetric and antisymmetric torsional oscillations of the wings, and plot the torsional mode corresponding to each.

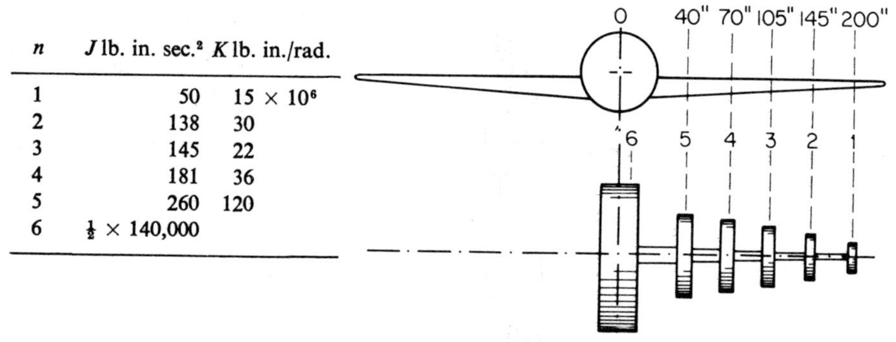

n	J lb. in. sec.^{2}	K lb. in./rad.
1	50	15×10^6
2	138	30
3	145	22
4	181	36
5	260	120
6	$\frac{1}{2} \times 140{,}000$	

Figure P. 7-28.

7-29 If a harmonic torque of 10,000 in. lb at $\omega = 150$ rad/sec is applied to disk 3 of the system shown in Example 7.9-1, determine the amplitude and phase of each disk.

7-30 A torsional system with a torsional damper is shown in Fig. P7-30. Determine the torque-frequency curve for the system.

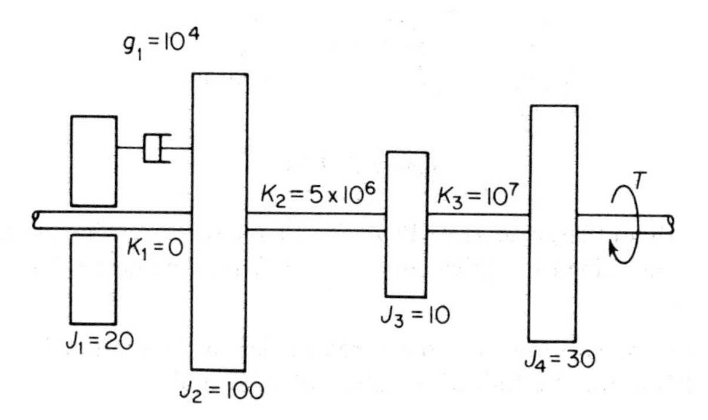

Figure P. 7-30.

7-31 For the system of Example 7.9-3, Fig. 7.9-11, determine the amplitude and phase of each disk at $\omega^2 = 600$ when a torque of 0.040×10^6 in. lb is impressed on disk 4.

7-32 Shown in Fig. P7-32 is a linear system with damping between mass 1 and 2. Carry out a computer analysis for numberical values assigned by the instructor, and determine the amplitude and phase of each mass at a specified frequency.

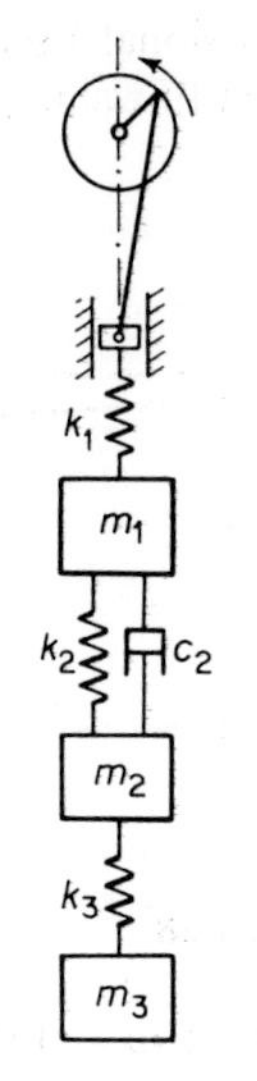

Figure P. 7-32.

7-33 Determine the equivalent torsional system for the geared system shown in Fig. P7-33, and find its natural frequency.

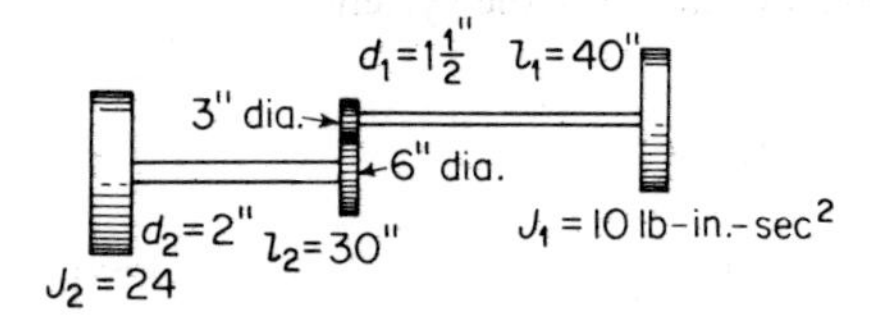

Figure P. 7-33.

7-34 If the small and large gears of Prob. 7-33 have the following inertias, $J' = 2$, $J'' = 6$, determine the equivalent single shaft system and establish the natural frequencies.

7-35 Determine the two lowest natural frequencies of the torsional system shown in Fig. P7-35 for the following values of J, K, and n

$$J_1 = 15 \text{ lb in. sec}^2 \qquad K_1 = 2 \times 10^6 \text{ lb in./rad}$$
$$J_2 = 10 \text{ lb in. sec}^2 \qquad K_2 = 1.6 \times 10^6 \text{ lb in./rad}$$
$$J_3 = 18 \text{ lb in. sec}^2 \qquad K_3 = 1 \times 10^6 \text{ lb in./rad}$$
$$J_4 = 6 \text{ lb in. sec}^2 \qquad K_4 = 4 \times 10^6 \text{ lb in./rad}$$

Speed ratio of drive shaft to axle = 4 to 1.

What are the amplitude ratios of J_2 to J_1 at the natural frequencies?

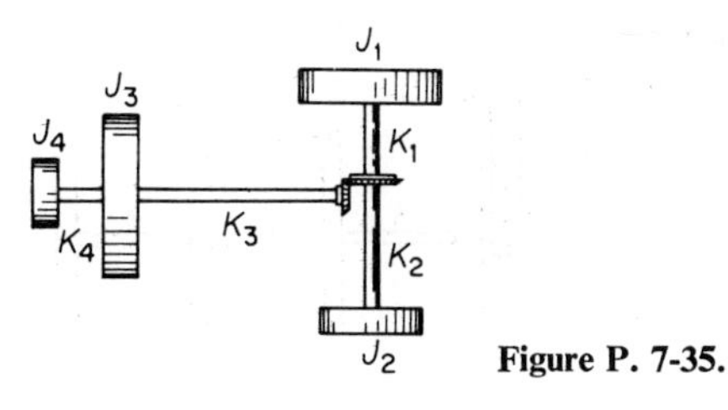

Figure P. 7-35.

7-36 Reduce the torsional system of the automobile shown in (a) to the equivalent torsional system shown in (b). The necessary information is given as follows

J of each rear wheel $= 9.2$ lb in. sec^2
J of flywheel $= 12.3$ lb in. sec^2
transmission speed ratio (drive shaft to engine speed) $= 1.0$ to 3.0
differential speed ratio (axle to drive shaft) $= 1.0$ to 3.5
axle dimensions $= 1\frac{1}{4}$ in. diameter, 25 in. long (each)
drive shaft dimensions $= 1\frac{1}{2}$ in. diameter, 74 in. long
stiffness of crankshaft between cylinders, measured experimentally
$= 6.1 \times 10^6$ lb in./rad
stiffness of crankshaft between cylinder 4 and flywheel $= 4.5 \times 10^6$ lb in./rad

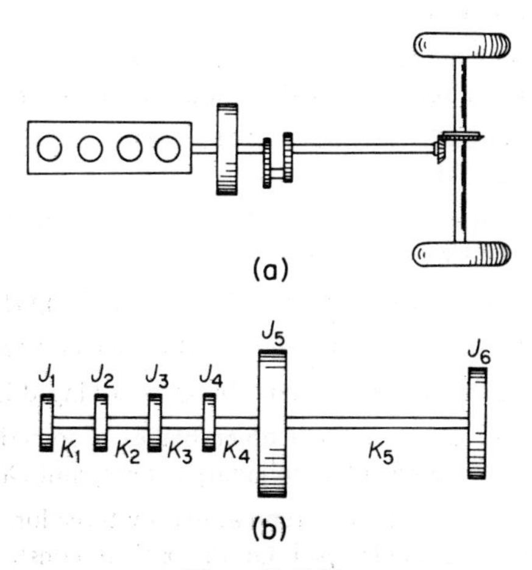

Figure P. 7-36.

7-37 Assume that the J of each cylinder of Prob. 7-36 $= 0.20$ lb in. sec^2 and determine the natural frequencies of the system.

7-38 Determine the equations of motion for the torsional system shown in Fig. P7-38, and arrange them into the matrix iteration form. Solve for the principal modes of oscillation.

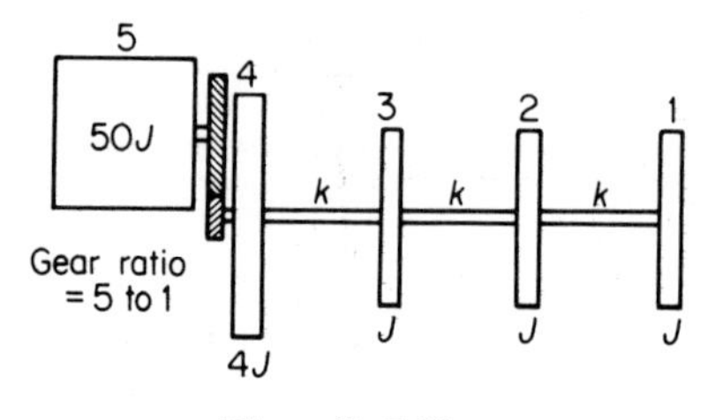

Figure P. 7-38.

7-39 Apply the matrix method to a cantilever beam of length l and mass m at the end, and show that the natural frequency equation is directly obtained.

7-40 Apply the matrix method to a cantilever beam with two equal masses spaced equally a distance l. Show that the boundary conditions of zero slope and deflection lead to the equation

$$\theta_1 = \frac{\frac{1}{2}m\omega^2 lK(5 + \frac{1}{6}m\omega^2 l^2 K)}{1 + \frac{1}{2}l^2 Km\omega^2} = \frac{1 + \frac{3}{2}m\omega^2 l^2 K + (\frac{1}{6}m\omega^2 l^2 K)^2}{2l + \frac{1}{6}m\omega^2 l^3 K}$$

where $K = l/EI$.

Obtain the frequency equation from the above relationship and determine the two natural frequencies.

7-41 Solve Prob. 7-39 by the method of Sec. 7.12(b) when the beam is rotated about an axis through the fixed end with an angular speed Ω.

7-42 Determine the natural frequencies of the cantilever beam of Prob. 7-17 by the method of Sec. 7.12(a).

7-43 From the boundary equation, Eq. (7.12-7), establish the boundary determinant $D(\omega)$ for a simply supported beam.

7-44 Determine the boundary determinant $D(\omega)$ for a clamped-clamped beam.

7-45 Determine the boundary determinant $D(\omega)$ for a clamped-hinged beam.

7-46 Determine the boundary determinant $D(\omega)$ for a hinged-free beam.

7-47 A rotating beam, such as a helicopter blade, is sometimes considered as pinned at the hub. Establish the boundary determinant $D(\omega)$ for such a case.

7-48 Assume a helicopter blade to be represented by three lumped masses equally spaced, with the hub end clamped. On the basis of constant bending stiffness determine the natural frequencies for rotational speed Ω.

7-49 Determine the flexure-torsion vibration for the system shown in Fig. P7-49.

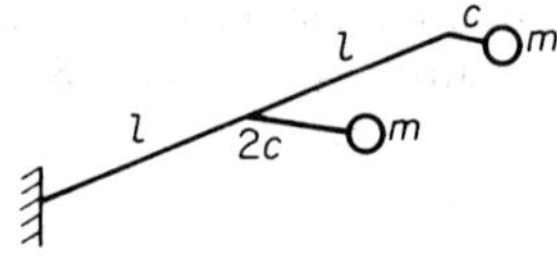

Figure P. 7-49.

7-50 Using the matrix formulation, establish the boundary conditions for the symmetric and antisymmetric bending modes for the system shown in Fig.

P7-50. Plot the boundary determinant against the frequency ω to establish the natural frequencies, and draw the first two mode shapes.

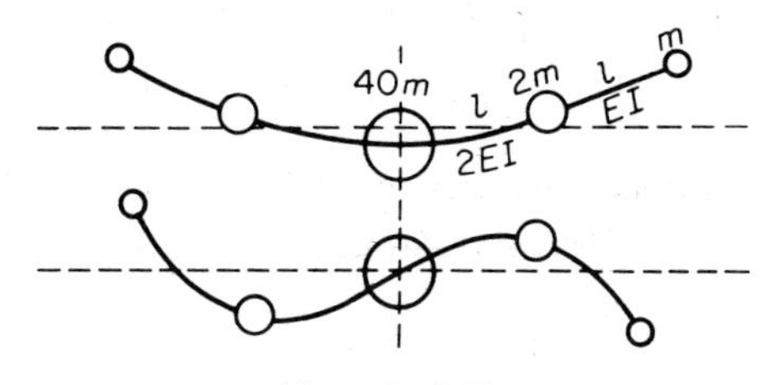

Figure P. 7-50.

7-51 Prove that the elements of the modal matrix [P] for a two degree of freedom system are

$$r_1 = \frac{\omega^2 m}{\mu_1 - 1} \qquad \text{and} \qquad r_2 = \frac{\omega^2 m}{\mu_2 - 1}$$

where the system section is a spring and a mass as in Fig. 7.13-1.

7-52 Show that for the system shown in Fig. P7-52, the natural frequency equation using the procedure of Sec. 7.13 reduces to

$$-\mu_1^n r_2 + \mu_2^n r_1 = 0$$

k m n sections

Figure P. 7-52.

7-53 Letting $\mu_1 = e^{\alpha}$ and $\mu_2 = e^{-\alpha}$ in Eq. (7.13-8), $(A + D)/2 = \cosh\alpha$. Develop the frequency equation in terms of this substitution.

7-54 Reduce the system of Fig. 7.13-3 to an equivalent of the system shown in Fig. 7.13-2.

7-55 Interchanging θ and y, Eq. 7.12-4 may be rearranged to the form

$$\begin{Bmatrix} \theta \\ y \\ M \\ V \end{Bmatrix}_n = \begin{bmatrix} 1 & 0 & \frac{l}{EI} & \frac{l^2}{2EI} \\ l & 1 & \frac{l^2}{2EI} & \frac{l^3}{6EI} \\ 0 & 0 & 1 & l \\ 0 & 0 & 0 & 1 \end{bmatrix} \begin{Bmatrix} \theta \\ y \\ M \\ V \end{Bmatrix}_{n-1}$$

$$= \begin{bmatrix} A' & B \\ 0 & A \end{bmatrix} \begin{Bmatrix} \theta \\ y \\ M \\ V \end{Bmatrix}_{n-1}$$

where B is symmetric about its diagonal. Letting $\delta = (\theta, y)'$ and $L = (M, V)'$, show that the stiffness matrix is

$$\begin{Bmatrix} L_{n-1} \\ L_n \end{Bmatrix} = \left[\begin{array}{c|c} B^{-1} & -B^{-1}A' \\ \hline AB^{-1} & -AB^{-1}A' \end{array}\right] \begin{Bmatrix} \delta_n \\ \delta_{n-1} \end{Bmatrix}$$

7-56 Evaluate the partitioned matrices of Problem 7-55 and show that they are in the form (primes indicate transpose)

$$\begin{Bmatrix} L_{n-1} \\ L_n \end{Bmatrix} = \left[\begin{array}{c|c} \mathcal{A} & \mathcal{B} \\ \hline -\mathcal{B}' & \mathcal{C} \end{array}\right] \begin{Bmatrix} \delta_n \\ \delta_{n-1} \end{Bmatrix}$$

which is expected due to the reciprocity theorem of Betti-Maxwell.*

7-57 Using the notation of Problem 7-56, rewrite Eq. 7.12-6 in the form

$$\begin{Bmatrix} \delta \\ L \end{Bmatrix}_n^R = \left[\begin{array}{c|c} A' & B \\ \hline Q & S \end{array}\right] \begin{Bmatrix} \delta \\ L \end{Bmatrix}_{n-1}^R$$

and show that the determinant of the transfer matrix is equal to unity.

7-58 Set up the difference equations for the torsional system shown in Fig. P7-58. Determine the boundary equations and solve for the natural frequencies.

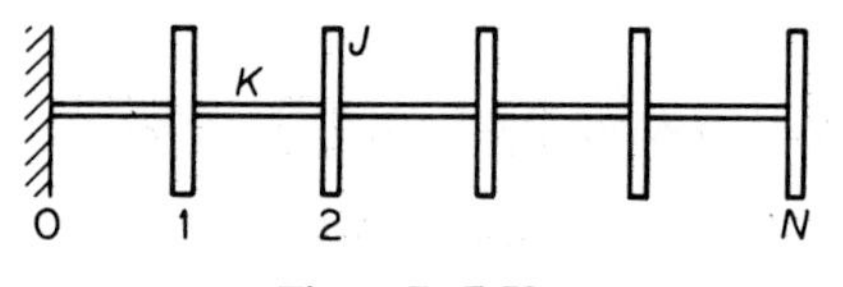

Figure P. 7-58.

7-59 Set up the difference equations for N equal masses on a string with tension T, as shown in Fig. P7-59. Determine the boundary equations and the natural frequencies.

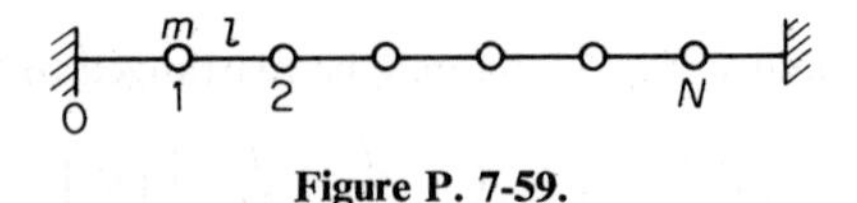

Figure P. 7-59.

7-60 Write the difference equations for the spring-mass system shown in Fig. P7-60 and find the natural frequencies of the system.

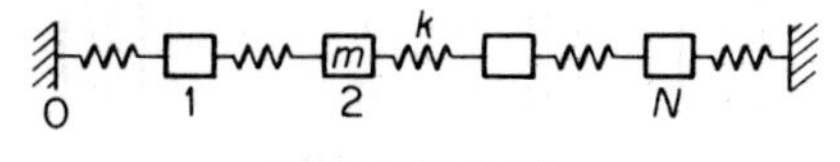

Figure P. 7-60.

*Pestel and Leckie, *Matrix Methods in Elastomechanics*, pp. 151, 215.

7-61 An N-mass pendulum is shown in Fig. P7-61. Determine the difference equations, boundary conditions, and the natural frequencies.

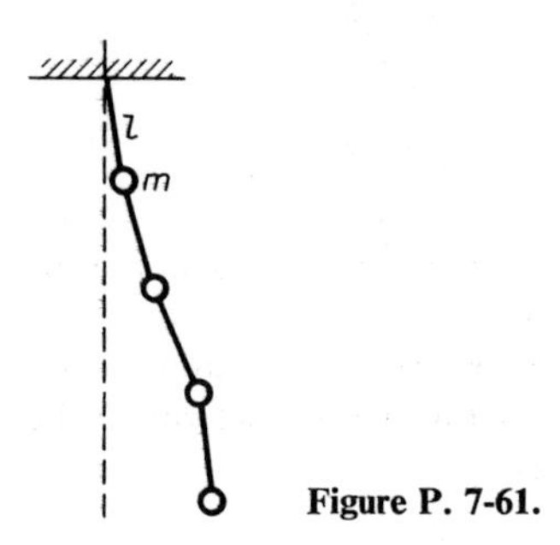

Figure P. 7-61.

7-62 If the left end of the system of Prob. 7-58 is connected to a heavy flywheel, as shown in Fig. P7-62, show that the boundary conditions lead to the equation

$$(-\sin N\beta \cos \beta + \sin N\beta)\left(1 + 4\frac{K}{K_a}\frac{J_a}{J}\sin^2\frac{\beta}{2}\right)$$
$$= -2\frac{J_a}{J}\sin^2\frac{\beta}{2}\sin\beta\cos N\beta$$

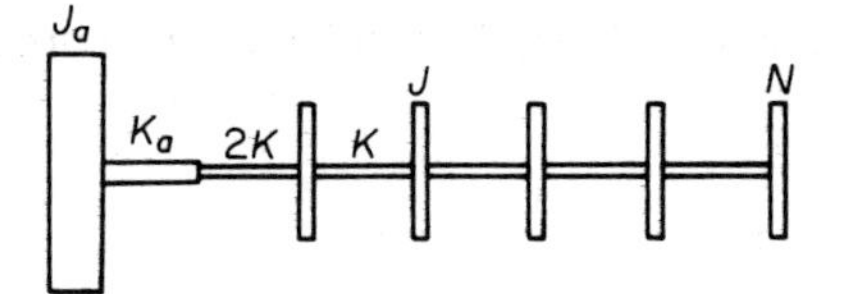

Figure P. 7-62.

7-63 If the top story of a building is restrained by a spring of stiffness K_N, as shown in Fig. P7-63, determine the natural frequencies of the N-story building.

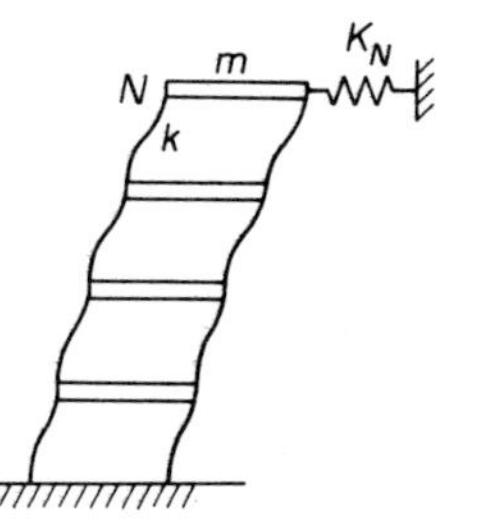

Figure P. 7-63.

7-64 A ladder-type structure is fixed at both ends, as shown in Fig. P7-64. Determine the natural frequencies.

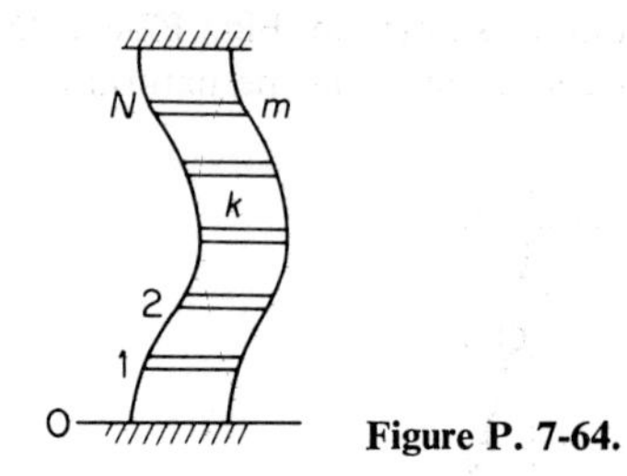

Figure P. 7-64.

7-65 If the base of an N-story building is allowed to rotate against a resisting spring K_θ, as shown in Fig. P7-65, determine the boundary equations and the natural frequencies.

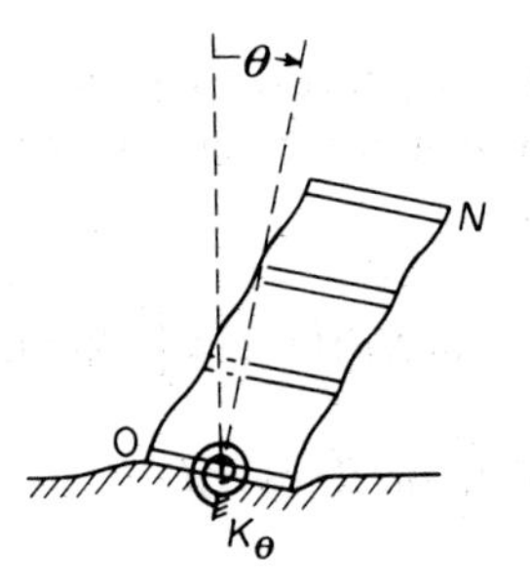

Figure P. 7-65.

7-66 Draw a flow diagram for Prob. 7-27 and write the Fortran program.

8

Continuous Systems

8.1 INTRODUCTION

The systems to be studied in this chapter have continuously distributed mass and elasticity. These bodies are assumed to be homogeneous and isotropic, obeying Hooke's law within the elastic limit. To specify the position of every particle in the elastic body, an infinite number of coordinates are necessary, and such bodies therefore possess an infinite number of degrees of freedom.

In general, the free vibration of these bodies is the sum of the principal modes as previously stated in Chapter 5. For the principal mode of vibration every particle of the body performs simple harmonic motion at the frequency corresponding to the particular root of the frequency equation, each particle passing simultaneously through its respective equilibrium position. If the elastic curve of the body under which the motion is started coincides exactly with one of the principal modes, only that principal mode will be produced. However, the elastic curve resulting from a blow or a sudden removal of forces seldom corresponds to that of a principal mode, and thus all modes are excited. In many cases, however, a particular principal mode can be excited by proper initial conditions.

In this chapter some of the simpler problems of vibration of elastic bodies are taken up. The solutions to these problems are treated in terms of the principal modes of vibration.

8.2 THE VIBRATING STRING

A flexible string of mass ρ per unit length is stretched under tension T. By assuming the lateral deflection y of the string to be small, the change in tension with deflection is negligible and can be ignored.

In Fig. 8.2-1, a free-body diagram of an elementary length dx of the string is shown. Assuming small deflections and slopes, the equation of motion in the y-direction is

$$T\left(\theta + \frac{\partial \theta}{\partial x}dx\right) - T\theta = \rho\, dx \frac{\partial^2 y}{\partial t^2}$$

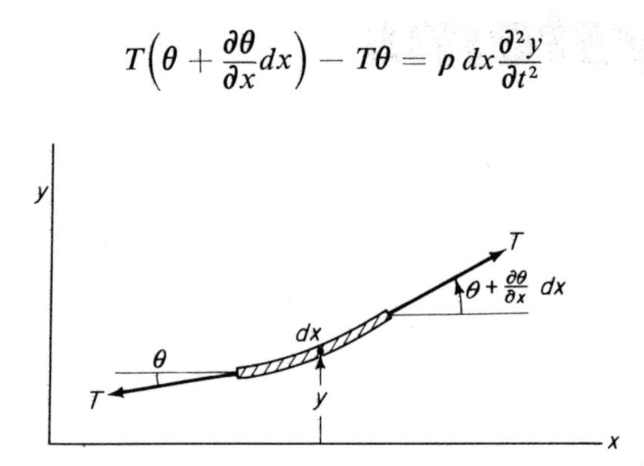

Figure 8.2-1. String element in lateral vibration.

or

$$\frac{\partial \theta}{\partial x} = \frac{\rho}{T}\frac{\partial^2 y}{\partial t^2} \tag{8.2-1}$$

Since the slope of the string is $\theta = \partial y/\partial x$, the above equation reduces to

$$\frac{\partial^2 y}{\partial x^2} = \frac{1}{c^2}\frac{\partial^2 y}{\partial t^2} \tag{8.2-2}$$

where $c = \sqrt{T/\rho}$ can be shown to be the velocity of wave propagation along the string.

The general solution of Eq. (8.2-2) can be expressed in the form

$$y = F_1(ct - x) + F_2(ct + x) \tag{8.2-3}$$

where F_1 and F_2 are arbitrary functions. Regardless of the type of function

F, the argument $(ct \pm x)$ upon differentiation leads to the equation

$$\frac{\partial^2 F}{\partial x^2} = \frac{1}{c^2}\frac{\partial^2 F}{\partial t^2} \tag{8.2-4}$$

and hence the differential equation is satisfied.

Considering the component $y = F_1(ct - x)$, its value is determined by the argument $(ct - x)$ and hence by a range of values of t and x. For example, if $c = 10$, the equation for $y = F_1(100)$ is satisfied by $t = 0$, $x = -100$; $t = 1$, $x = -90$; $t = 2$, $x = -80$, etc. Therefore, the wave profile moves in the positive x-direction with speed c. In a similar manner we can show that $F_2(ct + x)$ represents a wave moving toward the negative x-direction with speed c. We therefore refer to c as the velocity of wave propagation.

One method of solving partial differential equations is that of separation of variables. In this method the solution is assumed in the form

$$y(x, t) = Y(x)G(t) \tag{8.2-5}$$

By substitution into Eq. (8.2-2) we obtain

$$\frac{1}{Y}\frac{d^2 Y}{dx^2} = \frac{1}{c^2}\frac{1}{G}\frac{d^2 G}{dt^2} \tag{8.2-6}$$

Since the left side of this equation is independent of t, whereas the right side is independent of x, it follows that each side must be a constant. Letting this constant be $-(\omega/c)^2$ we obtain two ordinary differential equations

$$\frac{d^2 Y}{dx^2} + \left(\frac{\omega}{c}\right)^2 Y = 0 \tag{8.2-7}$$

$$\frac{d^2 G}{dt^2} + \omega^2 G = 0 \tag{8.2-8}$$

with the general solutions

$$Y = A \sin \frac{\omega}{c} x + B \cos \frac{\omega}{c} x \tag{8.2-9}$$

$$G = C \sin \omega t + D \cos \omega t \tag{8.2-10}$$

The arbitrary constants A, B, C, D depend on the boundary conditions and the initial conditions. For example, if the string is stretched between two fixed points with distance l between them, the boundary conditions are $y(0, t) = y(l, t) = 0$. The condition that $y(0, t) = 0$ will require that $B = 0$ so that the solution will appear as

$$y = (C \sin \omega t + D \cos \omega t) \sin \frac{\omega}{c} x \tag{8.2-11}$$

The condition $y(l, t) = 0$ then leads to the equation

$$\sin \frac{\omega l}{c} = 0$$

or

$$\frac{\omega_n l}{c} = \frac{2\pi l}{\lambda} = n\pi, \qquad n = 1, 2, 3 \cdots$$

and $\lambda = c/f$ is the wavelength and f is the frequency of oscillation. Each n represents a normal-mode vibration with natural frequency determined from the equation

$$f_n = \frac{n}{2l} c = \frac{n}{2l}\sqrt{\frac{T}{\rho}}, \qquad n = 1, 2, 3, \cdots \tag{8.2-12}$$

The mode shape is sinusoidal with the distribution

$$Y = \sin n\pi \frac{x}{l} \tag{8.2-13}$$

In the more general case of free vibration initiated in any manner, the solution will contain many of the normal modes and the equation for the displacement may be written as

$$y(x, t) = \sum_{n=1}^{\infty} (C_n \sin \omega_n t + D_n \cos \omega_n t) \sin \frac{n\pi x}{l} \tag{8.2-14}$$

$$\omega_n = \frac{n\pi c}{l}$$

Fitting this equation to the initial conditions of $y(x, 0)$ and $\dot{y}(x, 0)$, the C_n and D_n can be evaluated.

Example 8.2-1

A uniform string of length l is fixed at the ends and stretched under tension T. If the string is displaced into an arbitrary shape $y(x, 0)$ and released, determine C_n and D_n of Eq. (8.2-14).

Solution: At $t = 0$, the displacement and velocity are

$$y(x, 0) = \sum_{n=1}^{\infty} D_n \sin \frac{n\pi x}{l}$$

$$\dot{y}(x, 0) = \sum_{n=1}^{\infty} \omega_n C_n \sin \frac{n\pi x}{l} = 0$$

Multiplying each equation by $\sin k\pi x/l$ and integrating from $x = 0$ to $x = 1$ all of the terms on the right side will be zero, except the term $n = k$. Thus we arrive at the result

$$D_k = \frac{2}{l}\int_0^l y(x, 0)\sin\frac{k\pi x}{l}\,dx$$

$$C_k = 0 \qquad\qquad k = 1, 2, 3, \cdots$$

8.3 LONGITUDINAL VIBRATION OF RODS

The rod considered in this section is assumed to be thin and uniform along its length. Due to axial forces there will be displacements u along the rod which will be a function of both the position x and the time t. Since the rod has an infinite number of natural modes of vibration, the distribution of the displacement will differ with each mode.

Let us consider an element of this rod of length dx (Fig. 8.3-1). If u is

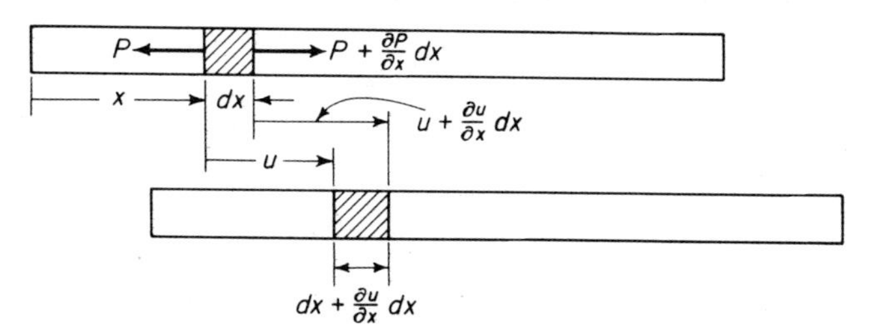

Figure 8.3-1. Displacement of rod element.

the displacement at x, the displacement at $x + dx$ will be $u + (\partial u/\partial x)\,dx$. It is evident then that the element dx in the new position has changed in length by an amount $(\partial u/\partial x)\,dx$, and thus the unit strain is $\partial u/\partial x$. Since from Hooke's law the ratio of unit stress to unit strain is equal to the modulus of elasticity E, we can write

$$\frac{\partial u}{\partial x} = \frac{P}{AE} \qquad (8.3\text{-}1)$$

where A is the cross-sectional area of the rod. Differentiating with respect to x

$$AE\frac{\partial^2 u}{\partial x^2} = \frac{\partial P}{\partial x} \qquad (8.3\text{-}2)$$

We now apply Newton's law of motion for the element and equate the unbalanced force to the product of the mass and acceleration of the element

$$\frac{\partial P}{\partial x} dx = \rho \frac{A\, dx}{g} \frac{\partial^2 u}{\partial t^2} \tag{8.3-3}$$

where ρ is the density of the rod in pounds per unit volume. Eliminating $\partial P/\partial x$ between Eqs. (8.3-2) and (8.3-3), we obtain the partial differential equation

$$\frac{\partial^2 u}{\partial t^2} = \left(\frac{Eg}{\rho}\right) \frac{\partial^2 u}{\partial x^2} \tag{8.3-4}$$

or

$$\frac{\partial^2 u}{\partial x^2} = \frac{1}{c^2} \frac{\partial^2 u}{\partial t^2} \tag{8.3-5}$$

which is similar to that of Eq. (8.2-2) for the string. The velocity of propagation of the displacement or stress wave in the rod is then equal to

$$c = \sqrt{Eg/\rho} \tag{8.3-6}$$

and a solution of the form

$$u(x, t) = U(x)G(t) \tag{8.3-7}$$

will result in two ordinary differential equations similar to Eqs. (8.2-7) and (8.2-8), with

$$U(x) = A \sin \frac{\omega}{c} x + B \cos \frac{\omega}{c} x \tag{8.3-8}$$

$$G(t) = C \sin \omega t + D \cos \omega t \tag{8.3-9}$$

Example 8.3-1

> Determine the natural frequencies and mode shapes of a free-free rod (a rod with both ends free).

Solution: For such a bar, the stress at the ends must be zero. Since the stress is given by the equation $E\, \partial u/\partial x$, the unit strain at the ends must also be zero; that is,

$$\frac{\partial u}{\partial x} = 0 \text{ at } x = 0, \text{ and } x = l$$

The two equations corresponding to the above boundary conditions are therefore

$$\left(\frac{\partial u}{\partial x}\right)_{x=0} = A\frac{\omega}{c}(C \sin \omega t + D \cos \omega t) = 0$$

$$\left(\frac{\partial u}{\partial x}\right)_{x=l} = \frac{\omega}{c}\left(A \cos \frac{\omega l}{c} - B \sin \frac{\omega l}{c}\right)(C \sin \omega t + D \cos \omega t) = 0$$

Since these equations must be true for any time t, A must be equal to zero from the first equation. Since B must be finite in order to have vibration, the second equation is satisfied when

$$\sin \frac{\omega l}{c} = 0$$

or

$$\frac{\omega_n l}{c} = \omega_n l\sqrt{\rho/Eg} = \pi, 2\pi, 3\pi, \ldots, n\pi$$

The frequency of vibration is thus given by

$$\omega_n = \frac{n\pi}{l}\sqrt{\frac{Eg}{\rho}}, \qquad f_n = \frac{n}{2l}\sqrt{\frac{Eg}{\rho}}$$

where n represents the order of the mode. The solution of the free-free rod with zero initial displacement can then be written as

$$u = u_0 \cos \frac{n\pi}{l} x \sin \frac{n\pi}{l}\sqrt{\frac{Eg}{\rho}}t$$

The amplitude of the longitudinal vibration along the rod is therefore a cosine wave having n nodes.

8.4 TORSIONAL VIBRATION OF RODS

The equation of motion of a rod in torsional vibration is similar to that of longitudinal vibration of rods discussed in the preceding section.

Letting x be measured along the length of the rod, the angle of twist in any length dx of the rod due to a torque T is

$$d\theta = \frac{T\,dx}{I_p G} \tag{8.4-1}$$

where $I_p G$ is the torsional stiffness given by the product of the polar moment of inertia I_P of the cross-sectional area and the shear modulus of elasticity G. The torque on the two faces of the element being T and $T + (\partial T/\partial x)\,dx$, as shown in Fig. 8.4-1, the net torque from Eq. (8.4-1) becomes

$$\frac{\partial T}{\partial x}dx = I_P G \frac{\partial^2 \theta}{\partial x^2}dx \tag{8.4-2}$$

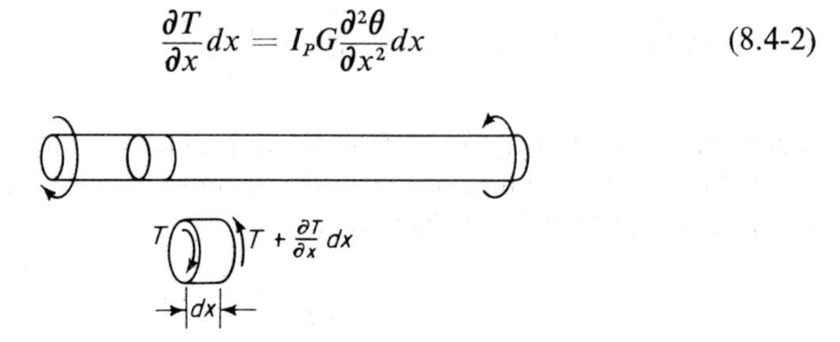

Figure 8.4-1. Torque acting on an element dx.

Equating this torque to the product of the mass moment of inertia $(\rho/g)I_P\,dx$ of the element and the angular acceleration $\partial^2\theta/\partial t^2$, where ρ is the density of the rod in pounds per unit volume, the differential equation of motion becomes

$$\frac{\rho}{g} I_P\,dx \frac{\partial^2 \theta}{\partial t^2} = I_P G \frac{\partial^2 \theta}{\partial x^2}dx, \qquad \frac{\partial^2 \theta}{\partial t^2} = \left(\frac{Gg}{\rho}\right)\frac{\partial^2 \theta}{\partial x^2} \tag{8.4-3}$$

This equation is of the same form as that of longitudinal vibration of rods where θ and Gg/ρ replace u and Eg/ρ, respectively. The general solution may hence be written immediately by comparison as

$$\theta = \left(A \sin \omega\sqrt{\frac{\rho}{Gg}}x + B \cos \omega\sqrt{\frac{\rho}{Gg}}x\right)(C \sin \omega t + D \cos \omega t) \tag{8.4-4}$$

Example 8.4-1

Determine the equation for the natural frequencies of a uniform rod in torsional oscillation with one end fixed and the other end free, as in Fig. 8.4-2.

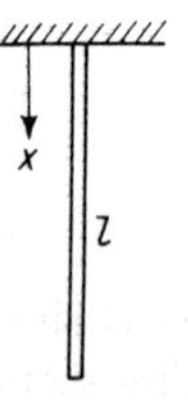

Figure 8.4-2.

Solution: Starting with equation

$$\theta = (A \sin \omega\sqrt{\rho/Gg}x + B \cos \omega\sqrt{\rho/Gg}x) \sin \omega t$$

apply the boundary conditions, which are

(1) when $x = 0$, $\theta = 0$,

(2) when $x = l$, torque $= 0$, or

$$\frac{\partial \theta}{\partial x} = 0$$

Boundary condition (1) results in $B = 0$.
Boundary condition (2) results in the equation

$$\cos \omega \sqrt{\rho/Gg}\, l = 0$$

which is satisfied by the following angles

$$\omega_n \sqrt{\frac{\rho}{Gg}}\, l = \frac{\pi}{2}, \frac{3\pi}{2}, \frac{5\pi}{2}, \ldots, \left(n + \frac{1}{2}\right)\pi$$

The natural frequencies of the bar are hence determined by the equation

$$\omega_n = \left(n + \frac{1}{2}\right)\frac{\pi}{l}\sqrt{\frac{Gg}{\rho}}$$

where $n = 0, 1, 2, 3, \cdots$

Example 8.4-2

The drill pipe of an oil well terminates at the lower end to a rod containing a cutting bit. Derive the expression for the natural frequencies, assuming the drill pipe to be uniform and fixed at the upper end and the rod and cutter to be represented by an end mass of moment of inertia J_0, as shown in Fig. 8.4-3.

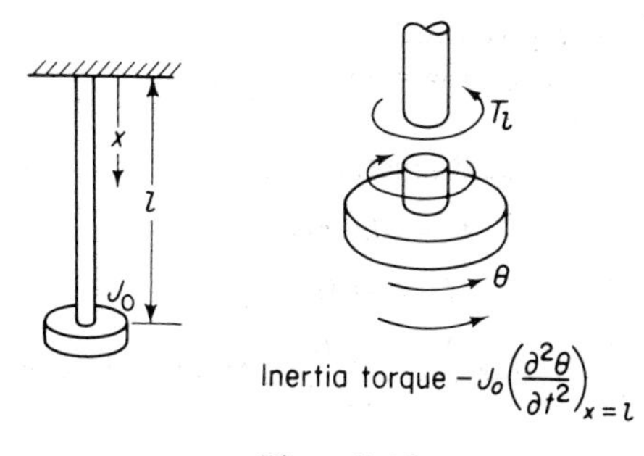

Figure 8.4-3.

Solution: The boundary condition at the upper end is $x = 0, \theta = 0$, which requires B to be zero in Eq. (8.4-4).

For the lower end, the torque on the shaft is due to the inertia torque of the end disk, as shown by the free-body diagram of Fig. 8.4-3. The inertia torque of the disk is $-J_0(\partial^2\theta/\partial^2 t)_{x=l} = J_0\omega^2(\theta)_{x=l}$, whereas the shaft torque from Eq. (8.4-1) is $T_l = GI_P(d\theta/dx)_{x=l}$. Equating the two, we have

$$GI_P\left(\frac{d\theta}{dx}\right)_{x=l} = J_0\omega^2(\theta)_{x=l}$$

Substituting from Eq. (8.4-4) with $B = 0$

$$GI_P\omega\sqrt{\frac{\rho}{Gg}}\cos\omega\sqrt{\frac{\rho}{Gg}}l = J_0\omega^2\sin\omega\sqrt{\frac{\rho}{Gg}}l$$

$$\tan\omega l\sqrt{\frac{\rho}{Gg}} = \frac{I_P}{\omega J_0}\sqrt{\frac{G\rho}{g}} = \frac{I_P\rho l}{gJ_0\omega l}\sqrt{\frac{Gg}{\rho}}$$

$$= \frac{J_{\text{rod}}}{J_0\omega l}\sqrt{\frac{Gg}{\rho}}$$

This equation is of the form

$$\beta\tan\beta = \frac{J_{\text{rod}}}{J_0}, \qquad \beta = \omega l\sqrt{\frac{\rho}{Gg}}$$

which can be solved graphically or from tables.*

Example 8.4-3

Using the frequency equation developed in the previous example, determine the first two natural frequencies of an oil-well drill pipe 500 ft long, fixed at the upper end and terminating at the lower end to a drill collar 120 ft long. The average values for the drill pipe and drill collar are given as

Drill pipe: outside diameter $= 4\frac{1}{2}$ in.

$$\text{inside diameter} = 3.83 \text{ in.}$$

$$I_P = 0.00094 \text{ ft}^4$$

$$l = 5000 \text{ ft}$$

$$J_{\text{rod}} = I_P\frac{\rho l}{g} = 0.00094 \times \frac{490}{32.2} \times 5000 = 71.4 \text{ lb ft sec}^2$$

Drill collar: outside diameter $= 7\frac{5}{8}$ in.

$$\text{inside diameter} = 2.0 \text{ in.}$$

$$J_0 = 0.244 \times 120 \text{ ft} = 29.3 \text{ lb ft sec}^2$$

*See Jahnke and Emde, *Tables of Functions*, 4th Ed. (Dover Publications, Inc., 1945), Table V, p. 32.

Solution: The equation to be solved is

$$\beta \tan \beta = \frac{J_{\text{rod}}}{J_0} = 2.44$$

From Table V, p. 32, Jahnke and Emde, $\beta = 1.135, 3.722, \ldots$

$$\beta = \omega l \sqrt{\frac{\rho}{Gg}} = 5000\omega \sqrt{\frac{490}{12 \times 10^6 \times 12^2 \times 32.2}} = 0.470\omega$$

Solving for ω, the first two natural frequencies are found to be

$$\omega_1 = \frac{1.135}{0.470} = 2.41 \text{ rad/sec} = 0.384 \text{ cps}$$

$$\omega_2 = \frac{3.722}{0.470} = 7.93 \text{ rad/sec} = 1.26 \text{ cps}$$

8.5 THE EULER EQUATION FOR THE BEAM

To determine the differential equation for the lateral vibration of beams, consider the forces and moments acting on an element of the beam shown in Fig. 8.5-1.

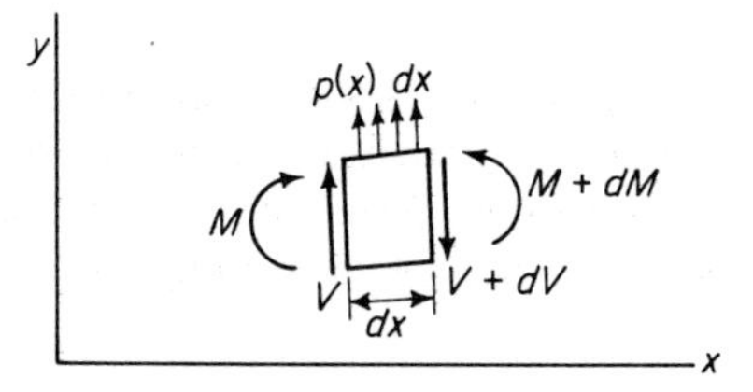

Figure 8.5-1.

V and M are shear and bending moments, respectively, and $p(x)$ represents the loading per unit length of the beam.

By summing forces in the y-direction

$$dV - p(x)\,dx = 0 \tag{8.5-1}$$

By summing moments about any point on the right face of the element

$$dM - V\,dx - \tfrac{1}{2}p(x)(dx)^2 = 0 \tag{8.5-2}$$

In the limiting process these equations result in the following important relationships

$$\frac{dV}{dx} = p(x), \qquad \frac{dM}{dx} = V \tag{8.5-3}$$

The first part of Eq. (8.5-3) states that the rate of change of shear along the length of the beam is equal to the loading per unit length, and the second states that the rate of change of the moment along the beam is equal to the shear.

From Eq. (8.5-3) we obtain the following

$$\frac{d^2M}{dx^2} = \frac{dV}{dx} = p(x) \tag{8.5-4}$$

The bending moment is related to the curvature by the flexure equation, which, for the coordinates indicated in Fig. 8.5-1, is

$$M = EI\frac{d^2y}{dx^2} \tag{8.5-5}$$

Substituting this relation into Eq. (8.5-4), we obtain

$$\frac{d^2}{dx^2}\left(EI\frac{d^2y}{dx^2}\right) = p(x) \tag{8.5-6}$$

For a beam vibrating about its static equilibrium position under its own weight, the load per unit length is equal to the inertia load due to its mass and acceleration. Since the inertia force is in the same direction as $p(x)$ as shown in Fig. 8.5-1, we have, by assuming harmonic motion

$$p(x) = \frac{w}{g}\omega^2 y \tag{8.5-7}$$

where w/g is the mass per unit length of the beam. Using this relation, the equation for the lateral vibration of the beam reduces to

$$\frac{d^2}{dx^2}\left(EI\frac{d^2y}{dx^2}\right) - \frac{w}{g}\omega^2 y = 0 \tag{8.5-8}$$

In the special case where the flexural rigidity EI is a constant, the above equation may be written as

$$EI\frac{d^4y}{dx^4} - \frac{w}{g}\omega^2 y = 0 \tag{8.5-9}$$

On substituting

$$\beta^4 = \frac{w}{g} \frac{\omega^2}{EI} \tag{8.5-10}$$

we obtain the fourth-order differential equation

$$\frac{d^4 y}{dx^4} - \beta^4 y = 0 \tag{8.5-11}$$

for the vibration of a uniform beam.

The general solution of Eq. (8.5-11) can be shown to be

$$y = A \cosh \beta x + B \sinh \beta x + C \cos \beta x + D \sin \beta x \tag{8.5-12}$$

To arrive at this result, we assume a solution of the form

$$y = e^{ax}$$

which will satisfy the differential equation when

$$a = \pm\, \beta, \text{ and } a = \pm\, i\beta$$

Since

$$e^{\pm \beta x} = \cosh \beta x \pm \sinh \beta x$$
$$e^{\pm i\beta x} = \cos \beta x \pm i \sin \beta x$$

the solution in the form of Eq. (8.5-12) is readily established.

The natural frequencies of vibration are found from Eq. (8.5-10) to be

$$\omega_n = \beta^2 \sqrt{gEI/w}$$

where the number β depends on the boundary conditions of the problem. The following table lists numerical values of $(\beta l)^2$ for typical end conditions.

Beam configuration	$(\beta_1 l)^2$ Fundamental	$(\beta_2 l)^2$ Second Mode	$(\beta_3 l)^2$ Third Mode
Simply supported	9.87	39.5	88.9
Cantilever	3.52	22.4	61.7
Free-free	22.4	61.7	121.0
Clamped-clamped	22.4	61.7	121.0
Clamped-hinged	15.4	50.0	104.0
Hinged-free	0	15.4	50.0

Example 8.5-1

Determine the natural frequencies of vibration of a uniform beam clamped at one end and free at the other.

Solution: The boundary conditions are

$$\text{at } x = 0 \begin{cases} y = 0 \\ \dfrac{dy}{dx} = 0 \end{cases}$$

$$\text{at } x = l \begin{cases} M = 0 \quad \text{or} \quad \dfrac{d^2y}{dx^2} = 0 \\ V = 0 \quad \text{or} \quad \dfrac{d^3y}{dx^3} = 0 \end{cases}$$

Substituting these boundary conditions in the general solution, we obtain

$$(y)_{x=0} = A + C = 0, \quad \therefore A = -C$$

$$\left(\frac{dy}{dx}\right)_{x=0} = \beta[A \sinh \beta x + B \cosh \beta x - C \sin \beta x + D \cos \beta x]_{x=0} = 0$$

$$\beta[B + D] = 0, \quad \therefore \quad B = -D$$

$$\left(\frac{d^2y}{dx^2}\right)_{x=l} = \beta^2[A \cosh \beta l + B \sinh \beta l - C \cos \beta l - D \sin \beta l] = 0$$

$$A(\cosh \beta l + \cos \beta l) + B(\sinh \beta l + \sin \beta l) = 0$$

$$\left(\frac{d^3y}{dx^3}\right)_{x=l} = \beta^3[A \sinh \beta l + B \cosh \beta l + C \sin \beta l - D \cos \beta l] = 0$$

$$A(\sinh \beta l - \sin \beta l) + B(\cosh \beta l + \cos \beta l) = 0$$

From the last two equations we obtain

$$\frac{\cosh \beta l + \cos \beta l}{\sinh \beta l - \sin \beta l} = \frac{\sinh \beta l + \sin \beta l}{\cosh \beta l + \cos \beta l}$$

which reduces to

$$\cosh \beta l \cos \beta l + 1 = 0$$

This last equation is satisfied by a number of values of βl, corresponding to each normal mode of oscillation, which for the first and second modes are 1.875 and 4.695, respectively. The natural frequency for the first mode is hence given as

$$\omega_1 = \frac{1.875^2}{l^2}\sqrt{\frac{gEI}{w}} = \frac{3.515}{l^2}\sqrt{\frac{gEI}{w}}$$

8.6 EFFECT OF ROTARY INERTIA AND SHEAR DEFORMATION

The Timoshenko theory accounts for both the rotary inertia and shear deformation of the beam. The free-body diagram and the geometry for the beam element are shown in Fig. 8.6-1. If the shear deformation is zero,

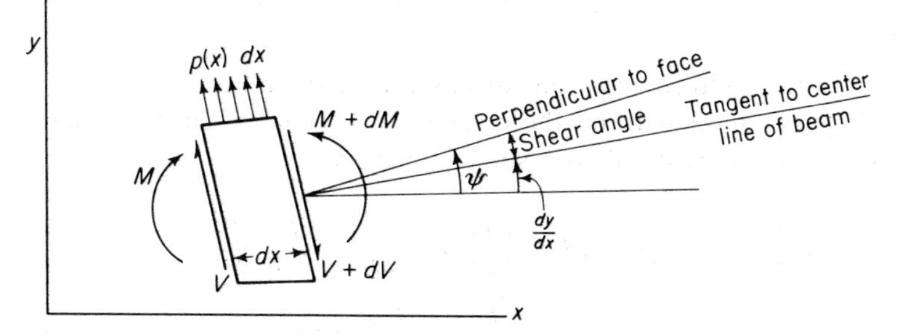

Figure 8.6-1. Effect of shear deformation.

the center line of the beam element will coincide with the perpendicular to the face of the cross section. Due to shear, the rectangular element tends to go into a diamond shape without rotation of the face and, the slope of the center line is diminished by the shear angle $(\psi - dy/dx)$. The following quantities can then be defined

$$y = \text{deflection of the center line of the beam}$$

$$\frac{dy}{dx} = \text{slope of the center line of the beam}$$

$$\psi = \text{slope due to bending}$$

$$\psi - \frac{dy}{dx} = \text{loss of slope, equal to the shear angle}$$

There are two elastic equations for the beam, which are

$$\psi - \frac{dy}{dx} = \frac{V}{kAG} \tag{8.6-1}$$

$$\frac{d\psi}{dx} = \frac{M}{EI} \tag{8.6-2}$$

where A is the cross-sectional area, G the shear modulus, k a factor depending on the shape of the cross section, and EI the bending stiffness. In addition, there are two dynamical equations

$$\text{(moment)} \quad J\ddot{\psi} = \frac{dM}{dx} - V \tag{8.6-3}$$

$$\text{(force)} \quad m\ddot{y} = -\frac{dV}{dx} + p(x, t) \tag{8.6-4}$$

where J and m are the rotary inertia and mass of the beam per unit length.

Substituting the elastic equations into the dynamical equations, we have

$$\frac{d}{dx}\left(EI\frac{d\psi}{dx}\right) + kAG\left(\frac{dy}{dx} - \psi\right) - J\ddot{\psi} = 0 \tag{8.6-5}$$

$$m\ddot{y} - \frac{d}{dx}\left[kAG\left(\frac{dy}{dx} - \psi\right)\right] - p(x, t) = 0 \tag{8.6-6}$$

which are the coupled equations of motion for the beam.

If ψ is eliminated and the cross section remains constant, these two equations can be reduced to a single equation

$$EI\frac{\partial^4 y}{\partial x^4} + m\frac{\partial^2 y}{\partial t^2} - \left(J + \frac{EIm}{kAG}\right)\frac{\partial^4 y}{\partial x^2 \partial t^2} + \frac{Jm}{kAG}\frac{\partial^4 y}{\partial t^4} = p(x, t)$$
$$+ \frac{J}{kAG}\frac{\partial^2 p}{\partial t^2} - \frac{EI}{kAG}\frac{\partial^2 p}{\partial x^2} \tag{8.6-7}$$

It is evident then that the Euler equation

$$EI\frac{\partial^4 y}{\partial x^4} + m\frac{\partial^2 y}{\partial t^2} = p(x, t)$$

is a special case of the general beam equation including the rotary inertia and the shear deformation.

8.7 VIBRATION OF MEMBRANES

A membrane has no bending stiffness, and the lateral load on it is resisted only by the tension in the membrane itself. Its equation of motion can be derived by a procedure similar to that used in the string but applied in two dimensions.

Assume that the membrane is under uniform tension, T lb per unit length, which is large so that its variation due to lateral deflection is small. Defining the equilibrium position of the membrane in the xy plane, and letting w be the lateral deflection, we examine the forces on an element $dx\,dy$ as shown in Fit. 8.7-1. The resultant force in the w-direction due to the tension on the edges dy is

$$T\,dy\left(\theta + \frac{\partial \theta}{\partial x}dx\right) - T\,dy\,\theta = T\frac{\partial \theta}{\partial x}dy\,dx \tag{8.7-1}$$

Similarly, the tension on the edges dx results in the component $T(\partial\phi/\partial y)dy\,dx$. Since the slopes in the x and y directions are $\theta = \partial w/\partial x$ and $\phi = \partial w/\partial y$,

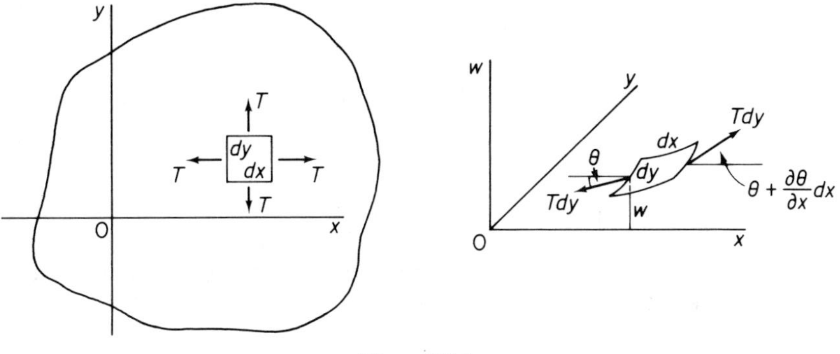

Figure 8.7-1.

the total lateral force due to the tension T is

$$T\left(\frac{\partial^2 w}{\partial x^2} + \frac{\partial^2 w}{\partial y^2}\right) dx\, dy \tag{8.7-2}$$

Letting ρ be the mass per unit area of the membrane and $p(x, y)$ the applied lateral pressure, the equation of motion becomes

$$\rho\, dx\, dy \frac{\partial^2 w}{\partial t^2} = T\left(\frac{\partial^2 w}{\partial x^2} + \frac{\partial^2 w}{\partial y^2}\right) dx\, dy + p(x, y)\, dx\, dy$$

or

$$\frac{\partial^2 w}{\partial t^2} = c^2 \nabla^2 w + \frac{1}{\rho} p(x, y) \tag{8.7-3}$$

where

$$\nabla^2 = \frac{\partial^2}{\partial x^2} + \frac{\partial^2}{\partial y^2}$$

$$c = \sqrt{\frac{T}{\rho}}$$

This equation also applies in other coordinates with appropriate expression for ∇^2.

For the normal mode type of vibration, $p(x, y) = 0$ and $\partial^2 w/\partial t^2 = -\omega^2 w$, and the differential equation reduces to

$$\nabla^2 w + \left(\frac{\omega}{c}\right)^2 w = 0 \tag{8.7-4}$$

For a rectangular membrane of dimensions $(x, y) = (a, b)$ shown in Fig. 8.7-2, the method of separation of variables may be used to arrive at the solution. Letting $w(x, y) = X(x)Y(y)$ and substituting into Eq. (8.7-4), it is

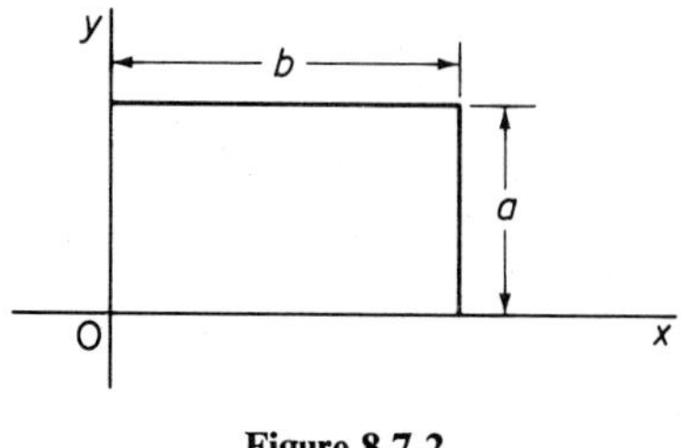

Figure 8.7-2.

easily shown that the solution is of the form

$$X(x) = C_1 \sin \alpha x + C_2 \cos \alpha x$$
$$Y(y) = C_3 \sin \beta y + C_4 \cos \beta y \quad (8.7\text{-}5)$$

where $\alpha^2 + \beta^2 = (\omega/c)^2$. The constants C_i in these equations must be determined from the boundary conditions.

8.8 DIGITAL COMPUTATION

In many problems analytical solutions are not possible, in which case approximate numerical methods must be used. There are several numerical methods available, and the choice as to the most suitable method depends on the problem itself. In this section we will briefly discuss two numerical methods which are extensively used.

Finite Difference. In this method the differential equations and their boundary conditions are replaced by the corresponding finite difference equations. This then reduces the problem to a set of simultaneous algebraic equations which can be solved by the digital computer.

Consider a function $y(x)$ which is shown in Fig. 8.8-1. At some point x_i the derivative is approximated by the equation

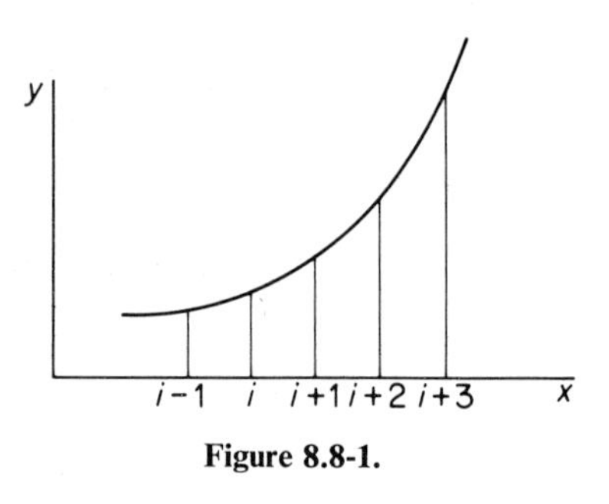

Figure 8.8-1.

$$\left(\frac{dy}{dx}\right)_i \cong \frac{1}{h}(y_{i+1} - y_i) = \frac{1}{h}\Delta y \qquad (8.8\text{-}1)$$

where $h = (x_{i+1} - x_i)$. The second derivative is

$$\left(\frac{d^2y}{dx^2}\right)_i = \frac{d}{dx}\left(\frac{dy}{dx}\right)_i \cong \frac{1}{h}\left[\frac{1}{h}(y_{i+2} - y_{i+1}) - \frac{1}{h}(y_{i+1} - y_i)\right]$$
$$= \frac{1}{h^2}(y_{i+2} - 2y_{i+1} + y_i) = \frac{1}{h^2}\Delta^2 y \qquad (8.8\text{-}2)$$

The above procedure can be repeated any number of times for higher order derivatives. The finite difference pattern up to the fourth derivative is shown in the following table

Finite Difference Table

x	y	$\frac{\Delta y}{\Delta x}$	$\frac{\Delta^2 y}{\Delta x^2}$	$\frac{\Delta^3 y}{dx^3}$	$\frac{\Delta^4 y}{\Delta x^4}$
x_1	y_1				
		$\frac{1}{h}(y_2 - y_1)$			
x_2	y_2		$\frac{1}{h_2}(y_3 - 2y_2 + y_1)$		
		$\frac{1}{h}(y_3 - y_2)$		$\frac{1}{h^3}(y_4 - 3y_3 + 3y_2 - y_1)$	
x_3	y_3		$\frac{1}{h^2}(y_4 - 2y_3 + y_2)$		$\frac{1}{h^4}(y_5 - 4y_4 + 6y_3 - 4y_2 + y_1)$
		$\frac{1}{h}(y_4 - y_3)$		$\frac{1}{h^3}(y_5 - 3y_4 + 3y_3 - y_2)$	
x_4	y_4		$\frac{1}{h^2}(y_5 - 2y_4 + y_3)$		$\frac{1}{h^4}(y_6 - 4y_5 + 6y_4 - 4y_3 + y_2)$
		$\frac{1}{h}(y_5 - y_4)$		$\frac{1}{h^3}(y_6 - 3y_5 + 3y_4 - y_3)$	
x_5	y_5		$\frac{1}{h^2}(y_6 - 2y_5 + y_4)$		
		$\frac{1}{h}(y_6 - y_5)$			
x_6	y_6				

Boundary Conditions. To satisfy the boundary conditions, fictitious points outside the structure must be chosen. The following examples of typical boundary conditions for beams are given.

Simply Supported Beam. As shown in Fig. 8.8-2, let the point on the left of station 1 be p. The boundary conditions at the left end of the beam are

$$y_1 = 0, \qquad \left(\frac{d^2y}{dx^2}\right)_1 = 0$$

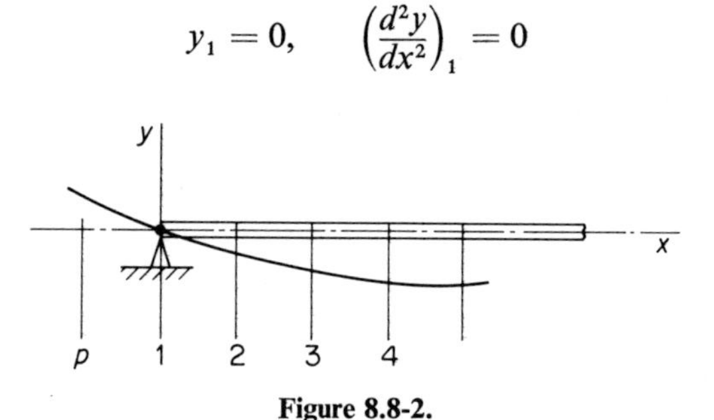

Figure 8.8-2.

Writing the difference equation for the second derivative at station 1, we have

$$\frac{1}{h^2}(y_2 - 2y_1 + y_p) = \frac{1}{h^2}(y_2 - 0 + y_p) = 0$$

Thus y_p must equal $-y_2$.

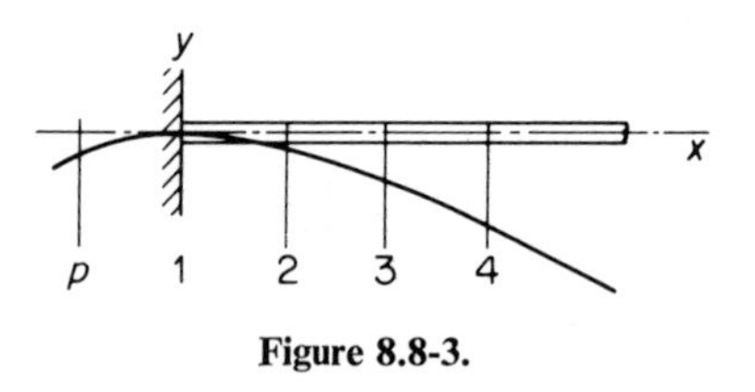

Figure 8.8-3.

Clamped End. At the clamped end the deflection and slope are both zero as shown in Fig. 8.8-3. Again letting y_p be the deflection at the left of station 1, we have, using an interval of $2h$

$$\left(\frac{dy}{dx}\right)_1 = \frac{1}{2h}(y_2 - y_p) = 0$$

Thus $y_p = y_2$ and the deflection curve is symmetrical about the wall.

Partially Restrained Beam. Consider next the case where the left end of the beam is partially restrained. We can represent this condition by a torsional spring of stiffness K lb in./rad as shown in Fig. 8.8-4. The moment at the boundary is $M_1 = -K\theta_1$, but

$$\theta_1 = \left(\frac{dy}{dx}\right)_1 = \frac{1}{2h}(y_2 - y_p)$$

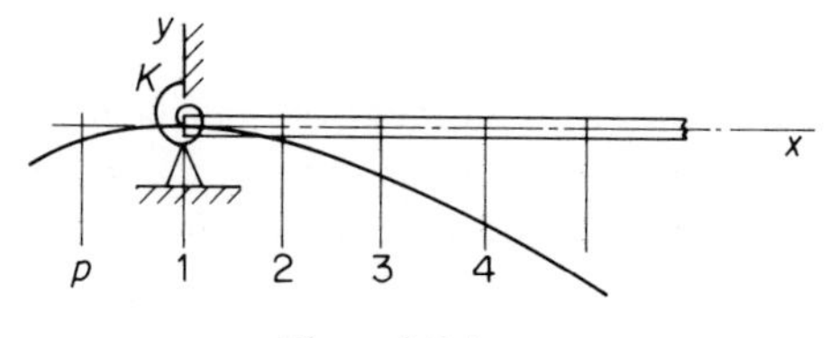

Figure 8.8-4.

and

$$M_1 = EI\left(\frac{d^2y}{dx^2}\right)_1 = \frac{EI}{h^2}(y_2 - 0 + y_p)$$

Substituting into $M_1 = -K\theta_1$, and solving for y_p, we obtain

$$y_p = -y_2\left(\frac{2EI + Kh}{2EI - Kh}\right)$$

Free End. At the free end of the beam, the moment and the shear must be zero. We introduce two fictitious points p and q, and an arbitrary number 4 for the station at the end, as shown in Fig. 8.8-5. Referring to the table of differences, for the moment, we have

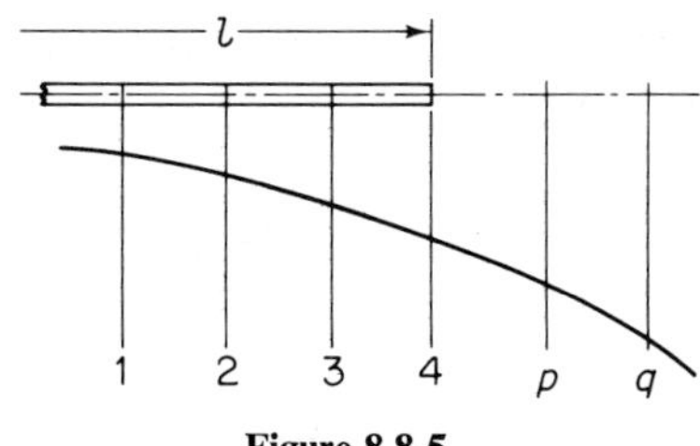

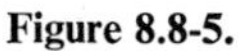

Figure 8.8-5.

$$\left(\frac{d^2y}{dx^2}\right)_4 = \frac{1}{h^2}(y_p - 2y_4 + y_3) = 0$$

or

$$y_p = 2y_4 - y_3$$

For the shear we will average the third derivatives at the end as follows. Generally greater accuracy is obtained in this way.

$$\left(\frac{d^3y}{dx^3}\right)_4 = \frac{1}{2}\left[\frac{1}{h^3}(y_q - 3y_p + 3y_4 - y_3) + \frac{1}{h^3}(y_p - 3y_4 + 3y_3 - y_2)\right]$$

$$= \frac{1}{2h^3}(y_q - 2y_p + 2y_3 - y_2) = 0$$

Thus

$$y_q = 4y_4 - 4y_3 + y_2$$

Example 8.8-1

A beam of non-uniform moment of inertia rests on an elastic foundation of stiffness k lb/in. as shown in Fig. 8.8-6. Its natural frequencies are to be found from its differential equation which is

$$\frac{d^2}{dx^2}\left(EI\frac{d^2 y}{dx^2}\right) + ky - \omega^2 my = 0 \qquad \text{(a)}$$

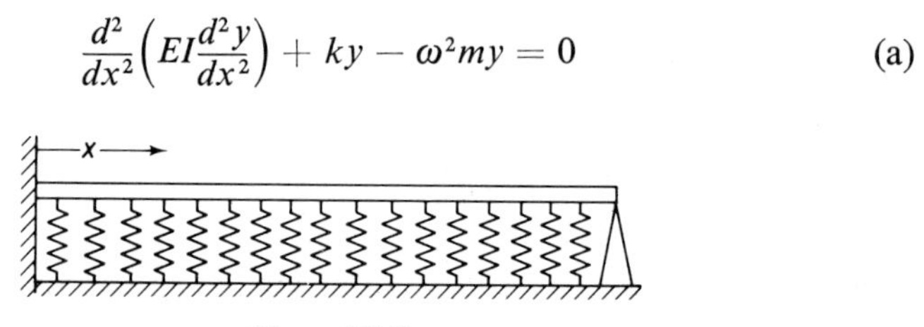

Figure 8.8-6.

To solve this problem by the finite difference method, we number the stations along the beam from 1 to n, and assign a new foundation stiffness for each section, which is k/h as shown in Fig. 8.8-7. Equation (a) is also rewritten as

$$EI\frac{d^4 y}{dx^4} + 2E\frac{d^3 y}{dx^3}\frac{dI}{dx} + E\frac{d^2 y}{dx^2}\frac{d^2 I}{dx^2} + (k - m\omega^2)y = 0 \qquad \text{(b)}$$

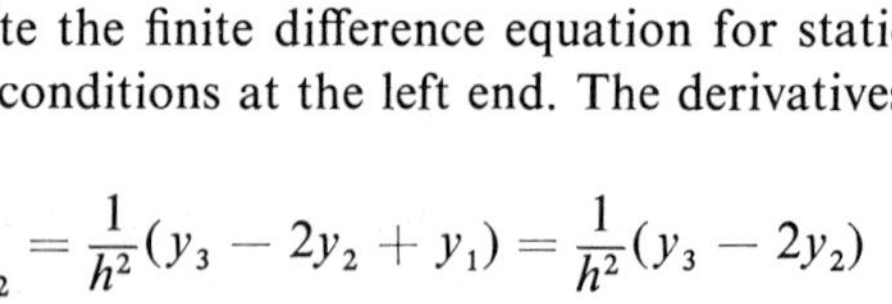

Figure 8.8-7.

We will now write the finite difference equation for station 2, taking note of the boundary conditions at the left end. The derivatives encountered are

$$\left(\frac{d^2 y}{dx^2}\right)_2 = \frac{1}{h^2}(y_3 - 2y_2 + y_1) = \frac{1}{h^2}(y_3 - 2y_2)$$

$$\left(\frac{d^3 y}{dx^3}\right)_2 = \frac{1}{2}\left[\frac{1}{h^3}(y_4 - 3y_3 + 3y_2 - y_1) + \frac{1}{h^3}(y_3 - 3y_2 + 3y_1 - y_p)\right]$$

$$= \frac{1}{2h^3}(y_4 - 2y_3 - y_p) = \frac{1}{2h^3}(y_4 - 2y_3 - y_2)$$

$$\left(\frac{d^4 y}{dx^4}\right)_2 = \frac{1}{h^4}(y_4 - 4y_3 + 6y_2 - 0 + y_2)$$

$$= \frac{1}{h^4}(y_4 - 4y_3 + 7y_2)$$

With these derivatives, the finite difference equation for station 2 becomes

$$\frac{EI_2}{h^4}(y_4 - 4y_3 + 7y_2) + \frac{2E}{2h^3}(y_4 - 2y_3 - y_2)\frac{1}{2h}(I_3 - I_1)$$
$$+ \frac{E}{h^2}(y_3 - 2y_2)\frac{1}{h^2}(I_3 - 2I_2 + I_1) + \left(\frac{k}{h} - m\omega^2\right)y_2 = 0 \quad \text{(c)}$$

In a similar manner, equations for other stations can be written. The boundary conditions at the right end must also be considered, and the resulting set of algebraic equation can be programmed for digital computation.

Runge-Kutta Method. The Runge-Kutta method is popular because it is self-starting and results in good accuracy. The error is of the order h^5.

To illustrate the procedure, we consider the beam with rotary inertia and shear terms, discussed in Sec. 8.6. The fourth order equation is first written in terms of four first order equations as follows

$$\begin{aligned}
\frac{d\psi}{dx} &= \frac{M}{EI} = F(x, \psi, y, M, V) \\
\frac{dy}{dx} &= \psi - \frac{V}{kAG} = G(x, \psi, y, M, V) \\
\frac{dM}{dx} &= V - \omega^2 J\psi = P(x, \psi, y, M, V) \\
\frac{dV}{dx} &= \omega^2 my = K(x, \psi, y, M, V)
\end{aligned} \qquad (8.8\text{-}3)$$

The Runge-Kutta procedure, discussed in Sec. 4.8 for a single coordinate, is now extended to the simultaneous solution of four variables listed below

$$\begin{aligned}
\psi &= \psi_1 + \frac{h}{6}(f_1 + 2f_2 + 2f_3 + f_4) \\
y &= y_1 + \frac{h}{6}(g_1 + 2g_2 + 2g_3 + g_4) \\
M &= M_1 + \frac{h}{6}(p_1 + 2p_2 + 2p_3 + p_4) \\
V &= V_1 + \frac{h}{6}(k_1 + 2k_2 + 2k_3 + k_4)
\end{aligned} \qquad (8.8\text{-}4)$$

where $h = \Delta x$.

The computation proceeds as follows

$$f_1 = F(x_1, \psi_1, y_1, M_1, V_1)$$
$$g_1 = G(x_1, \psi_1, y_1, M_1, V_1)$$

$$p_1 = P(x_1, \psi_1, y_1, M_1, V_1)$$
$$k_1 = K(x_1, \psi_1, y_1, M_1, V_1)$$

$$f_2 = F\left(x_1 + \frac{h}{2}, \psi_1 + f_1\frac{h}{2}, y_1 + g_1\frac{h}{2}, M_1 + p_1\frac{h}{2}, V_1 + k_1\frac{h}{2}\right)$$
$$g_2 = G\left(x_1 + \frac{h}{2}, \psi_1 + f_1\frac{h}{2}, y_1 + g_1\frac{h}{2}, M_1 + p_1\frac{h}{2}, V_1 + k_1\frac{h}{2}\right)$$
$$p_2 = P\left(x_1 + \frac{h}{2}, \psi_1 + f_1\frac{h}{2}, y_1 + g_1\frac{h}{2}, M_1 + p_1\frac{h}{2}, V_1 + k_1\frac{h}{2}\right)$$
$$k_2 = K\left(x_1 + \frac{h}{2}, \psi_1 + f_1\frac{h}{2}, y_1 + g_1\frac{h}{2}, M_1 + p_1\frac{h}{2}, V_1 + k_1\frac{h}{2}\right)$$

$$f_3 = F\left(x_1 + \frac{h}{2}, \psi_1 + f_2\frac{h}{2}, y_1 + g_2\frac{h}{2}, M_1 + p_2\frac{h}{2}, V_1 + k_2\frac{h}{2}\right)$$
$$g_3 = G\left(x_1 + \frac{h}{2}, \psi_1 + f_2\frac{h}{2}, y_1 + g_2\frac{h}{2}, M_1 + p_2\frac{h}{2}, V_1 + k_2\frac{h}{2}\right)$$
$$p_3 = P\left(x_1 + \frac{h}{2}, \psi_1 + f_2\frac{h}{2}, y_1 + g_2\frac{h}{2}, M_1 + p_2\frac{h}{2}, V_1 + k_2\frac{h}{2}\right)$$
$$k_3 = K\left(x_1 + \frac{h}{2}, \psi_1 + f_2\frac{h}{2}, y_1 + g_2\frac{h}{2}, M_1 + p_2\frac{h}{2}, V_1 + k_2\frac{h}{2}\right)$$

$$f_4 = F(x_1 + h, \psi_1 + f_3 h, y_1 + g_3 h, M_1 + p_3 h, V_1 + k_3 h)$$
$$g_4 = G(x_1 + h, \psi_1 + f_3 h, y_1 + g_3 h, M_1 + p_3 h, V_1 + k_3 h)$$
$$p_4 = P(x_1 + h, \psi_1 + f_3 h, y_1 + g_3 h, M_1 + p_3 h, V_1 + k_3 h)$$
$$k_4 = K(x_1 + h, \psi_1 + f_3 h, y_1 + g_3 h, M_1 + p_3 h, V_1 + k_3 h)$$

With these quantities substituted into Eq. (8.8-4), the dependent variables at the neighboring point x_2 are found, and the procedure is repeated for the point x_3, etc.

Returning to the beam equations, the boundary conditions at the beginning end x_1 provide a starting point. For example, in the cantilever beam with origin at the fixed end, the boundary conditions at the starting end are

$$\psi_1 = 0, \quad M_1 = M_1$$
$$y_1 = 0, \quad V_1 = V_1$$

These can be considered to be the linear combination of two boundary vectors as follows

$$\begin{Bmatrix} \psi_1 \\ y_1 \\ M_1 \\ V_1 \end{Bmatrix} = \begin{Bmatrix} 0 \\ 0 \\ 1 \\ 0 \end{Bmatrix} + \alpha \begin{Bmatrix} 0 \\ 0 \\ 0 \\ 1 \end{Bmatrix} = C_1 + \alpha D_1$$

Since the system is linear, we can start with each boundary vector separately. Starting with C_1, we obtain

$$C_N = \begin{Bmatrix} \psi_N \\ y_N \\ M_N \\ V_N \end{Bmatrix}_C$$

Starting with D_1, we obtain

$$\alpha D_N = \begin{Bmatrix} \psi_N \\ y_N \\ M_N \\ V_N \end{Bmatrix}_D$$

These must now add to satisfy the actual boundary conditions at the terminal end, which for a cantilever free end are

$$\begin{Bmatrix} \psi \\ y \\ M \\ V \end{Bmatrix}_N = \begin{Bmatrix} \psi \\ y \\ 0 \\ 0 \end{Bmatrix} = C_N + \alpha D_N$$

If the frequency chosen is correct, the above boundary equations lead to

$$M_{NC} + \alpha M_{ND} = 0$$

$$V_{NC} + \alpha V_{ND} = 0$$

$$\alpha = -\frac{M_{NC}}{M_{ND}} = -\frac{V_{NC}}{V_{ND}}$$

which is satisfied by the determinant

$$\begin{vmatrix} M_{NC} & V_{NC} \\ M_{ND} & V_{ND} \end{vmatrix} = 0$$

The iteration can be started with three different frequencies, which results in three values of the determinant. A parabola is passed through these three points and the zero of the curve is chosen for a new estimate of the frequency. When the frequency is close to the correct value, the new estimate may be made by a straight line between two values of the boundary determinant.

8.9 TRANSIENT SOLUTION BY LAPLACE TRANSFORMS

The response of continuous systems to arbitrarily prescribed boundary conditions can be examined with advantage using the Laplace transform technique. Since the problems of the string and the longitudinal and torsional motions of the slender rod all have the same differential equation, we can examine the equation

$$c^2 \frac{\partial^2 u}{\partial x^2} = \frac{\partial u^2}{\partial t^2} \tag{8.9-1}$$

with the initial conditions $u(x, 0) = u'(x, 0) = 0$ corresponding to the system being at rest initially.

We first take the Laplace transform $\bar{u}(x, s)$ in terms of the time t, reducing the equation to an ordinary differential equation with x as the independent variable

$$c^2 \frac{d^2 \bar{u}}{dx^2}(x, s) = s^2 \bar{u}(x, s) \tag{8.9-2}$$

The general solution of the above equation is

$$\bar{u}(x, s) = C_1 e^{(s/c)x} + C_2 e^{-(s/c)x} \tag{8.9-3}$$

where the constants C_1 and C_2 will depend on the boundary conditions. It is at this point that the physical problem must be defined in order for the boundary conditions to be consistent with reality.

The String. Consider a string of infinite length with arbitrarily prescribed motion of the end $x = 0$. The quantity $u(x, t)$ is then the lateral motion of the string and $c = \sqrt{T/\rho}$ is the propagation velocity of any disturbance along the string with T as tension and ρ as mass per unit length.

At the far end $x = l \rightarrow \infty$, the displacement must be zero which requires that $C_1 = 0$. At the end $x = 0$, the displacement is prescribed as $u(0, t)$ so that $C_2 = \bar{u}(0, s)$. The general solution then becomes

$$\bar{u}(x, s) = \bar{u}(o, s) e^{-(s/c)x}$$

Using the second shifting theorem (see Appendix B)

$$\mathfrak{L}^{-1} e^{-as} \bar{f}(s) = f(t - a) \mathfrak{U}(t - a)$$

the time solution for $u(x, t)$ becomes

$$u(x, t) = u\left(o, t - \frac{x}{c}\right) \mathfrak{U}\left(t - \frac{x}{c}\right)$$

which is interpreted as follows: The unit function $\mathcal{U}(t - x/c)$ is zero for $t < x/c$ so that the string x units from the origin remains at rest until time $t = x/c$. After $t = x/c$ the motion of the string at x undergoes the same motion as the prescribed motion of the end $x = 0$. It is evident then that the prescribed motion of the end $x = 0$ travels along the string at the propagation velocity c, as shown in Fig. 8.9-1.

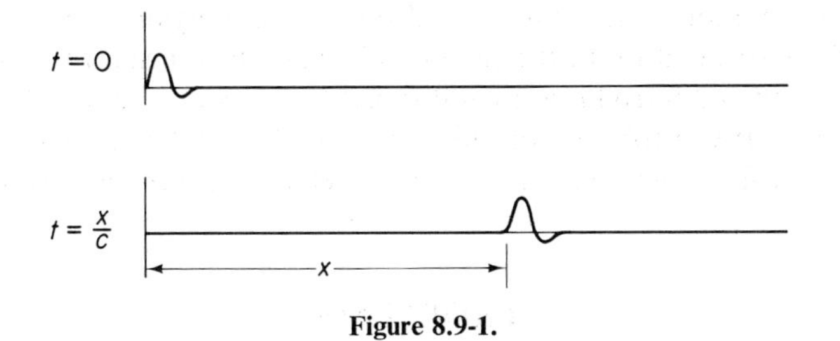

Figure 8.9-1.

Longitudinal Motion of a Rod. Consider here a rod fixed at $x = 0$ with a force $F(l, t)$ applied at the free end $x = l$. The longitudinal displacement is now $u(x, t)$ with $c = \sqrt{Eg/\rho}$ the propagation velocity of disturbances.

The boundary conditions are

$$\bar{u}(0, s) = C_1 + C_2 = 0$$

$$AE\frac{\partial \bar{u}}{\partial x}(l, s) = AE\frac{s}{C}(C_1 e^{sl/c} - C_2 e^{-sl/c}) = \bar{F}(l, s)$$

Solving for C_1 and C_2 we find

$$C_1 = -C_2 = \frac{c\bar{F}(l, s)}{2AEs \cosh \dfrac{sl}{c}}$$

and

$$\bar{u}(x, s) = \frac{\bar{F}(l, s) \sinh \dfrac{sx}{c}}{AE\dfrac{s}{c} \cosh \dfrac{sl}{c}}$$

Since the displacement is the time integral of the velocity, we can replace $\bar{u}(x, s)$ by $(1/s)\bar{v}(x, s)$ and obtain a general expression between the velocity at the end $x = l$ and the force $F(l, t)$ as follows

$$\begin{aligned}\bar{F}(l, s) &= \bar{v}(l, s)\left(\frac{AE}{c}\right) \coth\left(\frac{sl}{c}\right) \\ &= \bar{v}(l, s)\left(\frac{AE}{c}\right)[1 + 2e^{-2(sl/c)} + 2e^{-4(sl/c)} + \cdots]\end{aligned}$$

Again using the second shifting theorem we obtain

$$F(l, t) = \left(\frac{AE}{c}\right)\left[v(l, t) + 2v\left(l, t - \frac{2l}{c}\right)\mathfrak{u}\left(t - \frac{2l}{c}\right)\right.$$
$$\left. + 2v\left(l, t - \frac{4l}{c}\right)\mathfrak{u}\left(t - \frac{4l}{c}\right) + \cdots\right]$$

The above solution states that the end force is proportional to the velocity $v(l, t)$ of the free end up to the time $t = 2l/c$ at which instant the reflection from the fixed end introduces an additional term $2v(l, t - 2l/c)$, etc.

Many other problems of this type can be treated similarly by the Laplace transform method; the reader is referred to *Laplace Transformation.**

PROBLEMS

8-1 Find the wave velocity along a rope whose density is $\frac{1}{4}$ lb per foot when stretched to a tension of 100 lb.

8-2 Derive the equation for the natural frequencies of a uniform cord of length l fixed at the two ends. The cord is stretched to a tension T and its mass per unit length is ρ.

8-3 A cord of length l and mass per unit length ρ is under tension T with the left end fixed and the right end attached to a spring-mass system as shown in Fig. P8-3. Determine the equation for the natural frequencies.

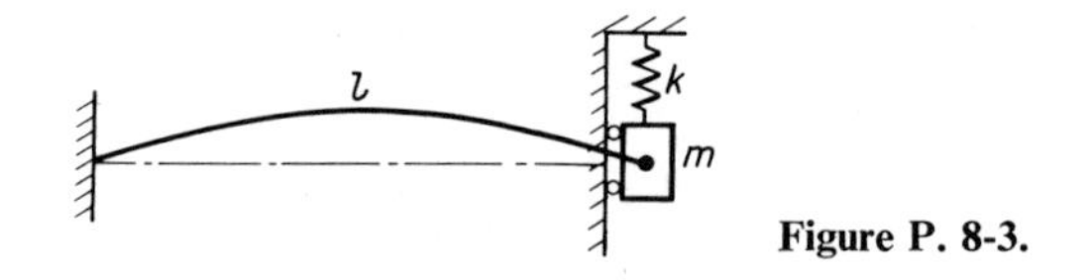

Figure P. 8-3.

8-4 A harmonic vibration has an amplitude that varies as a cosine function along the x-direction such that

$$y = a \cos kx \cdot \sin \omega t$$

Show that if another harmonic vibration of same frequency and equal amplitude displaced in space phase and time phase by a quarter wave length is added to the first vibration, the resultant vibration will represent a traveling wave with a propagation velocity equal to $c = \omega/k$.

8-5 Find the velocity of longitudinal waves along a thin steel bar. The modulus of elasticity and weight per unit volume of steel are 29×10^6 lb/in.^{2} and 0.282 lb/in.^{3}

*W.T. Thomson, *Laplace Transformation*, 2nd ed (Englewood Cliffs, N.J.: Prentice-Hall, Inc., 1960), Chapter 8.

8-6 A uniform bar of length l is fixed at one end and free at the other end. Show that the frequencies of normal longitudinal vibrations are $f = (n + \frac{1}{2})c/2l$, where $c = \sqrt{Eg/\rho}$ is the velocity of longitudinal waves in the bar, and $n = 0, 1, 2, \cdots$.

8-7 A uniform rod of length l and cross-sectional area A is fixed at the upper end and is loaded with a weight W on the other end. Show that the natural frequencies are determined from the equation

$$\omega l\sqrt{\frac{\rho}{Eg}} \tan \omega l\sqrt{\frac{\rho}{Eg}} = \frac{A\rho l}{W}$$

8-8 Show that the fundamental frequency for the system of Prob. 8-7 can be expressed in the form

$$\omega_1 = \beta_1\sqrt{k/rM}$$

where

$$n_1 l = \beta_1, \qquad r = \frac{M_{\text{rod}}}{M},$$

$$k = \frac{AE}{l}, \qquad M = \text{end mass}$$

Reducing the above system to a spring k and an end mass equal to $M + \frac{1}{3}M_{\text{rod}}$, determine an approximate equation for the fundamental frequency. Show that the ratio of the approximate to the exact frequency as found above is $(1/\beta_1)\sqrt{3r/(3 + r)}$.

8-9 The frequency of magnetostriction oscillators is determined by the length of the nickel alloy rod which generates an alternating voltage in the surrounding coils equal to the frequency of longitudinal vibration of the rod, as shown in Fig. P8-9. Determine the proper length of the rod clamped at the middle for a frequency of 20 kcps if the modulus of elasticity and density are given as $E = 30 \times 10^6$ lb/in.^{2} and $\rho = 0.31$ lb/in.^{3}

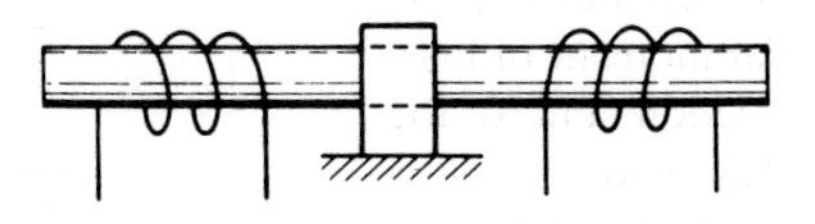

Figure P. 8-9.

8-10 Show that $c = \sqrt{Gg/\rho}$ is the velocity of propagation of torsional strain along the rod. What is the numerical value of c for steel?

8-11 Determine the expression for the natural frequencies of torsional oscillations of a uniform rod of length l clamped at the middle and free at the two ends.

8-12 Determine the natural frequencies of a torsional system consisting of a uniform shaft of mass moment of inertia J_s with a disk of inertia J_0 attached to each end. Check the fundamental frequency by reducing the uniform shaft to a torsional spring with end masses.

8-13 Determine the expression for the natural frequencies of a free-free bar in lateral vibration.

8-14 Determine the node position for the fundamental mode of the free-free beam by Rayleigh's method, assuming the curve to be $y = \sin(\pi x/l) - b$. By equating the momentum to zero, determine b. Substitute this value of b to find ω_1.

8-15 A concrete test beam $2 \times 2 \times 12$ in., supported at two points $0.224l$ from the ends, was found to resonate at 1690 cps. If the density of concrete is 153 lb/ft^3, determine the modulus of elasticity, assuming the beam to be slender.

8-16 Determine the natural frequencies of a uniform beam of length l clamped at both ends.

8-17 Determine the natural frequencies of a uniform beam of length l, clamped at one end and pinned at the other end.

8-18 A uniform beam of length l and weight W_b is clamped at one end and carries a concentrated weight W_0 at the other end. State the boundary conditions and determine the frequency equation.

8-19 The pinned end of a pinned-free beam is given a harmonic motion of amplitude y_0 perpendicular to the beam. Show that the boundary conditions result in the equation

$$\frac{y_0}{y_l} = \frac{\sinh \beta l \cos \beta l - \cosh \beta l \sin \beta l}{\sinh \beta l - \sin \beta l}$$

which for $y_0 \longrightarrow 0$, reduce to

$$\tanh \beta l = \tan \beta l$$

8-20 A uniform bar has these specifications: length l, mass density per unit volume ρ, and torsional stiffness $I_P G$ where I_P is the polar moment of inertia of the cross section and G the shear modulus. The end $x = 0$ is fastened to a torsional spring of stiffness K lb in./rad, while the end l is fixed as shown in Fig. P8-20. Determine the transcendental equation from which natural frequencies can be established. Verify the correctness of this equation by considering special cases for $K = 0$ and $K = \infty$.

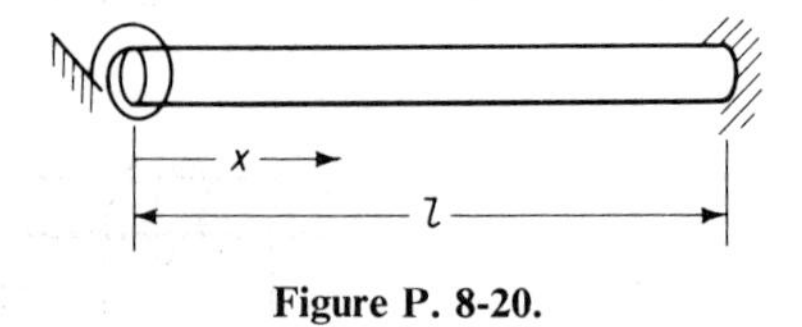

Figure P. 8-20.

8-21 A simply supported beam has an overhang of length l_2, as shown in Fig. P8-21. If the end of the overhang is free, show that boundary conditions require the deflection equation for each span to be

$$\phi_1 = C\left(\sin \beta x - \frac{\sin \beta l_1}{\sinh \beta l_1} \sinh \beta x\right)$$

$$\phi_2 = A\left\{\cos \beta x + \cosh \beta x - \left(\frac{\cos \beta l_2 + \cosh \beta l_2}{\sin \beta l_2 + \sinh \beta l_2}\right)(\sin \beta x + \sinh \beta x)\right\}$$

where x is measured from the left and right ends.

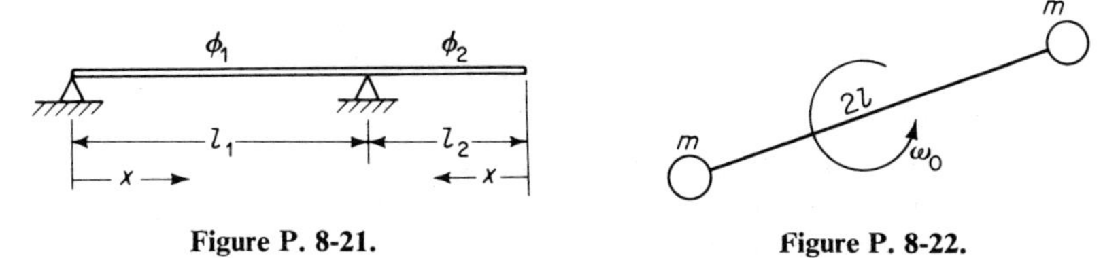

Figure P. 8-21.

Figure P. 8-22.

8-22 A particular satellite consists of two equal masses m each, connected by a cable of length $2l$ and mass density ρ, as shown in Fig. P8-22. The assembly rotates in space with angular speed ω_0. Show that if the variation in the cable tension is neglected, the differential equation of lateral motion of the cable is

$$\frac{\partial^2 y}{\partial x^2} = \frac{\rho}{m\omega_0^2 l}\left(\frac{\partial^2 y}{\partial t^2} - \omega_0^2 y\right)$$

and that its fundamental frequency of oscillation is

$$\omega^2 = \left(\frac{\pi}{2l}\right)^2\left(\frac{m\omega_0 l}{\rho}\right) - \omega_0^2.$$

8-23 Shown in Fig. P8-23 is a flexible cable supported at the upper end and free to oscillate under the influence of gravity. Show that the equation of lateral motion is

$$\frac{\partial^2 y}{\partial t^2} = g\left(x\frac{\partial^2 y}{\partial x^2} + \frac{\partial y}{\partial x}\right)$$

Figure P. 8-23.

8-24 In Prob. 23, assume a solution in the form $y = Y(x) \cos \omega t$ and show that $Y(x)$ can be reduced to a Bessel's differential equation

$$\frac{d^2 Y(z)}{dz^2} + \frac{1}{z}\frac{dY(z)}{dz} + Y(z) = 0$$

with solution

$$Y(z) = J_0(z) \quad \text{or} \quad Y(x) = J_0\left(2\omega\sqrt{\frac{x}{g}}\right)$$

by a change in variable $z^2 = 4\omega^2 x/g$.

8-25 A membrane is stretched with large tension T lb/in., so that its lateral deflection y does not increase T appreciably. Using polar coordinates, show that the differential equation of lateral vibration is

$$\frac{\partial^2 y}{\partial t^2} = \frac{T}{\rho}\left(\frac{\partial^2 y}{\partial r^2} + \frac{1}{r}\frac{\partial y}{\partial r} + \frac{1}{r^2}\frac{\partial^2 y}{\partial \theta^2}\right)$$

8-26 Apply the results of Prob. 25 to a circular membrane of radius a with the boundary conditions $y(a) = 0$. The deflection of the symmetric modes without radial node lines can be shown to be given by $J_0(r\sqrt{\rho\omega^2/T})$. For the general case of radial and circumferential nodes, the natural frequencies are evaluated from the boundary conditions at $r = a$ and $r = 0$, which result in an equation of the form

$$\omega = \frac{\alpha_{n,m}}{a}\sqrt{\frac{T}{\rho}}$$

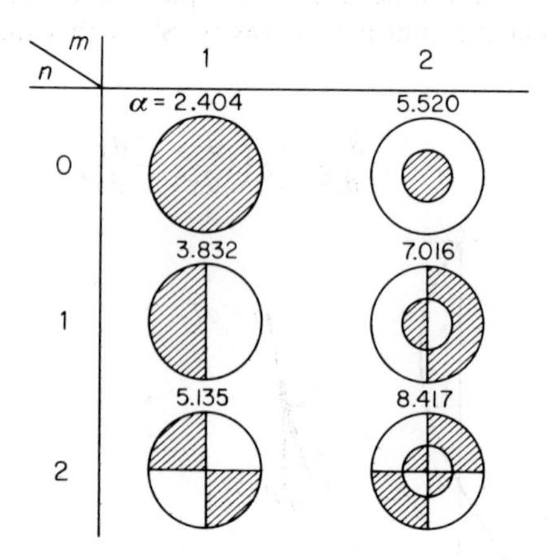

Figure P. 8-26. Deflection of membranes.

where n refers to the number of radial nodes, and m the number of circular nodes including that of the outer boundary. A few shapes are shown in Fig. P8-26.

8-27 The equation for the longitudinal oscillations of a slender rod with viscous damping is

$$m\frac{\partial^2 u}{\partial t^2} = AE\frac{\partial^2 u}{\partial x^2} - \alpha\frac{\partial u}{\partial t} + \frac{p_0}{l}p(x)f(t)$$

where the loading per unit length is assumed to be separable. Letting $u = \sum_i \phi_i(x)q_i(t)$ and $p(x) = \sum_i b_i\phi_i(x)$ show that

$$u = \frac{p_0}{ml\sqrt{1-\zeta^2}}\sum_j \frac{b_j\phi_j}{\omega_j}\int_0^t f(t-\tau)e^{-\zeta\omega_j\tau}\sin\omega_j\sqrt{1-\zeta^2}\,\tau\,d\tau$$

$$b_j = \frac{1}{l}\int_0^l p(x)\phi_j(x)\,dx$$

Derive the equation for the stress at any point x.

8-28 Assume that the edges of the rectangular membrane of Fig. 8.7-2 are clamped and show that its solution is

$$w(x, y, t) = \sum_{m=1}^{\infty}\sum_{n=1}^{\infty}\sin\frac{m\pi x}{b}\sin\frac{n\pi y}{a}(A_{mn}\sin\omega_{mn}t + B_{mn}\cos\omega_{mn}t)$$

8-29 Show that the natural frequencies of the membrane of Prob. 8-28 are given by the equation

$$\omega_{m,n}^2 = c^2\pi^2\left(\frac{m^2}{b^2} + \frac{n^2}{a^2}\right)$$

where $m, n = 1, 2, 3, \ldots$.

8-30 Describe the natural mode shapes for the square membrane with clamped edges.

8-31 When shear and rotary inertia are included, show that the differential equation of the beam may be expressed by the first order matrix equation

$$\frac{d}{dx}\begin{Bmatrix}\psi\\ y\\ M\\ V\end{Bmatrix} = \begin{bmatrix} 0 & 0 & \frac{1}{EI} & 0\\ 1 & 0 & 0 & \frac{-1}{kAG}\\ -\omega^2 J & 0 & 0 & 1\\ 0 & \omega^2 m & 0 & 0\end{bmatrix}\begin{Bmatrix}\psi\\ y\\ M\\ V\end{Bmatrix}$$

8-32 For the beam configuration shown in Fig. P8-32, determine the finite difference equation for station 2.

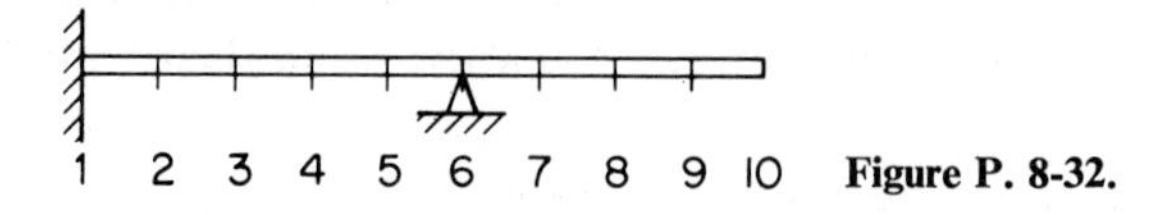

Figure P. 8-32.

8-33 For the beam of Prob. 32, establish the finite difference equations which apply to stations 5 and 7.

8-34 For the beam of Prob. 32, develop the finite difference equations for stations 9 and 10.

8-35 A string of length l, fixed at the ends, is under tension T. At $x=0$ the string is given an initial velocity

$$v(0, t)$$

Determine its motion.

8-36 A helical spring of length l and stiffness k lies unstrained on a horizontal frictionless plane. If the end $x = 0$ is given a prescribed velocity $v(0, t)$, determine the motion of any point x. What is the stress in the spring at point x?

9

Lagrange's Equation

9.1 INTRODUCTION

Lagrange* developed a general treatment of dynamical systems formulated from the scalar quantities of kinetic energy T, potential energy U, and work W. As the system becomes more complicated, the establishment of vector relationships required by Newton's laws becomes increasingly difficult, in which case the scalar approach based on energy and work offers considerable advantage. Furthermore, constraint forces of frictionless hinges and guides can be completely disregarded in Lagrange's formulation of the equations of motion.

9.2 GENERALIZED COORDINATES

The equations of motion of a system can be formulated in a number of different coordinate systems. However, n independent coordinates are

*Joseph L. C. Lagrange (1736-1813).

necessary to describe the motion of a system of n degrees of freedom. Such independent coordinates are called *generalized coordinates* and are usually denoted by the letters q_i.

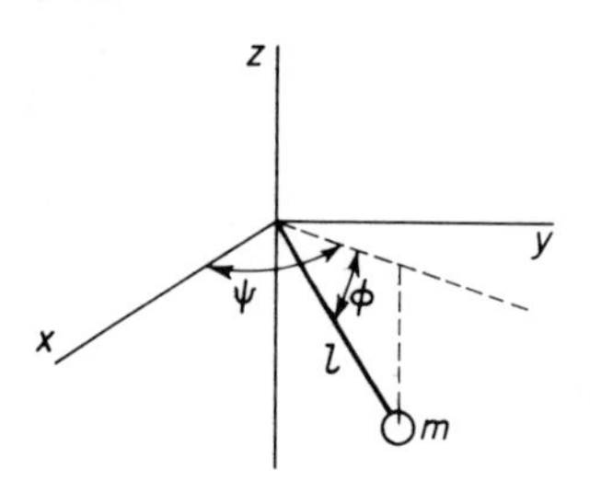

Figure 9.2-1.

Motion of bodies are not always free motions and are often constrained to move in a predetermined manner. As an example, Fig. 9.2-1 shows a spherical pendulum of length l. Its position can be completely defined by the two independent coordinates ψ and ϕ. Hence ψ and ϕ are generalized coordinates, and the spherical pendulum represents a system of two degrees of freedom.

The position of the spherical pendulum can also be described by the three rectangular coordinates x, y, z, which exceed the degrees of freedom of the system by one. The coordinates x, y, z are, however, not independent, because they are related by the *constraint equation*

$$x^2 + y^2 + z^2 - l^2 = 0 \tag{9.2-1}$$

One of the coordinates can be eliminated by the above equation, thereby reducing the number of necessary coordinates to two.

The excess coordinates exceeding the number of degrees of freedom of the system are called *superfluous coordinates*, and constraint equations equal in number to the superfluous coordinates are necessary for their elimination. Constraints are called *holonomic* if the excess coordinates can be eliminated through equations of constraint. Such constraints are in the form

$$C(q_1, q_2, \ldots q_n, t) = 0 \tag{9.2-2}$$

In nonholonomic systems, the constraints are not expressible in terms of the coordinates or coordinates and time, as in Eq. (9.2-2). *Nonholonomic constraints* are expressible only as relationships between the differentials as in the following equation

$$a_1 dq_1 + a_2 dq_2 + \cdots a_m dt = 0 \tag{9.2-3}$$

Elimination of the dependent coordinates are then not possible by algebraic means.

9.3 PRINCIPLE OF VIRTUAL WORK

A virtual displacement $\delta x, \delta\theta, \delta r$, etc., is an infinitesimal change in the coordinate which may be conceived in any manner irrespective of the time t, but without violating the constraints of the system.

Consider a system of particles acted upon by several forces. If the system is in static equilibrium, the resultant R_j of the forces acting on any particle j must be zero, and the work done by these forces in a virtual displacement δr_j is zero

$$\delta W = \sum_j \mathbf{R}_j \cdot \delta \mathbf{r}_j = 0 \tag{9.3-1}$$

If the force R_j is separated into an applied force F_j and a constraint force f_i, then F_j is balanced by f_i, and neither force is zero. Limiting our discussion to constraint forces that do no work, such as the reaction of a smooth floor, the virtual work equation reduces to

$$\delta W = \sum_j \mathbf{F}_j \cdot \delta \mathbf{r}_j = 0 \tag{9.3-1}$$

which expresses the principle of virtual work as presented by J. Bernoulli (1717). In summary, the above equation states that if a system is in static equilibrium, the work done by the *applied forces* in a virtual displacement compatible with the constraints is equal to zero.

Virtual Work in Terms of Generalized Coordinates. Consider an n degree of freedom system where the displacement r_j can be expressed by the n independent generalized coordinates q_i and the time t

$$\mathbf{r}_j = \mathbf{r}_j(q_1, q_2 \ldots q_n, t) \tag{9.3-3}$$

The virtual displacement of the coordinate r_j is

$$\delta \mathbf{r}_j = \sum_i \frac{\delta \mathbf{r}_j}{\delta q_i} \delta q_i \tag{9.3-4}$$

and the time t is not involved.

When the system is in equilibrium, the virtual work can now be expressed in terms of the generalized coordinates q_i by Eq. (9.3-4)

$$\delta W = \sum_j \mathbf{F}_j \cdot \delta \mathbf{r}_j = \sum_j \sum_i \mathbf{F}_j \cdot \frac{\delta \mathbf{r}_j}{\delta q_i} \delta q_i \tag{9.3-5}$$

Interchanging the order of summation and letting

$$Q_i = \sum_j \mathbf{F}_j \cdot \frac{\delta \mathbf{r}_j}{\delta q_i} \tag{9.3-6}$$

be defined as the *generalized force*, the virtual work for the system, expressed in terms of the generalized coordinates, becomes

$$\delta W = \sum_i Q_i \, \delta q_i \tag{9.3-7}$$

Example 9.3-1

To illustrate the method of virtual work, consider the problem of establishing the equilibrium position of the rigid bar constrained in its motion as shown in Fig. 9.3-1.

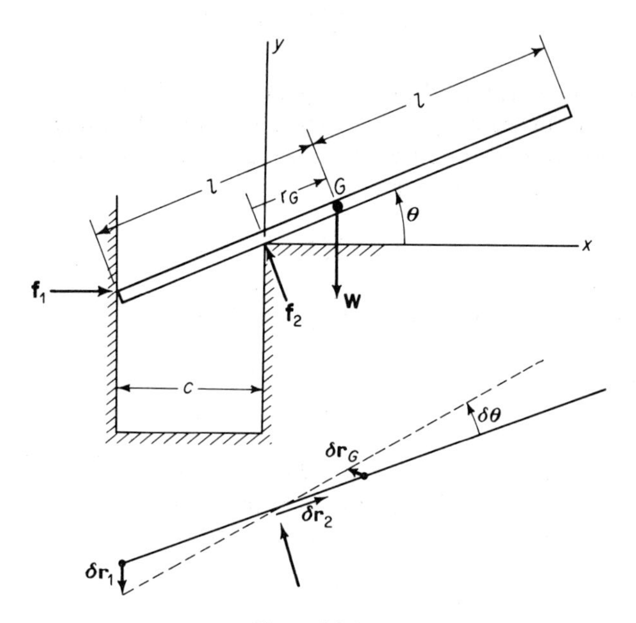

Figure 9.3-1.

The position of the bar is completely established by the coordinate θ, which can serve as the generalized coordinate. If the bar is given a virtual displacement $\delta\theta$, the corresponding displacements δr_1, δr_2 and δr_G of points 1, 2 and G must be compatible with the constraints of the system. They can all be expressed in terms of $\delta\theta$, which is the only independent quantity that can be assigned any arbitrary value.

There are two types of forces acting on the bar. The constraint forces are f_1 and f_2 whereas the gravity force w is an applied force. Assuming frictionless contacts, the constraint forces f_1 and f_2 are normal to the virtual displacements δr_1 and δr_2, respectively, and hence do no work when the bar undergoes a virtual displacement $\delta\theta$. Thus the virtual work of the system is due only to the applied force

$$\begin{aligned}\delta W &= \mathbf{f}_1\cdot\delta\mathbf{r}_1 + \mathbf{f}_2\cdot\delta\mathbf{r}_2 + w\cdot\delta\mathbf{r}_G \\ &= 0 + 0 + \mathbf{w}\cdot\delta\mathbf{r}_G\end{aligned}$$

Since $\mathbf{r}_G$ is some function of θ, we can write

$$\delta \mathbf{r}_G = \frac{\partial \mathbf{r}_G}{\partial \theta}\,\delta\theta$$

and the equation for the virtual work becomes

$$\delta W = \mathbf{w}\cdot\frac{\partial \mathbf{r}_G}{\partial \theta}\,\delta\theta = Q_\theta\,\delta\theta$$

In the above equation

$$Q_\theta = \mathbf{w}\cdot\frac{\partial \mathbf{r}_G}{\partial \theta}$$

is the generalized force associated with the generalized coordinate θ.

Using unit vectors $\mathbf{i}$ and $\mathbf{j}$ along the x and y axes (See Fig. 9.3-1), the equation for $\mathbf{r}_G$ is

$$\begin{aligned}\mathbf{r}_G &= r_G(\mathbf{i}\cos\theta + \mathbf{j}\sin\theta)\\ &= \left(l - \frac{c}{\cos\theta}\right)(i\cos\theta + j\sin\theta)\end{aligned}$$

Differentiating with respect to θ,

$$\delta \mathbf{r}_G = [(-l\sin\theta)\mathbf{i} + (l\cos\theta - c\sec^2\theta)\mathbf{j}]\delta\theta$$

and taking the dot product with $\mathbf{w} = -w\mathbf{j}$, the virtual work, which must be zero, becomes

$$\delta W = -w(l\cos\theta - c\sec^2\theta)\delta\theta = 0$$

The equation is satisfied by

$$(l\cos\theta - c\sec^2\theta) = 0$$

or

$$\cos\theta = \sqrt[3]{\frac{c}{l}}$$

which defines the equilibrium position of the bar. The student can verify the fact that the center of mass G occupies the lowest point in the above position and δr_G is a horizontal displacement.

9.4 DEVELOPMENT OF LAGRANGE'S EQUATION

The principle of virtual work, established for the case of static equilibrium, can be extended to dynamics by a reasoning advanced by D'Alembert (1743). D'Alembert reasoned that since the sum of the forces acting on a particle results in its acceleration $m_i\ddot{\mathbf{r}}_i$, the application of a force equal to $-m_i\ddot{\mathbf{r}}_i$ would produce a condition of equilibrium. The equation for the particle can then be written as

$$\mathbf{F}_i + \mathbf{f}_i - m_i\ddot{\mathbf{r}}_i = 0 \tag{9.4-1}$$

where $\mathbf{F}_i$ and $\mathbf{f}_i$ are the applied and constraint forces, respectively. It then follows from the principle of virtual work that for a system of particles

$$\sum_i (\mathbf{F}_i - m_i\ddot{\mathbf{r}}_i)\cdot\delta\mathbf{r}_i = 0 \tag{9.4-2}$$

where the work done by the constraint forces $\mathbf{f}_i$ is again zero. Thus, for a dynamical system, the principle of virtual work requires that the applied forces $\mathbf{F}_i$ be replaced by $(\mathbf{F}_i - m_i\ddot{\mathbf{r}}_i)$ which introduces a new term $\sum_i m_i\ddot{\mathbf{r}}_i\cdot\delta\mathbf{r}_i$. We will now show that this new term is related to the kinetic energy T by the equation

$$\sum_{k=1}^{n}\left[\frac{d}{dt}\left(\frac{\delta T}{\partial \dot{q}_k}\right) - \frac{\partial T}{\partial q_k}\right]\delta q_k$$

Considering a body to be representable by a system of particles, its kinetic energy is equal to

$$T = \sum_i \tfrac{1}{2}m_i\dot{r}_i^2 = \sum_i \tfrac{1}{2}m_i\dot{\mathbf{r}}_i\cdot\dot{\mathbf{r}}_i \tag{9.4-3}$$

For a system of n degrees of freedom the position of any particle can be expressed in terms of the n generalized coordinates $q_1, q_2, \ldots, q_n$ and in some cases time t,

$$\mathbf{r}_i = \mathbf{r}_i(q_1, q_2, \ldots, q_n, t) \tag{9.4-4}$$

and its velocity is

$$\dot{\mathbf{r}}_i = \frac{\partial \mathbf{r}_i}{\partial q_1}\dot{q}_1 + \frac{\partial \mathbf{r}_i}{\partial q_2}\dot{q}_2 + \cdots \frac{\partial \mathbf{r}_i}{\partial q_n}\dot{q}_n + \frac{\partial \mathbf{r}_i}{\partial t} \tag{9.4-5}$$

From these we form two important relationships. First, if we take the partial derivative of $\dot{\mathbf{r}}_i$ with respect to $\dot{q}_k$, it will be equal to the coefficient of $\dot{q}_k$

$$\frac{\partial \dot{\mathbf{r}}_i}{\partial \dot{q}_k} = \frac{\partial \mathbf{r}_i}{\partial q_k} \tag{9.4-6}$$

Second, the virtual displacement of $\mathbf{r}_i$ from Eq. (9.3-4) is

$$\delta\mathbf{r}_i = \frac{\partial \mathbf{r}_i}{\partial q_1}\delta q_1 + \frac{\partial \mathbf{r}_i}{\partial q_2}\delta q_2 + \cdots \frac{\partial \mathbf{r}_i}{\partial q_n}\delta q_n = \sum_{k=1}^{n} \frac{\partial \mathbf{r}_i}{\partial q_k}\delta q_k \tag{9.4-7}$$

where it should be noted that the time t does not enter into the equation (definition of virtual displacement, irrespective of time).

Making use of the above equation for $\delta\mathbf{r}_i$ we have

$$m_i\ddot{\mathbf{r}}_i \cdot \delta\mathbf{r}_i = \sum_{k=1}^{n} m_i\ddot{\mathbf{r}}_i \cdot \frac{\partial \mathbf{r}_i}{\partial q_k}\delta q_k \tag{9.4-8}$$

Next examine one of the terms of this summation

$$m_i\ddot{\mathbf{r}}_i \cdot \frac{\partial \mathbf{r}_i}{\partial q_k} = \frac{d}{dt}\left(m_i\dot{\mathbf{r}}_i \cdot \frac{\partial \mathbf{r}_i}{\partial q_k}\right) - m_i\dot{\mathbf{r}}_i \cdot \frac{d}{dt}\left(\frac{\partial \mathbf{r}_i}{\partial q_k}\right) \tag{9.4-9}$$

From Eq. (9.4-6), $\partial\mathbf{r}_i/\partial q_k$ in the first term can be replaced by $\partial\dot{\mathbf{r}}_i/\partial\dot{q}_k$, and the order of differentiation in the second term can be reversed so that

$$\frac{d}{dt}\left(\frac{\partial \mathbf{r}_i}{\partial q_k}\right) = \frac{\partial \dot{\mathbf{r}}_i}{\partial q_k}$$

The result is

$$\begin{aligned} m_i\ddot{\mathbf{r}}_i \cdot \frac{\delta \mathbf{r}_i}{\partial q_k} &= \frac{d}{dt}\left(m_i\dot{\mathbf{r}}_i \cdot \frac{\partial \dot{\mathbf{r}}_i}{\partial \dot{q}_k}\right) - m_i\dot{\mathbf{r}}_i \cdot \frac{\partial \dot{\mathbf{r}}_i}{\partial q_k} \\ &= \left[\frac{d}{dt}\frac{\partial}{\partial \dot{q}_k} - \frac{\partial}{\partial q_k}\right](\tfrac{1}{2}m_i\dot{\mathbf{r}}_i \cdot \dot{\mathbf{r}}_i) \end{aligned} \tag{9.4-10}$$

and

$$m_i\ddot{\mathbf{r}}_i \cdot \delta\mathbf{r}_i = \sum_{k=1}^{n}\left[\frac{d}{dt}\frac{\partial}{\partial \dot{q}_k} - \frac{\partial}{\partial q_k}\right](\tfrac{1}{2}m_i\dot{r}_i^2)\,\delta q_k \tag{9.4-11}$$

Summing over the i particles, we arrive at the result

$$\sum_i m_i\ddot{\mathbf{r}}_i \cdot \delta\mathbf{r}_i = \sum_{k=1}^{n}\left[\frac{d}{dt}\frac{\partial T}{\partial \dot{q}_k} - \frac{\partial T}{\partial q_k}\right]\delta q_k \tag{9.4-12}$$

where $T = \frac{1}{2}\sum_i m_i\dot{r}_i^2$ is the kinetic energy of the system.

To complete the development, the work done by the applied forces in the virtual displacement is written as

$$\begin{aligned} \delta W &= \sum_i \mathbf{F}_i \cdot \delta\mathbf{r}_i = \sum_i \mathbf{F}_i \cdot \sum_{k=1}^{n}\frac{\partial \mathbf{r}_i}{\partial q_k}\delta q_k \\ &= \sum_{k=1}^{n}\left(\sum_i \mathbf{F}_i \cdot \frac{\partial \mathbf{r}_i}{\partial q_k}\right)\delta q_k \\ &= \sum_{k=1}^{n} Q_k\,\delta q_k \end{aligned} \tag{9.4-13}$$

where

$$Q_k = \sum_i \mathbf{F}_i \cdot \frac{\partial \mathbf{r}_i}{\partial q_k} \tag{9.4-14}$$

is called the generalized force associated with the coordinate q_k. The dimensions of Q_k will depend on the dimensions of q_k, so that if q_k is an angle θ, the generalized force will be a moment.

We now put Eqs. (9.4-12) and (9.4-13) back into the original Eq. (9.4-2)

$$\sum_{k=1}^{n} \left(\frac{d}{dt} \frac{\partial T}{\partial \dot{q}_k} - \frac{\partial T}{\partial q_k} - Q_k \right) \delta q_k = 0 \tag{9.4-15}$$

Since the n δq_k corresponding to the n degrees of freedom are independent quantities, we can choose them in any manner we please. By singling out one of the $\delta q_i \neq 0$ and letting the remaining δq_s be zero, we obtain Lagrange's equation for the coordinate q_j

$$\frac{d}{dt} \frac{\partial T}{\partial \dot{q}_j} - \frac{\partial T}{\partial q_i} - Q_j = 0 \tag{9.4-16}$$

By repeating the procedure with other coordinates, a similar equation can be established for the n coordinates of the system.

There are a few variations of Lagrange's equation which can now be mentioned. If we have a conservative system, the work done is equal to the negative of the potential energy

$$W = -U(q_1, q_2, \ldots, q_n) \tag{9.4-17}$$

and the virtual work of Eq. (9.3-13) can be replaced by

$$\delta W = -\sum \frac{\partial U}{\partial q_k} \delta q_k \tag{9.4-18}$$

Thus in place of Q_k we use $-(\partial U/\partial q_k)$ and rewrite Lagrange's equation as

$$\frac{d}{dt} \frac{\partial T}{\delta \dot{q}_k} - \frac{\partial T}{\partial q_k} + \frac{\partial U}{\partial q_k} = 0 \tag{9.4-19}$$

The second variant results from recognizing that U is not a function of $\dot{q}$ so that if we define a Lagrangian L as

$$L = T - U \tag{9.4-20}$$

we can write Eq. (9.4-19) as

$$\frac{d}{dt} \frac{\partial L}{\partial \dot{q}_k} - \frac{\partial L}{\partial q_k} = 0 \tag{9.4-21}$$

When nonconservative forces exist in the system, the work done by them can be separated out in the form

$$\delta W = \sum_{k=1}^{n} \bar{Q}_k \, \delta q_k \tag{9.4-22}$$

in which case it is possible to present Lagrange's equation for a nonconservative system as

$$\begin{aligned} \frac{d}{dt}\frac{\partial L}{\partial \dot{q}_k} - \frac{\partial L}{\partial q_k} &= \bar{Q}_k \\ \frac{d}{dt}\frac{\partial T}{\partial \dot{q}_k} - \frac{\partial T}{\partial q_k} + \frac{\partial U}{\partial q_k} &= \bar{Q}_k \end{aligned} \tag{9.4-23}$$

These last forms enable us to extend the use of Lagrange's method to nonconservative systems and hence the method of Lagrange is applicable to all dynamical systems including damped vibrations.

9.5 GENERALIZED STIFFNESS AND MASS

In a conservative system, the forces can be derived from the potential energy U, which is a function of the generalized coordinates q_j. Expanding U in a Taylor series about the equilibrium position, we have, for a system of n degrees of freedom

$$U = U_0 + \sum_{j=1}^{n} \left(\frac{\partial U}{\partial q_j}\right)_0 q_j + \frac{1}{2} \sum_{j=1}^{n} \sum_{l=1}^{n} \left(\frac{\partial^2 U}{\partial q_j \, \partial q_l}\right)_0 q_j q_l + \cdots \tag{9.5-1}$$

In this expression U_0 is an arbitrary constant which we can set equal to zero. The derivatives of U are evaluated at the equilibrium position 0 and are constants when the q_j's are small quantities equal to zero at the equilibrium position. Since U is a minimum in the equilibrium position, the first derivative $(\partial U/\partial q_j)_0$ is zero, which leaves only $(\partial^2 U/\partial q_j \partial q_l)_0$ and higher order terms.

In the theory of small oscillations about the equilibrium position, terms beyond the second order are ignored and the equation for the potential energy reduces to

$$U = \frac{1}{2} \sum_{j=1}^{n} \sum_{l=1}^{n} \left(\frac{\partial^2 U}{\partial q_j \, \partial q_l}\right)_0 q_j q_l \tag{9.5-2}$$

The second derivative evaluated at 0 is a constant associated with the *generalized stiffness*

$$k_{jl} = \left(\frac{\partial^2 U}{\partial q_j \, \partial q_l}\right)_0 \tag{9.5-3}$$

and the potential energy is written as

$$U = \frac{1}{2}\sum_{j=1}^{n}\sum_{l=1}^{n} k_{jl}q_jq_l$$
$$= \frac{1}{2}\{q\}'[k]\{q\} \tag{9.5-4}$$

The kinetic energy is by definition equal to

$$T = \frac{1}{2}\sum_{i=1}^{n} m_i\,|\dot{r}_i|^2 \tag{9.5-5}$$

In terms of the generalized coordinates, the velocity $\dot{\mathbf{r}}_i$ is equal to

$$\mathbf{r}_i = \sum_{j=1}^{n} \frac{\partial \mathbf{r}_i}{\partial q_j}\dot{q}_j + \frac{\partial \mathbf{r}_i}{\partial t} \tag{9.5-6}$$

Considering a *scleronomic system* where the constraints are independent of time, the last term of the above equation is zero, and we have

$$|\dot{r}_i|^2 = \sum_{j=1}^{n}\sum_{l=1}^{n} \frac{\partial \mathbf{r}_i}{\partial q_j}\cdot\frac{\partial \mathbf{r}_i}{\partial q_l}\dot{q}_j\dot{q}_l \tag{9.5-7}$$

The kinetic energy equation thus becomes

$$T = \frac{1}{2}\sum_{i=1}^{n} m_i\left[\sum_{j=1}^{n}\sum_{l=1}^{n} \frac{\partial \mathbf{r}_i}{\partial q_j}\cdot\frac{\partial \mathbf{r}_i}{\partial q_l}\dot{q}_j\dot{q}_l\right] \tag{9.5-8}$$

We now interchange the order of summation and rewrite the above equation

$$T = \frac{1}{2}\sum_{j=1}^{n}\sum_{l=1}^{n} \dot{q}_j\dot{q}_l\left(\sum_{i}^{n} m_i\frac{\partial \mathbf{r}_i}{\partial q_j}\cdot\frac{\partial \mathbf{r}_i}{\partial q_l}\right) \tag{9.5-9}$$

Defining the *generalized mass* as

$$m_{jl} = \left(\sum_{i=1}^{n} m_i\frac{\partial \mathbf{r}_i}{\partial q_j}\cdot\frac{\partial \mathbf{r}_i}{\partial q_l}\right) \tag{9.5-10}$$

the kinetic energy can be written as

$$T = \frac{1}{2}\sum_{j=1}^{n}\sum_{l=1}^{n} m_{jl}\dot{q}_j\dot{q}_l$$
$$= \frac{1}{2}\{\dot{q}\}'[m]\{\dot{q}\} \tag{9.5-11}$$

It is clearly evident from Eqs. (9.5-3) and (9.5-10) that $k_{jl} = k_{lj}$ and

$m_{jl} = m_{lj}$. Thus the stiffness and mass matrices are symmetrical about the diagonal.

Substitution of T and U into Lagrange's equation leads to a set of equations which can be expressed by the following matrix equation

$$[m_{jl}]\{\ddot{q}\} + [k_{jl}]\{q\} = 0 \tag{9.5-11}$$

When the eigenvectors $\{q\}$ are normal (principal) coordinates, the off-diagonal terms of the matrix equation are zero and the equations of motion decouple to

$$m_{ii}\ddot{q}_i + k_{ii}q_i = 0 \tag{9.5-12}$$

The solution is then the normal mode vibration $q_i = A_i \sin \omega_i t$, whose substitution into the differential equation results in

$$k_{ii} = -\omega_i^2 m_{ii} \tag{9.5-13}$$

The use of normal coordinates will eliminate all k_{lj} and m_{lj} terms where $j \neq l$ and hence the kinetic and potential energy expressions simplify to the form

$$T = \tfrac{1}{2}\sum_i m_{ii}\dot{q}_i^2 = \tfrac{1}{2}\{\dot{q}\}'\lceil m \rfloor\{\dot{q}\} \tag{9.5-14}$$

$$U = \tfrac{1}{2}\sum_i k_{ii}q_i^2 = \tfrac{1}{2}\{q\}'\lceil k \rfloor\{q\} \tag{9.5-15}$$

where $\lceil \quad \rfloor$ denotes a diagonal matrix.

9.6 MODE SUMMATION METHOD

In Section 6.7 the equations of motion were decoupled by the model matrix to obtain the solution of forced vibration in terms of the normal coordinates of the system. In this section, we apply a similar technique to continuous systems by expanding the deflection in terms of the normal modes of the system.

Consider, for example, the general motion of a beam loaded by a distributed force $p(x, t)$, whose equation of motion is

$$[EIy''(x, t)]'' + m(x)\ddot{y}(x, t) = p(x, t) \tag{9.6-1}$$

The normal modes $\phi_i(x)$ of such a beam must satisfy the equation

$$(EI\phi_i'')'' - \omega_i^2 m(x)\phi_i = 0 \tag{9.6-2}$$

and its boundary conditions. The normal modes $\phi_i(x)$ are also orthogonal functions satisfying the relation

$$\int_0^l m(x)\phi_i\phi_j\,dx = \begin{cases} 0 \text{ for } j \neq i \\ M_i \text{ for } j = i \end{cases} \tag{9.6-3}$$

By representing the solution to the general problem in terms of $\phi_i(x)$

$$y(x, t) = \sum_i \phi_i(x) q_i(t) \tag{9.6-4}$$

the generalized coordinate $q_i(t)$ can be determined from Lagrange's equation by first establishing the kinetic and potential energies.

Recognizing the orthogonality relation, Eq. (9.6-3), the kinetic energy is

$$\begin{aligned} T &= \tfrac{1}{2}\int_0^l \dot{y}^2(x, t)m(x)\,dx = \tfrac{1}{2}\sum_i\sum_j \dot{q}_i\dot{q}_j \int_0^l \phi_i\phi_j m(x)\,dx \\ &= \tfrac{1}{2}\sum_i M_i\dot{q}_i^2 \end{aligned} \tag{9.6-5}$$

where the generalized mass M_i is defined as

$$M_i = \int_0^l \phi_i^2(x)m(x)\,dx \tag{9.6-6}$$

Similarly, the potential energy is

$$\begin{aligned} U &= \tfrac{1}{2}\int_0^l EIy''^2(x, t)\,dx = \tfrac{1}{2}\sum_i\sum_j q_i q_j \int_0^l EI\phi_i''\phi_j''\,dx \\ &= \tfrac{1}{2}\sum_i K_i q_i^2 = -\tfrac{1}{2}\sum \omega_i^2 M_i q_i^2 \end{aligned} \tag{9.6-7}^*$$

where the generalized stiffness is

$$K_i = \int_0^l EI[\phi_i''(x)]^2\,dx \tag{9.6-8}$$

In addition to T and U, we need the generalized force Q_i, which is determined from the work done by the applied force $p(x, t)dx$ in the virtual displacement δq_i.

$$\begin{aligned} \delta W &= \int_0^l p(x, t)\Big(\sum_i \phi_i \delta q_i\Big)\,dx \\ &= \sum_i \delta q_i \int_0^l p(x, t)\phi_i(x)\,dx \end{aligned} \tag{9.6-9}$$

*See Eq. (9.5-13).

or

$$Q_i = \int_0^l p(x, t)\phi_i(x)\,dx \tag{9.6-10}$$

Substituting into Lagrange's equation

$$\frac{d}{dt}\left(\frac{\partial T}{\partial \dot{q}_i}\right) - \frac{\partial T}{\partial q_i} + \frac{\partial U}{\partial q_i} = Q_i \tag{9.6-11}$$

the differential equation for $q_i(t)$ is found as

$$\ddot{q}_i + \omega_i^2 q_i = \frac{1}{M_i}\int_0^l p(x, t)\phi_i(x)\,dx \tag{9.6-12}$$

It is convenient at this point to consider the case where the loading per unit length $p(x, t)$ is separable in the form

$$p(x, t) = \frac{P_0}{l}p(x)f(t) \tag{9.6-13}$$

Eq. (9.6-12) then reduces to

$$\ddot{q}_i + \omega_i^2 q_i = \frac{P_0}{M_i}\Gamma_i f(t) \tag{9.6-14}$$

where

$$\Gamma_i = \frac{1}{l}\int_0^l p(x)\phi_i(x)\,dx \tag{9.6-15}$$

is defined as the mode participation factor for mode i. The solution of Eq. (9.6-14) is then

$$\begin{aligned} q_i(t) &= q_i(0)\cos\omega_i t + \frac{1}{\omega_i}\dot{q}_i(0)\sin\omega_i t \\ &\quad + \left(\frac{P_0\Gamma_i}{M_i\omega_i^2}\right)\omega_i\int_0^t f(\xi)\sin\omega_i(t-\xi)\,d\xi \end{aligned} \tag{9.6-16}$$

Since the ith mode statical deflection (with $\ddot{q}_i(t) = 0$) expanded in terms of $\phi_i(x)$ is $P_0\Gamma_i/M_i\omega_i^2$, the quantity

$$D_i(t) = \omega_i\int_0^t f(\xi)\sin\omega_i(t-\xi)d\xi \tag{9.6-17}$$

can be called the dynamic load factor for the ith mode.

Example 9.6-1

A simply supported uniform beam of mass M_0 is suddenly loaded by the force shown in Fig. 9.6-1. Determine the equation of motion.

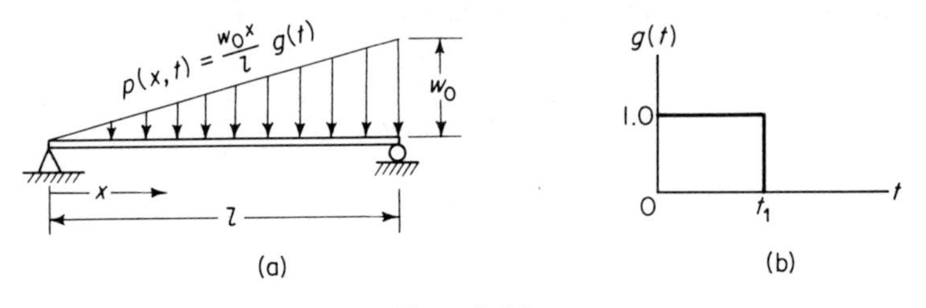

Figure 9.6-1.

Solution: The normal modes of the beam are

$$\phi_n(x) = \sqrt{2}\,\sin\frac{n\pi x}{l}$$

$$\omega_n = (n\pi)^2\sqrt{EI/M_0 l^3}$$

and the generalized mass is

$$M_n = \frac{M_0}{l}\int_0^l 2\sin^2\frac{n\pi x}{l}\,dx = M_0$$

The generalized force is

$$\begin{aligned}\int_0^l p(x,t)\phi_n\,dx &= g(t)\int_0^l \frac{w_0 x}{l}\sqrt{2}\,\sin\frac{n\pi x}{l}\,dx\\ &= g(t)\frac{w_0\sqrt{2}}{l}\left[\frac{\sin(n\pi x/l)}{(n\pi/l)^2} - \frac{x\cos(n\pi x/l)}{(n\pi/l)}\right]_0^l\\ &= -g(t)\frac{w_0\sqrt{2}\,l}{n\pi}\cos n\pi\\ &= -\frac{\sqrt{2}\,lw_0}{n\pi}g(t)(-1)^n\end{aligned}$$

where $g(t)$ is the time history of the load. The equation for q_n is then

$$\ddot{q}_n + \omega_n^2 q_n = -\frac{\sqrt{2}\,lw_0}{n\pi M_0}(-1)^n g(t)$$

which has the solution

$$q_n(t) = \frac{-\sqrt{2}\,lw_0}{n\pi M_0}\frac{(-1)^n}{\omega_n^2}(1 - \cos\omega_n t) \qquad 0 \le t \le t_1$$

$$= \frac{-\sqrt{2}\,lw_0}{n\pi M_0}\frac{(-1)^n}{\omega_n^2}(1 - \cos\omega_n t)$$

$$+ \frac{2\sqrt{2}\,lw_0(-1)^n}{n\pi M_0\omega_n^2}[1 - \cos\omega_n(t - t_1)] \qquad t_1 \le t \le \infty$$

Thus the deflection of the beam is expressed by the summation

$$y(x, t) = \sum_{n=1}^{\infty} q_n(t)\sqrt{2}\sin\frac{\pi n x}{l}$$

Example 9.6-2

A missile in flight is excited longitudinally by the thrust $F(t)$ of its rocket engine at the end $x = 0$. Determine the equation for the displacement $u(x, t)$ and the acceleration $\ddot{u}(x, t)$.

Solution: We assume the solution for the displacement to be

$$u(x, t) = \sum q_i(t)\varphi_i(x)$$

where $\varphi_i(x)$ are normal modes of the missile in longitudinal oscillation. The generalized coordinate q_i satisfies the differential equation

$$\ddot{q}_i + \omega_i^2 q_i = \frac{F(t)\varphi_i(0)}{M_i}$$

If, instead of $F(t)$, a unit impulse acted at $x = 0$, the above equation would have the solution $(\varphi_i(0)/M_i\omega_i)\sin\omega_i t$ for initial conditions $q_i(0) = \dot{q}_i(0) = 0$. Thus the response to the arbitrary force $F(t)$ is

$$q_i(t) = \frac{\varphi_i(0)}{M_i\omega_i}\int_0^t F(\xi)\sin\omega_i(t - \xi)\,d\xi$$

and the displacement at any point x is

$$u(x, t) = \sum_i \frac{\varphi_i(x)\varphi_i(0)}{M_i\omega_i}\int_0^t F(\xi)\sin\omega_i(t - \xi)\,d\xi$$

The acceleration $\ddot{q}_i(t)$ of mode i can be determined by rewriting the differential equation and substituting the former solution for $q_i(t)$

$$\ddot{q}_i(t) = \frac{F(t)\varphi_i(0)}{M_i} - \omega_i^2 q_i$$
$$= \frac{F(t)\varphi_i(0)}{M_i} - \frac{\varphi_i(0)\omega_i}{M_i}\int_0^t F(\xi)\sin\omega_i(t-\xi)\,d\xi$$

Thus the equation for the acceleration of any point x is found as

$$\ddot{u}(x, t) = \sum_i \ddot{q}_i(t)\varphi_i(x)$$
$$= \sum_i \left\{\frac{F(t)\varphi_i(0)\varphi_i(x)}{M_i} - \frac{\varphi_i(0)\varphi_i(x)\omega_i}{M_i}\int_0^t F(\xi)\sin\omega_i(t-\xi)\,d\xi\right\}$$

Example 9.6-3

Determine the response of a cantilever beam when its base is given a motion $y_b(t)$ normal to the beam axis as shown in Fig. 9.6-2.

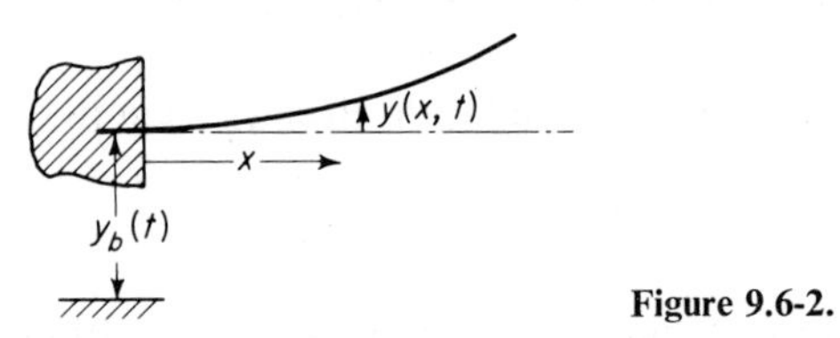

Figure 9.6-2.

Solution: The differential equation for the beam with base motion is

$$[EIy''(x, t)]'' + m(x)[\ddot{y}_b(t) + \ddot{y}(x, t)] = 0$$

which can be rearranged to

$$[EIy''(x, t)]'' + m(x)\ddot{y}(x, t) = -m(x)\ddot{y}_b(t)$$

Thus, instead of the force per unit length $F(x, t)$ we have the inertial force per unit length $-m(x)\ddot{y}_b(t)$. Assuming the solution in the form

$$y(x, t) = \sum_i q_i(t)\varphi_i(x)$$

the equation for the generalized coordinate q_i becomes

$$\ddot{q}_i + \omega_i^2 q_i = -\ddot{y}_b(t)\frac{1}{M_i}\int_0^l \varphi_i(x)\,dx$$

The solution for q_i then differs from that of a simple oscillator only by the factor $-1/M_i \int_0^l \varphi_i(x)\,dx$ so that for the initial conditions $y(0) = \dot{y}(0) = 0$

$$q_i(t) = \left\{-\frac{1}{M_i}\int_0^l \varphi_i(x)\,dx\right\}\frac{1}{\omega_i}\int_0^t \ddot{y}_b(\xi)\sin\omega_i(t-\xi)\,d\xi$$

9.7 BEAM ORTHOGONALITY, INCLUDING ROTARY INERTIA AND SHEAR DEFORMATION

The equations for the beam, including rotary inertia and shear deformation, were derived in Sec. 8.6. For such beams the orthogonality is no longer expressed by Eq. (9.6-3), but by the equation

$$\int [m(x)\varphi_j\varphi_i + J(x)\psi_j\psi_i]\,dx = \begin{cases} 0 & \text{if } j \neq i \\ M_i & \text{if } j = i \end{cases} \tag{9.7-1}$$

which can be proved in the following manner.

For convenience we will rewrite Eqs. (8.6-5) and (8.6-6), including a distributed moment per unit length $\mathfrak{M}(x, t)$

$$\frac{d}{dx}\left(EI\frac{d\psi}{dx}\right) + kAG\left(\frac{dy}{dx} - \psi\right) - J\ddot{\psi} - \mathfrak{M}(x, t) = 0 \tag{8.6-5}$$

$$m\ddot{y} - \frac{d}{dx}\left[kAG\left(\frac{dy}{dx} - \psi\right)\right] - p(x, t) = 0 \tag{8.6-6}$$

For the forced oscillation with excitation $p(x, t)$ and $\mathfrak{M}(x, t)$ per unit length of beam, the deflection $y(x, t)$ and the bending slope $\psi(x, t)$ can be expressed in terms of the generalized coordinates

$$\begin{aligned} y &= \sum_j q_j(t)\varphi_j(x) \\ \psi &= \sum_j q_j(t)\psi_j(x) \end{aligned} \tag{9.7-2}$$

With these summations substituted into the two beam equations, we obtain

$$\begin{aligned} J\sum_j \ddot{q}_j\psi_j &= \sum_j q_j\left\{\frac{d}{dx}(EI\psi_j') + kAG(\varphi_j' - \psi_j)\right\} + \mathfrak{M}(x, t) \\ m\sum_j \ddot{q}_j\varphi_j &= \sum_j q_j\frac{d}{dx}\{kAG(\varphi_j' - \psi_j)\} + p(x, t) \end{aligned} \tag{9.7-3}$$

However, normal-mode vibrations are of the form

$$\begin{aligned} y &= \varphi_j(x)e^{i\omega_j t} \\ \psi &= \psi_j(x)e^{i\omega_j t} \end{aligned} \tag{9.7-4}$$

which, when substituted into the beam equations with zero excitation, lead to

$$\begin{aligned} -\omega_j^2 J\psi_j &= \frac{d}{dx}(EI\psi_j') + kAG(\varphi_j' - \psi_j) \\ -\omega_j^2 m\varphi_j &= \frac{d}{dx}\{kAG(\varphi_j' - \psi_j)\} \end{aligned} \tag{9.7-5}$$

The right sides of this set of equations are the coefficients of the generalized coordinates q_j in the forced vibration equations, so that we can write Eqs. (9.7-3) as

$$\begin{aligned} J\sum_j \ddot{q}_j\psi_j &= -\sum_j q_j\omega_j^2 J\psi_j + \mathfrak{M}(x, t) \\ m\sum_j \ddot{q}_j\varphi_j &= -\sum_j q_j\omega_j^2 m\varphi_j + p(x, t) \end{aligned} \tag{9.7-6}$$

Multiplying these two equations by $\varphi_i\,dx$ and $\psi_i\,dx$, adding, and integrating, we obtain

$$\begin{aligned} \sum_j \ddot{q}_j \int_0^l (m\varphi_j\varphi_i + J\psi_j\psi_i)\,dx + \sum_j q_j\omega_j^2 \int_0^l (m\varphi_j\varphi_i + J\psi_j\psi_i)\,dx \\ = \int_0^l p(x, t)\varphi_i\,dx + \int_0^l \mathfrak{M}(x, t)\psi_i\,dx \end{aligned} \tag{9.7-7}$$

If the q's in these equations are generalized coordinates, they must be independent coordinates which satisfy the equation

$$\ddot{q}_i + \omega_i^2 q_i = \frac{1}{M_i}\left\{\int_0^l p(x, t)\varphi_i\,dx + \int_0^l \mathfrak{M}(x, t)\psi_i\,dx\right\} \tag{9.7-8}$$

We see then that this requirement is satisfied only if

$$\int_0^l (m\varphi_j\varphi_i + J\psi_j\psi_i)\,dx = \begin{cases} 0 & \text{if } j \neq i \\ M_i & \text{if } j = i \end{cases} \tag{9.7-9}$$

which defines the orthogonality for the beam, including rotary inertia and shear deformation.

9.8 NORMAL MODES OF CONSTRAINED STRUCTURES

When a structure is altered by the addition of a mass or a spring, we refer to it as a *constrained structure*. For example, a spring will tend to act as a constraint on the motion of the structure at the point of its application, and possibly increase the natural frequencies of the system. An added mass, on the other hand, may decrease the natural frequencies of the system. Such problems can be formulated in terms of generalized coordinates and the mode-summation technique.

Consider the forced vibration of any one dimensional structure (i.e., the points on the structure defined by one coordinate x) excited by a force per unit length $f(x, t)$ and moment per unit length $M(x, t)$. If we know the normal modes of the structure, ω_i and $\varphi_i(x)$, its deflection at any point x can be represented by

$$y(x, t) = \sum_i q_i(t)\varphi_i(x) \tag{9.8-1}$$

where the generalized coordinate q_i must satisfy the equation

$$\ddot{q}_i(t) + \omega_i^2 q_i(t) = \frac{1}{M_i}\left[\int f(x, t)\varphi_i(x)\,dx + \int M(x, t)\varphi_i'(x)\,dx\right] \tag{9.8-2}$$

The right side of this equation is $1/M_i$ times the generalized force Q_i, which can be determined from the virtual work of the applied loads as $Q_i = \delta W/\delta q_i$.

If, instead of distributed loads, we have a concentrated force $F(a, t)$ and a concentrated moment $M(a, t)$ at some point $x = a$, the generalized force for such loads is found from

$$\begin{aligned} \delta W &= F(a, t)\,\delta y(a, t) + M(a, t)\,\delta y'(a, t) \\ &= F(a, t)\sum_i \varphi_i(a)\,\delta q_i + M(a, t)\sum_i \varphi_i'(a)\,\delta q_i \\ Q_i &= \frac{\delta W}{\delta q_i} = F(a, t)\varphi_i(a) + M(a, t)\varphi_i'(a) \end{aligned} \tag{9.8-3}$$

Then, instead of Eq. (9.5-2), we obtain the equation

$$\ddot{q}_i(t) + \omega_i^2 q_i(t) = \frac{1}{M_i}[F(a, t)\varphi_i(a) + M(a, t)\varphi_i'(a)] \tag{9.8-4}$$

These equations form the starting point for the analysis of constrained structures, provided the constraints are expressible as external loads on the structure.

As an example, let us consider attaching a linear and torsional spring to the simply supported beam of Fig. 9.8-1. The linear spring exerts a force on the beam equal to

$$F(a, t) = -ky(a, t) = -k \sum_j q_j(t)\varphi_j(a) \tag{9.8-5}$$

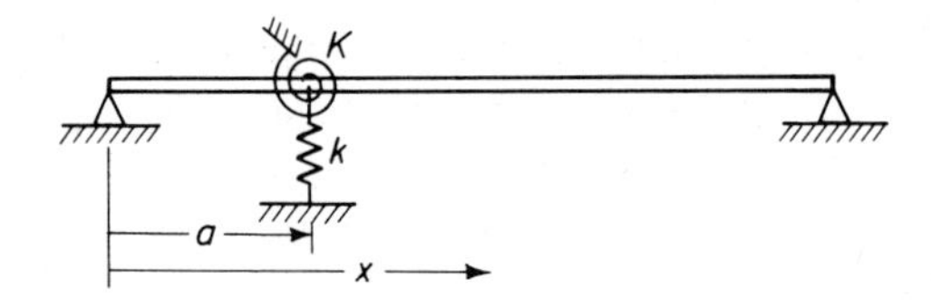

Figure 9.8-1.

whereas the torsional spring exerts a moment

$$M(a, t) = -Ky'(a, t) = -K \sum_j q_j(t)\varphi'_j(a) \tag{9.8-6}$$

Substituting these equations into Eq. (9.5-4), we obtain

$$\ddot{q}_i + \omega_i^2 q_i = \frac{1}{M_i}\left[-k\varphi_i(a) \sum_j q_j\varphi_j(a) - K\varphi'_i(a) \sum_j q_j\varphi'_j(a)\right] \tag{9.8-7}$$

The normal modes of the constrained modes are also harmonic and so we can write

$$q_i = \bar{q}_i e^{i\omega t}$$

The solution to the ith equation is then

$$\bar{q}_i = \frac{1}{M_i(\omega_i^2 - \omega^2)}\left[-k\varphi_i(a) \sum_j \bar{q}_j\varphi_j(a) - K\varphi'_i(a) \sum_j \bar{q}_j\varphi'_j(a)\right] \tag{9.8-8}$$

If we use n modes, there will be n values of $\bar{q}_j$ and n equations such as the one above. The determinant formed by the coefficients of the $\bar{q}_j$ will then lead to the natural frequencies of the constrained modes, and the mode shapes of the constrained structure are found by substituting the $\bar{q}_j$ into Eq. (9.8-1).

If, instead of springs, a mass m_0 is placed at a point $x = a$, as shown in Fig. 9.8-2, the force exerted by m_0 on the beam is

$$F(a, t) = -m_0\ddot{y}(a, t) = -m_0 \sum_j \ddot{q}_j\varphi_j(a) \tag{9.8-9}$$

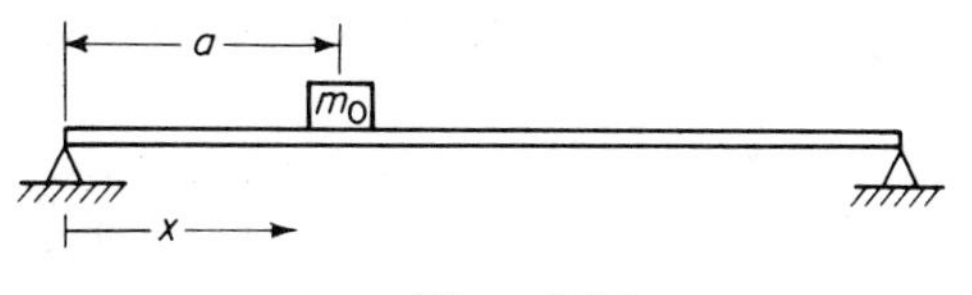

Figure 9.8-2.

Thus, in place of Eq. (9.8-8), we would obtain the equation

$$\bar{q}_i = \frac{1}{M_i(\omega_i^2 - \omega^2)}\left[\omega^2 m_0 \varphi_i(a) \sum_j \bar{q}_j \varphi_j(a)\right] \tag{9.8-10}$$

Example 9.8-1

Give a single mode approximation for the natural frequency of a simply supported beam when a mass m_0 is attached to it at $x = l/3$.

Solution: When only a single mode is used, Eq. (9.8-10) reduces to

$$M_1(\omega_1^2 - \omega^2) = \omega^2 m_0 \varphi_1^2(a)$$

Solving for ω^2, we obtain

$$\left(\frac{\omega}{\omega_1}\right)^2 = \frac{1}{1 + \dfrac{m_0}{M_1}\varphi_1^2(a)}$$

For the first mode of the unconstrained beam, we have

$$\omega_1 = \pi^2\sqrt{\frac{EI}{Ml^3}}, \qquad \varphi_1(x) = \sqrt{2}\,\sin\frac{\pi x}{l}$$

$$\varphi_1\left(\frac{l}{3}\right) = \sqrt{2}\,\sin\frac{\pi}{3} = \sqrt{2} \times 0.866$$

$$M_1 = M = \text{mass of the beam}$$

Thus its substitution into the above equation gives the one-mode approximation for the constrained beam the value

$$\left(\frac{\omega}{\omega_1}\right)^2 = \frac{1}{1 + 1.5\dfrac{m_0}{M}}$$

The same problem treated by the Dunkerley equation in Example 7.5-5

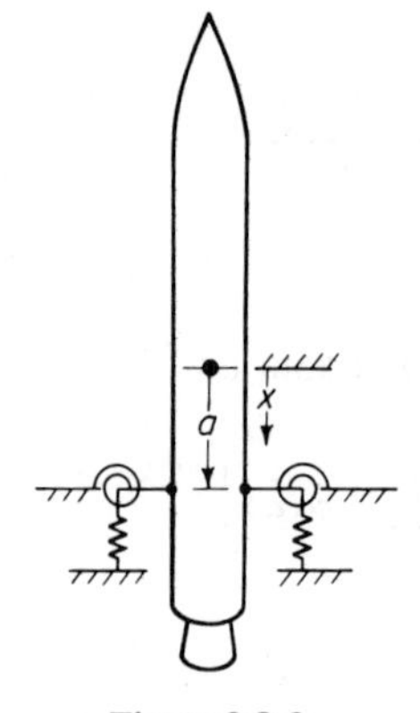

Figure 9.8-3.

gave, for this ratio, the result

$$\frac{1}{1 + 1.6\dfrac{m_0}{M}}$$

Example 9.8-2

A missile is constrained in a test stand by linear and torsional springs, as shown in Fig. 9.8-3. Formulate the inverse problem of determining its free-free modes from the normal modes of the constrained missile, which are designated as Φ_i and Ω_i.

Solution: The problem is approached in a manner similar to that of the direct problem where, in place of φ_i and ω_i, we use Φ_i and Ω_i. We now relieve the constraints at the supports by introducing opposing forces $-F(a)$ and $-M(a)$ equal to $ky(a)$ and $Ky'(a)$.

To carry out this problem in greater detail, we start with the equation

$$\bar{q}_i = \frac{-F(a)\Phi_i(a) - M(a)\Phi_i'(a)}{M_i\Omega_i^2[1 - (\omega/\Omega_i)^2]}$$

which replaces Eq. (9.8-8). Letting $D_i/\omega) = M_i\Omega_i^2[1 - (\omega/\Omega_i)^2]$, the displacement at $x = a$ is

$$y(a) = \sum_i \Phi_i(a)\bar{q}_i = \sum_i \frac{-F(a)\Phi_i^2(a) - M(a)\Phi_i'(a)\Phi_i(a)}{D_i(\omega)}$$

We now replace $-F(a)$ and $-M(a)$ with $ky(a)$ and $Ky'(a)$ and write

$$y(a) = \sum_i \frac{ky(a)\Phi_i^2(a) + Ky'(a)\Phi_i'(a)\Phi_i(a)}{D_i(\omega)}$$

$$y'(a) = \sum_i \frac{ky(a)\Phi_i'(a)\Phi_i(a) + Ky'(a)\Phi_i'^2(a)}{D_i(\omega)}$$

These equations may now be rearranged as

$$y(a)\left[1 - k\sum_i \frac{\Phi_i^2(a)}{D_i(\omega)}\right] = y'(a)K\sum_i \frac{\Phi_i'(a)\Phi_i(a)}{D_i(\omega)}$$

$$y(a)k\sum_i \frac{\Phi_i'(a)\Phi_i(a)}{D_i(\omega)} = y'(a)\left[1 - K\sum_i \frac{\Phi_i'^2(a)}{D_i(\omega)}\right]$$

The frequency equation then becomes

$$\left[1 - k\sum_i \frac{\Phi_i^2(a)}{D_i(\omega)}\right]\left[1 - K\sum_i \frac{\Phi_i'^2(a)}{D_i(\omega)}\right] - kK\left[\sum \frac{\Phi_i'(a)\Phi_i(a)}{D_i(\omega)}\right]^2 = 0$$

The slope to deflection ratio at $x = a$ is

$$\frac{y'(a)}{y(a)} = \frac{1 - k\sum_i \dfrac{\Phi_i^2(a)}{D_i(\omega)}}{K\sum_i \dfrac{\Phi_i'(a)\Phi_i(a)}{D_i(\omega)}}$$

The free-free mode shape is then given by

$$\frac{y(x)}{y(a)} = \sum_i \frac{k\Phi_i(a)\Phi_i(x) + K\dfrac{y'(a)}{y(a)}\Phi_i'(a)\Phi_i(x)}{D_i(\omega)}$$

Example 9.8-3

Determine the constrained modes of the missile of Fig. 9.8-3, using only the first free-free mode $\varphi_1(x), \omega_1$, together with translation $\varphi_T = 1$, $\Omega_T = 0$ and rotation $\varphi_R = x$, $\Omega_R = 0$, where x is measured positively toward the tail of the missile.

Solution: The generalized mass for each of the three modes is

$$M_T = \int dm = M$$

$$M_R = \int x^2\, dm = I = M\rho^2$$

$$M_1 = \int \varphi_1^2(x)\, dm = M$$

where the $\varphi_1(x)$ mode was normalized such that $M_1 = M =$ actual mass.

The frequency dependent factors D_i are

$$D_T = -M_T\omega^2 = -M\omega^2 = -M\omega_1^2\lambda$$

$$D_R = -M\rho^2\omega^2 = -M\rho^2\omega_1^2\lambda$$

$$D_1 = M\omega_1^2\left[1 - \left(\frac{\omega}{\omega_1}\right)^2\right] = M\omega_1^2(1 - \lambda)$$

$$\left(\frac{\omega}{\omega_1}\right)^2 = \lambda$$

The frequency equation for this problem is the same as that of Example 9.8-2, except that the minus k's are replaced by positive k's and $\varphi(x)$ and ω replace $\Phi(x)$ and Ω. Substituting the above quantities into the frequency equation, we have

$$\left\{1 - \frac{k}{M\omega_1^2}\left[\frac{1}{\lambda} + \frac{a^2}{\rho^2\lambda} - \frac{\varphi_1^2(a)}{(1-\lambda)}\right]\right\}\left\{1 - \frac{K}{M\omega_1^2}\left[\frac{1}{\rho^2\lambda} - \frac{\varphi_1'^2(a)}{(1-\lambda)}\right]\right\} - \frac{kK}{M^2\omega_1^4}\left\{\frac{-a}{\rho^2\lambda} + \frac{\varphi_1'(a)\varphi_1(a)}{(1-\lambda)}\right\}^2 = 0$$

which can be simplified to

$$\lambda^2(1-\lambda) + \left(\frac{k}{M\omega_1^2}\right)\left[\varphi_1^2(a) + \frac{K}{k}\varphi_1'^2(a)\right]\lambda^2 - \left(\frac{k}{M\omega_1^2}\right)\left[1 + \frac{a^2}{\rho^2} + \frac{K}{k\rho^2}\right]\lambda(1-\lambda)$$
$$+ \left(\frac{k}{M\omega_1^2}\right)^2 \frac{K}{k\rho^2}(1-\lambda) - \left(\frac{k}{M\omega_1^2}\right)^2 \frac{K}{k}\lambda\left\{\varphi_1'^2(a) + \frac{1}{\rho^2}[\varphi_1(a) - a\varphi_1'(a)]^2\right\} = 0$$

A number of special cases of the above equation are of interest, and we mention one of these. If $K = 0$, the frequency equation simplifies to

$$\lambda^2 - \left\{1 + \left(\frac{k}{M\omega_1^2}\right)\left[1 + \frac{a^2}{\rho^2} + \varphi_1^2(a)\right]\right\}\lambda + \left(\frac{k}{M\omega_1^2}\right)\left(1 + \frac{a^2}{\rho^2}\right) = 0$$

Here $x = a$ might be taken negatively so that the missile is hanging by a spring.

9.9 MODE-ACCELERATION METHOD

One of the difficulties encountered in any mode summation method has to do with the convergence of the procedure. If this convergence is poor, a large number of modes must be used, thereby increasing the order of the frequency determinant. The mode-acceleration method tends to overcome this difficulty by improving the convergence so that a fewer number of normal modes are needed.

The mode-acceleration method starts with the same differential equation for the generalized coordinate q_i, but rearranged in order. For example, we can start with Eq. (9.8-4) and write it in the order

$$q_i(t) = \frac{F(a,t)\varphi_i(a)}{M_i\omega_i^2} + \frac{M(a,t)\varphi_i'(a)}{M_i\omega_i^2} - \frac{\ddot{q}_i(t)}{\omega_i^2} \tag{9.9-1}$$

Substituting this into Eq. (9.8-1), we obtain

$$y(x,t) = \sum_i q_i(t)\varphi_i(x)$$
$$= F(a,t)\sum_i \frac{\varphi_i(a)\varphi_i(x)}{M_i\omega_i^2} + M(a,t)\sum_i \frac{\varphi_i'(a)\varphi_i(x)}{M_i\omega_i^2} - \sum_i \frac{\ddot{q}_i(t)\varphi_i(x)}{\omega_i^2} \tag{9.9-2}$$

We note here that, if $F(a, t)$ and $M(a, t)$ were static loads, the last term containing the acceleration would be zero. Thus the terms

$$\sum_i \frac{\varphi_i(a)\varphi_i(x)}{M_i\omega_i^2} = \alpha(a, x)$$
$$\sum \frac{\varphi_i'(a)\varphi_i(x)}{M_i\omega_i^2} = \beta(a, x) \tag{9.9-3}$$

must represent influence functions, where $\alpha(a, x)$ and $\beta(a, x)$ are the deflections at x due to a unit load and unit moment at a, respectively. We can therefore rewrite Eq. (9.9-2) as

$$y(x, t) = F(a, t)\alpha(a, x) + M(a, t)\beta(a, x) - \sum \frac{\ddot{q}_i(t)\varphi_i(x)}{\omega_i^2} \tag{9.9-4}$$

Because of ω_i^2 in the denominator of the terms summed, the convergence is improved over the mode-summation method.

In the forced vibration problem where $F(a, t)$ and $M(a, t)$ are excitations, Eq. (9.8-4) is first solved for $q_i(t)$ in the conventional manner, and then substituted into Eq. (9.9-4) for the deflection. For the normal modes of constrained structures, $F(a, t)$ and $M(a, t)$ are again the forces and moments exerted by the constraints, and the problem is treated in a manner similar to those of Sec. 9.8. However, because of the improved convergence, fewer number of modes will be found to be necessary.

Example 9.9-1

Using the mode-acceleration method, solve the problem of Fig. 9.8-2 of a concentrated mass m_0 attached to the structure.

Solution: Assuming harmonic oscillations

$$F(a, t) = \bar{F}(a)e^{i\omega t}$$
$$q_i(t) = \bar{q}_i e^{i\omega t}$$
$$y(x, t) = y(x)e^{i\omega t}$$

Substituting these equations into Eq. (9.9-4) and letting $x = a$,

$$\bar{y}(a) = \bar{F}(a)\alpha(a, a) + \omega^2 \sum_j \frac{\bar{q}_j\varphi_j(a)}{\omega_j^2}$$

Since the force exerted by m_0 on the structure is

$$\bar{F}(a) = m_0\omega^2\bar{y}(a)$$

we can eliminate $\bar{y}(a)$ between the above two equations, obtaining

$$\frac{\bar{F}(a)}{m_0\omega^2} = \bar{F}(a)x(a, a) + \omega^2 \sum_j \frac{\bar{q}_j\varphi_j(a)}{\omega_j^2}$$

or

$$\bar{F}(a) = \frac{\omega^2 \sum_j \frac{\bar{q}_j\varphi_j(a)}{\omega_j^2}}{\frac{1}{m_0\omega^2} - \alpha(a, a)}$$

If we now substitute this equation into Eq. (9.8-4) and assume harmonic motion, we obtain the equation

$$(\omega_i^2 - \omega^2)\bar{q}_i = \frac{\bar{F}(a)\varphi_i(a)}{M_i} = \frac{\omega^2\varphi_i(a)\sum_j \bar{q}_j \frac{\varphi_j(a)}{\omega_j^2}}{M_i\left[\frac{1}{m_0\omega^2} - \alpha(a, a)\right]}$$

Rearranging, we have

$$[1 - m_0\omega^2\alpha(a, a)](\omega_i^2 - \omega^2)\bar{q}_i = \frac{\omega^4 m_0\varphi_i(a)}{M_i} \sum_j \frac{\bar{q}_j\varphi_j(a)}{\omega_j^2}$$

which represents a set of linear equations in $\bar{q}_k$. The series represented by the summation will, however, converge rapidly because of ω_j^2 in the denominator. Offsetting this advantage of smaller number of modes is the disadvantage that these equations are now quartic rather than quadratic in ω.

9.10 COMPONENT MODE SYNTHESIS

The treatment of large structural systems may be simplified by breaking up the system into smaller subsystems which are related through the displacement and force conditions at their junction points. Each subsystem is represented by mode functions, the sum of which allows the satisfaction of the displacement and force conditions at the junctions. These functions need not be orthogonal or normal modes of the subsystem, and each mode used need not satisfy the junction conditions as long as their combined sum allows these conditions to be satisfied. Lagrange's equations, and in particular the method of superfluous coordinates, form the basis for the synthesis process.

To present the basic ideas of the method of modal synthesis, we will consider a simple beam with a 90° bend, an example which was used by

W. Hurty.* The beam, shown in Fig. 9.10-1, is considered to vibrate only in the plane of the paper.

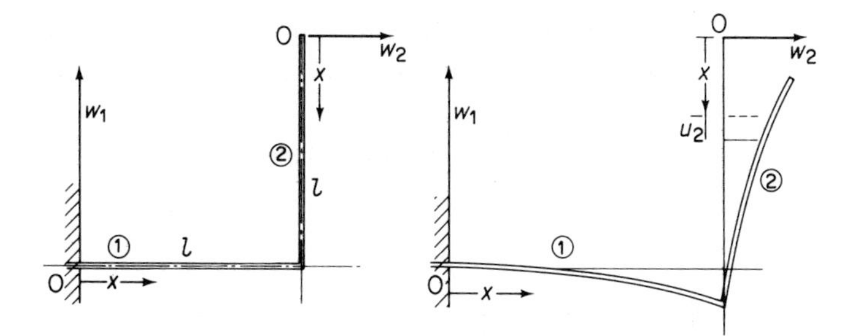

Figure 9.10-1. Beam sections 1 and 2 with their coordinates.

We separate the beam into two sections, ① and ②, whose coordinates are shown as w_1, x; w_2, x; and u_2, x. For part ① we assume the deflection to be

$$\begin{aligned} w_1(x, t) &= \phi_1(x)p_1(t) + \phi_2(x)p_2(t) + \cdots \\ &= \left(\frac{x}{l}\right)^2 p_1 + \left(\frac{x}{l}\right)^3 p_2 \end{aligned} \tag{9.10-1}$$

Note that the two mode functions satisfy the geometric and force conditions at the boundaries of section ① as follows

$$\begin{aligned} w_1(0) &= 0 & w_1(l) &= p_1 + p_2 \\ w'_1(0) &= 0 & w'_1(l) &= \frac{2}{l}p_1 + \frac{3}{l}p_3 \\ w''_1(0) &= \frac{M(0)}{EI} = \frac{2}{l^2}p_1 & w''_1(l) &= \frac{M(l)}{EI} = \frac{2}{l^2}p_1 + \frac{6}{l^2}p_2 \\ w'''_1(0) &= \frac{V(0)}{EI} = \frac{6}{l^3}p_2 & w'''_1(l) &= \frac{V(l)}{EI} = \frac{6}{l^3}p_2 \end{aligned} \tag{9.10-2}$$

Next consider part ② with the origin of the coordinates w_2, x at the free end. The following functions will satisfy the boundary conditions of beam section ②

$$\begin{aligned} w_2(x, t) &= \phi_3(x)p_3(t) + \phi_4(x)p_4(t) + \phi_5(x)p_5(t) + \cdots \\ &= 1p_3 + \left(\frac{x}{l}\right)p_4 + \left(\frac{x}{l}\right)^4 p_5 \end{aligned} \tag{9.10-3}$$

*Walter C. Hurty, "Vibrations of Structural Systems by Component Synthesis," *Jour. Engr. Mech. Div., Proc. of ASCE* (Aug. 1960), pp. 51–69.

$$
\begin{aligned}
u_2(x, t) &= \phi_6(x)p_6(t) + \cdots \\
&= 1\, p_6
\end{aligned}
\tag{9.10-4}
$$

where $u_2(x, t)$ is the displacement in the x direction.

The next step is to calculate the generalized mass from the equation

$$m_{ij} = \int_0^l m(x)\phi_i(x)\phi_j(x)\, dx$$

For subsection ① we have

$$m_{11} = \int_0^l m\phi_1\phi_1\, dx = \int_0^l m\left(\frac{x}{l}\right)^4 dx = 0.20ml$$

$$m_{12} = \int_0^l m\phi_1\phi_2\, dx = \int_0^l m\left(\frac{x}{l}\right)^5 dx = 0.166ml = m_{21}$$

$$m_{22} = \int_0^l m\phi_2\phi_2\, dx = \int_0^l m\left(\frac{x}{l}\right)^6 dx = 0.1428ml$$

The generalized mass for subsection ② is computed in a similar manner using ϕ_3 to ϕ_6

$$
\begin{aligned}
m_{33} &= 1.0ml \\
m_{34} &= 0.50ml = m_{43} \\
m_{35} &= 0.20ml = m_{53} \\
m_{44} &= 0.333ml \\
m_{45} &= 0.166ml = m_{54} \\
m_{55} &= 0.111ml \\
m_{66} &= 1.0ml
\end{aligned}
$$

Since there is no coupling between the longitudinal displacement u_2 and the lateral displacement w_2, $m_{63} = m_{64} = m_{65} = 0$.

The generalized stiffness is found from the equation

$$k_{ij} = \int_0^l EI\phi_i''\phi_j''\, dx$$

Thus

$$k_{11} = EI\int_0^l \phi_1''\phi_1''\, dx = EI\int_0^l \left(\frac{2}{l^2}\right)^2 dx = 4\frac{EI}{l^3}$$

$$k_{12} = k_{21} = EI\int_0^l \left(\frac{2}{l^2}\right)\left(\frac{6x}{l^3}\right) dx = 6\frac{EI}{l^3}$$

$$k_{22} = 12\frac{EI}{l^3}$$

$$k_{55} = 28.8\frac{EI}{l^3}$$

All other k_{ij} are zero.

The results computed for m_{ij} and k_{ij} can now be arranged in the mass and stiffness matrices partitioned as follows

$$[m] = ml\begin{bmatrix} 0.2000 & 0.1666 & 0 & 0 & 0 & 0 \\ 0.1666 & 0.1428 & 0 & 0 & 0 & 0 \\ 0 & 0 & 1.0000 & 0.5000 & 0.2000 & 0 \\ 0 & 0 & 0.5000 & 0.3333 & 0.1666 & 0 \\ 0 & 0 & 0.2000 & 0.1666 & 0.1111 & 0 \\ 0 & 0 & 0 & 0 & 0 & 1.0000 \end{bmatrix} \tag{9.10-5}$$

$$[k] = \frac{EI}{l^3}\begin{bmatrix} 4 & 6 & 0 & 0 & 0 & 0 \\ 6 & 12 & 0 & 0 & 0 & 0 \\ 0 & 0 & 0 & 0 & 0 & 0 \\ 0 & 0 & 0 & 0 & 0 & 0 \\ 0 & 0 & 0 & 0 & 28.8 & 0 \\ 0 & 0 & 0 & 0 & 0 & 0 \end{bmatrix} \tag{9.10-6}$$

where the upper left matrix refers to section ① and the remainder to section ②.

At the junction between sections ① and ② we have the following constraint equations

$$\begin{aligned} w_1(l) + u_2(l) &= 0 \qquad &\text{or} \qquad p_1 + p_2 + p_6 &= 0 \\ w_2(l) &= 0 & p_3 + p_4 + p_5 &= 0 \\ w_1'(l) - w_2'(l) &= 0 & 2p_1 + 3p_2 - p_4 - 4p_5 &= 0 \\ EI[w_1''(l) + w_2''(l)] &= 0 & 2p_1 + 6p_2 + 12p_5 &= 0 \end{aligned}$$

Arranged in matrix form, these are

$$\begin{bmatrix} 1 & 1 & 0 & 0 & 0 & 1 \\ 0 & 0 & 1 & 1 & 1 & 0 \\ 2 & 3 & 0 & -1 & -4 & 0 \\ 2 & 6 & 0 & 0 & 12 & 0 \end{bmatrix}\begin{Bmatrix} p_1 \\ p_2 \\ p_3 \\ p_4 \\ p_5 \\ p_6 \end{Bmatrix} = 0 \tag{9.10-7}$$

Since the total number of coordinates used are six and there are four constraint equations, the number of generalized coordinates for the system is two (i.e., there are four superfluous coordinates corresponding to the four constraint equations (See Sec. 9.2)). We can thus choose any two of the coordinates to be the generalized coordinates q. Let $p_1 = q_1$ and $p_6 = q_6$ be the generalized coordinates and express $p_1 \dots p_6$ in terms of q_1 and q_6. This is accomplished in the following steps.

Rearrange Eq. (9.10-7) by shifting columns 1 and 6 to the right side

$$\begin{bmatrix} 1 & 0 & 0 & 0 \\ 0 & 1 & 1 & 1 \\ 3 & 0 & -1 & -4 \\ 6 & 0 & 0 & 12 \end{bmatrix} \begin{Bmatrix} p_2 \\ p_3 \\ p_5 \\ p_4 \end{Bmatrix} = \begin{bmatrix} -1 & -1 \\ 0 & 0 \\ -2 & 0 \\ -2 & 0 \end{bmatrix} \begin{Bmatrix} q_1 \\ q_6 \end{Bmatrix} \tag{9.10-8}$$

In abbreviated notation the above equation is

$$[s]\{p_{2-5}\} = [Q]\{q_{1,6}\}$$

Premultiply by $[s]^{-1}$ to obtain

$$\{p_{2-5}\} = [s]^{-1}[Q]\{q_{1,6}\}$$

Supply the identity $p_1 = q_1$ and $p_6 = q_6$ and write

$$\{p_{1-6}\} = [C]\{q_{1,6}\}$$

The above constraint equation is now in terms of the generalized coordinates q_1 and q_6 as follows

$$\begin{Bmatrix} p_1 \\ p_2 \\ p_3 \\ p_4 \\ p_5 \\ p_6 \end{Bmatrix} = \begin{bmatrix} 1 & 0 \\ -1 & -1 \\ 2 & 4.50 \\ -2.333 & -5.0 \\ 0.333 & 0.50 \\ 0 & 1 \end{bmatrix} \begin{Bmatrix} q_1 \\ q_6 \end{Bmatrix} = [C]\begin{Bmatrix} q_1 \\ q_6 \end{Bmatrix} \tag{9.10-9}$$

Returning to the Lagrange equation for the system, which is

$$ml[m]\{\ddot{p}\} + \frac{EI}{l^3}[k]\{p\} = 0 \tag{9.10-10}$$

substitute for $\{p\}$ in terms of $\{q\}$ from the constraint equation (9.10-9)

$$ml[m][C]\{\ddot{q}\} + \frac{EI}{l^3}[k][C]\{q\} = 0$$

Premultiply by the transpose $[C]'$

$$ml[C]'[m][C]\{\ddot{q}\} + \frac{EI}{l^3}[C]'[k][C]\}q\} = 0 \qquad (9.10\text{-}11)$$

Comparing Eqs. (9.10-10) and (9.10-11), we note that in (9.10-10) the mass and stiffness matrices are 6×6 (See Eqs. 9.10-5 and 9.10-6), whereas the matrices $[C]'[m][C]$ and $[C]'[k][C]$ in Eq. (9.10-11) are 2×2. Thus we have reduced the size of the system from a 6×6 to a 2×2 problem.

Letting $\{\ddot{q}\} = -\omega^2\{q\}$, Eq. (9.10-11) is in the form

$$\left[-\omega^2 ml\begin{bmatrix} a_{11} & a_{12} \\ a_{21} & a_{22} \end{bmatrix} + \frac{EI}{l^3}\begin{bmatrix} b_{11} & b_{12} \\ b_{21} & b_{22} \end{bmatrix}\right]\begin{Bmatrix} q_1 \\ q_6 \end{Bmatrix} = 0 \qquad (9.10\text{-}12)$$

The numerical values of the matrix $[a_{ij}]$ and $[b_{ij}]$ from Eqs. (9.10-5), (9.10-6), and (9.10-9) are

$$[a_{ij}] = [C]'[m][C] = \begin{bmatrix} 1.1774 & 2.6614 \\ 2.6614 & 7.3206 \end{bmatrix}$$

$$[b_{ij}] = [C]'[k][C] = \begin{bmatrix} 7.200 & 10.800 \\ 10.800 & 19.200 \end{bmatrix}$$

Using these numerical results, we find the two natural frequencies of the system from the characteristic equation of Eq. (9.10-12)

$$\omega_1 = 1.172\sqrt{\frac{EI}{ml^4}}$$

$$\omega_2 = 3.198\sqrt{\frac{EI}{ml^4}}$$

Figure 9.10-2 shows the mode shapes corresponding to the above frequencies. Since Eq. (9.10-12) enables the solution of the eigenvectors only in terms of an arbitrary reference, q_6 can be solved with $q_1 = 1.0$. The coor-

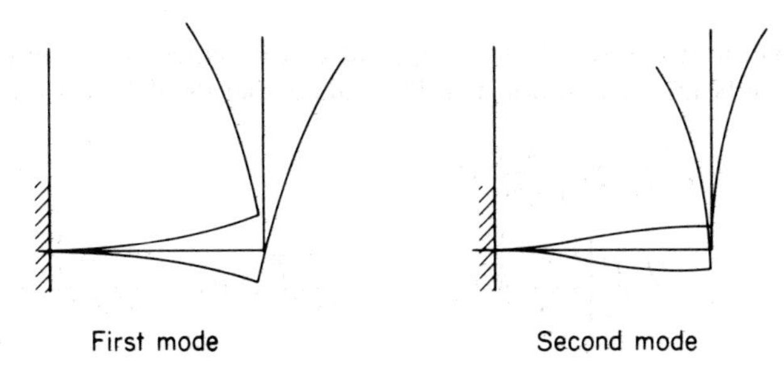

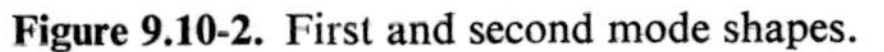

Figure 9.10-2. First and second mode shapes.

dinates p are then found from Eq. (9.10-9) and the mode shapes are obtained from Eqs. (9.10-1), (9.10-3), and (9.10-4).

PROBLEMS

9-1 Show that the dynamic load factor for a suddenly applied constant force reaches a maximum value of 2.0.

9-2 If a suddenly applied constant force is applied to a system for which the damping factor of the ith mode is $\zeta = c/c_{cr}$, show that the dynamic load factor is given approximately by the equation

$$D_i = 1 - e^{-\zeta \omega_i t} \cos \omega_i t$$

9-3 Determine the mode participation factor for a uniformly distributed force.

9-4 If a concentrated force acts at $x = a$, the loading per unit length corresponding to it can be represented by a delta function $l\,\delta(x - a)$. Show that the mode-participation factor then becomes $K_i = \varphi_i(a)$ and the deflection is expressible as

$$y(x, t) = \frac{P_0 l_3}{EI} \sum_i \frac{\varphi_i(a)\varphi_i(x)}{(\beta_i l)^4} D_i(t)$$

where $\omega_i^2 = (\beta_i l)^4 (EI/Ml^3)$ and $(\beta_i l)$ is the eigenvalue of the normal-mode equation.

9-5 For a couple of moment M_0 acting at $x = a$, show that the loading $p(x)$ is the limiting case of two delta functions shown in Fig. P9-5 as $\epsilon \longrightarrow 0$. Show also that the mode-participation factor for this case is

$$K_i = l \frac{d\varphi_i(x)}{dx}\bigg|_{x=a} = (\beta_i l)\varphi_i'(x)_{x=a}$$

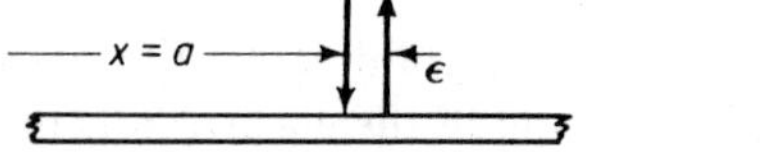

$\frac{l^2}{\epsilon}\delta(x-a)$ $\frac{l^2}{\epsilon}\delta(x-a-\epsilon)$ $x = a$ ϵ

Figure P. 9-5.

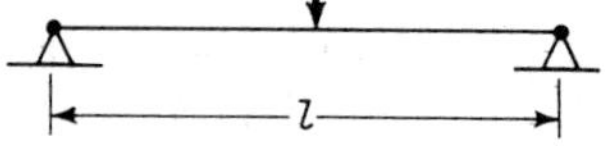

$P_0 f(t)$ l

Figure P. 9-6.

9-6 A concentrated force $P_0 f(t)$ is applied to the center of a simply supported uniform beam, as shown in Fig. P9-6. Show that the deflection is given by

$$y(x, t) = \frac{P_0 l^3}{EI} \sum_i \frac{K_i \varphi_i(x)}{(\beta_i l)^4} D_i$$

$$= \frac{2P_0 l^3}{EI} \left\{ \frac{\sin \pi \frac{x}{l}}{\pi^4} D_1(t) - \frac{\sin 3\pi \frac{x}{l}}{(3\pi)^4} D_3(t) + \frac{\sin 5\pi \frac{x}{l}}{(5\pi)^4} D_5(t) \ldots \right\}$$

9-7 A couple of moment M_0 is applied at the center of the beam of Prob. 9-6, as shown in Fig. P9-7. Show that the deflection at any point is given by the equation

$$y(x, t) = \frac{M_0 l^2}{EI} \sum_i \frac{\varphi_i'(a)\varphi_i(x)}{(\beta_i l)^3} D_i(t)$$

$$= \frac{2M_0 l^2}{.EI}\left\{ -\frac{\sin 2\pi \frac{x}{l}}{(2u)^3} D_2(t) + \frac{\sin 4\pi \frac{x}{l}}{(4\pi)^3} D_4(t) - \frac{\sin 6\pi \frac{x}{l}}{(6\pi)^3} D_6(t) \cdots \right\}$$

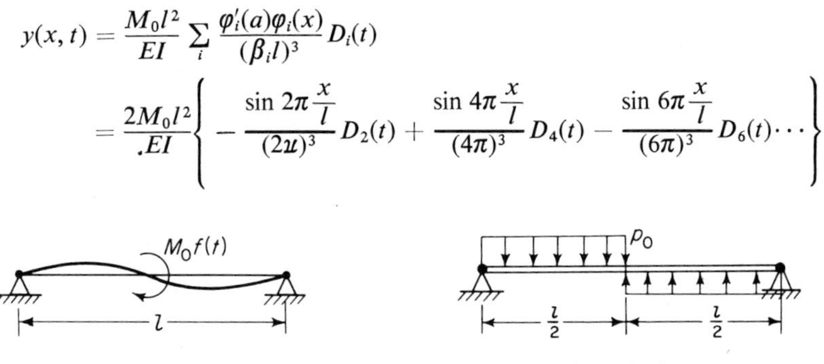

Figure P. 9-7.

Figure P. 9-8.

9-8 A simply supported uniform beam has suddenly applied to it the load distribution shown in Fig. P9-8, where the time variation is a step function. Determine the response $y(x, t)$ in terms of the normal modes of the beam. Indicate what modes are absent and write down the first two existing modes.

9-9 A slender rod of length l, free at $x = 0$ and fixed at $x = l$, is struck longitudinally by a time-varying force concentrated at the end $x = 0$. Show that all modes are equally excited (i.e., that the mode-participation factor is independent of the mode number), the complete solution being

$$u(x, t) = \frac{2F_0 l}{AE}\left\{ \frac{\cos \frac{\pi}{2}\frac{x}{l}}{\left(\frac{\pi}{2}\right)^2} D_1(t) + \frac{\cos \frac{3\pi}{2}\frac{x}{l}}{\left(\frac{3\pi}{2}\right)^2} D_3(t) + \cdots \right\}$$

9-10 If the force of Prob. 9-9 is concentrated at $x = l/3$, determine which modes will be absent in the solution.

9-11 In Prob. 9-10, determine the participation factor of the modes present and obtain a complete solution for an arbitrary time variation of the applied force.

9-12 Consider a uniform beam of mass M and length l supported on equal springs of total stiffness k, as shown in Fig. P9-12(a). Assume the deflection to be

$$y(x, t) = \varphi_1(x)q_1(t) + \varphi_2(x)\varphi_2(t)$$

and choose $\varphi_1 = \sin \frac{\pi x}{l}$ and $\varphi_2 = 1.0$.

Using Lagrange's equation, show that

$$\ddot{q}_1 + \frac{4}{\pi}\ddot{q}_2 + \omega_{11}^2 q_1 = 0$$

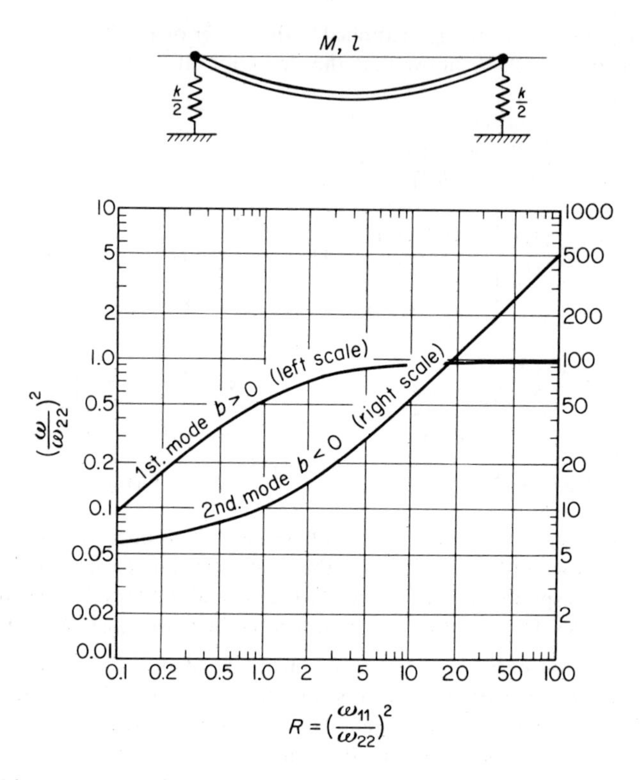

Figure P. 9-12. First two natural frequencies of the system of Fig. P. 9-12.

$$\frac{2}{\pi}\ddot{q}_1 + \ddot{q}_2 + \omega_{22}^2 q_2 = 0$$

where $\omega_{11}^2 = \pi^4(EI/Ml^3)$ = natural frequency of beam on rigid supports
$\omega_{22}^2 = k/M$ = natural frequency of rigid beam on springs

Solve these equations and show that

$$\omega^2 = \omega_{22}^2\frac{\pi^2}{2}\left\{\frac{(R+1) \pm \sqrt{(R-1)^2 + \frac{32}{\pi^2}R}}{\pi^2 - 8}\right\}$$

Let $y(x, t) = \left(b + \sin\frac{\pi x}{l}\right)q$ and use Rayleigh's method to obtain

$$\frac{q_2}{q_1} = b = \frac{\pi}{8}\left\{(R-1) \mp \sqrt{(R-1)^2 + \frac{32}{\pi^2}R}\right\}$$

$$R = \left(\frac{\omega_{11}}{\omega_{22}}\right)^2$$

A plot of the natural frequencies of the system is shown in Fig. P9-12(b).

9-13 A uniform beam, clamped at both ends, is excited by a concentrated force $P_0 f(t)$ at midspan, as shown in Fig. P9-13. Determine the deflection under the load and the resulting bending moment at the clamped ends.

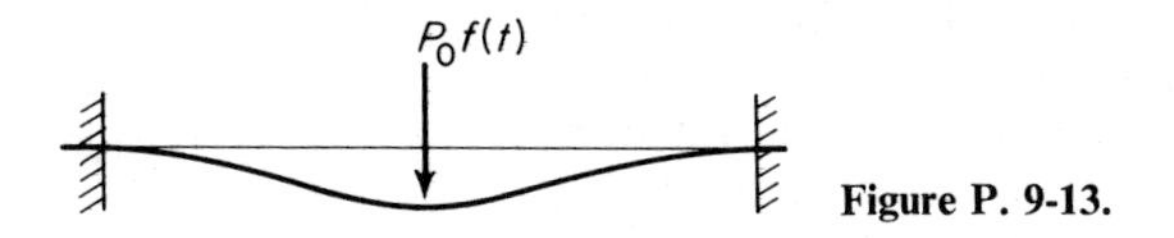

Figure P. 9-13.

9-14 If a uniformly distributed load of arbitrary time variation is applied to a uniform cantilever beam, determine the participation factor for the first three modes.

9-15 A spring of stiffness k is attached to a uniform beam, as shown in Fig. P9-15. Show that the one-mode approximation results in the frequency equation

$$\left(\frac{\omega}{\omega_1}\right)^2 = 1 + 1.5\left(\frac{k}{M}\right)\left(\frac{Ml^3}{\pi^4 EI}\right)$$

where

$$\omega_1^2 = \frac{\pi^4 EI}{Ml^3}$$

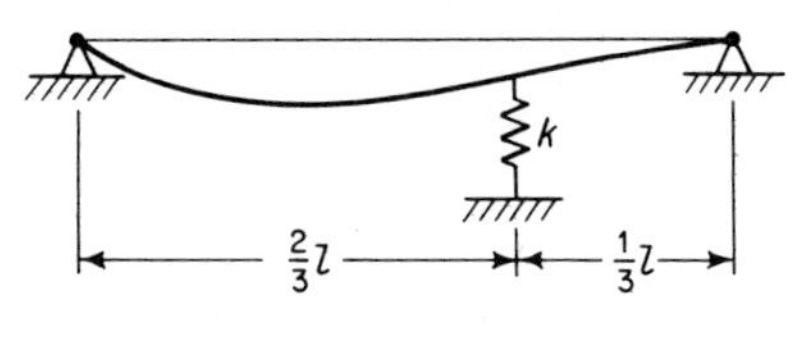

Figure P. 9-15.

9-16 Write the equations for the two-mode approximation of Prob. 9-15.

9-17 Repeat Prob. 9-16, using the mode-acceleration method.

9-18 Show that for the problem of a spring attached to any point $x = a$ of a beam, both the constrained-mode and the mode-acceleration methods result in the same equation when only one mode is used, this equation being

$$\left(\frac{\omega}{\omega_1}\right)^2 = 1 + \frac{k}{M\omega_1^2}\varphi_1^2(a)$$

9-19 The beam shown in Fig. P9-19 has a spring of rotational stiffness K lb in./rad at the left end. Using two modes in Eq. (9.8-8), determine the fundamental frequency of the system as a function of $K/M\omega_1^2$ where ω_1 is the fundamental frequency of the simply supported beam.

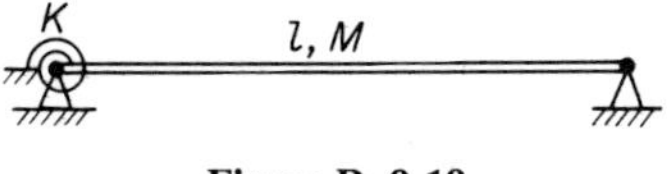

Figure P. 9-19.

9-20 If both ends of the beam of Fig. P9-19 are restrained by springs of stiffness K, determine the fundamental frequency. As K approaches infinity, the result should approach that of the clamped ended beam.

9-21 An airplane is idealized to a simplified model of a uniform beam of length l and mass per unit length m with a lumped mass M_0 at its center, as shown in Fig. P9-21. Using the translation of M_0 as one of the generalized coordinates, write the equations of motion and establish the natural frequency of the symmetric mode. Use first cantilever mode for the wing.

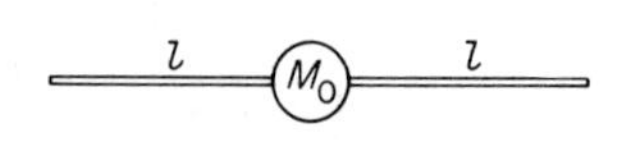

Figure P. 9-21.

9-22 For the system of Prob. 9-21, determine the antisymmetric mode by using the rotation of the fuselage as one the generalized coordinates.

9-23 If wing tip tanks of mass M_1 are added to the system of Prob. 9-21, determine the new frequency.

9-24 Using the method of constrained modes, show that the effect of adding a mass m_1 with moment of inertia J_1 to a point x_1 on the structure changes the first natural frequency ω_1 to

$$\omega_1' = \frac{\omega_1}{\sqrt{1 + \dfrac{m_1}{M_1}\varphi_1^2(x_1) + \dfrac{J_1}{M_1}\varphi_1'^2(x_1)}}$$

and the generalized mass and damping to

$$M_1' = M_1\left\{1 + \frac{m_1\varphi_1^2(x_1)}{M_1} + \frac{J_1}{M_1}\varphi_1'^2(x)\right\}$$

$$\zeta_1' = \frac{\zeta_1}{\sqrt{1 + \dfrac{m_1}{M_1}\varphi_1^2(x_1) + \dfrac{J_1}{M_1}\varphi_1'^2(x_1)}}$$

where a one-mode approximation is used for the inertia forces.

9-25 Formulate the vibration problem of the bent shown in Fig. P9-25 by the component mode synthesis. Assume the corners to remain at 90°.

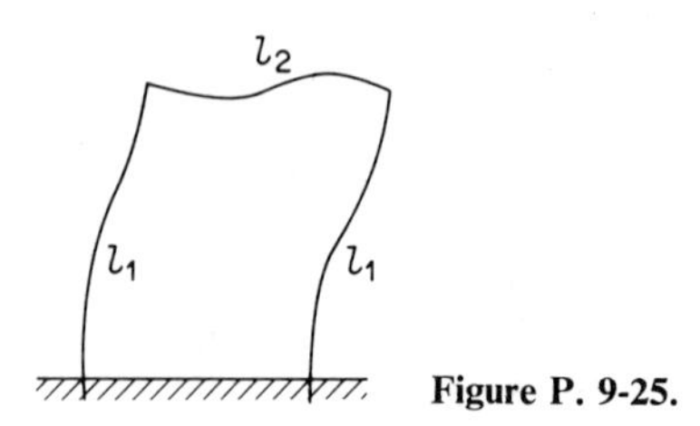

Figure P. 9-25.

9-26 A rod of circular cross-section is bent at right angles in a horizontal plane as shown in Fig. P9-26. Using component mode synthesis, set up the equations for the vibration perpendicular to the plane of the rod. Note that member 1 is in flexure and torsion. Assume its bending only in the vertical plane.

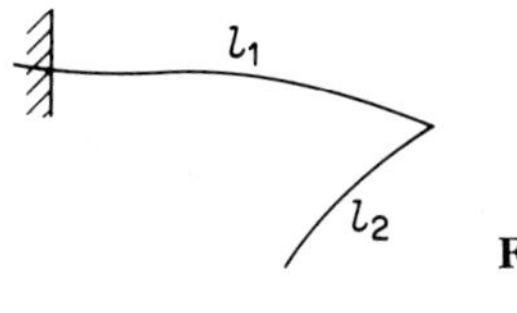

Figure P. 9-26.

10

Random Vibration

10.1 INTRODUCTION

In previous Chapters, we have dealt with the response of dynamical systems to deterministic excitations, describable by a mathematical function of time. The response to such excitation is also deterministic.

With the development of jet engines and high speed aircraft, a new aspect of vibration is encountered—that of excitation fluctuating in a random manner, as shown in Fig. 10.1-1. The characteristic of such random function is that its instantaneous value cannot be predicted in a deterministic sense.

In spite of their unpredictable fluctuations, many random phenomena exhibit some degree of statistical regularity which makes possible a statistical approach to the problem. For example, it is possible to predict the probability of finding the instantaneous value of the response within a specified range of values x to $x + \Delta x$. Other quantities, such as the mean or the mean square values, can be established by averaging, and the frequency content of the variable in question can be determined by various methods based on Fourier analysis.

Figure 10.1-1. A record of random time function.

In any statistical method, a large amount of data is necessary to establish reliability. For example, to establish the statistics of the pressure fluctuation due to air turbulance over a certain air route, an airplane may collect hundreds of records of the type shown in Fig. 10.1-2. Each record is called a *sample*, and the total collection of samples the *ensemble*. We can compute the average of the instantaneous pressures at time t_1. We can also multiply the instantaneous pressures in each sample at times t_1 and $t_1 + \tau$, and average these results for the ensemble. If such averages do not differ as we choose different values of t_1, then the random process described by the above ensemble is said to be *stationary*.

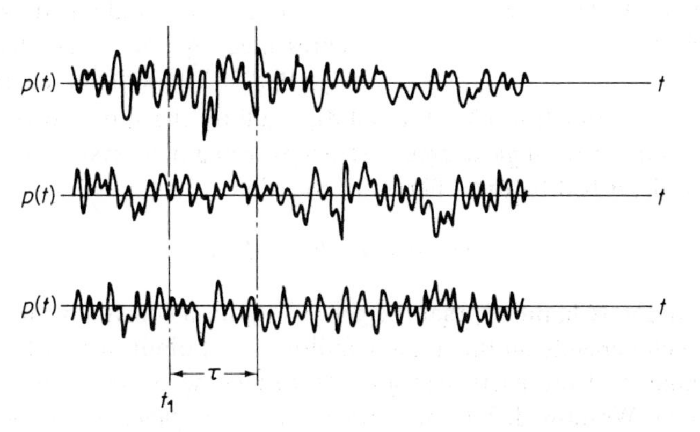

Figure 10.1-2. An ensemble of random time functions.

If the ensemble averages are replaced next by time averages, and if the results computed from each sample are the same as those of any other sample and equal to the ensemble average, then the random process is said to be *ergodic*. This chapter will treat only this class of random functions for which time averaging can be adopted with assured stationarity.

Throughout the chapter, we will encounter the concept of time averaging over a long time interval. The most common notations for this operation are defined by the following equation in which $x(t)$ is the variable.

$$\overline{x(t)} = \langle x(t) \rangle = \lim_{T\to\infty} \frac{1}{T} \int_0^T x(t)\, dt \tag{10.1-1}$$

The above number is also equal to the expected value of $x(t)$, or $E[x(t)]$, which is defined as the average or mean value of a quantity sampled over a large number, or a long time. In the case of discrete variables x_i, the expected value is given by the equation

$$E[x] = \lim_{n\to\infty} \frac{1}{n} \sum_{i=1}^{n} x_i \tag{10.1-2}$$

These averaging operations can be applied to any variable such as $x^2(t)$ or $x(t) \cdot y(t)$, and expected value (or expectation) is associated with the probability distribution of the variable.

10.2 THE FREQUENCY RESPONSE FUNCTION

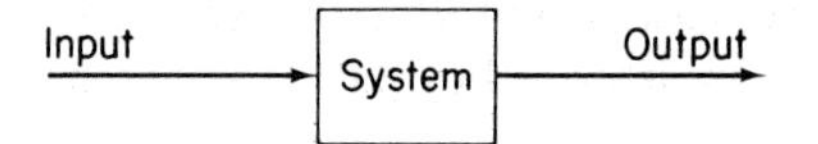

Figure 10.2-1. Block diagram of a linear system.

In any linear system, there is a direct linear relationship between the input and the output. This relationship, which also holds for random functions, is represented by the block diagram of Fig. 10.2-1. The system, characterized by its transfer function (See Eq. 4.4-4), modifies the input to the output.

Considering a single degree of freedom spring-mass system with viscous damping, described by the differential equation,

$$m\ddot{x} + c\dot{x} + kx = F(t) \tag{10.2-1}$$

we have found (Chapter 3) that the general solution consists of the transient term, which depends on the initial conditions and diminishes with time due to damping, and the particular solution depending on the excitation. (See Eq. 3.2-11). We now define the *frequency response function* as: *the ratio of the output to the input under steady state conditions, with the input equal to a harmonic time function of unit amplitude.* The transient solution is thus excluded in this consideration.

Letting the input be

$$F(t) = e^{i\omega t} \tag{10.2-2}$$

and the steady state output be

$$x = H(\omega) e^{i\omega t} \tag{10.2-3}$$

their substitution into Eq. (10.2-1) results in the frequency response function $H(\omega)$ which is

$$H(\omega) = \frac{1}{k - m\omega^2 + i\omega c} = \frac{1}{k}\frac{1}{1 - \left(\frac{\omega}{\omega_n}\right)^2 + i2\zeta\left(\frac{\omega}{\omega_n}\right)} \tag{10.2-4}^*$$

Note that $H(\omega)$ is a complex function of ω/ω_n and the damping factor ζ, and has the dimensions of displacement over force.† The absolute value of this quantity is given by Eq. (3.2-7) and its variation with frequency and damping is plotted in Fig. 3.2-3. For small damping, its peak occurs at $\omega/\omega_n \cong 1.0$ and the sharpness of the resonance curve is defined by $Q = \frac{1}{2}\zeta$.

Mean Square Value. In random vibrations, the initial conditions and the phase ϕ have little meaning, and are therefore ignored. We are mainly concerned with the average energy, which we can associate with the mean square value of x. The *mean square value*, designated by the notation $\overline{x^2}$, is found by integrating x^2 over a time interval T and taking its average value according to the equation

$$\overline{x^2} = \lim_{T\to\infty} \frac{1}{T}\int_0^T x^2\,dt \tag{10.2-5}$$

This equation can, of course, be applied to the exciting force or the response. For example, if we have a harmonic force $F = F_0 \sin \omega t$, its mean square value is

$$\overline{F^2} = \lim_{T\to\infty} \frac{1}{T}\int_0^T \frac{F_0^2}{2}(1 - \cos 2\omega t)\,dt = \frac{F_0^2}{2} \tag{10.2-6}$$

To determine the relationship between the mean square response and the mean square excitation, we start with the response equation

$$x = Re\{F_0 H(\omega)e^{i\omega t}\} \tag{10.2-7}$$

*Often the dimensional factor ($1/k$ in Eq. 10.2-4) is considered together with the force, leaving the frequency response function a nondimensional quantity

$$H(\omega) = \frac{1}{1 - \left(\frac{\omega}{\omega_n}\right)^2 + i2\zeta\left(\frac{\omega}{\omega_n}\right)}$$

†In Example 10.6-3, the frequency response function is also shown to be the Fourier transform of the impulse response function.

recognizing that we are interested in the real part of the above expression. Since for any complex number the real part is equal to one half the sum of the number and its complex conjugate, denoted by *, we can rewrite Eq. (10.2-7) as

$$x = \tfrac{1}{2}F_0(He^{i\omega t} + H^*e^{-i\omega t}) \tag{10.2-8}$$

Thus by squaring and substituting into Eq. (10.2-5), the mean square value of x is

$$\begin{aligned}\overline{x^2} &= \frac{F_0^2}{4}\lim_{T\to\infty}\frac{1}{T}\int_0^T (H^2e^{i2\omega t} + 2HH^* + H^*e^{-i2\omega t})\,dt \\ &= \frac{F_0^2}{2}H(\omega)H^*(\omega) = \overline{F^2}\,|H(\omega)|^2\end{aligned} \tag{10.2-9}$$

In the above evaluation, the first and last terms become zero because of $T \longrightarrow \infty$ in the denominator, whereas the middle term is independent of T. Equation (10.2-9) indicates that the mean square value of the response is equal to the mean square excitation multiplied by the square of the absolute value of the system response function.

10.3 SPECTRAL DENSITY

Random vibrations contain frequencies in a continuous distribution over a wide range. The energy represented at the various frequencies is a quantity that is of interest in random vibration. We approach this problem by first considering a periodic function $F(t)$ which contains many discrete frequencies. It can be represented by the real part of the series

$$F(t) = Re \sum_n F_n e^{in\omega_0 t} \tag{10.3-1}$$

where F_n is a complex number, and Re stands for the real part of the series. We write this equation in terms of its complex conjugate as

$$F(t) = \frac{1}{2}\left\{\sum_n F_n e^{in\omega_0 t} + \sum_n F_n^* e^{-in\omega_0 t}\right\} \tag{10.3-2}$$

and determine its mean square value as

$$\begin{aligned}\bar{F}^2 &= \lim_{T\to\infty}\frac{1}{T}\int_0^T \frac{1}{4}\left\{\sum_n F_n e^{in\omega_0 t} + \sum_n F_m^* e^{-in\omega_0 t}\right\}^2 dt \\ &= \sum_n \frac{F_n F_n^*}{2} = \sum \frac{1}{2}|F_n|^2 = \sum_n \overline{F_m^2}\end{aligned} \tag{10.3-3}$$

Thus the mean square value of the multifrequency wave is simply the sum of the mean square values of each harmonic component present, the result being a discrete frequency spectrum shown in Fig. 10.3-1.†

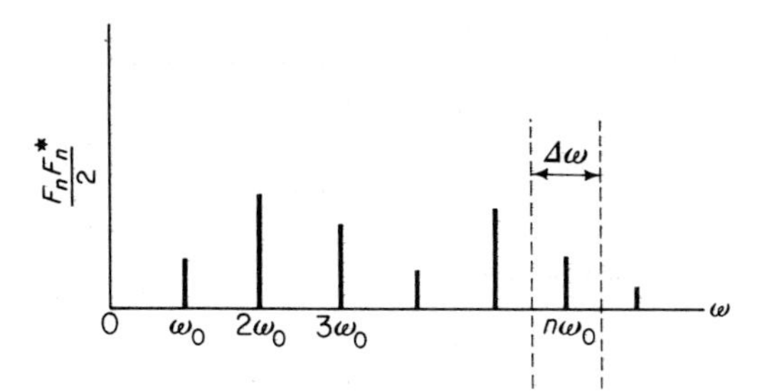

Figure 10.3-1. Discrete frequency spectrum of a periodic function.

We next examine the mean square contribution in the frequency interval $\Delta\omega$. By letting $S(n\omega_0)$ be the *density* of the mean square value in the interval $\Delta\omega$ at the frequency $n\omega_0$, we obtain

$$S(n\omega_0)\Delta\omega = \frac{F_n F_n^*}{2} \tag{10.3-4}$$

and the discrete spectral density becomes

$$S(n\omega_0) = \frac{F_n F_n^*}{2\Delta\omega} \tag{10.3-5}$$

It is evident that when $F(t)$ contains a very large number of frequency components, the discrete density function $S(n\omega_0)$ becomes more nearly a continuous spectral density function $S(\omega)$, such as the one shown in Fig. 10.3-2. The mean square value of F is then

$$\overline{F^2} = \int_0^\infty S(\omega)\,d\omega \tag{10.3-6}$$

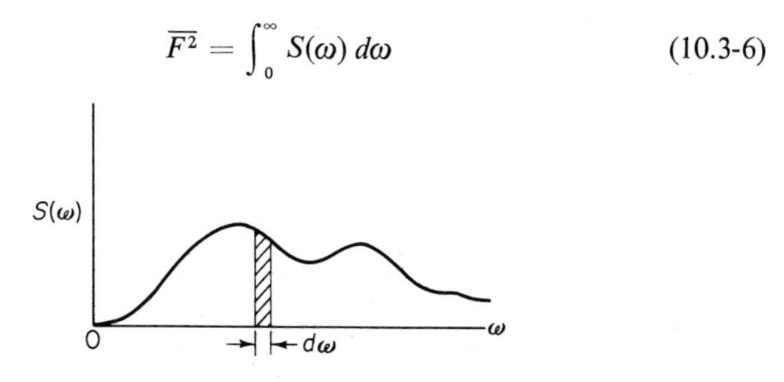

Figure 10.3-2. Continuous spectral density curve.

†In Eq. (10.3-1) we have specified the real part of the series which is expressed in Eq. (10.3-2). Thus n is a positive integer and the spectrum of Fig. 10.3-1 is defined over real frequencies, rather than over both the positive and negative frequencies.

or

$$\overline{F^2} = \int_0^\infty S(f)\, df \tag{10.3-6'}$$

If an exciting force $F_n e^{in\omega_0 t}$ acts on a system with frequency response function $H(\omega)$, its response, from Eq. (10.2-7), is

$$x = Re\{F_n H(n\omega_0) e^{in\omega_0 t}\} \tag{10.3-7}$$

and its mean square response is

$$\overline{x^2} = \frac{F_n F_n^*}{2} H(n\omega_0) H^*(n\omega_0) \tag{10.3-8}$$

Thus for a multifrequency input, the mean square response is the superposition of all such values or

$$\begin{aligned} \overline{x^2} &= \sum_n \frac{F_n F_n^*}{2} H(n\omega_0) H^*(n\omega_0) \\ &= \sum S_F(n\omega_0) H(n\omega_0) H^*(n\omega_0) \Delta\omega \end{aligned} \tag{10.3-9}$$

By defining the spectral density of the response as

$$S_x(n\omega_0) = S_F(n\omega_0) H(n\omega_0) H^*(n\omega_0) \tag{10.3-10}$$

we find that $S_x(n\omega_0)$ is also a discrete spectrum, equal to the spectral density of the excitation modified by the frequency response function. Thus $S_F(n\omega_0)$ and $S_x(n\omega_0)$ may appear as in Fig. 10.3-3. For a continuous specturm, the summation of Eq. (10.3-9) is replaced by an integral, and the mean square response is given by the equation

$$\overline{x^2} = \int_0^\infty S(\omega) H(\omega) H^*(\omega)\, d\omega \tag{10.3-11}$$

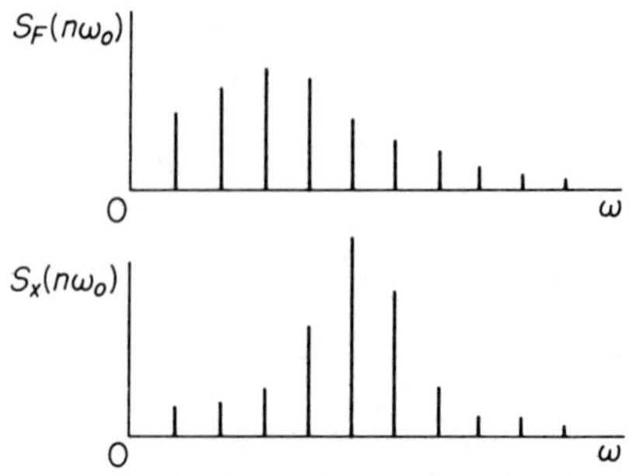

Figure 10.3-3. Spectral density of excitation and response.

In practice, the spectral density function is generally given in terms of the frequency $f = \omega/2\pi$ cps and hence the equation becomes

$$\overline{x^2} = \int_0^\infty S(f)H(f)H^*(f)\,df \tag{10.3-12}$$

where

$$S(f) = 2\pi S(\omega)$$

$$H(f) = \frac{\frac{1}{k}}{[1 - (f/f_n)^2] + i[2\zeta(f/f_n)]}$$

For a lightly damped system, the response function $H(f)$ is peaked steeply at resonance, and if the spectral density of the excitation is broad, as in Fig. 10.3-4, the mean square response can be approximated by the equation

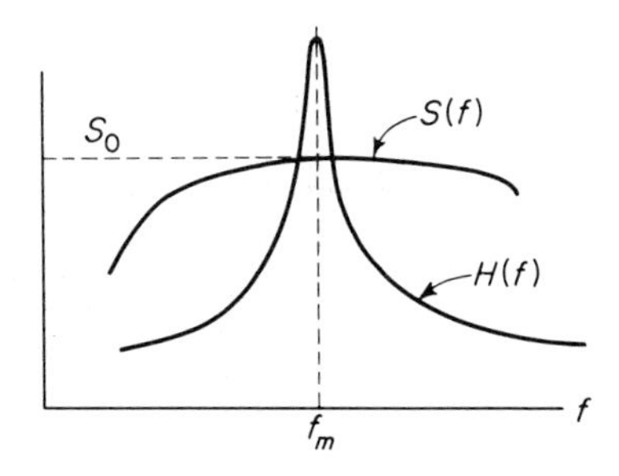

Figure 10.3-4. $S(f)$ and $H(f)$ leading to $\bar{x}^2$ of Equation 10.3-12.

$$\overline{x^2} \cong f_n S(f_n) \frac{\pi}{4\zeta} \tag{10.3-13}$$

Typical spectral density functions for two common types of random records are shown in Figs. 10.3-5 and 10.3-6. The first is a wide-band noise-

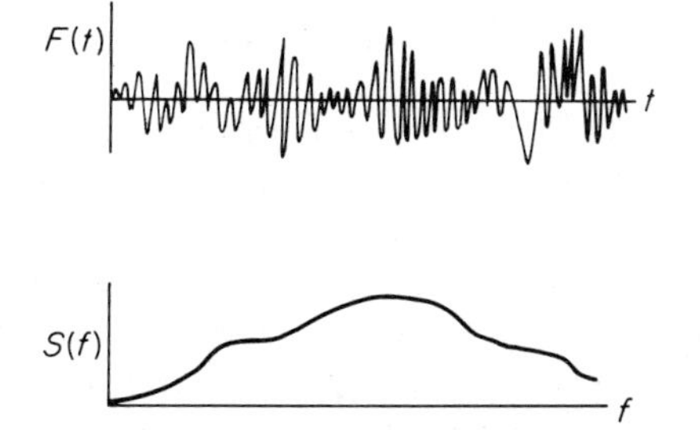

Figure 10.3-5. Wide band record and its spectral density.

type of record which has a broad spectral density function. The second is a narrow-band random record which is typical of a response of a sharply resonant system to a wide-band input. Its spectral density function is concentrated around the frequency of the instantaneous variation within the envelope.

The spectral density of a given record can be measured electronically by the circuit of Fig. 10.3-7. Here the spectral density is noted as the con-

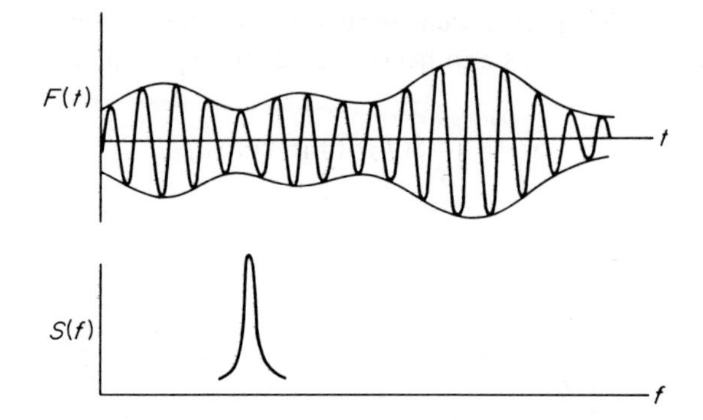

Figure 10.3-6. Narrow band record and its spectral density.

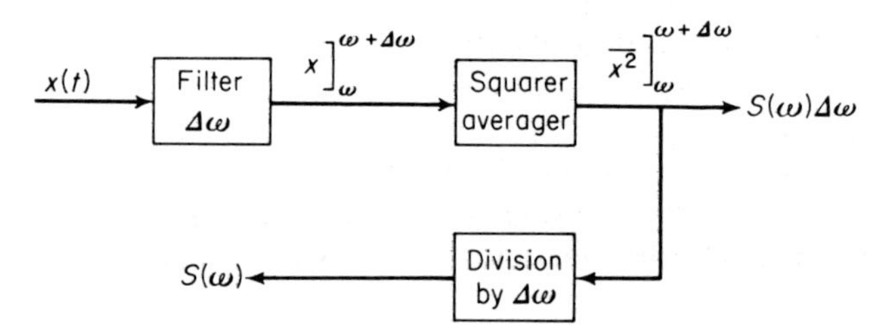

Figure 10.3-7. Power spectral density analyzer.

tribution of the mean square value in the frequency interval $\Delta\omega$ divided by $\Delta\omega$.

$$S(\omega) = \lim_{\Delta\omega \to 0} \frac{\Delta(\bar{x}^2)}{\Delta\omega} \tag{10.3-14}$$

The band-pass filter of pass band $B = \Delta\omega$ passes $x(t)$ in the frequency interval ω to $\omega + \Delta\omega$, and the output is squared, averaged, and divided by $\Delta\omega$.

For high resolution, $\Delta\omega$ should be made as narrow as possible; however, the pass band of the filter cannot be reduced indefinitely without losing the reliability of the measurement. Also, a long record is required for the true estimate of the mean square value, but actual records are always of finite length. It is evident now that a parameter of importance is the product of the record length and the band width, $2BT$, which must be sufficiently large.*

Example 10.3-1

A single degree of freedom system with natural frequency $\omega_n = \sqrt{k/m}$ and damping $\zeta = 0.20$ is excited by the force

*See Bendat, J. S, and A. G Piersol "Random Data" Wiley Interscience, N. Y. (1971) p. 96.

$$F(t) = F\cos\tfrac{1}{2}\omega_n t + F\cos\omega_n t + F\cos\tfrac{3}{2}\omega_n t$$
$$= \sum_{m=1/2,\,1,\,3/2} F\cos m\omega_n t$$

Determine the mean square response and compare the output spectrum with that of the input.

Solution: The response of the system is simply the sum of the response of the single degree of freedom system to each of the harmonic components of the exciting force.

$$x(t) = \sum_{m=1/2,\,1,\,3/2} |H(n\omega)|\,F\cos(m\omega_n t - \phi_m)$$

where

$$|H(\tfrac{1}{2}\omega_n)| = \frac{\frac{1}{k}}{\sqrt{\frac{9}{16} + (0.20)^2}} = \frac{1.29}{k}$$

$$|H(\omega_n)| = \frac{\frac{1}{k}}{\sqrt{4(0.20)^2}} = \frac{2.50}{k}$$

$$|H(\tfrac{3}{2}\omega_n)| = \frac{\frac{1}{k}}{\sqrt{\frac{25}{16} + 9(0.02)^2}} = \frac{0.72}{k}$$

$$\phi_{1/2} = \tan^{-1}\frac{4\zeta}{3} = 0.083\pi$$

$$\phi_1 = \tan^{-1}\infty = 0.50\pi$$

$$\phi_{3/2} = \tan^{-1}\frac{-12\zeta}{5} = -0.142\pi$$

Substituting these values into $x(t)$, we obtain the equation

$$x(t) = \frac{F}{k}\,[1.29\cos(0.5\omega_n - 0.083\pi)$$
$$+ 2.50\cos(\omega_n t - 0.50\pi)$$
$$+ 0.72\cos(1.5\omega_n t + 0.142\pi)]$$

The mean square response is then

$$\overline{x^2} = \frac{F^2}{2k^2}[(1.29)^2 + (2.50)^2 + (0.72)^2]$$

Figure 10.3-8 shows the input and output spectra for the problem. The components of the mean square input are the same for each frequency and equal to $F^2/2$. The output spectra is modified by the system frequency response function.

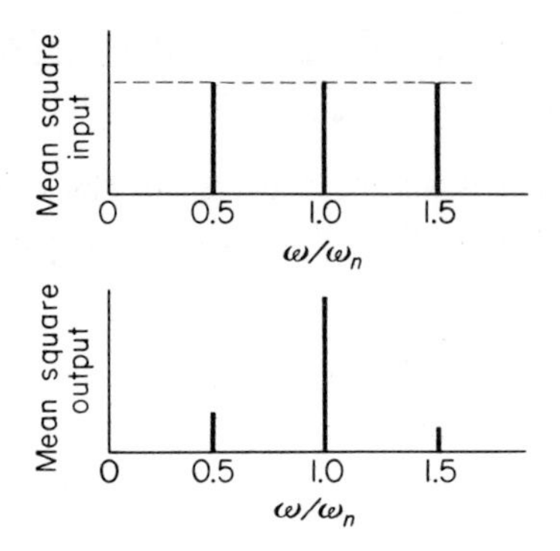

Figure 10.3-8. Input and output spectra with discrete frequencies.

Example 10.3-2

Determine the Fourier coefficients C_n and the power spectral density of the periodic function shown in Fig. 10.3-9.

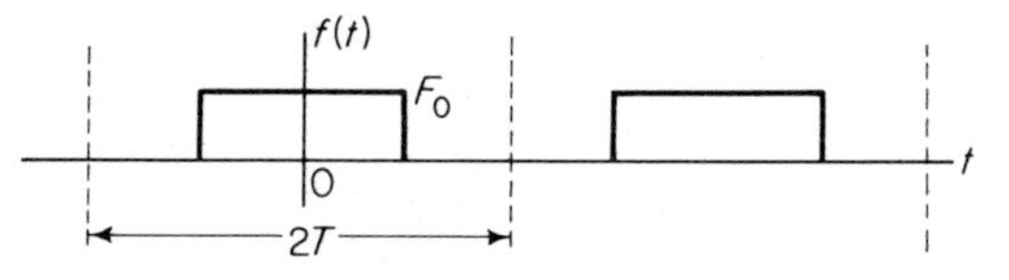

Figure 10.3-9.

Solution: The period is $2T$ and C_n are

$$C_0 = \frac{1}{2T}\int_{-T/2}^{T/2} F_0\, d\xi = \frac{F_0}{2}$$

$$C_n = \frac{1}{2T}\int_{-T/2}^{T/2} F_0 e^{-in\omega_0\xi}\, d\xi = \frac{F_0}{2}\left(\frac{\sin(n\pi/2)}{n\pi/2}\right)$$

Numerical values of C_n are computed as follows and plotted in Fig. 10.3-10.

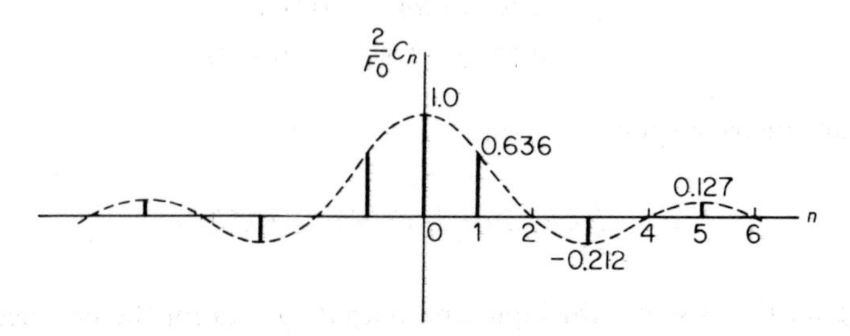

Figure 10.3-10. Fourier coefficients versus n.

n	$\frac{n\pi}{2}$	$\sin\frac{n\pi}{2}$	C_n
0	0	0	$\frac{F_0}{2} = 1.0\frac{F_0}{2}$
1	$\frac{\pi}{2}$	1	$\left(\frac{2}{\pi}\right)\frac{F_0}{2} = 0.636\frac{F_0}{2}$
2	π	0	0
3	$3\frac{\pi}{2}$	-1	$\left(-\frac{2}{3\pi}\right)\frac{F_0}{2} = -0.212\frac{F_0}{2}$
4	2π	0	0
5	$5\frac{\pi}{2}$	1	$\left(\frac{2}{5\pi}\right)\frac{F_0}{2} = 0.127\frac{F_0}{2}$

The mean square value is determined from the equation

$$\overline{f^2} = \lim_{T\to\infty}\frac{1}{2T}\int_{-T}^{T} f^2(t)\,dt = \lim_{T\to\infty}\frac{1}{2T}\int_{-T}^{T}\frac{1}{4}\left\{\sum_n (C_n e^{in\omega_0 t} + C_n^* e^{-in\omega_0 t})\right\}^2 dt$$

$$= \sum_{n=1}^{\infty}\frac{C_n C_n^*}{2}$$

and since $\bar{f}_2 = \int_0^\infty S_f(\omega)\,d\omega$, the spectral density function can be represented by a series of delta functions as

$$S_f(\omega) = \sum_{n=1}^{\infty}\frac{C_n C_n^*}{2}\delta(\omega - n\omega_0)$$

10.4 PROBABILITY DISTRIBUTION

Referring to the random time function of Fig. 10.4-1, what is the probability of its instantaneous value being less than (more negative than) some specified value x_1? To answer this question, we draw a horizontal line at the specified value x_1 and sum the time intervals Δt_i during which $x(t)$ is less than x_1. This sum divided by the total time then represents the fraction of the total time that $x(t)$ is less than x_1, which is the probability that $x(t)$ will be found less than x_1.

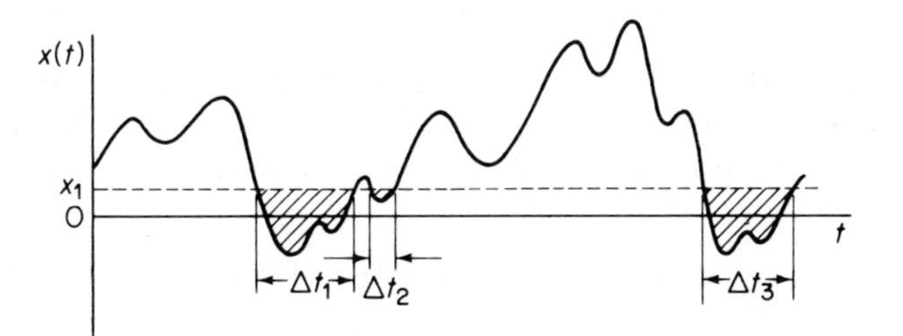

Figure 10.4-1. Calculation of cumulative probability.

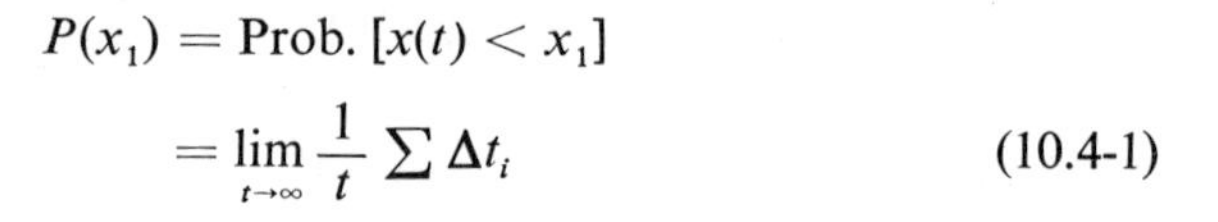

$$P(x_1) = \text{Prob. } [x(t) < x_1]$$
$$= \lim_{t \to \infty} \frac{1}{t} \sum \Delta t_i \qquad (10.4\text{-}1)$$

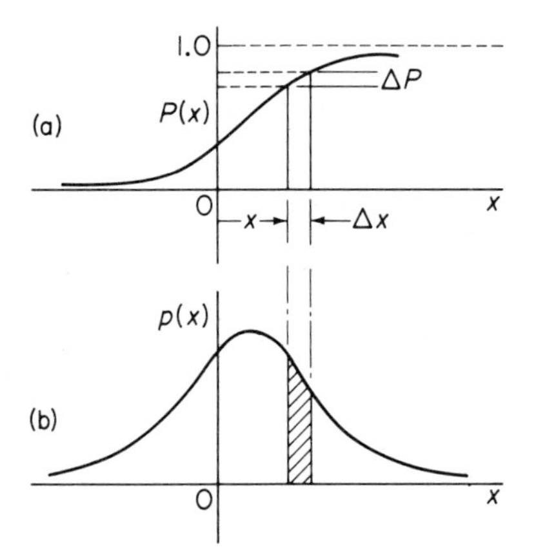

Figure 10.4-2. (a)-Cumulative probability (b)-Probability density.

If a large negative number is chosen for x_1, none of the curve will extend negatively beyond x_1, and hence $P(x_1 \to -\infty) = 0$. As the horizontal line corresponding to x_1 is moved up, more of $x(t)$ will extend negatively beyond x_1, and the fraction of the total time in which $x(t)$ extends below x_1 must increase as shown in Fig. 10.4-2(a). As $x \to \infty$, all $x(t)$ will lie in the region less than $x = \infty$, and hence the probability of $x(t)$ being less than $x = \infty$ is certain, or $P(x = \infty) = 1.0$. Thus the curve of Fig. 10.4-2(a) which is cumulative towards positive x must increase monotonically from zero at $x = -\infty$ to 1.0 at $x = +\infty$. The curve is called the cumulative probability distribution function $P(x)$.

If next we wish to determine the probability of $x(t)$ lying between the values x_1 and $x_1 + \Delta x$, all we need to do is subtract $P(x_1)$ from $P(x_1 + \Delta x)$, which is also proportional to the time occupied by $x(t)$ in the zone x_1 to $x_1 + \Delta x$.

We now define the *probability density function* $p(x)$ as

$$p(x) = \lim_{\Delta x \to 0} \frac{P(x + \Delta x) - P(x)}{\Delta x} = \frac{dP(x)}{dx} \qquad (10.4\text{-}2)$$

and it is evident from Fig. 10.4-2(b) that $p(x)$ is the slope of the cumulative probability distribution $P(x)$. From the above equation we can also write

$$P(x_1) = \int_{-\infty}^{x_1} p(x)\, dx \qquad (10.4\text{-}3)$$

The area under the probability density curve of Fig. 10.4-2(b) between two values of x represents the probability of the variable being in this interval. Since the probability of $x(t)$ being between $x = \pm\infty$ is certain

$$P(\infty) = \int_{-\infty}^{+\infty} p(x)\, dx = 1.0 \qquad (10.4\text{-}4)$$

and the total area under the $p(x)$ curve must be unity.

Figure 10.4-3 shows a block diagram of a circuit which will perform the calculation for the probability density function electronically. With $x(t)$ as the variable, the analyzer measures the cumulative time during which $x(t)$ remains within a set interval Δx. The probability density is found by dividing this quantity by Δx and t.

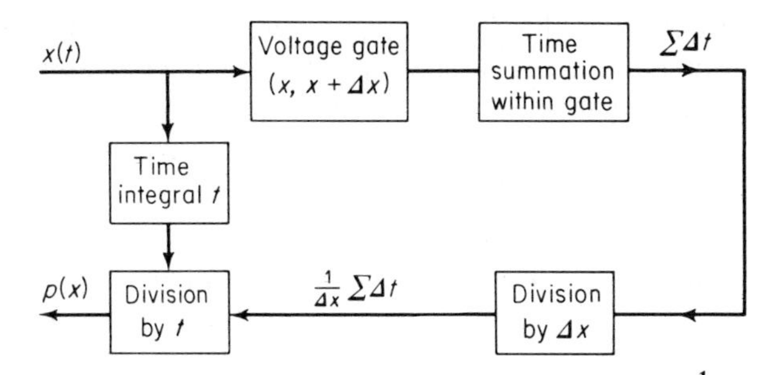

Figure 10.4-3. Probability density analyzer $p(x) = \lim\limits_{t\to\infty} \lim\limits_{\Delta x\to 0} \dfrac{1}{t\Delta x} \sum \Delta t$.

The mean and the mean square value, previously defined in terms of the time average, are related to the probability density function in the following manner. The mean value $\bar{x}$ coincides with the centroid of the area under the probability density curve $p(x)$, as shown in Fig. 10.4-4. It can therefore be determined by the first moment

$$\bar{x} = \int_{-\infty}^{\infty} xp(x)\,dx \quad (10.4\text{-}5)$$

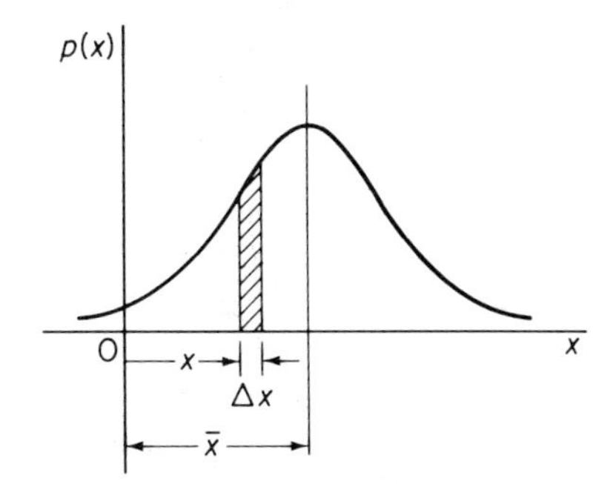

Figure 10.4-4. First and second moments of $p(x)$.

Likewise the mean square value is determined from the second moment

$$\overline{x^2} = \int_{-\infty}^{\infty} x^2 p(x)\,dx \qquad (10.4\text{-}6)$$

which is analogous to the moment of inertia of the area under the probability density curve about $x = 0$.

The *variance* σ^2 is defined as the mean square value about the mean, or

$$\begin{aligned}
\sigma^2 &= \int_{-\infty}^{\infty} (x - \bar{x})^2 p(x)\,dx \\
&= \int_{-\infty}^{\infty} x^2 p(x)\,dx - 2\bar{x}\int_{-\infty}^{\infty} xp(x)\,dx + (\bar{x})^2 \int_{-\infty}^{\infty} p(x)\,dx \\
&= \overline{x^2} \qquad - 2(\bar{x})^2 \qquad + (\bar{x})^2 \\
&= \overline{x^2} \qquad - (\bar{x})^2
\end{aligned} \qquad (10.4\text{-}7)$$

The *standard deviation* σ is the positive square root of the variance. When the mean value is zero, $\sigma = \sqrt{\overline{x^2}}$ and the standard deviation is equal to the root mean square (rms) value.

Gaussian and Rayleigh Distribution. Certain distributions which occur frequently in nature are the *Gaussian* (or normal) distribution and the *Rayleigh* distribution, both of which can be expressed mathematically. The Gaussian distribution is a bell shaped curve, symmetric about the mean value (which will be assumed to be zero) with the following equation.

$$p(x) = \frac{1}{\sigma\sqrt{2\pi}} e^{-x^2/2\sigma^2} \tag{10.4-8}$$

The standard deviation σ is a measure of the spread about the mean value; the smaller the value of σ, the narrower the $p(x)$ curve (remember that the total area = 1.0), as shown in Fig. 10.4-5(a).

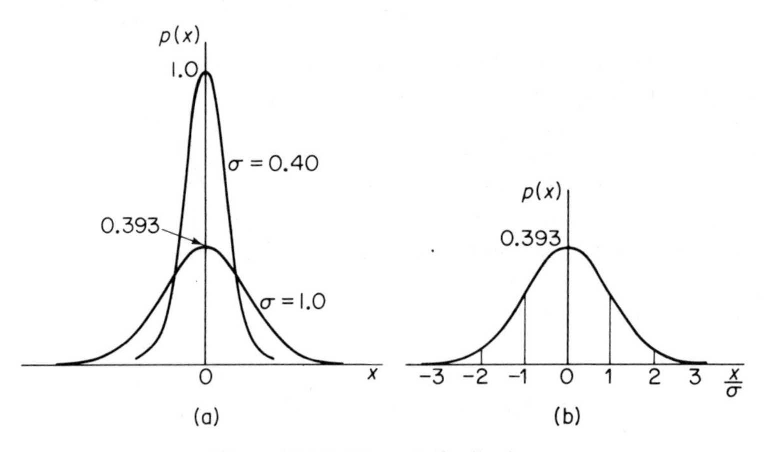

Figure 10.4-5. Normal distribution.

In Fig. 10.4-5(b) the Gaussian distribution is plotted nondimensionally in terms of x/σ. The probability of $x(t)$ being between $\pm\lambda\sigma$ where λ is any positive number is found from the equation

$$\text{Prob}\,[-\lambda\sigma \leq x(t) \leq \lambda\sigma] = \frac{1}{\sigma\sqrt{2\pi}} \int_{-\lambda\sigma}^{\lambda\sigma} e^{-x^2/2\sigma^2}\,dx \tag{10.4-9}$$

The following table presents numerical values associated with $\lambda = 1$, 2, and 3.

λ	Prob $[-\lambda\sigma \leq x(t) \leq \lambda\sigma]$	Prob $[\lvert x \rvert > \lambda\sigma]$
1	68.3%	31.7%
2	95.4%	4.6%
3	99.7%	0.3%

The probability of $x(t)$ lying outside $\pm\lambda\sigma$ is the probability of $|x|$ exceeding $\lambda\sigma$, which is 1.0 minus the above values, or the equation

$$\text{Prob}\,[|x| > \lambda\sigma] = \frac{2}{\sigma\sqrt{2\pi}}\int_{\lambda\sigma}^{\infty} e^{-x^2/2\sigma^2}\,dx = erfc\left(\frac{\lambda}{\sqrt{2}}\right) \qquad (10.4\text{-}10)$$

Random variables restricted to positive values, such as the absolute value A of the amplitude often tend to follow the Rayleigh distribution, which is defined by the equation

$$p(A) = \frac{A}{\sigma^2}e^{-A^2/2\sigma^2} \qquad A > 0 \qquad (10.4\text{-}11)$$

The probability density $p(A)$ is zero here for $A < 0$ and has the shape shown in Fig. 10.4-6.

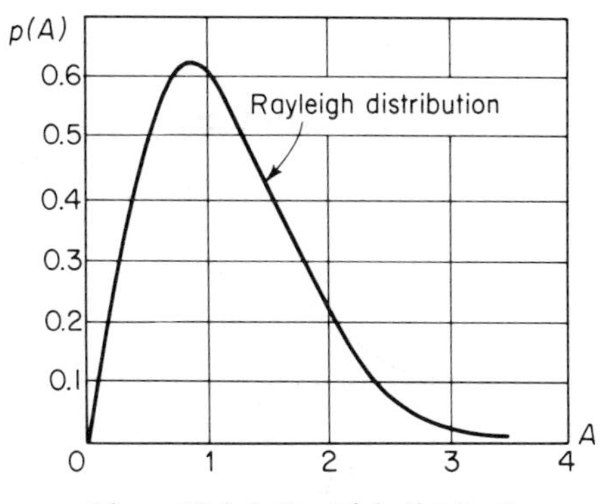

Figure 10.4-6. Rayleigh distribution.

The mean and mean square values for the Rayleigh distribution can be found from the first and second moments to be

$$\begin{aligned} \bar{A} &= \int_0^{\infty} Ap(A)\,dA = \int_0^{\infty}\frac{A^2}{\sigma^2}\,e^{-A^2/2\sigma^2}\,dA = \sqrt{\frac{\pi}{2}}\sigma \\ \overline{A^2} &= \int_0^{\infty} A^2p(A)\,dA = \int_0^{\infty}\frac{A^3}{\sigma^2}\,e^{-A^2/2\sigma^2}\,dA = 2\sigma^2 \end{aligned} \qquad (10.4\text{-}12)$$

The variance associated with the Rayleigh distribution is

$$\sigma_A^2 = \overline{A^2} - (\bar{A})^2 = \left(\frac{4-\pi}{2}\right)\sigma^2 \cong \frac{2}{3}\sigma \qquad (10.4\text{-}13)$$

Also, the probability of A exceeding a specified value $\lambda\sigma$ is

$$\text{Prob}\,[A > \lambda\sigma] = \int_{\lambda\sigma}^{\infty}\frac{A}{\sigma^2}e^{-A^2/2\sigma^2}\,dA \qquad (10.4\text{-}14)$$

which has the following numerical values

λ	$P[A > \lambda\sigma]$
0	100%
1	60.7%
2	13.5%
3	1.2%

Three important examples of time records frequently encountered in practice are shown in Fig. 10.4-7 where the mean value is arbitrarily chosen to be zero. The cumulative probability distribution for the sine wave is easily shown to be

$$P(x) = \tfrac{1}{2} + \frac{1}{\pi}\sin^{-1}\frac{x}{A}$$

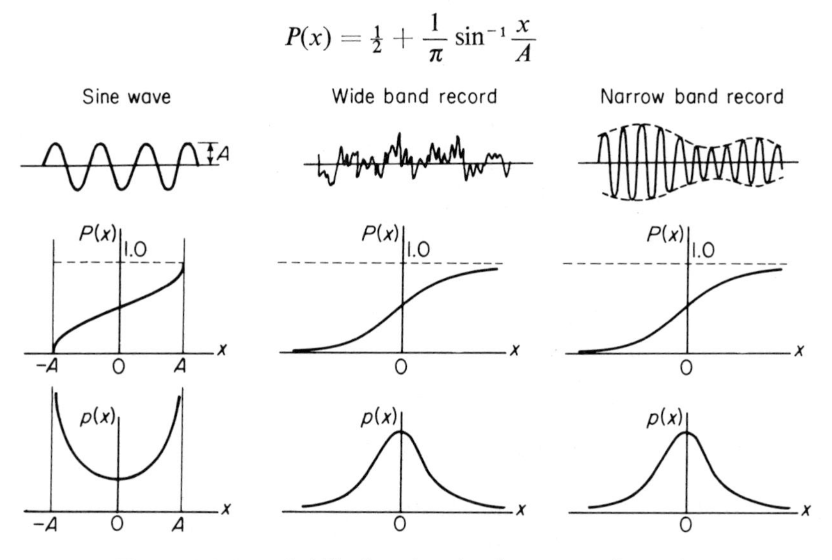

Figure 10.4-7. Probability functions for three types of records.

and its probability density, by differentiation, is

$$\begin{aligned} p(x) &= \frac{1}{\pi\sqrt{A^2 - x^2}} \qquad |x| < A \\ &= 0 \qquad\qquad\qquad |x| > A \end{aligned}$$

For the wide-band record, the amplitude, phase, and frequency all vary randomly and an analytical expression is not possible for its instantaneous value. Such functions are encountered in radio noise, jet engine pressure fluctuation, atmospheric turbulence, etc., and a most likely probability distribution for such records is the *Gaussian distribution.*

When a wide-band record is put through a narrow-band filter, or a resonance system where the filter bandwidth is small compared to its central frequency f_0, we obtain the third type of wave which is essentially a constant frequency oscillation with slowly varying amplitude and phase. The probability distribution for its instantaneous values is the same as that for the wide band random function. However, the absolute values of its peaks, corresponding to the envelope, will have a Rayleigh distribution.

Another quantity of great interest is the distribution of the peak values. Rice* shows that the distribution of the peak values depends on a quantity $N_0/2M$ where N_0 is the number of zero crossings and $2M$ is the number of positive and negative peaks. For a sine wave or a narrow band, N_0 is equal to $2M$ so that the ratio $N_0/2M = 1$. For a wide-band random record, the number of peaks will greatly exceed the number of zero crossings so that $N_0/2M$ tends to approach zero. When $N_0/2M = 0$, the probability density distribution of peak values turns out to be Gaussian, whereas when $N_0/2M = 1$, as in the narrow band case, the probability density distribution of the peak values tends to a *Rayleigh distribution.*

Example 10.4-1

A random vibration test specification calls for
Mean value of acceleration $= 0$
Acceleration density, 0.025 g^2/cps
Frequency range, 20 to 2000 cps
Determine the rms acceleration.

Solution: The rms acceleration is found by multiplying the acceleration density by the bandwidth and taking the square root of this number.

$$\text{rms accel} = \sqrt{0.025 \times (2000 - 20)} = \sqrt{49.5} = 7.03\, g$$

Example 10.4-2

A random signal has a spectral density which is constant

$$S(f) = 0.004 \text{ in.}^2/\text{cps}$$

between 20 to 1200 cps, and zero outside this frequency range. Its mean value is 2.0 in. Determine its standard deviation and its rms value.

Solution: If the mean value is not zero, we must use Eq. (10.4-7)

$$\overline{x^2} = (\bar{x})^2 + \sigma^2$$

*See Ref. 8.

The standard deviation is found from

$$\sigma^2 = \int_0^\infty S(f)\,df = \int_{20}^{1200} 0.004\,df = 4.72$$

and the mean-square value is

$$\overline{x^2} = 2^2 + 4.72 = 8.72$$

The rms value is then

$$\text{rms} = \sqrt{\overline{x^2}} = \sqrt{8.72} = 2.95 \text{ in.}$$

The problem is graphically displayed by Fig. 10.4-8.

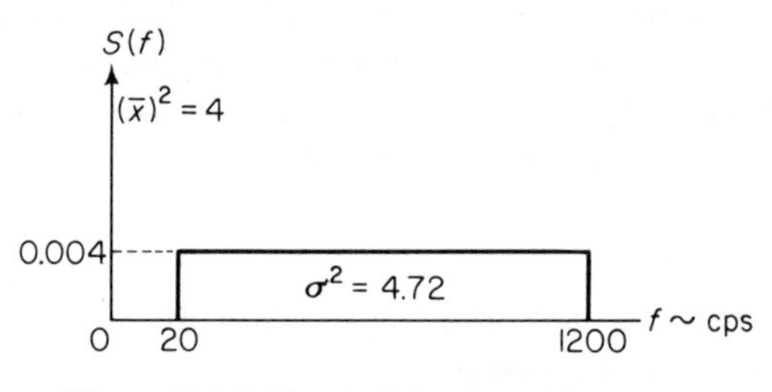

Figure 10.4-8. Spectral density vs. frequency.

Example 10.4-3

The response of any structure to a single point random excitation can be computed by a simple numerical procedure, provided the spectral density of the excitation and the frequency response curve of the structure are known. For example, consider the structure of Fig. 10.4-9(a) whose base is subjected to a random acceleration input with the power spectral density function shown in Fig. 10.4-9(b). It is desired to compute the response of the point p and establish the probability of exceeding any specified acceleration.

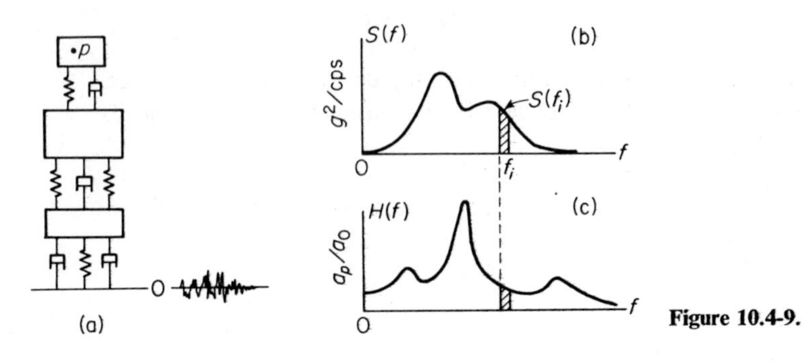

Figure 10.4-9.

The frequency response function $H(f)$ for the point p may be obtained experimentally by applying to the base a variable frequency sinusoidal shaker with a constant acceleration input a_0, and measuring the acceleration response at p. Dividing the measured acceleration by a_0, $H(f)$ may appear as in Fig. 10.4-9(c).

The mean-square response $\overline{a_p^2}$ at p is calculated numerically from the equation

$$\overline{a_p^2} = \sum_i S(f_i)\,|H(f_i)|^2 \Delta f_i$$

The following numerical table illustrates the computational procedure.

Numerical Example

f cps	Δf cps	$S(f_i)$ g^2/cps	$\|H(f_i)\|$ Nondimensional	$\|H(f_i)\|^2\,\Delta f$ cps	$S(f_i)\,\|H(f_i)\|^2\,\Delta f$ g^2 units
0	10	0	1.0	10	0
10	10	0	1.0	10	0
20	10	0.2	1.1	12.1	2.4
30	10	0.6	1.4	19.6	11.8
40	10	1.2	2.0	40	48.0
50	10	1.8	1.3	16.9	30.5
60	10	1.8	1.3	16.9	30.5
70	10	1.1	2.0	40	44.0
80	10	0.9	3.7	137	123
90	10	1.1	5.4	291	320
100	10	1.2	2.2	48.4	57.7
110	10	1.1	1.3	16.9	18.6
120	10	0.8	0.8	6.4	5.1
130	10	0.6	0.6	3.6	2.2
140	10	0.3	0.5	2.5	0.8
150	10	0.2	0.6	3.6	0.7
160	10	0.2	0.7	4.9	0.1
170	10	0.1	1.3	16.9	1.7
180	10	0.1	1.1	12.1	1.2
190	10	0.5	0.7	4.9	2.3
200	10	0	0.5	2.5	0
210	10	0	0.4	1.6	0

$$\overline{a^2} = 700.6g^2$$
$$\sigma = \sqrt{700.6g^2} = 26.6g$$

The probability of exceeding specified accelerations are

$$p[|a| > 26.6\,g] = 31.7\%$$
$$p[a_{\text{peak}} > 26.6\,g] = 60.7\%$$

$$p[|a| > 79.8\ g] = 0.3\%$$
$$p[a_{\text{peak}} > 79.8\ g] = 1.2\%$$

10.5 CORRELATION

Correlation is a measure of the similarity between two quantities. Suppose we have two records, $x_1(t)$ and $x_2(t)$, as shown in Fig. 10.5-1. The *correlation*

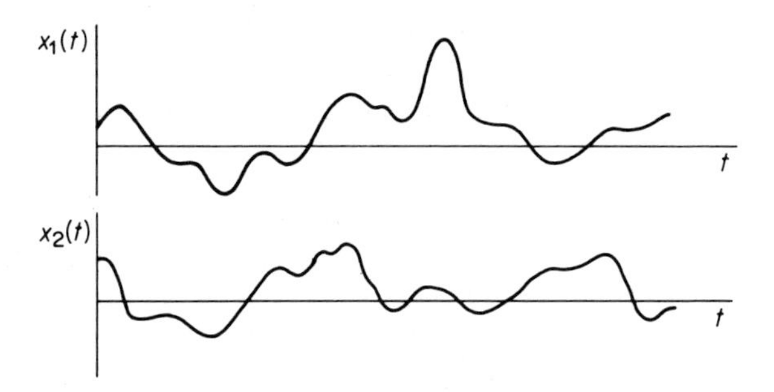

Figure 10.5-1. Correlation between $x_1(t)$ and $x_2(t)$.

between them is computed by multiplying the ordinates of the two records at each time t and determining the average value $\langle x_1(t)x_2(t)\rangle$ by dividing the sum of the products by the number of products. It is evident that the correlation so found will be largest when the two records are similar or identical. For dissimilar records, some of the products will be positive and others will be negative, so their sum will be smaller.

Consider next the case where $x_2(t)$ is identical to $x_1(t)$ but shifted to the left by a time τ as shown in Fig. 10.5-2. Then at time t, when x_1 is $x(t)$, the value of x_2 is $x(t + \tau)$, and the correlation will be given by $\langle x(t)x(t + \tau)\rangle$. Here, if $\tau = 0$, we have complete correlation; as τ increases the correlation will decrease.

It is evident that the above result can be computed from a single record by multiplying the ordinates at times t and $t + \tau$ and determining the average. We then call this result the *autocorrelation* and designate it by $R(\tau)$. It is also

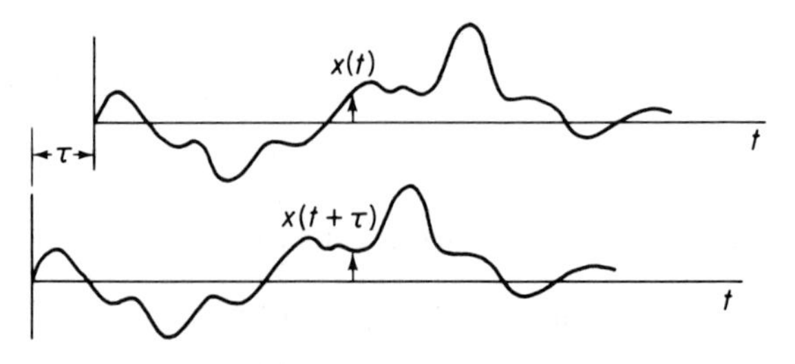

Figure 10.5-2. Function $x(t)$ shifted by τ.

the expected value of the product $x(t)x(t+\tau)$, or

$$\begin{aligned} R(\tau) &= E[x(t)x(t+\tau)] = \langle x(t)x(t+\tau)\rangle \\ &= \lim_{T\to\infty}\frac{1}{T}\int_{-T/2}^{T/2} x(t)x(t+\tau)\,dt \end{aligned} \tag{10.5-1}$$

When $\tau = 0$, the above definition reduces to the mean square value

$$R(0) = \overline{x^2} = \sigma^2 \tag{10.5-2}$$

Since the second record of Fig. 10.5-2 can be considered to be delayed with respect to the first record, or the first record advanced with respect to the second record, it is evident that $R(\tau) = R(-\tau)$ is symmetric about the origin $\tau = 0$ and is always less than $R(0)$.

Highly random functions, such as the one shown in Fig. 10.5-3, soon lose their similarity within a short time shift. Its autocorrelation, therefore, is a sharp spike at $\tau = 0$ that drops off rapidly with $\pm\tau$ as shown.

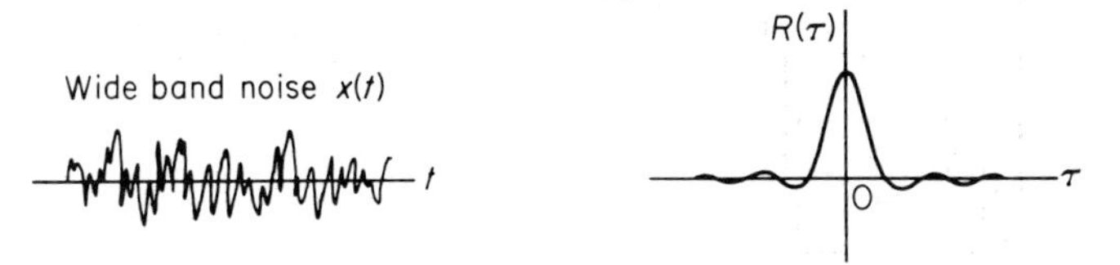

Figure 10.5-3. Highly random function and its autocorrelation.

For the special case of a periodic wave, the autocorrelation must be periodic of the same period since shifting the wave one period brings the wave into coincidence again. Fig. 10.5-4 shows a sine wave and its autocorrelation.

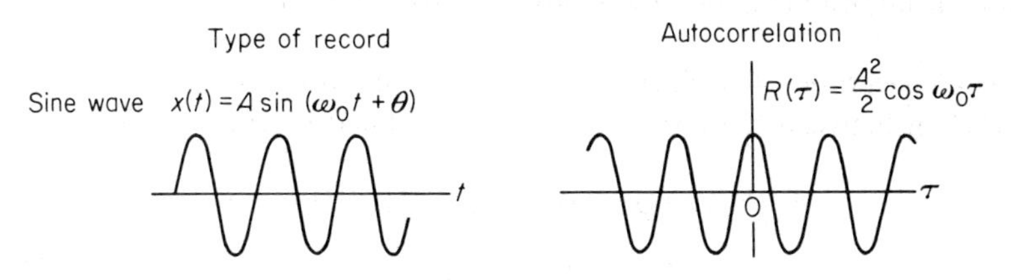

Figure 10.5-4. Sine wave and its autocorrelation.

The autocorrelation for the wide band noise is a peaked curve at $\tau = 0$, which tends to approach zero very quickly on either side. It implies that wide-band random records have little or no correlation except near $\tau = 0$.

For the narrow-band record shown in Fig. 10.5-5, the autocorrelation has some of the characteristics found for the sine wave in that it is again an even function with a maximum at $\tau = 0$ and frequency ω_0 corresponding to

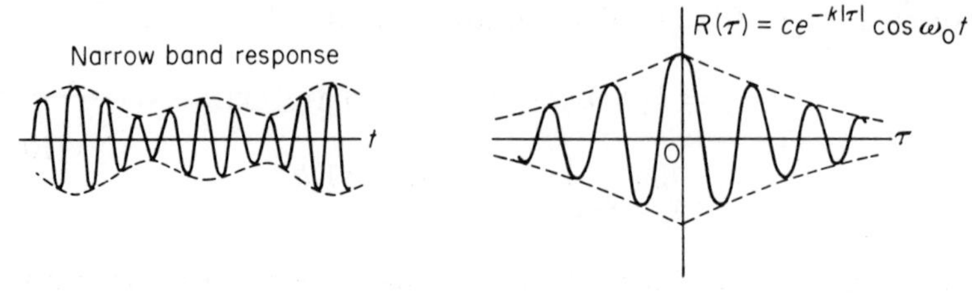

Figure 10.5-5. Autocorrelation for three types of records.

the dominant or central frequency. The difference appears in the fact that $R(\tau)$ approaches zero for large τ for the narrow band record. It is evident then that hidden periodicities in a random record can be detected by determining $R(\tau)$ for large values of τ.

Further properties of the autocorrelation function will be deferred to a later section; we now show a block diagram in Fig. 10.5-6 which presents

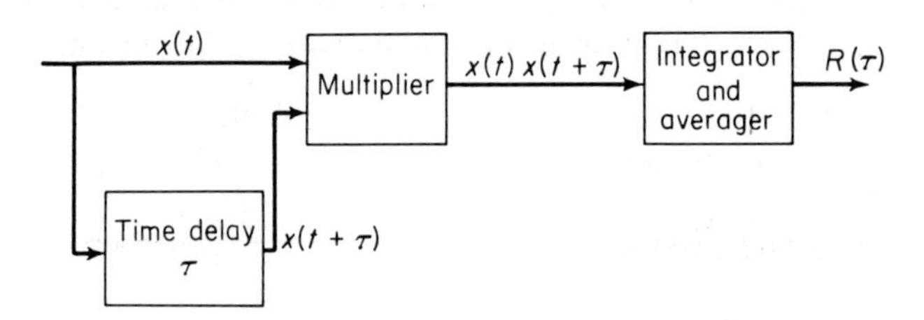

Figure 10.5-6. Block diagram of the autocorrelation analyzer.

the underlying operation for the determination of the autocorrelation. The signal $x(t)$ is delayed by τ and multiplied, after which it is integrated and averaged. The delay time τ is fixed during each run and is changed in steps or is continuously changed by a slow sweeping technique.

Cross Correlation. Consider two random quantities $x(t)$ and $y(t)$. The correlation between these two quantities is defined by the equation

$$\begin{aligned} R_{xy}(\tau) &= E[x(t)y(t+\tau)] = \langle x(t)y(t+\tau) \rangle \\ &= \lim_{T\to\infty} \frac{1}{T} \int_{-\infty}^{\infty} x(t)y(t+\tau)dt \end{aligned} \tag{10.5-3}$$

that can also be called the *cross correlation* between the quantities x and y.

Such quantities often arise in dynamical problems. For example, let $x(t)$ be the deflection at the end of a beam due to a load $F_1(t)$ at some specified point. $y(t)$ is the deflection at the same point, due to a second load $F_2(t)$ at a different point than the first, as illustrated in Fig. 10.5-7. The deflection due to both loads is then $z(t) = x(t) + y(t)$, and the autocorrelation of $z(t)$ as a

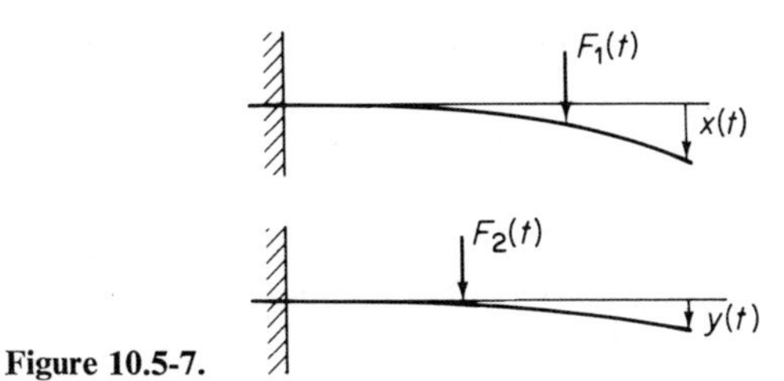

Figure 10.5-7.

result of the two loads is

$$\begin{aligned}R_z(\tau) &= \langle z(t)z(t+\tau)\rangle \\ &= \langle [x(t)+y(t)][x(t+\tau)+y(t+\tau)]\rangle \\ &= \langle x(t)x(t+\tau)\rangle + \langle x(t)y(t+\tau)\rangle \\ &\quad + \langle y(t)x(t+\tau)\rangle + \langle y(t)y(t+\tau)\rangle \\ &= R_x(\tau) + R_{xy}(\tau) + R_{yx}(\tau) + R_y(\tau)\end{aligned} \tag{10.5-4}$$

Thus the autocorrelation of a deflection at a given point due to separate loads $F_1(t)$ and $F_2(t)$ cannot be determined simply by adding the autocorrelations $R_x(\tau)$ and $R_y(\tau)$ resulting from each load acting separately. $R_{xy}(\tau)$ and $R_{yx}(\tau)$ are here referred to as *cross correlation*, and, in general, they are not equal.

10.6 FOURIER TRANSFORM

In Sec. 10.3 we examined the frequency content of periodic time functions which resulted in discrete frequency spectra. The concept of the spectral density was then introduced as the mean square contribution in the frequency interval divided by the frequency interval, this quantity approaching a continuous variation as the period approached a large value. When the spectral density function is known, the task of determining the mean square value reduces to that of summation over the frequency according to Eq. (10.3-6).

In general, random vibrations are not periodic, and then frequency analysis requires the use of the Fourier integral. The Fourier integral, however, can be viewed as a limiting case of the Fourier series as the period is extended to infinity. Fourier transforms, which result from the Fourier integral, enable a more extensive treatment of the random vibration problem.

We begin with a periodic function $x(t)$ which is a real quantity expressible by the equation

$$x(t) = \tfrac{1}{2}\sum_{n=0}^{\infty}(C_n e^{in\omega_0 t} + C_n^* e^{-in\omega_0 t}) \tag{10.6-1}$$

In this expression, the coefficient C_n is given by the equation

$$C_n = \frac{1}{T} \int_{-T/2}^{T/2} x(\xi) e^{-in\omega_0 \xi} \, d\xi \tag{10.6-2}$$

and T is the period. Equation (10.6-2) indicates that $C_n^* = C_{-n}$ and so Eq. (10.6-1) can be rewritten as

$$\begin{aligned} x(t) &= \sum_{n=-\infty}^{\infty} C_n e^{in\omega_0 t} \\ &= \sum_{n=-\infty}^{\infty} \frac{1}{T} \int_{-T/2}^{T/2} x(\xi) e^{-in\omega_0 \xi} e^{in\omega_0 t} \, d\xi \end{aligned} \tag{10.6-3}$$

The frequency $\omega = n\omega_0$ is specified here in discrete intervals, and hence the increment of the frequency is

$$\Delta\omega = (n+1)\omega_0 - n\omega_0 = \omega_0 = \frac{2\pi}{T}$$

From this expression, we replace $1/T$ by $\Delta\omega/2\pi$ and note that as $T \longrightarrow \infty$, $\Delta\omega \longrightarrow d\omega$ and $n\omega_0 \longrightarrow \omega$. Thus in the limiting case, Eq. (10.6-3) becomes

$$x(t) = \frac{1}{2\pi} \int_{-\infty}^{\infty} \left\{ \int_{-\infty}^{\infty} x(\xi) e^{-i\omega\xi} \, d\xi \right\} e^{i\omega t} \, d\omega \tag{10.6-4}$$

which is the Fourier integral.

Since the quantity within the inner braces is a function only of $i\omega$, we can rewrite this equation in two parts as

$$X(i\omega) = \int_{-\infty}^{\infty} x(\xi) e^{-i\omega\xi} \, d\xi \tag{10.6-5}$$

and

$$x(t) = \frac{1}{2\pi} \int_{-\infty}^{\infty} X(i\omega) e^{i\omega t} \, d\omega \tag{10.6-6}$$

The quantity $X(i\omega)$ is the Fourier transform of $x(t)$, and the two equations above are referred to as the Fourier transform pair. Equation (10.6-5) resolves the function $x(t)$ into harmonic components $X(i\omega)$, whereas Eq. (10.6-6) synthesizes these harmonic components to the original function $x(t)$.

In actual measurements, it is more convenient to deal with the frequency f rather than the circular frequency ω. Mathematically this also has the advantage of reducing the Fourier transform pair into the symmetric form below.

$$X(f) = \int_{-\infty}^{\infty} x(\xi)e^{-i2\pi f\xi}\, d\xi \tag{10.6-7}$$

and

$$x(t) = \int_{-\infty}^{\infty} X(f)e^{i2\pi ft}\, df \tag{10.6-8}$$

We will henceforth use this symmetric form of the Fourier transform whenever possible.

Parseval's Theorem. Parseval's theorem is a useful tool for converting time integration into frequency integration. If $X_1(f)$ and $X_2(f)$ are Fourier transforms of real time functions $x_1(t)$ and $x_2(t)$ respectively, Parseval's theorem states that

$$\begin{aligned} \int_{-\infty}^{\infty} x_1(t)x_2(t)\, dt &= \int_{-\infty}^{\infty} X_1(f)X_2(-f)\, df \\ &= \int_{-\infty}^{\infty} X_1(-f)X_2(f)\, df \end{aligned} \tag{10.6-9}$$

This relationship may be proved using the Fourier transform as follows

$$\begin{aligned} x_1(t)x_2(t) &= x_2(t)\int_{-\infty}^{\infty} X_1(f)e^{i2\pi ft}\, df \\ \int_{\infty}^{\infty} x_1(t)x_2(t)\, dt &= \int_{-\infty}^{\infty} x_2(t)\int_{-\infty}^{\infty} X_1(f)e^{i2\pi ft}\, df\, dt \\ &= \int_{-\infty}^{\infty} X_1(f)\left[\int_{-\infty}^{\infty} x_2(t)e^{i2\pi ft}\, dt\right] df \\ &= \int_{-\infty}^{\infty} X_1(f)X_2(-f)\, df \end{aligned}$$

All of the previous formulas for the mean square value, autocorrelation, and cross correlation can now be expressed in terms of the Fourier transform by Parseval's theorem.

Example 10.6-1

Express the mean square value in terms of the Fourier transform.

Letting $x_1(t) = x_2(t) = x(t)$, and averaging over T, which is allowed to go to ∞, we obtain

$$\overline{x^2} = \lim_{T\to\infty} \frac{1}{T}\int_{-\infty}^{\infty} x^2(t)\, dt = \int_{-\infty}^{\infty} \lim_{T\to\infty} \frac{1}{T} X(f)X^*(f)\, df$$

Comparing this with Eq. (10.3-6′), we obtain the relationship

$$S(f) = \lim_{T\to\infty} \frac{1}{T} X(f)X^*(f) \tag{10.6-10}$$

for the spectral density function.

Example 10.6-2

Express the auto-correlation in terms of the Fourier transform. We begin with the Fourier transform of $x(t + \tau)$

$$x(t+\tau) = \int_{-\infty}^{\infty} X(f)e^{i2\pi f(t+\tau)}\, df$$

Substituting this into the expression for the autocorrelation, we obtain

$$\begin{aligned} R(\tau) &= \lim_{T\to\infty} \frac{1}{T} \int_{-\infty}^{\infty} x(t)x(t+\tau)\, dt \\ &= \lim_{T\to\infty} \frac{1}{T} \int_{-\infty}^{\infty} x(t) \int_{-\infty}^{\infty} X(f)e^{i2\pi ft}e^{i2\pi f\tau}\, df\, dt \\ &= \int_{-\infty}^{\infty} \lim_{T\to\infty} \frac{1}{T}\left\{\int_{-\infty}^{\infty} x(t)e^{i2\pi ft}\, dt\right\} X(f)e^{i2\pi f\tau}\, df \\ &= \int_{-\infty}^{\infty} \left\{\lim_{T\to\infty} \frac{1}{T} X^*(f)X(f)\right\} e^{i2\pi f\tau}\, df \end{aligned}$$

Substituting from Eq. (10.6-10) the above equation becomes

$$R(\tau) = \int_{-\infty}^{\infty} S(f)e^{i2\pi f\tau}\, df \tag{10.6-11}$$

The inverse of the above equation is also available from the Fourier transform

$$S(f) = \int_{-\infty}^{\infty} R(\tau)e^{-i2\pi f\tau}\, d\tau \tag{10. 6-12}$$

Since $R(\tau)$ is symmetric about $\tau = 0$, the last equation can also be written as

$$S(f) = 2\int_{0}^{\infty} R(\tau)\cos 2\pi f\tau\, d\tau \tag{10.6-13}$$

These are the *Wiener-Kinchin* equations, and they state that the spectral density function can be determined from the autocorrelation function.

As a parallel to the Wiener-Khinchin equations, we can define the cross correlation between two quantities $x(t)$ and $y(t)$ as

$$R_{xy}(\tau) = \langle x(t)y(t+\tau)\rangle = \lim_{T\to\infty} \frac{1}{T}\int_{-T/2}^{T/2} x(t)y(t+\tau)\,dt$$

$$= \int_{-\infty}^{\infty} \lim_{T\to\infty} \frac{1}{T} X^*(f)Y(f)e^{i2\pi f\tau}\,df \tag{10.6-14}$$

$$R_{xy}(\tau) = \int_{-\infty}^{\infty} S_{xy}(f)e^{i2\pi f\tau}\,df$$

where the cross spectral density is defined as

$$S_{xy}(f) = \lim_{T\to\infty} \frac{1}{T} X^*(f)Y(f) \qquad -\infty \le f \le \infty$$

$$= \lim_{T\to\infty} \frac{1}{T} X(f)Y^*(f) \tag{10.6-15}$$

$$= S_{xy}^*(f) = S_{xy}(-f)$$

Its inverse from the Fourier transform is

$$S_{xy}(f) = \int_{-\infty}^{\infty} R_{xy}(\tau)e^{-i2\pi f\tau}\,d\tau \tag{10.6-16}$$

which is the parallel to Eq. (10.6-12). Unlike the autocorrelation, the cross correlation and the cross spectral density functions are, in general, not even functions; hence, the limits $-\infty$ to $+\infty$ are retained.

Example 10.6-3

Show that the frequency response function $H(\omega)$ is the Fourier transform of the impulse response function $g(t)$.

Solution: From the convolution integral, Eq. (4.3-1), the response equation in terms of the impulse response function is

$$x(t) = \int_{-\infty}^{t} f(\xi)g(t-\xi)\,d\xi$$

where the lower limit has been extended to $-\infty$ to account for all past excitations. By letting $\tau = (t-\xi)$, the above integral becomes

$$x(t) = \int_{0}^{\infty} f(t-\tau)g(\tau)\,d\tau$$

For a harmonic excitation $f(t) = e^{i\omega t}$, the above equation becomes

$$x(t) = \int_0^\infty e^{i\omega(t-\tau)} g(\tau)\, d\tau$$
$$= e^{i\omega t} \int_0^\infty g(\tau) e^{-i\omega\tau}\, d\tau$$

Comparison of this result with Eq. (10.2-3) shows that the frequency response function is

$$H(\omega) = \int_0^\infty g(\tau) e^{-i\omega\tau}\, d\tau \tag{10.6-17}$$

Spectral Density from Laplace Transformation of Autocorrelation. So far we have used the Fourier transforms assuming that they exist for the record in question. For the Fourier transforms

$$f(t) = \frac{1}{2\pi} \int_{-\infty}^{\infty} F(\omega) e^{i\omega t}\, d\omega$$
$$F(\omega) = \int_{-\infty}^{\infty} f(t) e^{-i\omega t}\, dt$$

the integration is along the real axis from $-\infty$ to $+\infty$.

Suppose we change the integration path to a line parallel to the real axis but below it by γ as shown in Fig. 10.6-1a. Then the limits of integration

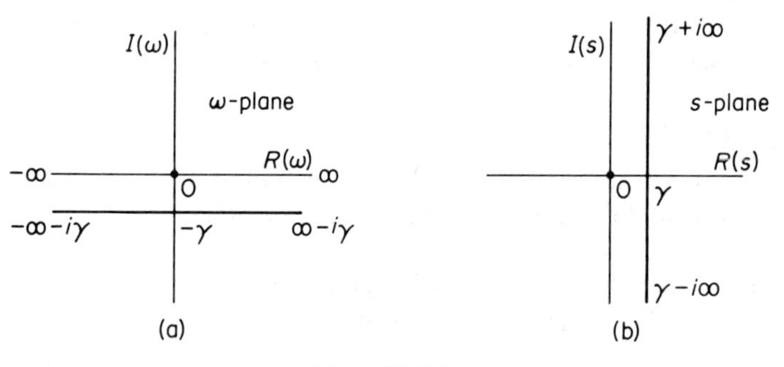

Figure 10.6-1.

are $\omega = -\infty - i\gamma$ to $\omega = +\infty - i\gamma$, and the Fourier integral is extended to include functions for which the previous equations might not be valid.

$$f(t) = \frac{1}{2\pi} \int_{-\infty - i\gamma}^{+\infty - i\gamma} F(\omega) e^{i\omega t}\, d\omega$$

If now we introduce $s = i\omega$, it is evident that points on Fig. 10.6-1(a) are rotated 90° as in Fig. 10.6-1(b), and the integration path becomes a vertical line at a distance γ to the right of the origin. The Fourier transforms now become

$$f(t) = \frac{1}{2\pi i}\int_{\gamma - i\infty}^{\gamma + i\infty} F(s)e^{st}\,ds$$

$$F(s) = \int_{-\infty}^{\infty} f(t)e^{-st}\,dt$$

and are thus converted to the *two-sided Laplace transform pair.*

Consider next Eq. (10.6-12) for the power spectral density function.

$$S(f) = \int_{-\infty}^{\infty} R(\tau)e^{-i2\pi f\tau}\,d\tau$$

With $i2\pi f = i\omega = s$, the above equation is recognized as the two-sided Laplace transform

$$S(s) = \int_{-\infty}^{\infty} R(\tau)e^{-s\tau}\,d\tau \tag{10.6-18}$$

Since $R(\tau)$ is a symmetric function, the lower limit can be changed to 0 and the value of the integral doubled.

$$S(s) = 2\int_{0}^{\infty} R(\tau)e^{-s\tau}\,d\tau$$

$$s = i2\pi f \tag{10.6-19}$$

Thus the spectral density function may be found from the Laplace transform of the autocorrelation function. For autocorrelation functions which do not have a Fourier transform, the above equation offers an alternative procedure for the evaluation of the spectral density function $S(i2\pi f)$.

10.7 RESPONSE OF CONTINUOUS STRUCTURES TO RANDOM EXCITATION

We consider here the problem of determining the mean square response $\overline{y^2(x, t)}$ of a continuous elastic structure excited by a distributed random force $f(x, t)$. By treating the problem from the normal mode summation approach

$$y(x, t) = \sum_j \phi_j(x)q_j(t) \tag{10.7-1}$$

where $\phi_j(x)$ are the normal modes of the structure, we can make use of our previous knowledge of the response of the single degree of freedom system, discussed in Sec. 10.2. For this we must assume proportional damping defined by

$$\int_0^l c(x)\phi_j(x)\phi_k(x)\,dx = 0 \qquad \text{for} \quad j \neq k$$

which leads to uncoupled normal mode equations of the form

$$\ddot{q}_j + 2\zeta_j\omega_j\dot{q}_j + \omega_j^2 q_j = \frac{1}{M_j}F_j(t) \tag{10.7-2}$$

where

$$M_j = \int_0^l \phi_j^2(x)\,dm \qquad = \text{generalized mass}$$

$$F_j(t) = \int_0^l f(x,t)\phi_j(x)\,dx = \text{generalized force.}$$

In forming the mean square response of $y(x, t)$, the following summation must be considered.

$$\overline{y^2(x,t)} = \lim_{T\to\infty} \frac{1}{T} \int_{-T/2}^{T/2} y^2(x,t)\,dt$$

$$= \sum_j \sum_k \phi_j(x)\phi_k(x) \lim_{T\to\infty} \frac{1}{T} \int_{-T/2}^{T/2} q_j(t)q_k(t)\,dt$$

We note here that we are involved with the cross correlation of $q_j(t)$ and $q_k(t)$ which from Parseval's theorem can be replaced by the frequency integration of the Fourier transforms

$$\lim_{T\to\infty} \frac{1}{T} \int_{-T/2}^{T/2} q_j(t)q_k(t)\,dt = \int_{-\infty}^{\infty} \lim_{T\to\infty} \frac{1}{T} Q_j(f)Q_k^*(f)\,df$$

where the capital letters stand for the F.T. of the corresponding quantities in lower case letters. We further note from Eq. (10.6-15) that

$$S_{q_jq_k}(f) = \lim_{T\to\infty} \frac{1}{T} Q_j(f)Q_k^*(f)$$

is the cross spectral density of the generalized coordinates, which is related to the cross spectral density of the exciting force, $S_{F_jF_k}(f)$ (See Eq. 10.3-10).

$$S_{q_jq_k}(f) = H_j(f)H_k^*(f)S_{F_jF_k}(f)$$

Putting these quantities together, the equation for the mean square response becomes

$$\overline{y^2(x, t)} = \sum_j \sum_k \phi_j(x)\phi_k(x) \int_{-\infty}^{\infty} S_{F_jF_k}(f)H_j(f)H_k^*(f)\, df \tag{10.7-3}$$

It is frequently necessary to work strictly in the time domain where the differential equation of motion is of the form

$$L(x, \dot{x}, \ddot{x}) = F(t) \tag{10.7-4}$$

where $F(t)$ is assumed to be a random time function and $L(x, \dot{x}, \ddot{x})$ is a differential equation that may be nonlinear. The solution to such an equation would most likely be obtained on the digital or analog computer, the result being a random response $x(t)$.

If the response spectrum for the above problem is desired, the first step will be to form the autocorrelation function

$$R(\tau) = \langle x(t)x(t + \tau)\rangle$$

The response spectrum $S(f)$ can then be obtained from the Wiener-Khinchin relation, Eq. (10.6-13).

Example 10.7-1

The height of ocean waves are generally distributed in a Rayleigh distribution, with a frequency spectrum known as the Bretschneider* sea spectrum. For a given sea state, Fig. 10.7-1 may represent such a spectrum.

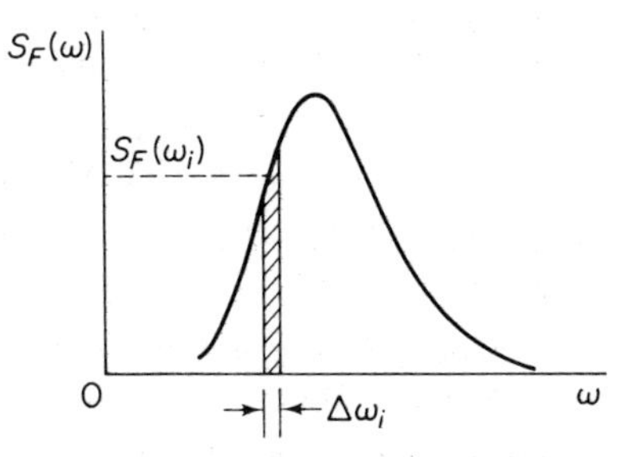

Figure 10.7-1. Typical spectrum for ocean waves.

In determining the response of an ocean structure to such excitation, one approach would be to assume harmonic wave forces of the form

$$F(t) = C \sum_i a_i \cos(\omega_i t + \phi_i) \tag{10.7-5}$$

where to comply with the wave spectrum the amplitudes a_i are chosen for each frequency ω_i by the relation

$$\tfrac{1}{2}a_i^2 = S_F(\omega_i)\Delta\omega_i \tag{10.7-6}$$

*C. L., Bretschneider, "Wave Variability and Wave Spectra for Wind-Generated Gravity Waves." T. M. No. 118 Beach Erosion Board, U.S. Army Corps of Engineers.

The phase ϕ_i can be assumed to have an equal probability between 0 to 2π and hence can be chosen by spinning a roulette wheel (or by using the Monte Carlo method with random numbers). When summed over all frequencies corresponding to the wave spectrum the excitation $F(t)$ is a random time function.

Applying $F(t)$ to the differential equation of the system under consideration, the response $x(t)$ is obtained by a computer. From the response $x(t)$ the autocorrelation $R(\tau)$ is computed and the response spectrum is obtained from the equation (10.6-13) or (10.6-19)

$$S(f) = 2\int_0^\infty R(\tau)\cos 2\pi f\tau d\tau$$

$$= 2\int_0^\infty R(\tau)e^{-s\tau}\,d\tau \qquad (s = i2\pi f)$$

REFERENCES

1. Bendat, J. S. *Principles and Applications of Random Noise Theory.* New York: John Wiley & Sons, Inc., 1958.
2. Bendat, J. S., and Piersol, A. G. *Measurement and Analysis of Random Data.* New York: John Wiley & Sons, Inc., 1966.
3. Blackman, R. B., and Tukey, J. W. *The Measurement of Power Spectra.* New York: Dover Publications, Inc., 1958.
4. Clarkson, B. L. "The Effect of Jet Noise on Aircraft Structures." *Aeronautical Quarterly*, Vol. 10, Part 2, May 1959.
5. Cramer, H. *The Elements of Probability Theory.* New York: John Wiley & Sons, Inc., 1955.
6. Crandall, S. H. *Random Vibration.* Cambridge, Mass.: The Technology Press of M.I.T., 1948.
7. Crandall, S. H. *Random Vibration*, Vol. 2. Cambridge, Mass.: The Technology Press of M.I.T., 1963.
8. Rice, S. O. *Mathematical Analysis of Random Noise.* New York: Dover Publications, Inc., 1954.
9. Robson, J. D. *Random Vibration.* Edinburgh University Press, (1964).
10. Thomson, W. T., and Barton, M. V. "The Response of Mechanical Systems to Random Excitation." *J. Appld. Mech.*, June 1957, pp. 248–51.

PROBLEMS

10-1 Give examples of random data and indicate classifications for each example.

10-2 Discuss the differences between nonstationary, stationary, and ergodic data.

10-3 Discuss what we mean by the expected value. What is the expected number of heads when 8 coins are thrown 100 times; 1000 times? What is the probability for tails?

10-4 Throw a coin 50 times, recording 1 for head and 0 for tail. Determine the probability of obtaining heads by dividing the cumulative heads by the number of throws and plot this number as a function of the number of throws. The curve should approach 0.5.

10-5 For the series of triangular waves shown in Fig. P10-5, determine the mean and the mean square values.

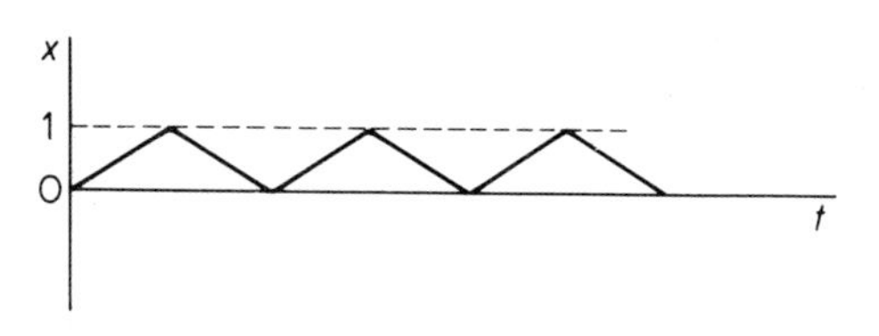

Figure P. 10-5.

10-6 A sine wave with a steady component has the equation

$$x = A_0 + A_1 \sin \omega t$$

Determine the expected values $E(x)$ and $E(x^2)$.

10-7 Determine the mean and the mean square values for the rectified sine wave.

10-8 The sharpness of the frequency response curve near resonance is often expressed in terms of $Q = \frac{1}{2}\zeta$. Points on either side of resonance where the response falls to a value $1/\sqrt{2}$ are called half-power points. Determine the respective frequencies of the half-power points in terms of ω_n and Q.

10-9 Show that

$$\int_0^\infty \frac{d\eta}{[1 - \eta^2]^2 + [2\zeta\eta]^2} = \frac{\pi}{4\zeta} \quad \text{for} \quad \zeta \ll 1$$

10-10 The differential equation of a system with structural damping is given as

$$m\ddot{x} + k(1 + i\gamma)x = F(t)$$

Determine the frequency response function.

10-11 A single degree of freedom system with natural frequency ω_n and damping factor $\zeta = 0.10$ is excited by the force

$$F(t) = F \cos (0.5\omega_n t - \theta_1) + F \cos (\omega_n t - \theta_2) + F \cos (2\omega_n t - \theta_3)$$

Show that the mean square response is

$$\overline{y^2} = (1.74 + 25.0 + 0.110) \frac{1}{2}\left(\frac{F}{k}\right)^2 = 13.43\left(\frac{F}{k}\right)^2$$

10-12 Derive the equation for the coefficients C_n of the periodic function

$$f(t) = Re \sum_{n=0}^{\infty} C_n e^{in\omega_0 t}$$

10-13 Show that for Prob. 12, $C_{-n} = C_n^*$, and that $f(t)$ can be written as

$$f(t) = \sum_{n=-\infty}^{\infty} C_n e^{in\omega_0 t}$$

10-14 Determine the Fourier series for the saw tooth wave shown in Fig. P10-14, and plot its spectral density.

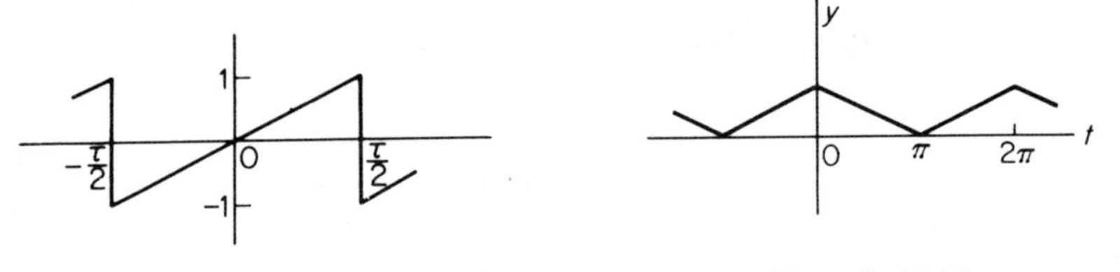

Figure P. 10-14.

Figure P. 10-15.

10-15 Determine the complex form of the Fourier series for the wave shown in Fig. P10-15, and plot its spectral density.

10-16 Determine the complex form of the Fourier series for the rectangular wave shown in Fig. P10-16, and plot its spectral density.

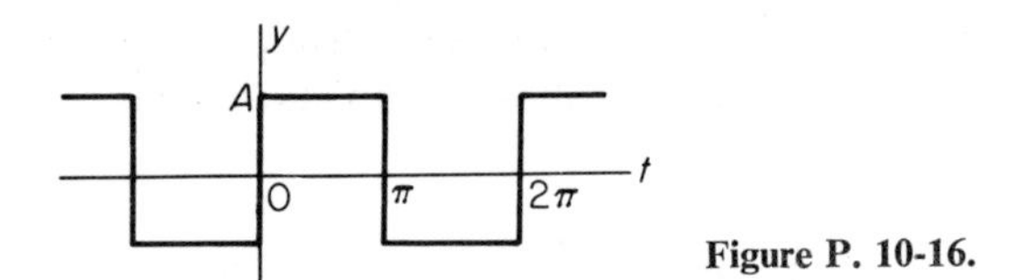

Figure P. 10-16.

10-17 Discuss why the probability distribution of the peak values of a random function should follow the Rayleigh distribution or one similar in shape to it.

10-18 Show that for the Gaussian probability distribution $p(x)$ the central moments are given by

$$E(x^n) = \int_{-\infty}^{\infty} x^n p(x)\,dx = \begin{cases} 0 \text{ for } n \text{ odd} \\ 1\cdot 3\cdot 5\cdots(n-1)\sigma^n \text{ for } n \text{ even} \end{cases}$$

10-19 Derive the equations for the cumulative probability and the probability density functions of the sine wave. Plot these results.

10-20 What would the cumulative probability and the probability density curves look like for the rectangular wave shown in Fig. P10–20.

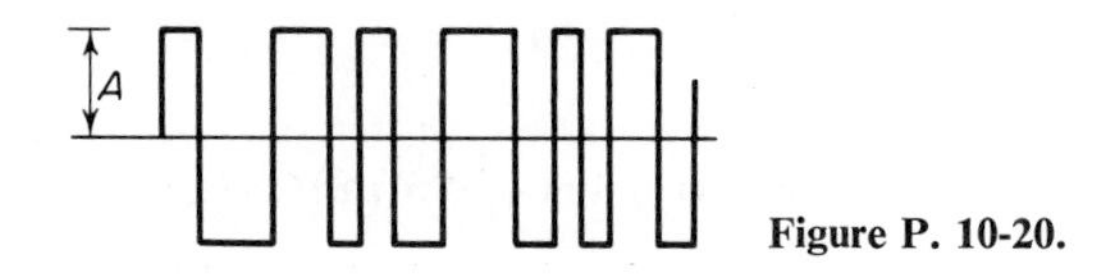

Figure P. 10-20.

10-21 A random signal is found to have a constant spectral density of $S(f) = 0.002$ in.^{2}/cps between 20 cps and 2000 cps. Outside this range, the spectral density is zero. Determine the standard deviation and the rms value if the mean value is 3.0 in. Plot this result.

10-22 In Example 10.4-3, what is the probability of the instantaneous acceleration exceeding a value $53.2g$? Of the peak value exceeding this value?

10-23 Determine the autocorrelation of a cosine wave $x(t) = A \cos t$, and plot it against τ.

10-24 Determine the autocorrelation of the rectangular wave of Prob. 10-16.

10-25 Determine the autocorrelation of a rectangular pulse and plot it against τ.

10-26 Determine the autocorrelation of the binary sequence shown in Fig. P10-26. Suggestion: Trace the above wave on transparent graph paper and shift it through τ.

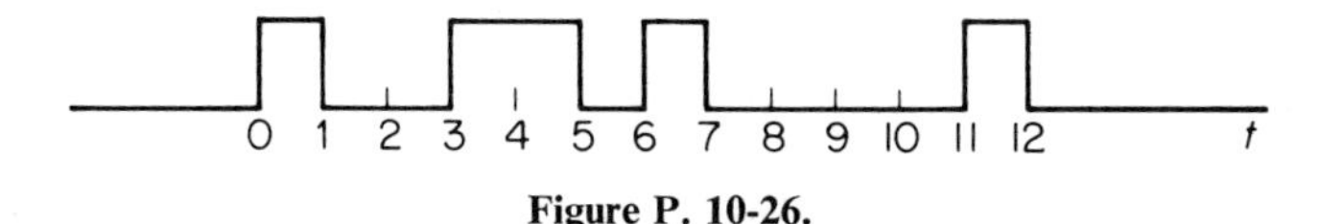

Figure P. 10-26.

10-27 Determine the autocorrelation of the triangular wave shown in Fig. P10-27

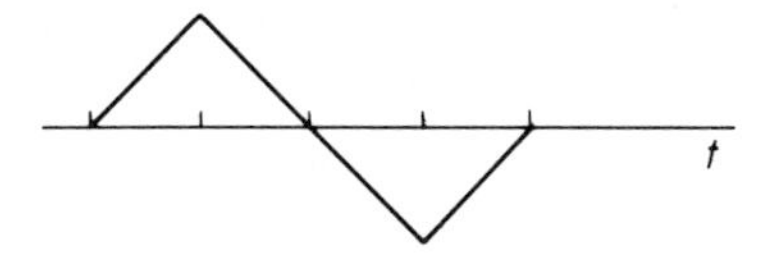

Figure P. 10-27.

10-28 Starting with the relationship

$$x(t) = \int_0^\infty f(t - \xi)g(\xi)\, d\xi$$

and using the Fourier transform technique, show that

$$X(i\omega) = F(i\omega)H(i\omega)$$

and

$$\overline{x^2} = \int_0^\infty S_F(\omega)\,|H(i\omega)|^2\, d\omega$$

where

$$S_F(\omega) = \lim_{T\to\infty} \frac{1}{2\pi T} F(i\omega)F^*(i\omega)$$

10-29 Starting with the relationship

$$H(i\omega) = |H(i\omega)|\, e^{i\phi(\omega)}$$

show that

$$\frac{H(i\omega)}{H^*(i\omega)} = e^{i2\phi(\omega)}$$

10-30 Starting with the equations

$$S_{FX}(\omega) = \lim_{T\to\infty} \frac{1}{2\pi T} F^*(i\omega)X(i\omega) = \lim_{T\to\infty} \frac{1}{2\pi T} F^*(FH) = S_F H$$

and

$$S_{XF}(\omega) = \lim_{T\to\infty} \frac{1}{2\pi T} X^*F = \lim_{T\to\infty} \frac{1}{2\pi T} (F^*H^*)F = S_F H^*$$

show that

$$\frac{S_{FX}(\omega)}{S_{XF}(\omega)} = e^{i2\phi(\omega)}$$

and

$$\frac{S_F(\omega)}{S_{XF}(\omega)} = \frac{S_{FX}(\omega)}{S_F(\omega)} = H(i\omega)$$

10-31 The differential equation for the longitudinal motion of a uniform slender rod is

$$\frac{\partial^2 u}{\partial t^2} = c^2 \frac{\partial^2 u}{\partial x^2}$$

Show that for an arbitrary axial force at the end $x = 0$, with the other end $x = l$ free, the Laplace transform of the response is

$$\bar{u}(x, s) = \frac{-c\bar{F}(s)e^{-s(l/c)}}{sAE(1 - e^{-2s(l/c)})}\{e^{(s/c))x-l)} + e^{-(s/c)(x-l)}\}$$

10-32 If the force in Prob. 10-31 is harmonic and equal to $F(t) = F_o\, e^{i\omega t}$, show that

$$u(x, t) = \frac{cF_o e^{i\omega t} \cos[(\omega l/c)(x/l - 1)]}{\omega AE \sin(\omega l/c)}$$

and

$$\sigma(x, t) = \frac{-\sin[(\omega l/c)(x/l - 1)]}{\sin(\omega l/c)} \frac{F_o}{A} e^{i\omega t}$$

where σ is the stress.

10-33 With $S(\omega)$ as the spectral density of the excitation stress at $x = 0$, show that the mean square stress in Prob. 10-31 is

$$\overline{\sigma^2} \cong \frac{2\pi}{\gamma} \sum_n \frac{c}{n\pi l} S(\omega_n) \sin^2 n\pi \frac{x}{l}$$

where structural damping is assumed. The normal modes of the problem are

$$\varphi_n(x) = \sqrt{2} \cos n\pi(x/l - 1), \quad \omega_n = n\pi(c/l), \quad c = \sqrt{AE/m}.$$

11

Nonlinear Vibrations

11.1 INTRODUCTION

Linear system analysis serves to explain much of the behavior of oscillatory systems. However, there are a number of oscillatory phenomena which cannot be predicted or explained by the linear theory.

In the linear systems which we have studied, cause and effect are related linearly; i.e., if we double the load, the response is doubled. In a nonlinear system this relationship between cause and effect is no longer proportional. For example, the center of an oil can may move proportionally to the force for small loads, but at a certain critical load it will snap over to a large displacement. The same phenomenon is also encountered in the buckling of columns, electrical oscillations of circuits containing inductance with an iron core, and vibration of mechanical systems with nonlinear restoring forces.

The differential equation describing a nonlinear oscillatory system may have the general form

$$\ddot{x} + f(\dot{x}, x, t) = 0 \tag{11.1-1}$$

Such equations are distinguished from linear equations in that the principle of superposition does not hold for their solution.

Analytical procedures for the treatment of nonlinear differential equations are difficult and require extensive mathematical study. Exact solutions that are known are relatively few, and a large part of the progress in the knowledge of nonlinear systems comes from approximate and graphical solutions and from studies made on computing machines. Much can be learned about a nonlinear system, however, by using the state space approach and studying the motion presented in the phase plane.

11.2 THE PHASE PLANE

In an *autonomous* system, the time t does not appear explicitly in the differential equation of motion. Thus only the differential of time, dt, will appear in such an equation.

We will first study an automonous system with the differential equation

$$\ddot{x} + f(x, \dot{x}) = 0 \tag{11.2-1}$$

where $f(x, \dot{x})$ may be a nonlinear function of x and $\dot{x}$. In the method of *state space*, we express the above equation in terms of two first order equations as follows

$$\begin{aligned} \dot{x} &= y \\ \dot{y} &= -f(x, y) \end{aligned} \tag{11.2-2}$$

If x and y are Cartesian coordinates, the xy plane is called the *phase plane*. The *state* of the system is defined by the coordinate x and $y = \dot{x}$, which represents a point on the phase plane. As the state of the system changes, the point on the phase plane moves, thereby generating a curve which is called the *trajectory*.

Another useful concept is the *state speed* V defined by the equation

$$V = \sqrt{\dot{x}^2 + \dot{y}^2} \tag{11.2-3}$$

When the state speed is zero, an *equilibrium state* is reached in that both the velocity $\dot{x}$ and the acceleration $\dot{y} = \ddot{x}$ are zero.

Dividing the second of Eq. (11.2-2) by the first we obtain the relation

$$\frac{dy}{dx} = -\frac{f(x, y)}{y} = \phi(x, y) \tag{11.2-4}$$

Thus for every point x, y in the phase plane for which $\phi(xy)$ is not indeterminate, there is a unique slope of the trajectory.

If $y = 0$ (i.e., points along the x-axis) and $f(x, y) \neq 0$, the slope of the trajectory is infinite. Thus all trajectories corresponding to such points must cross the x-axis at right angles.

If $y = 0$ and $f(x, y) = 0$, the slope is indeterminate. We define such points as *singular points*. Singular points correspond to a state of equilibrium in that both the velocity $y = \dot{x}$ and the force $\ddot{x} = \dot{y} = -f(x, y)$ are zero. Further discussion is required to establish whether the equilibrium represented by the singular point is stable or unstable.

Example 11.2-1

Determine the phase plane of a single degree of freedom oscillator

$$\ddot{x} + \omega^2 x = 0$$

Solution: With $y = \dot{x}$, the above equation is written in terms of two first order equations

$$\dot{y} = -\omega^2 x$$
$$\dot{x} = y$$

Dividing, we obtain

$$\frac{dy}{dx} = -\frac{\omega^2 x}{y}$$

Separating variables and integrating

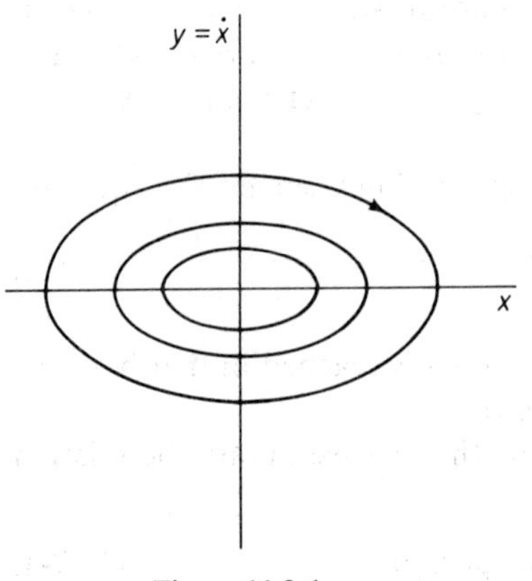

Figure 11.2-1.

$$y^2 + \omega^2 x^2 = C$$

which is a series of ellipses, the size of which is determined by C. The above equation is also that of conservation of energy

$$\tfrac{1}{2}m\dot{x}^2 + \tfrac{1}{2}kx^2 = C'$$

Since the singular point is at $x = y = 0$, the phase plane plot appears as in Fig. 11.2-1. If y/ω is plotted in place of y, the ellipses of Fig. 11.2-1 reduce to circles.

11.3 CONSERVATIVE SYSTEMS

In a conservative system the total energy remains constant. Summing the kinetic and potential energies per unit mass, we have

$$\tfrac{1}{2}\dot{x}^2 + U(x) = E = \text{constant} \tag{11.3-1}$$

Solving for $y = \dot{x}$, the ordinate of the phase plane is given by the equation

$$y = \dot{x} = \pm\sqrt{2[E - U(x)]} \tag{11.3-2}$$

It is evident from this equation that the trajectories of a conservative system must be symmetric about the x-axis.

The differnetial equation of motion for a conservative system can be shown to have the form

$$\ddot{x} = f(x) \tag{11.3-3}$$

Since $\ddot{x} = \dot{x}(d\dot{x}/dx)$, the above equation can be written as

$$\dot{x}\,d\dot{x} - f(x)\,dx = 0 \tag{11.3-4}$$

Integrating, we have

$$\frac{\dot{x}^2}{2} - \int_0^x f(x)\,dx = E \tag{11.3-5}$$

and by comparison with Eq. (11.3-1) we find

$$\begin{aligned} U(x) &= -\int_0^x f(x)\,dx \\ f(x) &= -\frac{dU}{dx} \end{aligned} \tag{11.3-6}$$

Thus for a conservative system the force is equal to the negative gradient of the potential energy.

With $y = \dot{x}$, Eq. (11.3-4) in the state-space becomes

$$\frac{dy}{dx} = \frac{f(x)}{y} \tag{11.3-7}$$

We note from this equation that singular points correspond to $f(x) = 0$ and $y = \dot{x} = 0$, and hence are equilibrium points. Equation (11.3-6) then indicates that at the equilibrium points the slope of the potential energy curve $U(x)$ must be zero. It can be shown that the minima of $U(x)$ are stable equilibrium positions, whereas the saddle points corresponding to the maxima of $U(x)$ are positions of unstable equilibrium.

Stability of Equilibrium. Examining Eq. (11.3-2) the value of E is determined by the initial conditions $x(0)$ and $y(0) = \dot{x}(0)$. If the initial conditions are large, E will also be large. For every position x, there is a potential $U(x)$; for motion to take place, E must be greater than $U(x)$. Otherwise, Eq. (11.3-2) shows that the velocity $y = \dot{x}$ is imaginary.

Figure 11.3-1 shows a general plot of $U(x)$ and the trajectory y vs. x for various values of E computed from Eq. (11.3-2).

For $E = 7$, $U(x)$ lies below $E = 7$ only between $x = 0$ to 1.2, $x = 3.8$ to 5.9, and $x = 7$ to 8.7. The trajectories corresponding to $E = 7$ are closed curves and the period associated with them can be found from Eq. (11.3-2) by integration

Computation of Phase Plane for $U(x)$ given in Fig. 11.3-1

$$y = \pm\sqrt{2[E - U(x)]}$$

x	$U(x)$	$\pm y$ at $(E = 7)$	$\pm y$ at $(E = 8)$	$\pm y$ at $(E = 9)$	$\pm y$ at $(E = 11)$
0	5.0	2.0	2.45	2.83	3.46
1.0	6.3	1.18	1.84	2.32	
1.5	8.0	imag	0	1.41	2.45
2.0	9.6	imag	imag	imag	
3.0	10.0	imag	imag	imag	1.41
3.5	8.0	imag	0	1.41	2.45
4.0	6.5	1.0	1.73	2.24	
5.0	5.0	2.0	2.45	2.83	3.46
5.5	5.7	1.61	2.24	2.57	
6.0	7.2	imag	1.26	1.90	
6.5	8.0	imag	0	1.41	2.45
7.0	7.0	0	1.41	2.0	
7.5	6.0	1.41	2.0	2.45	3.16
8.0	6.3	1.18	1.84	2.32	
9.0	7.4	imag	1.09	1.79	
9.5	8.0	imag	0	1.41	
10.0	8.8	imag	imag	0.63	
11.5					0

$$\tau = 2\int_{x_1}^{x_2} \frac{dx}{\sqrt{2[E - U(x)]}}$$

where x_1 and x_2 are extreme points of the trajectory on the x-axis.

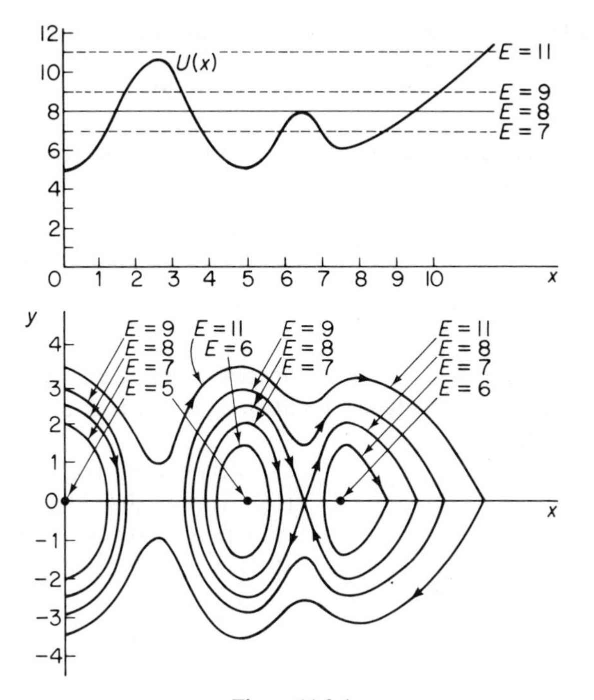

Figure 11.3-1.

For smaller initial conditions, these closed trajectories become smaller. For $E = 6$ the trajectory about the equilibrium point $x = 7.5$ contracts to a point, while the trajectory about the equilibrium point $x = 5$ is a closed curve between $x = 4.2$ to 5.7.

For $E = 8$ one of the maxima of $U(x)$ at $x = 6.5$ is tangent to $E = 8$ and the trajectory at this point has four branches. The point $x = 6.5$ is a saddle point for $E = 8$ and the motion is unstable. The saddle point trajectories are called *separatrices*.

For $E > 8$ the trajectories may or may not be closed. $E = 9$ shows a closed trajectory between $x = 3.3$ to 10.2. Note that at $x = 6.5$, $dU/dx = -f(x) = 0$ and $y = \dot{x} \neq 0$ for $E = 9$, and hence equilibrium does not exist.

11.4 STABILITY OF EQUILIBRIUM

Expressed in the general form

$$\frac{dy}{dx} = \frac{P(x, y)}{Q(x, y)} \tag{11.4-1}$$

the singular points (x_s, y_s) of the equation are identified by

$$P(x_s, y_s) = Q(x_s, y_s) = 0 \tag{11.4-2}$$

Equation (11.4-1), of course, is equivalent to the two equations

$$\begin{aligned} \frac{dx}{dt} &= Q(x, y) \\ \frac{dy}{dt} &= P(x, y) \end{aligned} \tag{11.4-3}$$

from which the time dt has been eliminated. A study of these equations in the neighborhood of the singular point provides us with answers as to the stability of equilibrium.

Recognizing that the slope dy/dx of the trajectories does not vary with translation of the coordinate axes, we will translate the u, v axes to one of the singular points to be studied, as shown in Fig. 11.4-1. We then have

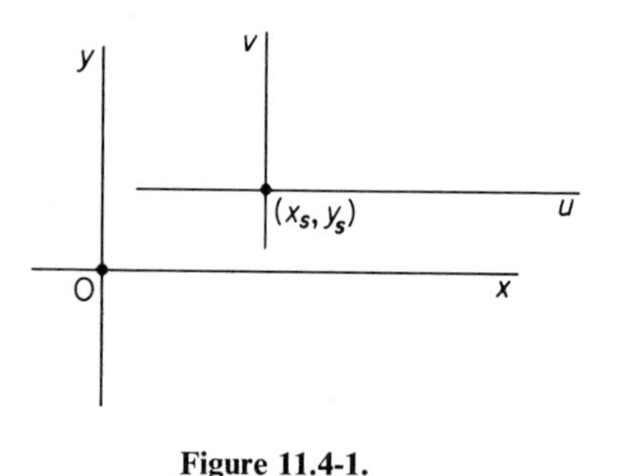

Figure 11.4-1.

$$\begin{aligned} x &= x_s + u \\ y &= y_s + v \\ \frac{dy}{dx} &= \frac{dv}{du} \end{aligned} \tag{11.4-4}$$

If $P(x, y)$ and $Q(x, y)$ are now expanded in terms of the Taylor series about the singular point (x_s, y_s), we obtain for $Q(x, y)$

$$Q(x, y) = Q(x_s, y_s) + \left(\frac{\partial Q}{\partial u}\right)_s u + \left(\frac{\partial Q}{\partial v}\right)_s v + \left(\frac{\partial^2 Q}{\partial u^2}\right)_s u^2 + \cdots \tag{11.4-5}$$

and a similar equation for $P(x, y)$. Since $Q(x_s y_s)$ is zero and $(\partial Q/\partial u)_s$ and $(\partial Q/\partial v)_s$ are constants, Eq. (11.4-1) in the region of the singularity becomes

$$\frac{dv}{du} = \frac{cu + ev}{au + bv} \tag{11.4-6}$$

where the higher order derivatives of P and Q have been omitted. Thus a study of the singularity at (x_s, y_s) is possible by studying Eq. (11.4-6) for small u and v.

Returning to Eq. (11.4-3) and taking note of Eq. (11.4-4) and (11.4-5), Eq. (11.4-6) is seen to be equivalent to

$$\frac{du}{dt} = au + bv$$
$$\frac{dv}{dt} = cu + ev \tag{11.4-7}$$

which may be rewritten in the matrix form

$$\begin{Bmatrix} \dot{u} \\ \dot{v} \end{Bmatrix} = \begin{bmatrix} a & b \\ c & e \end{bmatrix} \begin{Bmatrix} u \\ v \end{Bmatrix} \tag{11.4-8}$$

It was shown in Chapter 6, Sec. 6.7 that if the eigenvalues and eigenvectors of a matrix equation such as Eq. (11.4-8) are known, a transformation

$$\begin{Bmatrix} u \\ v \end{Bmatrix} = [P] \begin{Bmatrix} \xi \\ \eta \end{Bmatrix} = \left[\begin{Bmatrix} u_1 \\ v_1 \end{Bmatrix} \begin{Bmatrix} u_2 \\ v_2 \end{Bmatrix} \right] \begin{Bmatrix} \xi \\ \eta \end{Bmatrix} \tag{11.4-9}$$

where $[P]$ is a modal matrix of the eigenvector columns will decouple the equation to the form

$$\begin{Bmatrix} \dot{\xi} \\ \dot{\eta} \end{Bmatrix} = [\Lambda] \begin{Bmatrix} \xi \\ \eta \end{Bmatrix} = \begin{bmatrix} \lambda_1 & 0 \\ 0 & \lambda_2 \end{bmatrix} \begin{Bmatrix} \xi \\ \eta \end{Bmatrix} \tag{11.4-10}$$

Since Eq. (11.4-10) has the solution

$$\xi = e^{\lambda_1 t}$$
$$\eta = e^{\lambda_2 t} \tag{11.4-11}$$

the solution for u and v are

$$u = u_1 e^{\lambda_1 t} + u_2 e^{\lambda_2 t}$$
$$v = v_1 e^{\lambda_1 t} + v_2 e^{\lambda_2 t} \tag{11.4-12}$$

It is evident, then, that the stability of the singular point depends on the eigenvalues λ_1 and λ_2 determined from the characteristic equation

$$\begin{vmatrix} (a - \lambda) & b \\ c & (e - \lambda) \end{vmatrix} = 0$$

or

$$\lambda_{1,2} = \left(\frac{a + e}{2}\right) \pm \sqrt{\left(\frac{a + e}{2}\right)^2 - (ac - bc)} \tag{11.4-13}$$

Thus,

if $(ae - bc) > \left(\frac{a+e}{2}\right)^2$, the motion is oscillatory;

if $(ae - bc) < \left(\frac{a+e}{2}\right)^2$, the motion is aperiodic;

if $(a + e) > 0$, the system is unstable;
if $(a + e) < 0$, the system is stable.

The type of trajectories in the neighborhood of the singular point can be determined by first examining Eq. (11.4-10) in the form

$$\frac{d\xi}{d\eta} = \frac{\lambda_1}{\lambda_2}\frac{\xi}{\eta} \tag{11.4-14}$$

which has the solution

$$\xi = (\eta)^{\lambda_1/\lambda_2}$$

and using the transformation of Eq. (11.4-9) to plot v vs. u.

11.5 METHOD OF ISOCLINES

Consider the autonomous system with the equation

$$\frac{dy}{dx} = -\frac{f(x, y)}{y} = \phi(x, y) \tag{11.5-1}$$

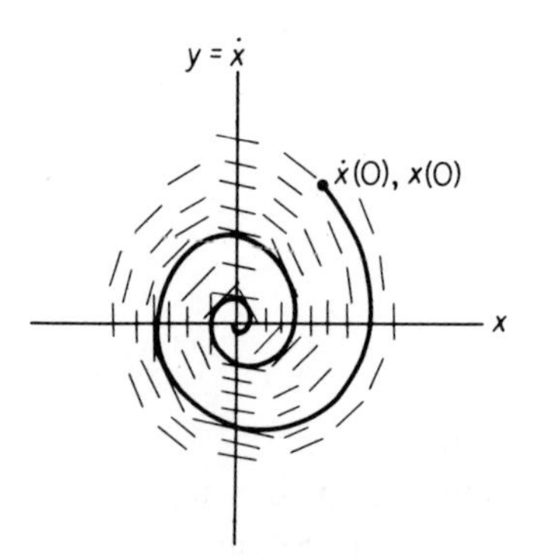

Figure 11.5-1.

that was discussed in Sec. 11.2, Eq. (11.2-4). In the method of isoclines we fix the slope dy/dx by giving it a definite number α, and solve for the curve

$$\phi(x, y) = \alpha \tag{11.5-2}$$

With a family of such curves drawn, it is possible to sketch in a trajectory starting at any point x, y as shown in Fig. 11.5-1.

Example 11.5-1

Determine the isoclines for the simple pendulum.

Solution: The equation for the simple pendulum is

$$\ddot{\theta} + \frac{g}{l}\sin\theta = 0 \tag{a}$$

Letting $x = \theta$ and $y = \dot{\theta} = \dot{x}$ we obtain

$$\frac{dy}{dx} = -\frac{g}{l}\frac{\sin x}{y} \tag{b}$$

Thus for $dy/dx = \alpha$, a constant, the equation for the isocline, Eq. (11.5-2), becomes

$$y = -\left(\frac{g}{l\alpha}\right)\sin x \tag{c}$$

It is evident from Eq. (b) that the singular points lie along the x-axis at $x = 0, \pm\pi, \pm 2\pi$ etc. Figure 11.5-2 shows isoclines in the first quadrant that correspond to negative values of α. Starting at an arbitrary point $x(0)$, $y(0)$, the trajectory can be sketched in by proceeding tangentially to the slope segments.

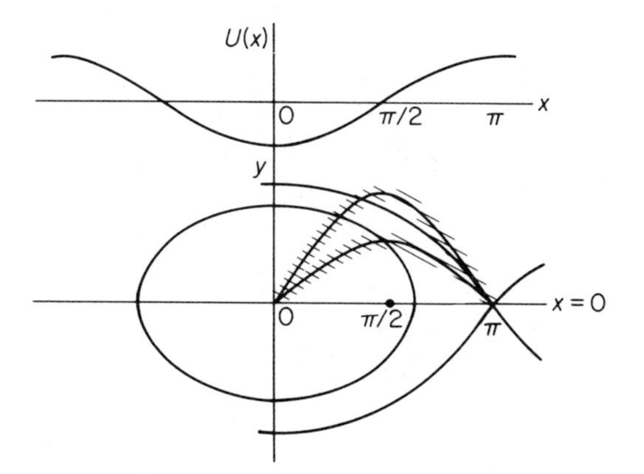

Figure 11.5-2. Isocline curves for the simple pendulum.

In this case the integral of Eq. (a) is readily availabe as

$$\frac{y^2}{2} - \frac{g}{l}\cos x = E$$

where E is a constant of integration corresponding to the total energy (see Eq. (11.3-1)). We also have $U(x) = -g/l \cos x$ and the discussions of Sec. 11.3 apply. For the motion to exist, E must be greater than $-g/l$. $E = g/l$ corresponds to the separatrix and for $E > g/l$ the trajectory does not close. This means that the initial conditions are large enough to cause the pendulum to continue past $\theta = 2\pi$.

Example 11.5-2

One of the interesting nonlinear equations which has been studied extensively is the *van der Pol equation*

$$\ddot{x} - \mu\dot{x}(1 - x^2) + x = 0$$

The equation somewhat resembles that of free vibration of a spring-mass system with viscous damping; however, the damping term of this equation is nonlinear in that it depends on both the velocity and the displacement. For small oscillations ($x < 1$) the damping is negative, and the amplitude will increase with time. For $x > 1$ the damping is positive, and the amplitude will diminish with time. If the system is initiated with $x(0)$ and $\dot{x}(0)$, the amplitude will increase or decrease, depending on whether x is small or large, and it will finally reach a stable state known as the *limit cycle*, graphically displayed by the phase plane plot of Fig. 11.5-3.

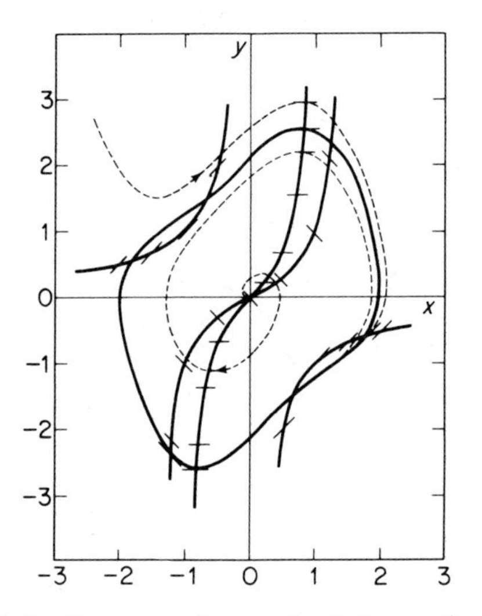

Figure 11.5-3. Isocline curves for van der Pol's equation with $\mu = 1.0$.

11.6 THE DELTA METHOD

The *delta method*, proposed by L. S. Jacobsen*, is a graphical method for the solution of the equation

$$\ddot{x} + f(\dot{x}, x, t) = 0 \tag{11.6-1}$$

where $f(\dot{x}, x, t)$ must be continuous and single valued. The equation is first rewritten by adding and subtracting a term $\omega_0^2 x$

$$\ddot{x} + f(\dot{x}, x, t) - \omega_0^2 x + \omega_0^2 x = 0 \tag{11.6-2}$$

*L. S. Jacobsen, "On a General Method of Solving Second Order Ordinary Differential Equations by Phase Plane Displacements," *J. Appl. Mech.* 19 (December 1952), pp. 543–53.

Introducing new variables τ and y defined by

$$\tau = \omega_0 t \qquad \text{and} \qquad y = \frac{dx}{d\tau} = \frac{\dot{x}}{\omega_0} \tag{11.6-3}$$

and letting

$$\delta(x, y, \tau) = \frac{1}{\omega_0^2}[f(\dot{x}, x, t) - \omega_0^2 x] \tag{11.6-4}$$

Eq. (11.6-2) may be written as

$$\frac{dy}{dx} = \frac{-(x + \delta)}{y} \tag{11.6-5}$$

The function $\delta(y, x, \tau)$ given in Eq. (11.6-4) depends on the variables y, x, and τ; however, for small changes in these variables, it may be assumed to remain constant. With δ a constant, Eq. (11.6-5) can be integrated to give

$$(x + \delta)^2 + y^2 = \rho^2 = \text{constant} \tag{11.6-6}$$

The above equation is that of a circle of radius ρ with its center at $y = 0$ and $x = -\delta$. Thus for small increments of τ, the solution corresponds to a small arc of a circle as shown in Fig. 11.6-1.

The procedure for the drawing of the trajectory is as follows

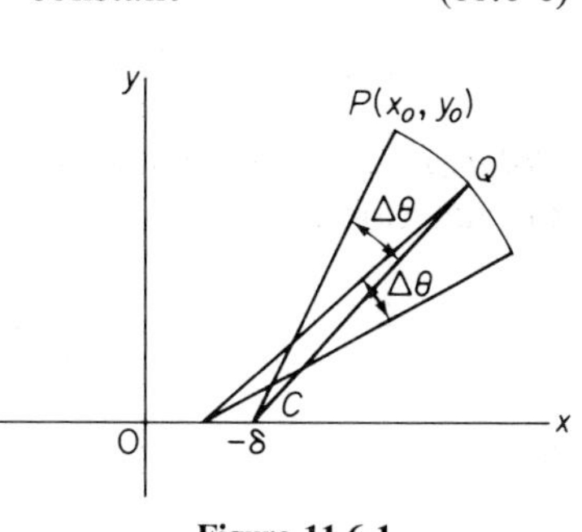

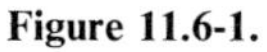

Figure 11.6-1.

(1) Locate the initial point $P(x_0 y_0)$ in the xy phase plane.

(2) From Eq. (11.6-4), calculate the initial value of $\delta(x_0, y_0, 0)$ and locate the point $-\delta$ on the x-axis.

(3) With the center $(-\delta, 0)$ draw a short arc through point $P(x_0 y_0)$. The length of the arc over which the solution is valid depends on the variation of δ, which is assumed to be constant.

(4) For an assumed value of $\Delta\theta$, locate the next point Q on the trajectory, and using the new values of x and y coorresponding to point Q, compute a new δ.

(5) Draw a line through the new $-\delta$ and the point Q and measure $\Delta\theta$ from this line for the location of the third point as shown in Fig. 11.6-1. The procedure is then repeated.

The relationship between the angular rotation $d\theta$ of the line CP and the time increment $d\tau$ can be found as follows

$$d\theta = \frac{ds}{\rho} = \frac{\sqrt{1 + \left(\frac{dy}{dx}\right)^2}\, dx}{\sqrt{y^2 + (x + \delta)^2}}$$

Substituting for dy/dx from Eq. (11.6-5) and noting from Eq. (11.6-3) that $dx = yd\tau$, we obtain

$$d\theta = \frac{dx}{y} = d\tau \tag{11.6-7}$$

It is also evident from Eq. (11.6-5) that the slope of the trajectory is negative in the first quadrant and that $d\theta$ proceeds in the clockwise direction.

Example 11.6-1

Determine the phase plane trajectory of the equation

$$\ddot{x} + \mu|\dot{x}|\dot{x} + \omega^2 x = 0$$

with $\omega = 10$ and $x(0) = 4$, $\dot{x}(0) = 0$.

Solution: We first make the substitution $\tau = \omega t$ and $dx/d\tau = y$ to reduce the equation to the form

$$\frac{dy}{d\tau} + \mu|y|y + x = 0$$

or

$$\frac{dy}{dx} = \frac{-(\mu|y|y + x)}{y}$$

Thus δ for the delta method is

$$\delta = \mu|y|y$$

which for a given value of μ is a parabola as plotted in Fig. 11.6-2. If the delta method is applied to the problem, the center $-\delta$ of the trajectory arc for any point P is M. Thus in a half cycle, the point M moves from the origin to some extreme point corresponding to the maximum value of $-\delta$ and back again to 0.

Instead of using the step-by-step delta method we will consider here the average δ method, which will enable one to draw the trajectory for the half cycle as a circle with a center at the average value of $-\delta$.

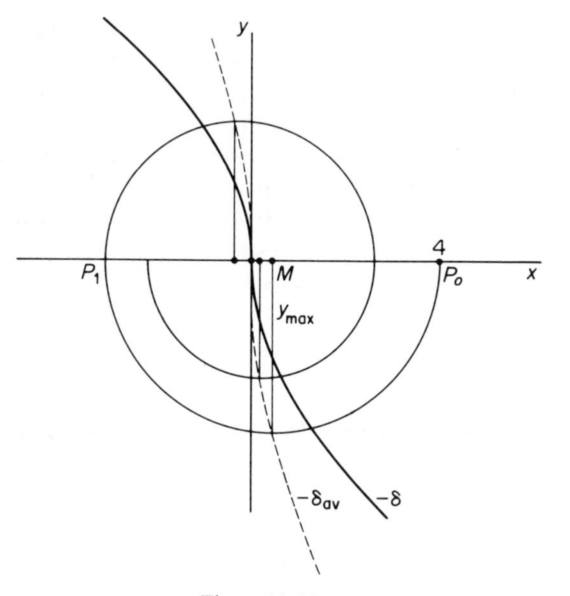

Figure 11.6-2.

The average $-\delta$ for this problem can be found for any y by integration

$$\delta_{av} = \frac{1}{y}\int_0^y \mu y^2\, dy = \mu\frac{y^2}{3}$$

Thus the average δ curve of $\frac{1}{3}$ the original delta can be drawn on the phase plane, and the trajectory with the center at $-\delta_{av}$ must cut this curve with the same $x = -\delta_{av}$. This is easily done with a compass, employing trial and error, such that $MP_0 = MP_1 = My_{max}$.

11.7 LIENARD'S METHOD

In the case where the damping term is some nonlinear function of the velocity $f(\dot{x})$, the phase plane trajectory may be graphically determined by a method proposed by Lienard.* The form of the differential equation to be treated is

$$\ddot{x} + f(\dot{x}) + x = 0 \tag{11.7-1}$$

With $y = \dot{x}$ the equation may be written as

$$\frac{dy}{dx} = -\frac{f(y) + x}{y} \tag{11.7-2}$$

*A. Lienard, "Études des Oscillations Entretenues," *Rev. Gen. de l'Élec.*, 23 (1928), pp. 901–12, 946–54.

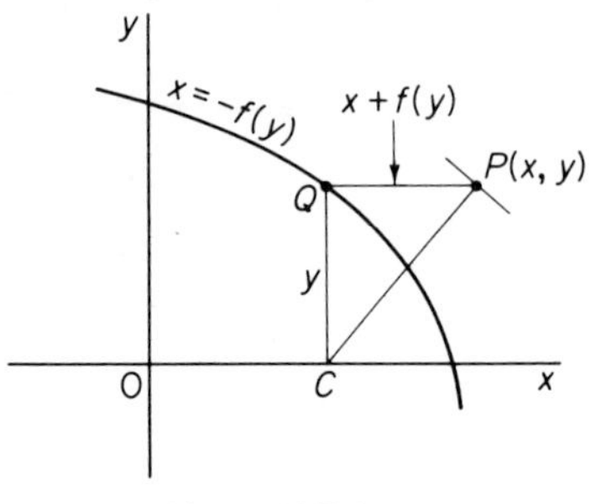

Figure 11.7-1.

As a first step the curve $x = -f(y)$ is plotted in the x, y plane as shown in Fig. (11.7-1). Starting with any point $P(x, y)$, draw a line from the point P parallel to the x-axis until it intersects the curve $x = -f(y)$ at Q. From Q draw a line *perpendicular* to x intersecting it at C. The slope of the line CP is then

$$\frac{y}{x + f(y)} = -\frac{dx}{dy}$$

and hence CP is perpendicular to the trajectory at P with slope dy/dx.

Example 11.7-1

Determine the phase plane plot of a spring-mass system with Coulomb damping.

Solution: The differential equation of motion is

$$\ddot{x} + f(\dot{x}) + \omega^2 x = 0$$

where

$$f(\dot{x}) = \begin{cases} +\mu g \text{ for } \dot{x} > 0 \\ -\mu g \text{ for } \dot{x} < 0 \end{cases}$$

Let $\tau = \omega t$, $y = dx/d\tau$ and rewrite the above equation as

$$\frac{dx}{dy} = -\frac{f_1(y) + x}{y}$$

where

$$f_1(y) = \begin{cases} +b \text{ for } y > 0 \\ -b \text{ for } y < 0 \\ \left(b = \dfrac{\mu g}{\omega^2}\right) \end{cases}$$

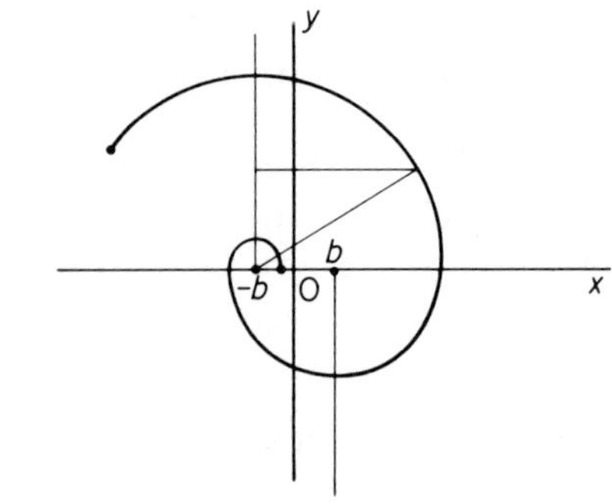

Figure 11.7-2.

Thus the curve $x = -f_1(y)$ is $-b$ for positive y and $+b$ for negative y as shown in Fig. 11.7-2. It is evident then that the point $(-b, 0)$ remains the center for all points above the x-axis and the point $(+b, 0)$ is the center for the curve below the x-axis. Thus in each half plane the curves are circles. The motion stops when the curve intersects the x-axis between the points $\pm b$.

*11.8 SLOPELINE NUMERICAL INTEGRATION**

The slopeline method is based on the step-by-step integration procedure from which a graphical construction is developed. We will present the basic principles in terms of a single first-order differential equation. The method is also applicable to second-order differential equations expressed in terms of two first-order equations.

Considering the differential equation to be

$$\frac{dx}{dt} = f(t) \tag{11.8-1}$$

the following incremental relation holds

$$\Delta x = f_{av}(t) \cdot \Delta t \tag{11.8-2}$$

where

$$\begin{aligned} f_{av}(t) &= \frac{1}{\Delta t}\int_{t_i}^{t_{i+1}} f(t)\,dt \\ \Delta t &= (t_{i+1} - t_i) \\ \Delta x &= (x_{i+1} - x_i) \end{aligned} \tag{11.8-3}$$

The average value of $f(t)$ is now replaced by its arithmetic mean

$$f_{av}(t) = \tfrac{1}{2}(f_i + f_{i+1}) \tag{11.8-4}$$

where

$$f_i = f(t_i) \quad \text{and} \quad f_{i+1} = f(t_{i+1})$$

Equation (11.8-2) then becomes

$$\Delta x = [f(t_i) + f(t_i + \Delta t)]\frac{\Delta t}{2} \tag{11.8-5}$$

In converting this procedure into a graphical method we start with the initial values x_0, t_0, and $f(t_0)$ plotted as point P_0 in the x, $f(t)$ plane as shown in Fig. 11.8-1. Assuming a time incre-

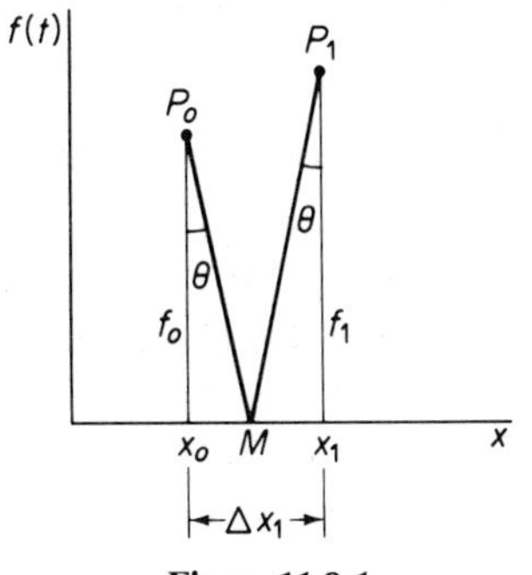

Figure 11.8-1.

*See Ref. 12.

ment Δt, draw a straight line P_0M with angle

$$\tan\theta = \frac{\Delta t}{2} \tag{11.8-6}$$

From M draw another straight line MP_1 with the same angle θ. The ordinate of P_1 is $f_1 = f(t_0 + \Delta t)$, and the increment Δx_1 is established as

$$\Delta x_1 = f_0 \tan\theta + f_1 \tan\theta = (f_0 + f_1)\frac{\Delta t}{2}$$

which is Eq. (11.8-5).

Figure 11.8-2 represents a convenient procedure for carrying out the above process. The function $f(t)$ is plotted on the right half plane with co-

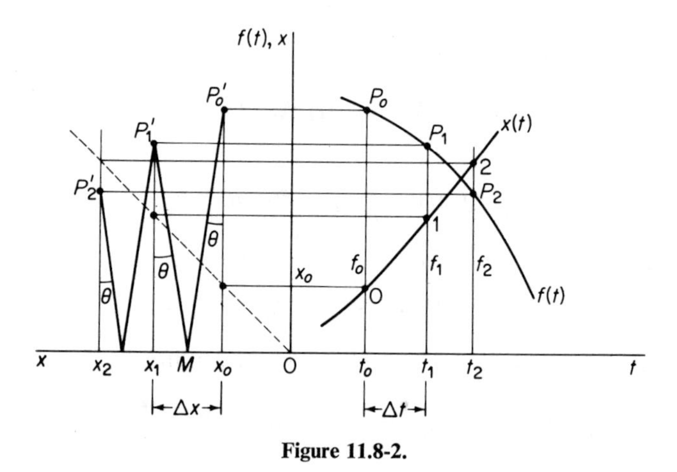

Figure 11.8-2.

ordinates t and $f(t)$. The coordinates of the left half plane are x and $f(t)$, and the dotted 45° line merely serves to transfer the x values from the horizontal to the vertical axis. The sequence of the graphical procedure is as follows

(1) Start with the initial point $P_0 = f(t_0)$.
(2) From P_0 draw a horizontal line to P'_0 located by x_0.
(3) Draw lines P'_0M and MP'_1 with angle θ based on Eq. (11.8-6). Having chosen Δt, the point P_1 on the curve $f(t)$ is known, which establishes P'_1.
(4) The value of x_1 is now known and is transferred to the right half plane at t_1 by the 45° line.
(5) From P'_1 again draw the slope line at angle θ and repeat the process. The curve passing through points 0, 1, 2, etc., is the solution $x(t)$.

The procedure can also be applied to first-order equations of the form

$$\frac{dx}{dt} = F(x) \tag{11.8-7}$$

since the initial conditions x_0, t_0 enable one to locate the starting point P_0.

The method described here can, with some modifications, be applied to second-order equations by noting that the following simultaneous first-order equations

$$\begin{aligned} \frac{dx}{dt} + g(x) - y = 0 \\ \frac{dy}{dt} + h(y) + x = 0 \end{aligned} \tag{11.8-8}$$

are equivalent to the second-order equation

$$\frac{d^2x}{dt^2} + \frac{dg}{dx}\frac{dx}{dt} + h\left(\frac{dx}{dt} + g\right) + x = 0 \tag{11.8-9}$$

For the graphical solution of the simultaneous first-order equations (Eqs. 11.8-8) in the x, y plane, the functions $g(x)$ and $-h(y)$ are first plotted against the x and y axes, respectively, as shown in Fig. 11.8-3. The reason for this is evident if Eqs. (11.8-8) are written as

$$\begin{aligned} \frac{dx}{dt} &= y - g(x) \\ \frac{dy}{dt} &= -[x + h(y)] \end{aligned}$$

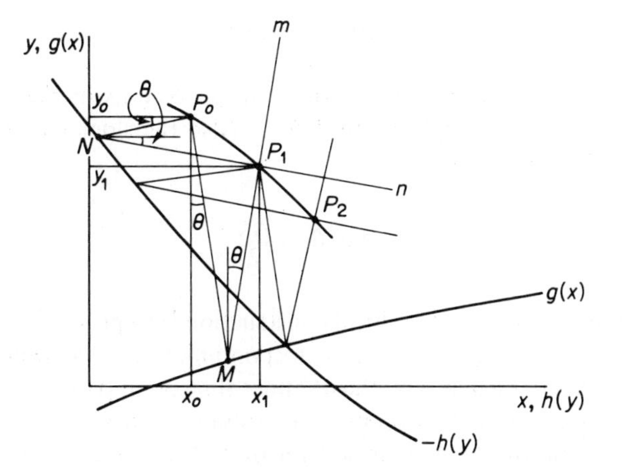

Figure 11.8-3.

and compared to Eq. (11.8-1) or to Eq. (11.8-7). The ordinates $y - g(x)$ and $x - [-h(y)]$ now replace $f(t)$ or $f(x)$.

The graphical construction proceeds as follows

(1) Locate the starting point $P_0(x_0, y_0)$ corresponding to the initial time t_0.
(2) From P_0 draw the slopeline P_0M to the curve $g(x)$ at angle $\theta = \tan^{-1}(\Delta t/2)$ with the vertical. From M draw the second slopeline Mm at the same angle θ.
(3) From P_0 draw the slopeline P_0N to the curve $-h(y)$ at angle $\theta = \tan^{-1}(\Delta t/2)$ with the horizontal, and another slopeline Nn.
(4) The intersection $P_1(x_1 y_1)$ of the lines Mm and Nn is the solution on the phase plane xy at time $t_1 = t_0 + \Delta t$.
(5) The process is now repeated from point P_1.

Example 11.8-1

Using the slopeline method, solve the equation

$$\frac{dx}{dt} = \phi(x)$$

Solution: Rearranging the equation in the form

$$dt = \frac{1}{\phi(x)}\,dx$$

and comparing with Eq. (11.8-1) written as

$$dx = f(t)\,dt$$

it is evident that the graphical solution can proceed as in Fig. 11.8-2 with t replaced by x and $f(t)$ replaced by $1/\phi(x)$. Also the slope equation must be modified to

$$\tan\theta = \frac{\Delta x}{2}$$

Fig. 11.8-4 shows how the graphical construction is to proceed. Starting from P_0 corresponding to x_0, locate P_0' corresponding to t_0. The time axis can be projected to the vertical axis by the 45° line; thus point 0 on the right quarter plane x, t represents the first point in the solution. Choosing a Δx, locate P_1 and P_1', and from the 45° line draw a horizontal to point 1. The procedure is repeated with Δx to locate P_2, etc.

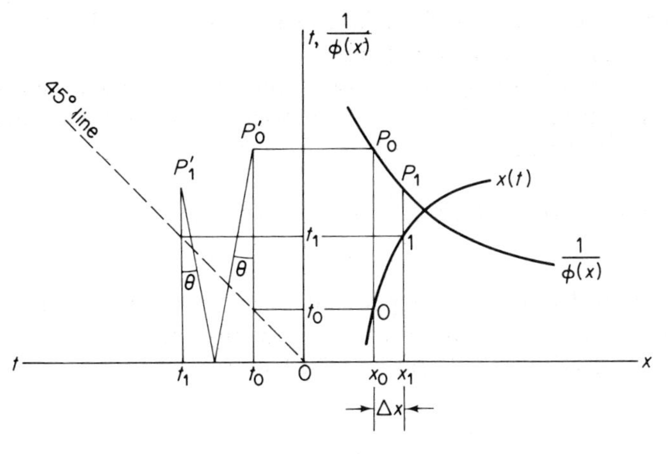

Figure 11.8-4.

11.9 THE PERTURBATION METHOD

The *perturbation method* is applicable to problems where a small parameter μ is associated with the nonlinear term of the differential equation. The solution is formed in terms of a series of the perturbation parameter μ, the result being a development in the neighborhood of the solution of the linearized problem. If the solution of the linearized problem is periodic, and if μ is small, we can expect the perturbed solution to be periodic also. We can reason from the phase plane that the periodic solution must represent a closed trajectory. The period which depends on the initial conditions is then a function of the amplitude of vibration.

Consider the free oscillation of a mass on a nonlinear spring which is defined by the equation

$$\ddot{x} + \omega_n^2 x + \mu x^3 = 0 \tag{11.9-1}$$

with initial conditions $x(0) = A$ and $\dot{x}(0) = 0$. When $\mu = 0$, the frequency of oscillation is that of the linear system $\omega_n = \sqrt{k/m}$.

We seek a solution in the form of an infinite series of the perturbation parameter μ as follows

$$x = x_0(t) + \mu x_1(t) + \mu^2 x_2(t) + \cdots \tag{11.9-2}$$

Furthermore, we know that the frequency of the nonlinear oscillation will depend on the amplitude of oscillation as well as on μ. We express this fact also in terms of a series in μ.

$$\omega^2 = \omega_n^2 + \mu\alpha_1 + \mu^2\alpha_2 + \cdots \tag{11.9-3}$$

where the α_i are as yet undefined functions of the amplitude, and ω is the frequency of the nonlinear oscillations.

We will consider only the first two terms of Eqs. (11.9-2) and (11.9-3), which will adequately illustrate the procedure. Substituting these into Eq. (11.9-1), we obtain

$$\ddot{x}_0 + \mu\ddot{x}_1 + (\omega^2 - \mu\alpha_1)(x_0 + \mu x_1) + \mu(x_0^3 + 3\mu x_0^2 x_1 + \cdots) = 0 \tag{11.9-4}$$

Since the perturbation parameter μ could have been chosen arbitrarily, the coefficients of the various powers of μ must be equated to zero. This leads to a system of equations which can be solved successively

$$\begin{aligned} &\ddot{x}_0 + \omega^2 x_0 = 0 \\ &\ddot{x}_1 + \omega^2 x_1 = \alpha_1 x_0 - x_0^3 \\ &\vdots \end{aligned} \tag{11.9-5}$$

The solution to the first equation, subject to the initial conditions $x(0) = A$, $\dot{x}(0) = 0$, is

$$x_0 = A \cos \omega t \tag{11.9-6}$$

which is called the *generating solution*. Substituting this into the right side of the second equation in Eq. (11.9-5) we obtain

$$\begin{aligned} \ddot{x}_1 + \omega^2 x_1 &= \alpha_1 A \cos \omega t - A^3 \cos^3 \omega t \\ &= \left(\alpha_1 - \frac{3}{4}A^2\right) A \cos \omega t - \frac{A^3}{4} \cos 3\omega t \end{aligned} \tag{11.9-7}$$

where $\cos^3 \omega t = \frac{3}{4} \cos \omega t + \frac{1}{4} \cos 3\omega t$ has been used. We note here that the forcing term $\cos \omega t$ would lead to a secular term $t \cos \omega t$ in the solution for x_1 (i.e., we have a condition of resonance). Such terms violate the initial stipulation that the motion is to be periodic; hence, we impose the condition

$$(\alpha_1 - \tfrac{3}{4}A^2) = 0$$

Thus α_1, which we stated earlier to be some function of the amplitude A, is evaluated to be equal to

$$\alpha_1 = \tfrac{3}{4}A^2 \tag{11.9-8}$$

With the forcing term $\cos \omega t$ eliminated from the right side of the equation, the general solution for x_1 is

$$x_1 = C_1 \sin \omega t + C_2 \cos \omega t + \frac{A^3}{32\omega^2} \cos 3\omega t$$
$$\omega^2 = \omega_n^2 + \frac{3}{4} \mu A^2 \tag{11.9-9}$$

Imposing the initial conditions $x_1(0) = \dot{x}_1(0) = 0$, the constants C_1 and C_2 are evaluated as

$$C_1 = 0 \quad C_2 = -\frac{A^3}{32\omega^2}$$

Thus

$$x_1 = \frac{A^3}{32\omega^2} (\cos 3\omega t - \cos \omega t) \tag{11.9-10}$$

and the solution at this point from Eq. (11.9-2) becomes

$$x = A \cos \omega t + \mu \frac{A^3}{32\omega^2} (\cos 3\omega t - \cos \omega t)$$
$$\omega = \omega_n \sqrt{1 + \frac{3}{4} \frac{\mu A^2}{\omega_0^2}} \tag{11.9-11}$$

The solution is thus found to be periodic, and the fundamental frequency ω is found to increase with the amplitude, as expected for a hardening spring.

Mathieu Equation. Consider the nonlinear equation

$$\ddot{x} + \omega_n^2 x + \mu x^3 = F \cos \omega t \tag{a}$$

and assume a perturbation solution

$$x = x_1(t) + \xi(t) \tag{b)*}$$

Substituting Eq. (b) into (a), we obtain the following two equations

$$\ddot{x}_1 + \omega_n^2 x_1 + \mu x_1^3 = F \cos \omega t \tag{c}$$
$$\ddot{\xi} + (\omega_n^2 + \mu 3 x_1^2)\xi = 0 \tag{d}$$

If μ is assumed to be small, we can let

$$x_1 \cong A \sin \omega t \tag{e}$$

*See Ref. 5, pp. 259–73.

and substitute it into Eq. (d), which becomes

$$\ddot{\xi} + \left[\left(\omega_n^2 + \frac{3\mu}{2}A^2\right) - \frac{3\mu}{2}A^2 \cos 2\omega t\right]\xi = 0 \tag{f}$$

This equation is of the form

$$\frac{d^2y}{dz^2} + (a - 2b \cos 2z)y = 0 \tag{g}$$

which is known as the *Mathieu equation*. The stable and unstable regions of the Mathieu equation depend on the parameters a and b, and are shown in Fig. 11.9-1.

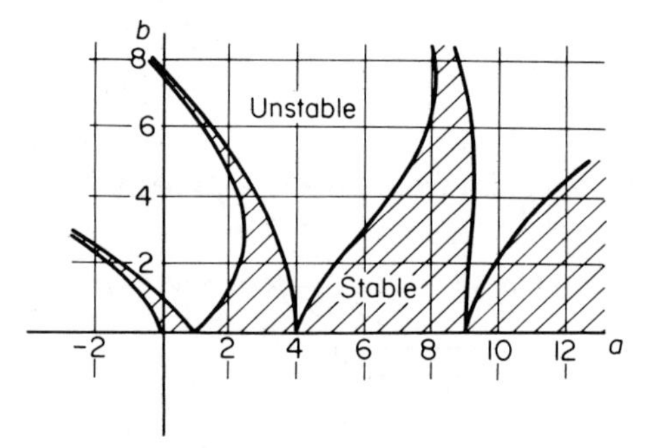

Figure 11.9-1. Stable region of Mathieu equation indicated by shaded area, which is symmetric about the horizontal axis.

11.10 METHOD OF ITERATION

Duffing* made an exhaustive study of the equation

$$m\ddot{x} + c\dot{x} + kx \pm \mu x^3 = F \cos \omega t$$

which represents a mass on a cubic spring, excited harmonically. The $\pm$ sign signifies a hardening or softening spring. The equation is nonautonomous in that the time t appears explicitly in the forcing term.

In this section we wish to examine a simpler equation where damping is zero, written in the form

$$\ddot{x} + \omega_n^2 x \pm \mu x^3 = F \cos \omega t \tag{11.10-1}$$

We seek only the steady state harmonic solution by the *method of iteration*, which is essentially a process of *successive approximation*. An assumed solu-

*See Ref. 7.

tion is substituted into the differential equation, which is integrated to obtain a solution of improved accuracy. The procedure may be repeated any number of times to achieve the desired accuracy.

For the first assumed solution, let

$$x_0 = A \cos \omega t \tag{11.10-2}$$

and substitute into the differential equation

$$\begin{aligned}\ddot{x} &= -\omega_n^2 A \cos \omega t \mp \mu A^3(\tfrac{3}{4} \cos \omega t + \tfrac{1}{4} \cos 3\omega t) + F \cos \omega t \\ &= (-\omega_n^2 A \mp \tfrac{3}{4}\mu A^3 + F) \cos \omega t \mp \tfrac{1}{4}\mu A^3 \cos 3\omega t\end{aligned}$$

In integrating this equation it is necessary to set the constants of integration to zero if the solution is to be harmonic with period $\tau = 2\pi/\omega$. Thus, we obtain for the improved solution

$$x_1 = \frac{1}{\omega^2}\left(\omega_n^2 A \pm \frac{3}{4}\mu A^3 - F\right) \cos \omega t \mp \cdots \tag{11.10-3}$$

where the higher harmonic term is ignored.

The procedure may be repeated but we will not go any further. Duffing reasoned at this point that if the first and second approximations are reasonable solutions to the problem, then the coefficients of $\cos \omega t$ in the two equations (11.10-2) and (11.10-3) must not differ greatly. Thus, by equating these coefficients we obtain

$$A = \frac{1}{\omega^2}\left(\omega_n^2 A \pm \frac{3}{4}\mu A^3 - F\right) \tag{11.10-4}$$

which may be solved for ω^2

$$\omega^2 = \omega_n^2 \pm \frac{3}{4}\mu A^2 - \frac{F}{A} \tag{11.10-5}$$

It is evident from this equation that if the nonlinear parameter is zero, we obtain the exact result for the linear system

$$A = \frac{F}{\omega_n^2 - \omega^2}$$

For $\mu \neq 0$, the frequency ω is a function of μ, F and A. It is evident that when $F = 0$, we obtain the frequency equation for free vibration

$$\frac{\omega^2}{\omega_n^2} = 1 \pm \frac{3}{4}\mu\frac{A^2}{\omega_n^2}$$

discussed in the previous section. Here we see that the frequency increases with the amplitude for the hardening spring (+), and decreases for the softening spring (−).

For $\mu \neq 0$ and $F \neq 0$, it is convenient to hold both μ and F constant and plot $|A|$ against ω/ω_n. In the construction of these curves, it is helpful to rearrange Eq. (11.10-5) to

$$\frac{3}{4}\mu\frac{A^3}{\omega_n^2} = \left(1 - \frac{\omega^2}{\omega_n^2}\right)A - \frac{F}{\omega_n^2} \tag{11.10-6}$$

each side of which can be plotted against A as shown in Fig. 11.10-1. The left side of this equation is a cubic, whereas the right side is a straight line of slope $(1 - \omega^2/\omega_n^2)$ and intercept $-F/\omega_n^2$. For $\omega/\omega_n < 1$, the two curves intersect at three points 1, 2, 3, which are also shown in the amplitude-frequency plot. As ω/ω_n increases towards unity, points 2 and 3 approach each other, after which only one value of the amplitude will satisfy Eq. (11.10-6). When $\omega/\omega_n = 1$, or when $\omega/\omega_n > 1$, these points are 4 or 5.

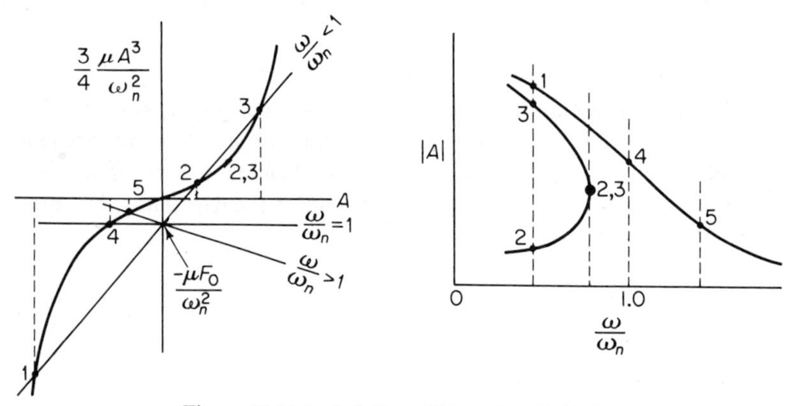

Figure 11.10-1. Solution of Equation 11.10-6.

The Jump Phenomenon. In problems of this type, it is found that the amplitude A undergoes a sudden discontinuous jump near resonance. The *jump phenomenon* can be described as follows. For the softening spring, with increasing frequency of excitation, the amplitude gradually increases until point "a" in Fig. 11.10-2 is reached. It then suddenly jumps to a larger value indicated by the point b, and diminishes along the curve to its right. In decreasing the frequency from some point c, the amplitude continues to increase beyond b to point d, and suddenly drops to a smaller value at e. The shaded region in the amplitude-frequency plot is unstable; the extent of unstableness depends on a number of factors such as the amount of damping present, the rate of change of the exciting frequency, etc. if a hardening spring had been chosen instead of the softening spring, the same type of analysis would be applicable and the result would be a curve of the type shown in Fig. 11.10-3.

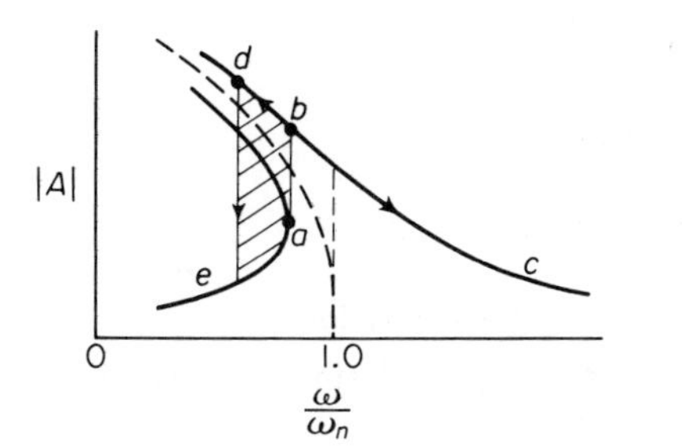

Figure 11.10-2. The jump phenomenon for the softening spring.

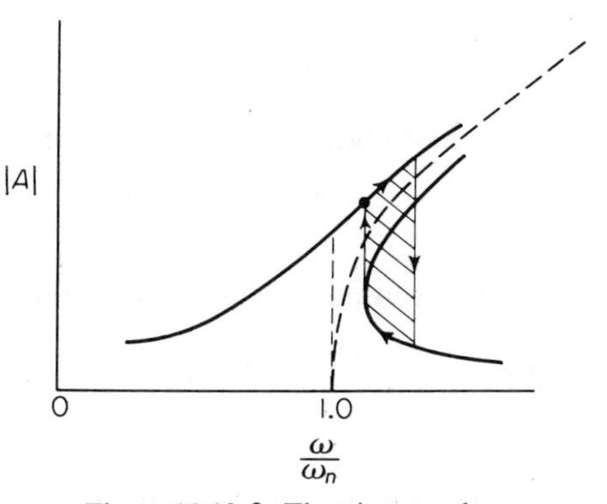

Figure 11.10-3. The jump phenomenon for the hardening spring.

Effect of Damping. In the undamped case the amplitude-frequency curves approach the backbone curve (shown dotted) asymptotically. This is also the case for the linear system where the backbone curve is the vertical line at $\omega/\omega_n = 1.0$.

With a small amount of damping present, the behavior of the system cannot differ appreciably from that of the undamped system. The upper end of the curve, instead of approaching the backbone curve asymptotically, will cross over in a continuous curve as shown in Fig. 11.10-4. The jump phenomenon is also present here but damping generally tends to reduce the size of the unstable region.

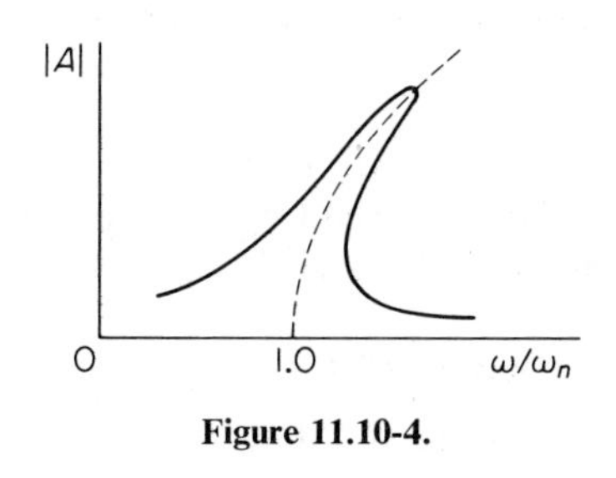

Figure 11.10-4.

The method of successive approximation is also applicable to the damped vibration case. The major difference in its treatment lies in the phase angle between the force and the displacement, which is no longer 0° or 180° as in the undamped problem. It is found that by introducing the phase in the force term rather than the displacement, the algebraic work is somewhat simplified. The differential equation can then be written as

$$\begin{aligned}\ddot{x} + c\dot{x} + \omega_n^2 x + \mu x^3 &= F\cos(\omega t + \phi)\\ &= A_0 \cos\omega t - B_0 \sin\omega t \qquad (11.10\text{-}7)\end{aligned}$$

where the magnitude of the force is

$$F = \sqrt{A_0^2 + B_0^2} \qquad (11.10\text{-}8)$$

and the phase can be determined from

$$\tan \phi = \frac{B_0}{A_0}$$

Assuming the first approximation to be

$$x_0 = A \cos \omega t$$

its substitution into the differential equation results in

$$[(\omega_n^2 - \omega^2)A + \tfrac{3}{4}\mu A^3] \cos \omega t - c\omega A \sin \omega t + \tfrac{1}{4}\mu A^3 \cos 3\omega t$$
$$= A_0 \cos \omega t - B_0 \sin \omega t \qquad (11.10\text{-}9)$$

We again ignore the cos $3\omega t$ term and equate coefficients of cos ωt and sin ωt to obtain

$$\begin{aligned} (\omega_n^2 - \omega^2)A + \tfrac{3}{4}\mu A^3 &= A_0 \\ c\omega A &= B_0 \end{aligned} \qquad (11.10\text{-}10)$$

Squaring and adding these results, the relationship between the frequency, amplitude and force becomes

$$F^2 = [(\omega_n^2 - \omega^2)A + \tfrac{3}{4}A^3]^2 + [c\omega A]^2 \qquad (11.10\text{-}11)$$

By fixing μ, c and F, the frequency ratio ω/ω_n can be computed for assigned values of A.

Example 11.10-1

Using the iteration method, solve for the period of the linear equation

$$\ddot{x} + \omega_n^2 x = 0 \qquad \text{(a)}$$

with the initial conditions $x(0) = A$ and $\dot{x}(0) = 0$.

*Solution:** Assume for the first solution $x = 1$ and substitute into the differential equation

$$\ddot{x} = -\omega_n^2 1$$

Integrate to obtain

$$\dot{x} = -\omega_n^2 \int_0^{\xi} 1 \, d\xi = -\omega_n^2 \xi$$

*See Ref. 2.

and

$$\int_A^x dx = -\omega_n^2 \int_0^t \xi \, d\xi$$
$$x(t) = A - \omega_n^2 \frac{t^2}{2} \tag{b}$$

Letting $t = t_1$ at a quarter cycle and noting that $x(t_1) = 0$, the above equation may be written as

$$x = A\left(1 - \frac{t^2}{t_1^2}\right) \tag{c}$$

We now substitute Eq. (c) into Eq. (a) and repeat the process

$$\begin{aligned}\dot{x}(t) &= -\omega_n^2 A \int_0^\xi \left(1 - \frac{\xi^2}{t_1^2}\right) d\xi \\ &= -\omega_n^2 A\left(\xi - \frac{\xi^3}{3t_1^2}\right) \\ x(t) &= A - \omega_n^2 A \int_0^t \left(\xi - \frac{\xi^3}{3t_1^2}\right) d\xi \\ &= A - \omega_n^2 A\left(\frac{t^2}{2} - \frac{t^4}{12t_1^2}\right)\end{aligned} \tag{d}$$

Next let $t = t_1$ and $x(t_1) = 0$.

$$0 = A\left[1 - \omega_n^2\left(\frac{t_1^2}{2} - \frac{t_1^2}{12}\right)\right]$$

Solving for t_1 we obtain

$$t_1 = \frac{1}{\omega_n}\sqrt{\frac{12}{5}} = \frac{\tau}{2\pi}\sqrt{\frac{12}{5}} = \frac{\tau}{4.05}$$

and we find that after two iterations, the value of t_1 is found to be nearly equal to the exact value of $\tau/4$.

11.11 SELF-EXCITED OSCILLATIONS

Oscillations which depend on the motion itself are called self-excited. The shimmy of automobile wheels, the flutter of airplane wings, and the oscillations of the van der Pol equation are some examples.

Self-excited oscillations may occur in a linear or a nonlinear system. The motion is induced by an excitation that is some function of the velocity or of the velocity and the displacement. If the motion of the system tends to increase the energy of the system, the amplitude will increase, and the system may become unstable.

As an example, consider a viscously damped single degree of freedom linear system excited by a force which is some function of the velocity. Its equation of motion is

$$m\ddot{x} + c\dot{x} + kx = F(\dot{x}) \tag{11.11-1}$$

Rearranging the equation to the form

$$m\ddot{x} + (c\dot{x} - F(\dot{x})) + kx = 0 \tag{11.11-2}$$

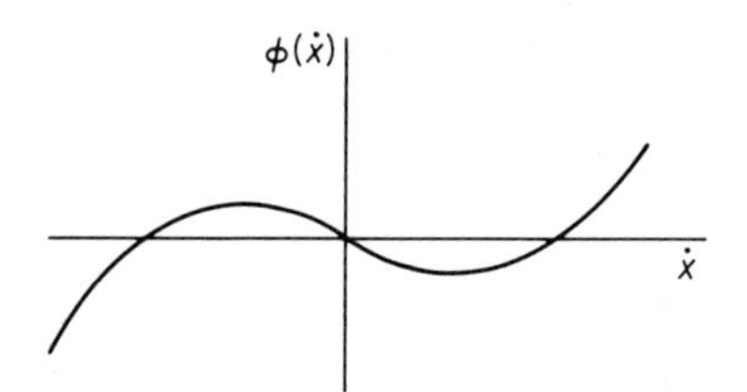

Figure 11.11-1. System with apparent damping $\phi(\dot{x}) = c\dot{x} - F(\dot{x})$.

we can recognize the possibility of negative damping if $F(\dot{x})$ becomes greater than $c\dot{x}$.

Suppose that $\phi(\dot{x}) = c\dot{x} - F(\dot{x})$ in the above equations varies as in Fig. 11.11-1. For small velocities the apparent damping $\phi(\dot{x})$ is negative, and the amplitude of oscillation will increase. For large velocities the opposite is true, and hence the oscillations will tend to a limit cycle.

Example 11.11-1

The coefficient of kinetic friction μ_k is generally less than the coefficient of static friction μ_s, this difference increasing somewhat with the velocity. Thus if the belt of Fig. 11.11-2 is started, the mass will move with the belt until the spring force is balanced by the static friction.

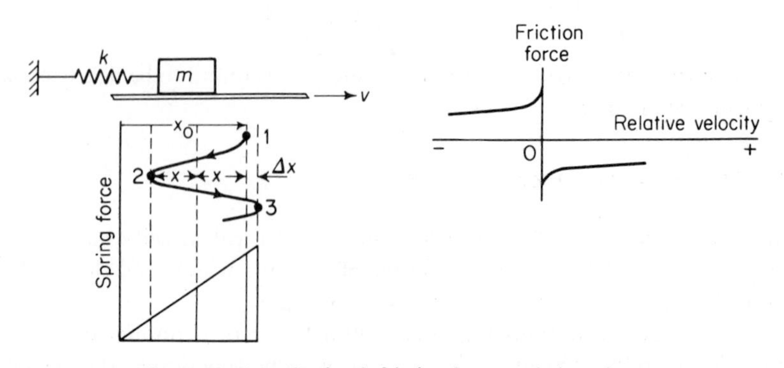

Figure 11.11-2. Coulomb friction between belt and mass.

$$kx_0 = \mu_s mg \tag{a}$$

At this point the mass will start to move back to the left, and the forces will again be balanced on the basis of kinetic friction when

$$k(x_0 - x) = \mu_{kl} mg$$

From these two equations, the amplitude of oscillation is found to be

$$x = x_0 - \mu_{kl}\frac{mg}{k} = \frac{(\mu_s - \mu_{kl})g}{\omega_n^2} \tag{b}$$

While the mass is moving to the left, the relative velocity between it and the belt is greater than when it is moving to the right, thus μ_{kl} is less than μ_{kr} where the subscripts l and r refer to left and right. It is evident then that the work done by the friction force while moving to the right is greater than that while moving to the left, so that more energy is put into the spring-mass system than taken out. This then represents one type of self-excited oscillation and the amplitude will continue to increase.

The work done by the spring from 2 to 3 is

$$-\tfrac{1}{2}k[(x_0 + \Delta x) + (x_0 - 2x)](2x + \Delta x)$$

The work done by friction from 2 to 3 is

$$\mu_{kr} mg(2x + \Delta x)$$

Equating the net work done between 2 and 3 to the change in kinetic energy which is zero,

$$-\tfrac{1}{2}k(2x_0 - 2x + \Delta x) + \mu_{kr} mg = 0 \tag{c}$$

Substituting (a) and (b) into (c), the increase in amplitude per cycle of oscillation is found to be

$$\Delta x = \frac{2g(\mu_{kr} - \mu_{kl})}{\omega_n^2} \tag{d}$$

11.12 ANALOG COMPUTER CIRCUITS FOR NONLINEAR SYSTEMS

Many nonlinear systems can be studied by the use of the electronic analog computer. Presented in this section are some of the circuit diagrams associated with nonlinear systems with a brief discussion as to their working principles.

System with Hysteresis Damping. Figure 11.12-1 shows a typical variation for the spring force leading to hysteresis damping, together with an integrator circuit proposed by T. K. Caughey* that limits the output voltage by means of diodes. The circuit works in the following manner. Assuming the voltage across the capacitor C to be initially zero, we apply a positive velocity $\dot{x}$ to

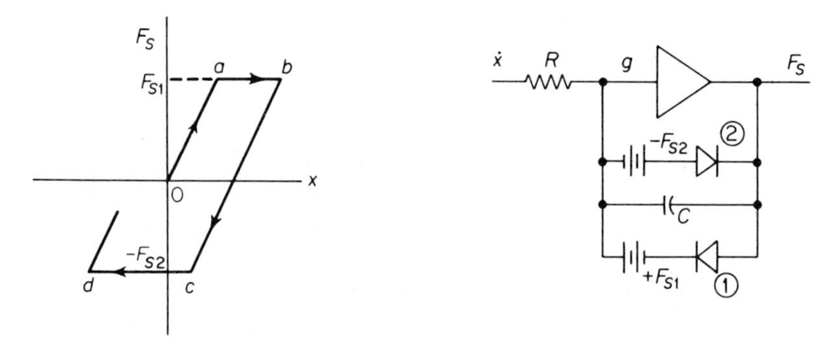

Figure 11.12-1. Bilinear-hysteresis.

its input. The circuit being that of an integrator, the output voltage begins to build up according to the equation

$$F_s = \frac{1}{RC}\int_0^t \dot{x}\,dt = kx$$

Noting that the potential of the grid g is essentially zero, this voltage appears across C. During this time diode ② cannot conduct and hence circuit ② appears as open. Diode ① can conduct only when its voltage from the output side exceeds the bias voltage $+F_{s1}$ which is set to the level indicated by a in the force displacement curve. When diode ① conducts, the voltage across C is limited to $+F_{s1}$ until $\dot{x}$ becomes negative at b, at which time diode ① becomes nonconducting and the circuit appears as if only C is present across the amplifier. When the output voltage reaches some negative value set by $-F_{s2}$ at c, the diode ② conducts and limits the negative voltage on C until again the velocity $\dot{x}$ becomes positive at d.

Figure 11.12-2 shows how the limited integrator is incorporated into the circuit to solve the equation

$$\frac{d^2x}{d\tau^2} + F_s = F(t) \tag{11.12-1}$$

By inserting an additional capacitor C_2 (shown dotted), it is possible to give the line ab and cd of the stiffness curve a slope other than zero.

*See Ref. 4.

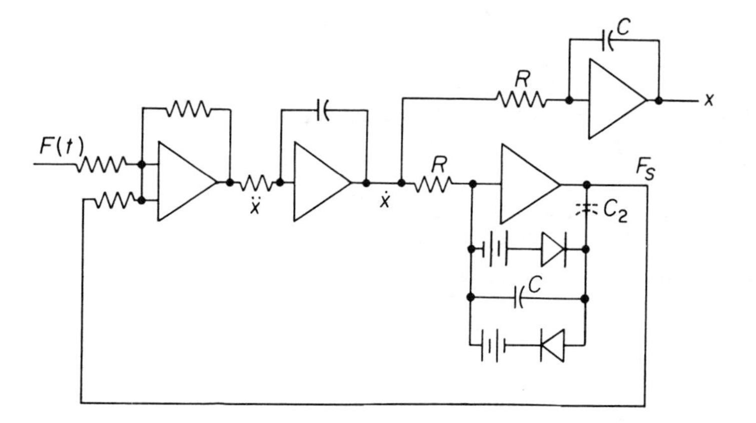

Figure 11.12-2. Circuit for Equation 11.12-1.

11.13 THE RUNGE-KUTTA METHOD

The Runge-Kutta method discussed in Chapters 4 and 8 can be used to solve nonlinear differential equations. We will consider the nonlinear equation

$$\frac{d^2x}{d\tau^2} + 0.4\frac{dx}{d\tau} + x + 0.5x^3 = 0.5 \cos (0.5\tau) \qquad (11.13\text{-}1)$$

and rewrite it in the first order form by letting $y = dx/d\tau$ as follows

$$\frac{dy}{d\tau} = 0.5 \cos (0.5\tau) - x - 0.5x^3 - 0.4y = F(\tau, x, y)$$

The computational equations to be used are programmed for the digital computer in the following order. From these results the values of x and y

τ	x	y	F
$t_1 = \tau_1$	$k_1 = x_1$	$g_1 = y_1$	$f_1 = F(t_1, k_1, g_1)$
$t_2 = \tau_1 + h/2$	$k_2 = x_1 + g_1h/2$	$g_2 = y_1 + f_1h/2$	$f_2 = F(t_2, k_2, g_2)$
$t_3 = \tau_1 + h/2$	$k_3 = x_1 + g_2h/2$	$g_3 = y_1 + f_2h/2$	$f_3 = F(t_3, k_3, g_3)$
$t_4 = \tau_1 + h$	$k_4 = x_1 + g_3h$	$g_4 = y_1 + f_3h$	$f_4 = F(t_4, k_4, g_4)$

are determined from the following recurrence equations, where $h = \Delta t$.

$$x_{i+1} = x_i + \frac{h}{6}(g_1 + 2g_2 + 2g_3 + g_4) \qquad (11.13\text{-}2)$$

$$y_{i+1} = y_i + \frac{h}{6}(f_1 + 2f_2 + 2f_3 + f_4) \qquad (11.13\text{-}3)$$

Thus with $i = 1$, x_2 and y_2 are found, and with $\tau_2 = \tau_1 + \Delta\tau$, the previous table of t, k, g, and f is computed and again substituted into the recurrence equations to find x_3 and y_3.

The error in the Runge-Kutta method is of order $h^5 = (\Delta\tau)^5$. Also, the method avoids the necessity of calculating derivatives and hence excellent accuracy is obtained.

Equation (11.13-1) was solved on the digital computer with the Runge-Kutta program and with $h = \Delta\tau = 0.1333$. The results were plotted out by the machine for the phase plane plot y vs. x in Fig. 11.13-1. It is evident that the limit cycle was reached in less than two cycles.

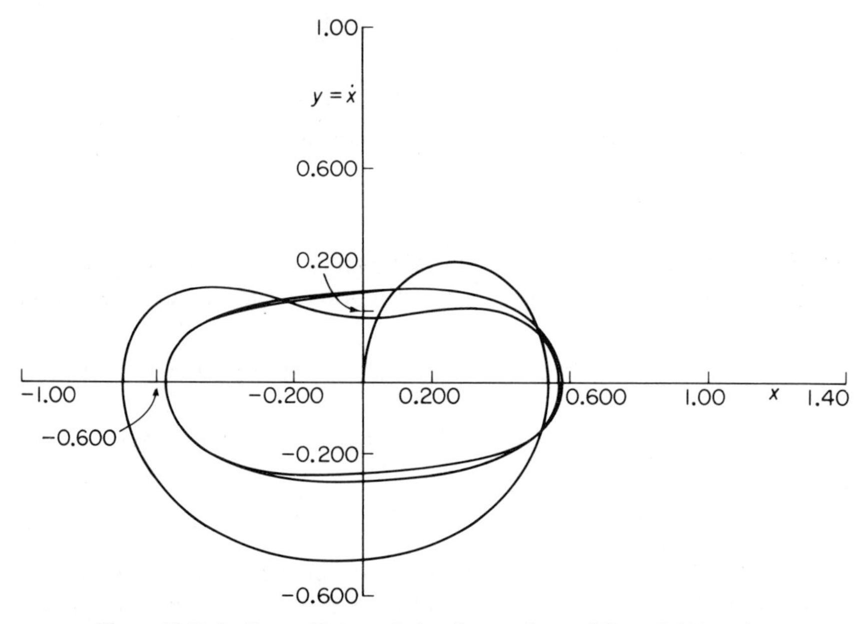

Figure 11.13-1. Runge-Kutta solution for nonlinear differential Equation 11.13-1.

In contrast to the Runge-Kutta method, the problem was also tried on the simple numerical program of Sec. 4.7, and Fig. 11.13-2 shows that the computation becomes unstable. In the numerical program of Sec. 4.7 the error is of order $h^3 = (\Delta\tau)^3$, which proved to be inadequate for the problem at hand.

Using the digital computer, the van der Pol equation

$$\ddot{x} - \mu\dot{x}(1 - x^2) + x = 0$$

was solved by the Runge-Kutta method for $\mu = 0.2, 0.7, 1.5, 3$ and 4 with a small initial displacement. Both the phase plane and the time plots were automatically plotted.

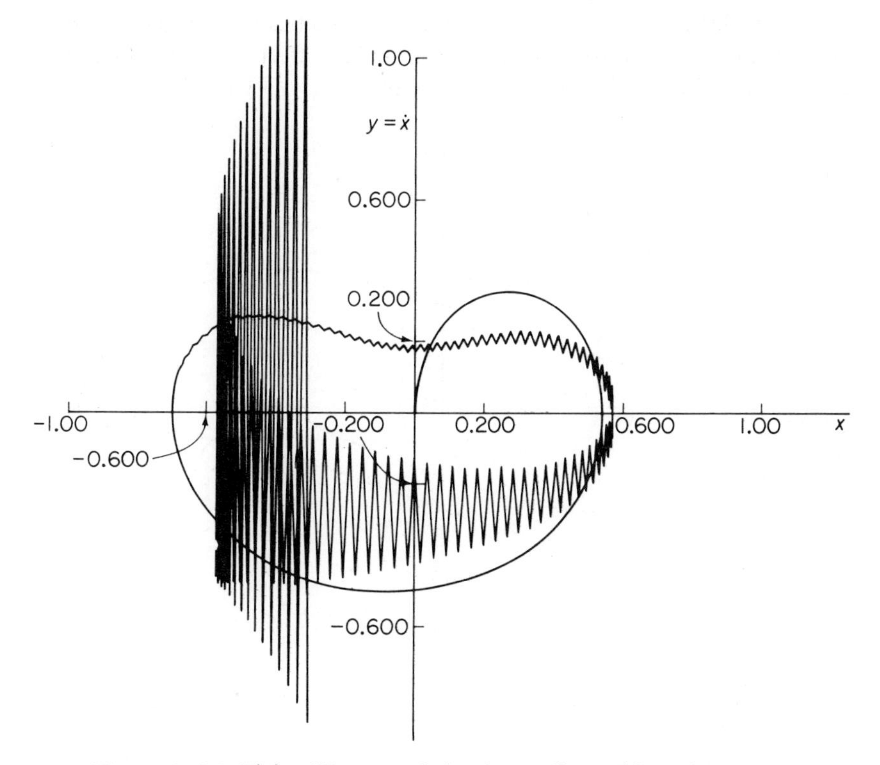

Figure 11.13-2. Finite difference solution for nonlinear differential Equation 11.13-1.

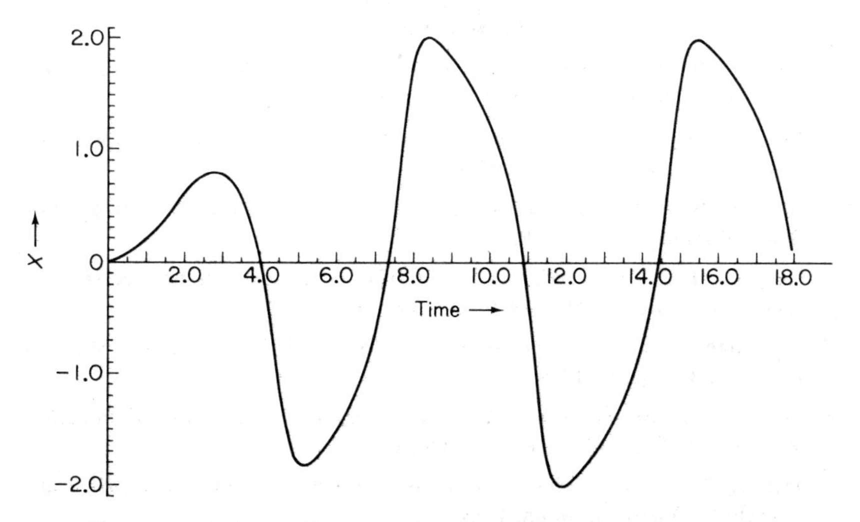

Figure 11.13-3. Runge-Kutta solution of van der Pol's Equation with $\mu = 1.5$.

For the case $\mu = 0.2$ the response is practically sinusoidal and the phase plane plot is nearly an elliptic spiral. The effect of the nonlinearity is quite evident for $\mu = 1.5$ which is shown in Fig. 11.13-3 and 11.13-4.

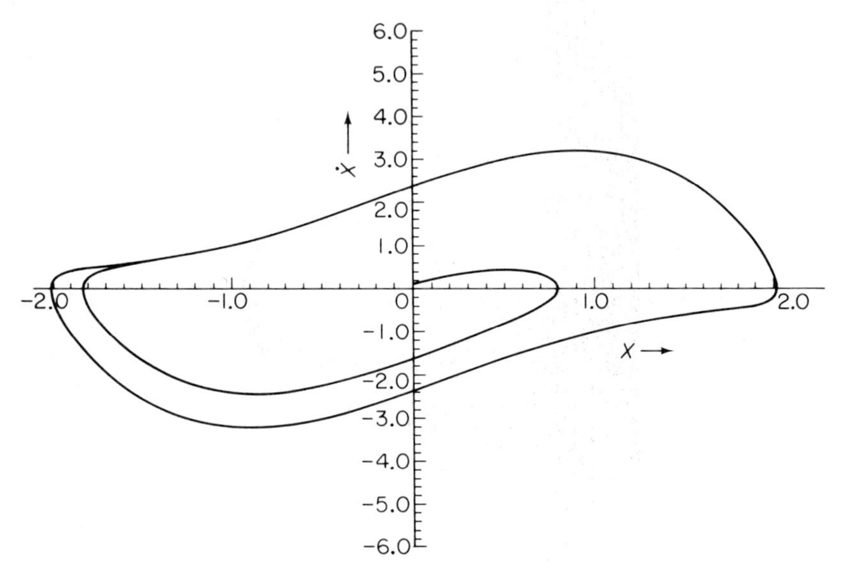

Figure 11.13-4. Runge-Kutta solution of van der Pol's Equation with $\mu = 1.5$.

REFERENCES

1. Bellman, R. *Perturbation Techniques in Mathematics, Physics and Engineering.* New York: Holt, Rinehart & Winston, Inc., 1964.
2. Brock, J. E. "An Iterative Numerical Method for Nonlinear Vibrations." *Jour. Appl'd. Mech.*, (March 1951), pp. 1–11.
3. Butenin, N. V. *Elements of the Theory of Nonlinear Oscillations.* New York: Blaisdell Publishing Co., 1965.
4. Caughey, T. K. "Sinusoidal Excitation of a System with Bilinear Hysteresis." *Jour. Appl'd. Mech.*, (Dec. 1960), pp. 640–43.
5. Cunningham, W. J. *Introduction to Nonlinear Analysis.* New York: McGraw-Hill Book Company, 1958.
6. Davis, H. T. *Introduction to Nonlinear Differential and Integral Equations.* Washington, D.C.: Govt. Printing Office, 1956.
7. Duffing, G. *Erwugene Schwingungen bei veranderlicher Eigenfrequenz.* Braunschweig: F. Vieweg u. Sohn, 1918.

8. Hayashi, C. *Forced Oscillations in Nonlinear Systems.* Osaka, Japan: Nippon Printing & Publishing Co., Ltd., 1953.
9. Jacobsen, L. S. "On a General Method of Solving Second Order Ordinary Differential Equations by Phase Plane Displacements." *Jour. Appl'd. Mech.*, (Dec. 1952), pp. 543–53.
10. Malkin, I. G. *Some Problems in the Theory of Nonlinear Oscillations*, Books I and II. Washington, D.C.: Dept. of Commerce, 1959.
11. Minorsky, N. *Nonlinear Oscillations.* Princeton, N.J.: D. Van Nostrand Co., Inc., 1962.
12. Nishikawa, Y. *A Contribution of the Theory of Nonlinear Oscillations.* Osaka, Japan: Nippon Printing & Publishing Co. Ltd., 1964.
13. Rauscher, M. "Steady Oscillations of Systems with Nonlinear and Unsymmetrical Elasticity." *Jour. Appl'd. Mech.*, (Dec. 1938), pp. A169–77.
14. Stoker, J. J. *Nonlinear Vibrations.* New York: Interscience Publishers, Inc., 1950.

PROBLEMS

11-1 Using the nonlinear equation

$$x + x^3 = 0$$

show that if $x_1 = \varphi_1(t)$ and $x_2 = \varphi_2(t)$ are solutions satisfying the differential equation, their superposition $(x_1 + x_2)$ is not a solution.

11-2 A mass is attached to the midpoint of a string of length $2l$ as shown in Fig. P11-2. Determine the differential equation of motion for large deflection. Assume string tension to be T.

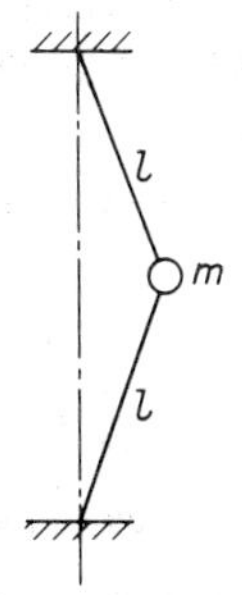

Figure P. 11-2.

11-3 A buoy is composed of two cones of diameter $2r$ and height h as shown in Fig. P11-3. A weight attached to the bottom allows it to float in the equilibrium position x_0. Establish the differential equation of motion for vertical oscillation.

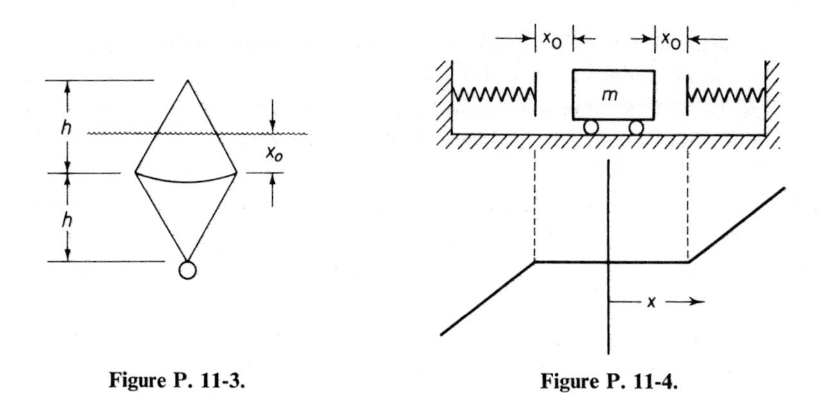

Figure P. 11-3. **Figure P. 11-4.**

11-4 Determine the differential equation of motion for the spring-mass system with the discontinuous stiffness resulting from the free gaps of Fig. P11-4.

11-5 The cord of a simple pendulum is wrapped around a fixed cylinder of radius R such that its length is l when in the vertical position as shown in Fig. P11-5. Determine the differential equation of motion.

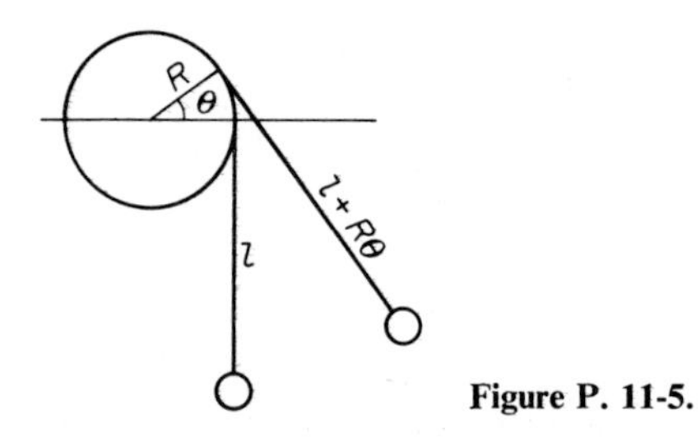

Figure P. 11-5.

11-6 Plot the phase plane trajectory for the undamped spring-mass system including the potential energy curve $U(x)$. Discuss the initial conditions associated with the plot.

11-7 From the plot of $U(x)$ vs. x of Prob. 11-6, determine the period from the equation

$$\tau = 4 \int_0^{x_{max}} \frac{dx}{\sqrt{2[E - U(x)]}}$$

(Remember that E in the text was for a unit mass.)

11-8 For the undamped spring-mass system with initial conditions $x(0) = A$ and $\dot{x}(0) = 0$ determine the equation for the state speed V and state under what condition the system is in equilibrium.

11-9 The solution to a certain linear differential equation is given as

$$x = \cos \pi t + \sin 2\pi t$$

Determine $y = \dot{x}$ and plot a phase plane diagram.

11-10 Determine the phase plane equation for the damped spring-mass system

$$\ddot{x} + 2\zeta\omega_n\dot{x} + \omega_n^2 x = 0$$

and plot one of the trajectories with $v = y/\omega_n$ and x as coordinates.

11-11 If the potential energy of a simple pendulum is given with the positive sign

$$U(\theta) = +\frac{g}{l}\cos\theta$$

determine which of the singular points are stable or unstable and explain their physical implications. Compare the phase plane with Fig. 11.5-2.

11-12 Given the potential $U(x) = 8 - 2\cos \pi x/4$, plot the phase plane trajectories for $E = 6, 7, 8, 10, 12$, and discuss the curves.

11.13 Determine the eigenvalues and eigenvectors of the equations

$$\dot{x} = 5x - y$$
$$\dot{y} = 2x + 2y$$

11-14 Determine the modal transformation of the equations of Prob. 11.13 which will decouple them to the form

$$\dot{\xi} = \lambda_1 \xi$$
$$\dot{\eta} = \lambda_2 \eta$$

11-15 Plot the ξ, η phase plane trajectories of Prob. 11.14 for $\lambda_1/\lambda_2 = 0.5$ and 2.0.

11.16 For $\lambda_1/\lambda_2 = 2.0$ in Prob. 11.15, plot the trajectory y vs. x.

11.17 If λ_1 and λ_2 of Prob. 11.14 are complex conjugates $-\alpha \pm i\beta$, show that the equation in the u, v plane becomes

$$\frac{dv}{du} = \frac{\beta u + \alpha v}{\alpha u - \beta v}$$

11-18 Using the transformation $u = \rho\cos\theta$ and $v = \rho\sin\theta$ show that the phase plane equation for Prob. 11.17 becomes

$$\frac{d\rho}{\rho} = \frac{\alpha}{\beta}\,d\theta$$

with the trajectories identified as logarithmic spirals

$$\rho = e^{(\alpha/\beta)\theta}$$

11-19 Near a singular point in the x, y plane, the trajectories appear as shown in Fig. P11-19.

Determine the form of the phase plane equation and the corresponding trajectories in the $\xi - \eta$ plane.

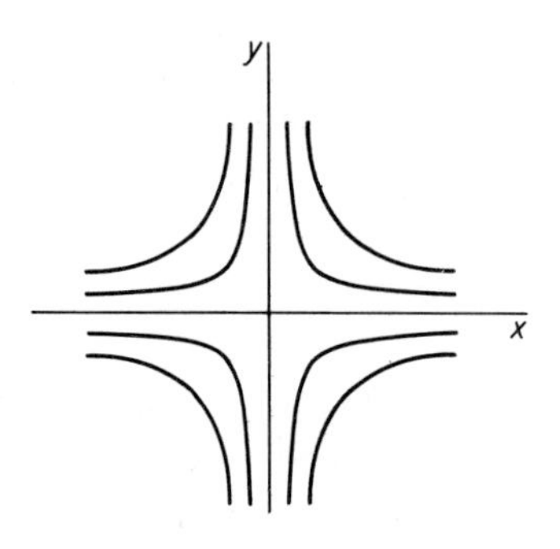

Figure P. 11-19.

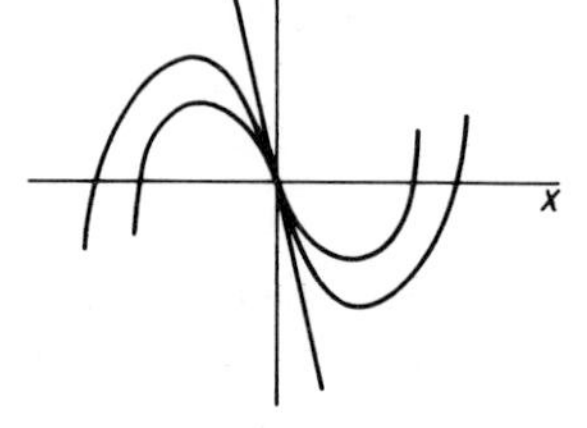

Figure P. 11-20.

11-20 The phase plane trajectories in the vicinity of a singularity of an overdamped system ($\zeta > 1$) is shown in Fig. P11-20.
Identify the phase plane equation and plot the corresponding trajectories in the $\xi - \eta$ plane.

11-21 Show that the solution of the equation

$$\frac{dy}{dx} = \frac{-x-y}{x+3y}$$

is $x^2 + 2xy + 3y^2 = C$, which is a family of ellipses with axes rotated from the x, y coordinates. Determine the rotation of the semimajor axis and plot one of the ellipses.

11-22 Show that the isoclines of the linear differential equation of second order are straight lines.

11-23 Draw the isoclines for the equation

$$\frac{dy}{dx} = xy(y-2)$$

11-24 Consider the nonlinear equation

$$\ddot{x} + \omega_n^2 x + \mu x^3 = 0$$

Replacing $\ddot{x}$ by $y(dy/dx)$ where $y = \dot{x}$, its integral becomes

$$y^2 + \omega_n^2 x^2 + \tfrac{1}{2}\mu x^4 = 2E$$

With $y = 0$ when $x = A$, show that the period is available from

$$\tau = 4\int_0^A \frac{dx}{\sqrt{2[E - U(x)]}}$$

11-25 What do the isoclines of Prob. 11-24 look like?

11-26 Plot the isoclines of the van der Pol's equation

$$\ddot{x} - \mu\dot{x}(1 - x^2) + x = 0$$

for $\mu = 2.0$ and $dy/dx = 0, -1$ and $+1$

11-27 The equation for the free oscillation of a damped system with hardening spring is given as

$$m\ddot{x} + c\dot{x} + kx + \mu x^3 = 0$$

Express this equation in the phase plane form for the delta method.

11-28 The following numerical values are given for the equation in Prob. 11.27

$$\omega_n^2 = \frac{k}{m} = 25, \qquad \frac{c}{m} = 2\zeta\omega_n = 2.0, \qquad \frac{\mu}{m} = 5$$

Plot the phase trajectory for the initial conditions $x(0) = 4.0$, $\dot{x}(0) = 0$, using the delta method.

11.29 Using the delta method, plot the phase plane trajectory for the simple pendulum with the initial conditions $\theta(0) = 60°$ and $\dot{\theta}(0) = 0$.

11-30 Determine the period of the pendulum of Prob. 11-29 and compare with that of the linear system.

11-31 The equation of motion for a spring-mass system with constant Coulomb damping can be written as

$$\ddot{x} + \omega_n^2 x + C\, sgn\,(\dot{x}) = 0$$

where $sgn\,(\dot{x})$ signifies either a positive or negative sign equal to that of the sign of $\dot{x}$. Express this equation in a form suitable for Lienard's method.

11-32 A system with Coulomb damping has the following numerical values: $k = 3.60$ lb/in., $m = 0.10$ lb sec² in.⁻¹ $\mu = 0.20$. Using Lienard's method, plot the trajectory for $x(0) = 20''$, $\dot{x}(0) = 0$.

11-33 Discuss the use of Lineard's method for a spring-mass system with viscous damping.

11-34 Discuss the similarity of the delta method and the Lienard's method for a spring-mass system with Coulomb damping.

11-35 Consider the motion of a simple pendulum with viscous damping and determine the singular points. With the aid of Fig. 11.5-2, and the knowledge that the trajectories must spiral into the origin, sketch some approximate trajectories.

11-36 Solve the problem of the spring-mass system with Coulomb damping by the slope-line method.

11-37 Van der Pol's equation may be written in terms of two first order equations as given Eq. (11.8-8), where

$$g(x) = \mu(-x + \tfrac{1}{3}x^3)$$
$$h(y) = 0$$

Determine the solution starting from $x = 0$, $y = -0.05$ using the slope-line method. A Δt of 0.20 is suggested.

11-38 If in the slope-line method the time increment Δt is kept constant, the displacement-time curve can be easily drawn. Draw the x vs. t curve from the graphical solution of Prob. 11-37.

11-39 Show that when $g(x) = c$ and $h(y) = ky$ in Eq. (11.8-8) where c and k are constants, the equivalent second order equation is

$$\ddot{x} + k\dot{x} + x + kc = 0$$

11-40 Apply the perturbation method to the simple pendulum with $\sin\theta$ replaced by $\theta - \frac{1}{6}\theta^3$. Use only the first two terms of the series for x and ω.

11-41 From the perturbation method, what is the equation for the period of the simple pendulum as a function of the amplitude.

11-42 For a given system the numerical values of Eq. (11.10-7) are given as

$$\ddot{x} + 0.15\dot{x} + 10x + x^3 = 5\cos(\omega t + \varphi)$$

Plot A vs. ω from Eq. (11.10-11) by first assuming a value of A and solving for ω^2.

11-43 Determine the phase angle φ vs. ω for Prob. 11-42.

11-44 The supporting end of a simple pendulum is given a motion as shown in Fig. P11-44. Show that the equation of motion is

$$\ddot{\theta} + \left(\frac{g}{l} - \frac{\omega^2 y_0}{l}\cos 2\omega t\right)\sin\theta = 0$$

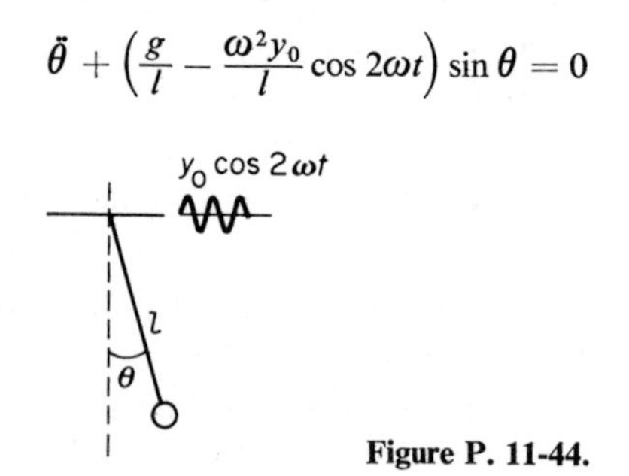

Figure P. 11-44.

11-45 For a given value of g/l, determine the frequencies of the excitation for which the simple pendulum of Prob. 11.44 with a stiff arm 1 will be stable in the inverted position.

11-46 Determine the perturbation solution for the system shown in Fig. P11-46, leading to a Mathieu equation. Use initial conditions $\dot{x}(0) = 0$, $x(0) = A$.

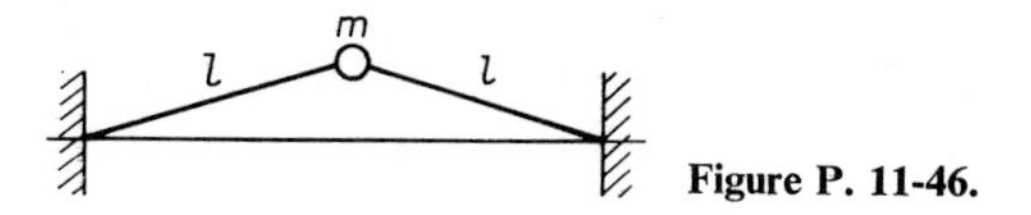

Figure P. 11-46.

11-47 A circuit which simulates a dead zone in the spring stiffness is shown in Fig. P11-47. Complete the analogue circuit to solve Prob. 11-4.

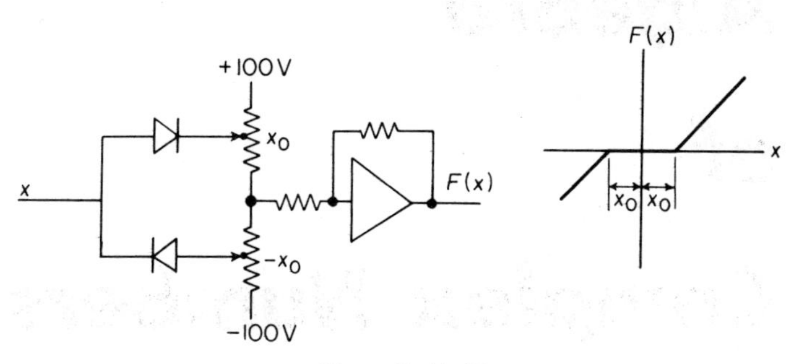

Figure P. 11-47.

Algebra of Complex Numbers

A complex number $z = a + ib$ can be displayed graphically by the *Argand diagram* of Fig. A-1, where the horizontal axis represents the real component and the vertical axis the imaginary component. The magnitude $A = \sqrt{a^2 + b^2}$ is called the *modulus* of z and the angle θ is called the *argument* of z.

The complex quantity z can also be expressed in polar form by *Euler's equation*

$$e^{i\theta} = \cos\theta + i\sin\theta \tag{1}$$

Thus another expression for z is

$$z = Ae^{i\theta} = (A\cos\theta) + i(A\sin\theta) \tag{2}$$

Of particular interest is the case where the modulus A is constant and the argument θ is linearly proportional to the time t; namely, $\theta = \omega t$, where ω is the angular rate which is also constant. The complex quantity z is then

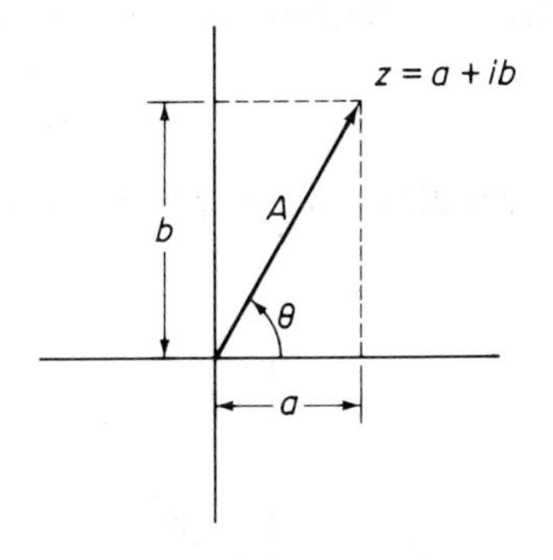

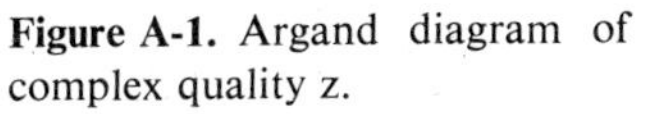

Figure A-1. Argand diagram of complex quality z.

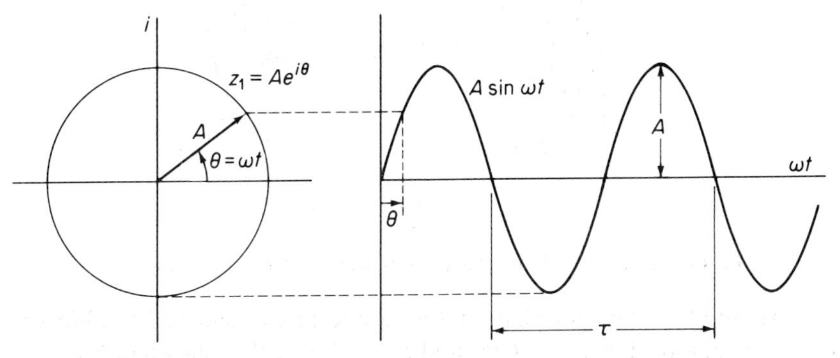

Figure A-2. Harmonic motion as component of a phasor.

called a *phasor*. Fig. A-2 shows a phasor

$$z_1 = Ae^{i\theta} - (A\cos\theta) + i(A\sin\theta) \tag{3}$$

or a point moving along a circle of radius A at a constant angular rate ω. It is evident that its components must vary sinusoidally when plotted against time, or

$$A\cos\theta = Re\,z \tag{4}$$

$$iA\sin\theta = Im\,z \tag{5}$$

A phasor may also rotate in a negative direction, in which case the exponent is negative

$$z_2 = Ae^{-i\theta} = (A\cos\theta) - i(A\sin\theta) \tag{6}$$

Thus z_2 is the complex conjugate of z_1, or $z_2 = z_1^*$. By adding z_1 and z_1^* and dividing by 2, we obtain the real component

$$Re\,z = \tfrac{1}{2}(z_1 + z_1^*) = A\cos\theta \tag{7}$$

Likewise, the imaginary component is obtained by half the difference

$$Im\, z = \tfrac{1}{2}(z_1 - z_1^*) = iA \sin\theta \tag{8}$$

These operations are graphically displayed in Fig. A-3.

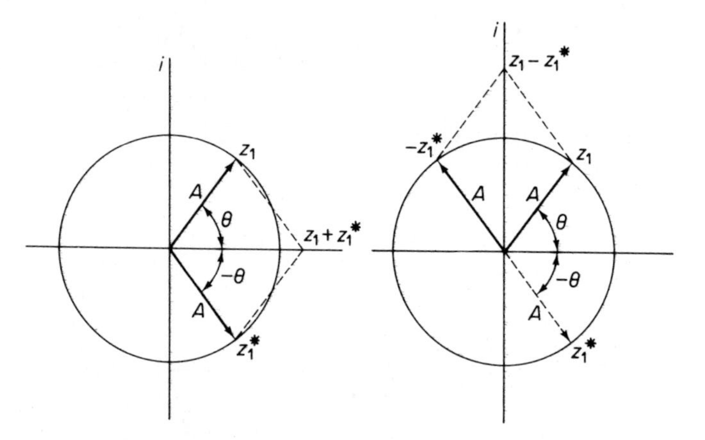

Figure A-3. Real and imaginary components in terms of z_1 and z_2*.

To add two different phasors, the second phasor should be referenced to the first phasor. Let $z_1 = A_1 e^{i\omega t}$ and $z_2 = A_2 e^{i(\omega t + \phi)}$ as shown in Fig. A-4. Resolve z_2 into components parallel and perpendicular to z_1 as follows

$$z_2 = A_2 e^{i\phi} e^{i\omega t} = (A_2 \cos\phi + iA_2 \sin\phi) e^{i\omega t} \tag{9}$$

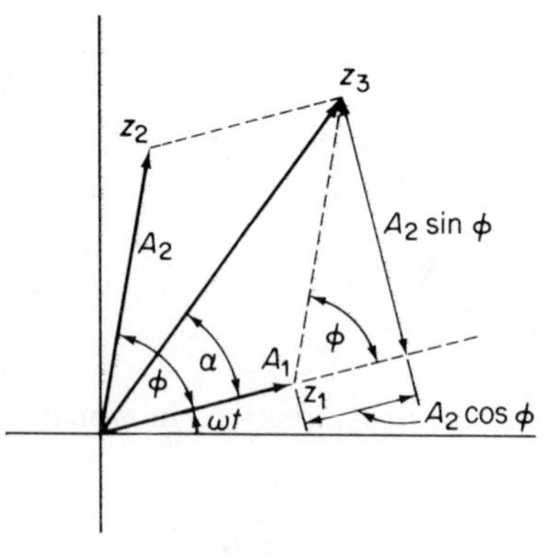

Figure A-4. Addition of two phasors.

Adding to z_1, we obtain

$$\begin{aligned} z_1 + z_2 &= [(A_1 + A_2 \cos\phi) + i(A_2 \sin\phi)] e^{i\omega t} \\ &= A_3 e^{i(\omega t + \alpha)} = (A_3 e^{i\alpha}) e^{i\omega t} \end{aligned} \tag{10}$$

where

$$A_3 = \sqrt{(A_1 + A_2 \cos \phi)^2 + (A_2 \sin \phi)^2}$$

$$\tan \alpha = \frac{A_2 \sin \phi}{A_1 + A_2 \cos \phi}$$

When two phasors of different frequencies are added, one phasor will move relative to the other with angular speed $(\omega_1 - \omega_2)$. Thus the resultant of the two phasors

$$A_1 e^{i\omega_1 t} + A_2 e^{i\omega_2 t}$$

will vary in amplitude between $A_1 + A_2$ and $A_1 - A_2$.

Other operations of phasors result from the rules of multiplication, division, and powers of complex numbers in the exponential form. They are summarized as follows

$$z_1 = A_1 e^{i\theta_1}$$

$$z_2 = A_2 e^{i\theta_2}$$

Multiplication

$$z_1 z_2 = A_1 A_2 e^{i(\theta_1 + \theta_2)} \tag{11}$$

Division

$$\frac{z_1}{z_2} = \frac{A_1}{A_2} e^{i(\theta_1 - \theta_2)} \tag{12}$$

Powers

$$(z_1)^n = A_1^n e^{in\theta} \tag{13}$$

$$z_1^{1/n} = A_1^{1/n} e^{i\theta/n} \tag{14}$$

Introduction to Laplace Transformation

Definition

If $f(t)$ is a known function of t for values of $t > 0$, its Laplace transform $\bar{f}(s)$ is defined by the equation

$$\bar{f}(s) = \int_0^\infty e^{-st} f(t)\, dt = \mathcal{L} f(t) \tag{1}$$

where s is a complex variable. The integral exists for the real part of $s > 0$ provided $f(t)$ is an absolutely integrable function of t in the time interval 0 to ∞.

Example 1. Let $f(t)$ be a constant c for $t > 0$. Its L.T. is

$$\mathcal{L}c = \int_0^\infty c e^{-st}\, dt = -\frac{ce^{-st}}{s}\Bigg]_0^\infty = \frac{c}{s}$$

which exists for $R(s) > 0$.

Example 2. Let $f(t) = t$. Its L.T. is found by integration by parts, letting

$$u = t \qquad du = dt$$

$$dv = e^{-st}\,dt \qquad v = -\frac{e^{-st}}{s}$$

The result is

$$\mathcal{L}t = -\frac{te^{-st}}{s}\bigg]_0^\infty + \frac{1}{s}\int_0^\infty e^{-st}\,dt = \frac{1}{s^2} \qquad R(s) > 0$$

Laplace Transform of Derivatives. If $\mathcal{L}f(t) = \bar{f}(s)$ exists, where $f(t)$ is continuous, then $f(t)$ tends to $f(0)$ as $t \longrightarrow 0$ and the L.T. of its derivative $f'(t) = df(t)/dt$ is equal to

$$\mathcal{L}f'(t) = s\bar{f}(s) - f(0) \tag{2}$$

The above relation is found by integration by parts

$$\int_0^\infty e^{-st}f'(t)\,dt = e^{-st}f(t)\bigg]_0^\infty + \frac{1}{s}\int_0^s e^{-st}f(t)\,dt$$
$$= -f(0) + s\bar{f}(s)$$

Similarly the L.T. of the second derivative can be shown to be

$$\mathcal{L}f''(t) = s^2\bar{f}(s) - sf(0) - f'(0) \tag{3}$$

Shifting Theorem

Consider the L.T. of the function $e^{at}x(t)$.

$$\mathcal{L}e^{at}x(t) = \int_0^\infty e^{-st}[e^{at}x(t)]\,dt = \int_0^\infty e^{-(s-a)t}x(t)\,dt$$

We conclude from this expression that

$$\mathcal{L}e^{at}x(t) = \bar{x}(s - a) \tag{4}$$

where $\mathcal{L}x(t) = \bar{x}(s)$. Thus, the multiplication of $x(t)$ by e^{at} shifts the transform by a, where a may be any number, real or complex.

Transformation of Ordinary Differential Equations

Consider the differential equation

$$m\ddot{x} + c\dot{x} + kx = F(t) \tag{5}$$

Its L.T. is

$$m[s^2\bar{x}(s) - sx(0) - \dot{x}(0)] + c[s\bar{x}(s) - x(0)] + k\bar{x}(s) = \bar{F}(s)$$

which can be rearranged to

$$\bar{x}(s) = \frac{\bar{F}(s)}{ms^2 + cs + k} + \frac{(ms + c)x(0) + m\dot{x}(0)}{ms^2 + cs + k} \tag{6}$$

The above equation is called the subsidiary equation of the differential equation. The response $x(t)$ is found from the inverse transformation, the first term representing the forced response and the second term the response due to the initial conditions.

For the more general case, the subsidiary equation can be written in the form

$$\bar{x}(s) = \frac{A(s)}{B(s)} \tag{7}$$

where $A(s)$ and $B(s)$ are polynomials. $B(s)$ is in general of higher order than $A(s)$.

Transforms Having Simple Poles

Considering the subsidiary equation

$$\bar{x}(s) = \frac{A(s)}{B(s)}$$

we examine the case where $B(s)$ is factorable in terms of n roots a_k which are distinct (simple poles).

$$B(s) = (s - a_1)(s - a_2) \cdots (s - a_n)$$

The subsidiary equation can then be expanded in the following partial fractions

$$\bar{x}(s) = \frac{A(s)}{B(s)} = \frac{C_1}{s - a_1} + \frac{C_2}{s - a_2} + \cdots + \frac{C_n}{s - a_n} \tag{8}$$

To determine the constants C_k, we multiply both sides of the above equation by $(s - a_k)$ and let $s = a_k$. Every term on the right will then be zero except C_k and we arrive at the result

$$C_k = \lim_{s \to a_k} (s - a_k)\frac{A(s)}{B(s)} \tag{9}$$

Since $\mathcal{L}^{-1}C_k/(s - a_k) = C_k e^{a_k t}$ the inverse transform of $\bar{x}(s)$ becomes

$$x(t) = \sum_{k=1}^{n} \lim_{s \to a_k} (s - a_k)\frac{A(s)}{B(s)} e^{a_k t} \tag{10}$$

Another expression for the above equation becomes apparent by noting that

$$B(s) = (s - a_k)B_1(s)$$
$$B'(s) = (s - a_k)B_1'(s) + B_1(s)$$
$$\lim_{s \to a_k} B'(s) = B_1(a_k)$$

Since $(s - a_k)A(s)/B(s) = A(s)/B_1(s)$, it is evident that

$$x(t) = \sum_{k=1}^{n} \frac{A(a_k)}{B'(a_k)} e^{a_k t} \tag{11}$$

Transforms Having Poles of Higher Order

If in the subsidiary equation

$$\bar{x}(s) = \frac{A(s)}{B(s)}$$

a factor in $B(s)$ is repeated m times, we say that $\bar{x}(s)$ has an mth order pole. Assuming that there is an mth order pole at a_1, $B(s)$ will have the form

$$B(s) = (s - a_1)^m (s - a_2)(s - a_3) \text{———}$$

The partial fraction expansion of $\bar{x}(s)$ then becomes

$$\bar{x}(s) = \frac{C_{11}}{(s - a_1)^m} + \frac{C_{12}}{(s - a_1)^{m-1}} + \cdots$$
$$+ \frac{C_{1m}}{(s - a_1)} + \frac{C_2}{(s - a_2)} + \frac{C_3}{(s - a_3)} + \cdots \tag{12}$$

The coefficient C_{11} is determined by multiplying both sides of the equation by $(s - a_1)^m$ and letting $s = a_1$

$$(s - a_1)^m \bar{x}(s) = C_{11} + (s - a_1)C_{12} + \cdots$$
$$+ (s - a_1)^{m-1} C_{1m} + \frac{(s - a_1)^m}{(s - a_2)} C_2 + \cdots$$
$$\therefore \quad C_{11} = [(s - a_1)^m \bar{x}(s)]_{s=a_1} \tag{13}$$

The coefficient C_{12} is determined by differentiating the equation for $(s - a_1)^m \bar{x}(s)$ with respect to s and then letting $s = a_1$

$$C_{12} = \left[\frac{d}{ds}(s - a_1)^m \bar{x}(s)\right]_{s=a_1} \tag{14}$$

It is evident then that

$$C_{1n} = \frac{1}{(n-1)!}\left[\frac{d^{n-1}}{ds^{n-1}}(s - a_1)^m \bar{x}(s)\right]_{s=a_1} \tag{15}$$

The remaining coefficients C_2, C_3, etc., are evaluated as in the previous section for simple poles.

Since by the shifting theorem

$$\mathfrak{L}^{-1}\frac{1}{(s-a_1)^n} = \frac{t^{n-1}}{(n-1)!}a_1 t$$

the inverse transform of $\bar{x}(s)$ becomes

$$x(t) = \left[C_{11}\frac{t^{m-1}}{(m-1)!} + C_{12}\frac{t^{m-2}}{(m-2)!} + \cdots\right]e^{a_1 t} + C_2 e^{a_2 t} + C_3 e^{a_3 t} + \cdots \tag{16}$$

Most ordinary differential equations can be solved by the elementary theory of L.T. The tables here give the L.T. of simple functions. The table is also used to establish the inverse L.T., since if

$$\mathfrak{L}f(t) = \bar{f}(s)$$

then

$$f(t) = \mathfrak{L}^{-1}\bar{f}(s).$$

Short Table of Laplace Transforms

	$\bar{f}(s)$	$f(t)$
(1)	1	$\delta(t)$ = unit impulse at $t = 0$
(2)	$\frac{1}{s}$	$\mathfrak{U}(t)$ = unit step function at $t = 0$
(3)	$\frac{1}{s^n}(n = 1, 2, \cdots)$	$\frac{t^{n-1}}{(n-1)!}$
(4)	$\frac{1}{s+a}$	e^{-at}

Short Table of Laplace Transforms (Continued)

	$f(s)$	$f(t)$
(5)	$\frac{1}{(s+a)^2}$	te^{-at}
(6)	$\frac{1}{(s+a)^n} (n = 1, 2, \cdots)$	$\frac{1}{(n-1)!} t^{n-1} e^{-at}$
(7)	$\frac{1}{s(s+a)}$	$\frac{1}{a}(1 - e^{-at})$
(8)	$\frac{1}{s^2(s+a)}$	$\frac{1}{a^2}(e^{-at} + at - 1)$
(9)	$\frac{s}{s^2 + a^2}$	$\cos at$
(10)	$\frac{s}{s^2 - a^2}$	$\cosh at$
(11)	$\frac{1}{s^2 + a^2}$	$\frac{1}{a} \sin at$
(12)	$\frac{1}{s^2 - a^2}$	$\frac{1}{a} \sinh at$
(13)	$\frac{1}{s(s^2 + a^2)}$	$\frac{1}{a^2}(1 - \cos at)$
(14)	$\frac{1}{s^2(s^2 + a^2)}$	$\frac{1}{a^3}(at - \sin at)$
(15)	$\frac{1}{(s^2 + a^2)^2}$	$\frac{1}{2a^3}(\sin at - at \cos at)$
(16)	$\frac{s}{(s^2 + a^2)^2}$	$\frac{t}{2a} \sin at$
(17)	$\frac{s^2 - a^2}{(s^2 + a^2)^2}$	$t \cos at$
(18)	$\frac{1}{s^2 + 2\zeta\omega_0 s + \omega_0^2}$	$\frac{1}{\omega_0\sqrt{1 - \zeta^2}} e^{-\zeta\omega_0 t} \sin \omega_0\sqrt{1 - \zeta^2}t$

REFERENCE

Thomson, W. T. *Laplace Transformation*, 2nd Ed. Englewood Cliffs, N. J.: Prentice-Hall, Inc., 1960.

C

Determinant and Matrices

I DETERMINANT

A determinant of the second order and its numerical evaluation are defined by the following notation and operation

$$D = \begin{vmatrix} a & b \\ c & d \end{vmatrix} = ad - bc$$

An nth order determinant has n rows and n columns, and in order to identify the position of its elements, the following notation is adopted

$$\begin{vmatrix} a_{11} & a_{12} & a_{13} & \cdots & a_{1n} \\ a_{21} & a_{22} & a_{23} & \cdots & a_{2n} \\ \cdot & & & & \\ \cdot & & & & \\ \cdot & & & & \\ a_{n1} & a_{n2} & a_{n3} & \cdots & a_{nn} \end{vmatrix}$$

Minors

A minor M_{ij} of the element a_{ij} is a determinant formed by deleting the ith row and the jth column from the original determinant.

Cofactor

The cofactor C_{ij} of the element a_{ij} is defined by the equation

$$C_{ij} = (-1)^{i+j} M_{ij}$$

Example: Given the third order determinant

$$\begin{vmatrix} 2 & 1 & 5 \\ 4 & 2 & 1 \\ 2 & 0 & 3 \end{vmatrix}$$

The minor of the term $a_{21} = 4$ is

$$M_{21} \text{ of } \begin{vmatrix} 2 & 1 & 5 \\ 4 & 2 & 1 \\ 2 & 0 & 3 \end{vmatrix} = \begin{vmatrix} 1 & 5 \\ 0 & 3 \end{vmatrix} = 3$$

and its cofactor is

$$C_{21} = (-1)^{2+1} 3 = -3$$

Expansion of a Determinant

The order of a determinant can be reduced by one by expanding any row or column in terms of its cofactors.

Example: The determinant of the previous example is expanded in terms of the second row as

$$D = \begin{vmatrix} 2 & 1 & 5 \\ 4 & 2 & 1 \\ 2 & 0 & 3 \end{vmatrix} = 1(-1)^{1+2} \begin{vmatrix} 4 & 1 \\ 2 & 3 \end{vmatrix} + 2(-1)^{2+2} \begin{vmatrix} 2 & 5 \\ 2 & 3 \end{vmatrix}$$

$$+ 0(-1)^{3+2} \begin{vmatrix} 2 & 5 \\ 4 & 1 \end{vmatrix}$$

$$= -10 - 8 = -18$$

Properties of Determinants

The following properties of determinants are stated without proof.

(1) Interchange of any two columns or rows changes the sign of the determinant.

(2) If two rows or two columns are identical, the determinant is zero.

(3) Any row or column may be multiplied by a constant and added to another row or column without changing the value of the determinant.

II MATRICES

Matrix. A rectangular array of terms arranged in m rows and n columns is called a matrix. For example

$$A = \begin{bmatrix} a_{11} & a_{12} & a_{13} & a_{14} \\ a_{21} & a_{22} & a_{23} & a_{24} \\ a_{31} & a_{32} & a_{33} & a_{34} \end{bmatrix}$$

is a 3×4 matrix.

Square Matrix. A square matrix is one in which the number of rows is equal to the number of columns. It is referred to as an $n \times n$ matrix or a matrix of order n.

Symmetric Matrix. A square matrix is said to be symmetric if the elements on the upper right half can be obtained by flipping the matrix about the diagonal

$$A = \begin{bmatrix} 2 & 1 & 3 \\ 1 & 5 & 0 \\ 3 & 0 & 1 \end{bmatrix} = \text{symmetric matrix}$$

Trace. The sum of the diagonal elements of a square matrix is called the trace. For the matrix above

$$\text{Trace } A = 2 + 5 + 1 = 8$$

Singular Matrix. If the determinant of a matrix is zero, the matrix is said to be singular.

Row Matrix. A row matrix has $m = 1$.

$$B = [b_1 \; b_2 \; b_3]$$

Column Matrix. A column matrix has $n = 1$.

$$C = \begin{Bmatrix} C_1 \\ C_2 \\ C_3 \end{Bmatrix}$$

Zero Matrix. The zero matrix is defined as one in which all elements are zero.

$$0 = \begin{bmatrix} 0 & 0 & 0 \\ 0 & 0 & 0 \end{bmatrix}$$

Unit Matrix. The unit matrix

$$I = \begin{bmatrix} 1 & 0 & 0 \\ 0 & 1 & 0 \\ 0 & 0 & 1 \end{bmatrix}$$

is a square matrix in which the diagonal elements from the top right to the bottom right are unity with all other elements equal to zero.

Diagonal Matrix. A square matrix having elements a_{ii} along the diagonal with all other elements equal to zero is a diagonal matrix

$$[a_{ii}] = \begin{bmatrix} a_{11} & 0 & 0 \\ 0 & a_{22} & 0 \\ 0 & 0 & a_{33} \end{bmatrix}$$

Transpose. The transpose A' of a matrix A is one in which the rows and columns are interchanged. For example

$$A = \begin{bmatrix} a_{11} & a_{12} & a_{13} \\ a_{21} & a_{22} & a_{23} \end{bmatrix} \qquad A' = \begin{bmatrix} a_{11} & a_{21} \\ a_{12} & a_{22} \\ a_{13} & a_{23} \end{bmatrix}$$

The transpose of a column matrix is a row matrix

$$X = \begin{Bmatrix} x_1 \\ x_2 \\ x_3 \end{Bmatrix} \qquad X' = [x_1 x_2 x_3]$$

Minor. A minor M_{ij} of a matrix A is formed by deleting the ith row and the jth column from the determinant of the original matrix

$$\text{Let } A = \begin{bmatrix} a_{11} & a_{12} & a_{13} \\ a_{21} & a_{22} & a_{23} \\ a_{31} & a_{32} & a_{33} \end{bmatrix}$$

$$M_{12} = \begin{vmatrix} a_{11} & a_{12} & a_{13} \\ a_{21} & a_{22} & a_{23} \\ a_{31} & a_{32} & a_{33} \end{vmatrix} = \begin{vmatrix} a_{21} & a_{23} \\ a_{31} & a_{33} \end{vmatrix}$$

Cofactor. The cofactor C_{ij} is equal to the signed minor $(-1)^{i+j}M_{ij}$. From the previous example

$$C_{12} = (-1)^{1+2}M_{12} = -M_{12}$$

Adjoint Matrix. An adjoint matrix of a square matrix A is a transpose of the matrix of cofactors of A.

Let cofactor matrix of A be

$$[C_{ij}] = \begin{bmatrix} C_{11} & C_{12} & C_{13} \\ C_{21} & C_{22} & C_{23} \\ C_{31} & C_{32} & C_{33} \end{bmatrix}$$

$$adj\ A = [C_{ij}]' = [C_{ji}] = \begin{bmatrix} C_{11} & C_{21} & C_{31} \\ C_{12} & C_{22} & C_{32} \\ C_{13} & C_{23} & C_{33} \end{bmatrix}$$

Inverse Matrix. The inverse A^{-1} of a matrix A satisfies the relationship

$$A^{-1}A = AA^{-1} = I$$

Orthogonal Matrix. An orthogonal matrix A satisfies the relationship

$$A'A = AA' = I$$

From the definition of an inverse matrix it is evident that for an orthogonal matrix $A' = A^{-1}$.

Symmetric Matrix. A square matrix is said to be symmetric if the elements on the upper right half can be obtained by flipping the matrix about the diagonal

$$A = \begin{bmatrix} 2 & 1 & 3 \\ 1 & 5 & 0 \\ 3 & 0 & 1 \end{bmatrix} = \text{symmetric matrix}$$

Trace. The sum of the diagonal elements of a square matrix is called the trace. For the matrix above,

$$\text{Trace } A = 2 + 5 + 1 = 8$$

Singular Matrix. If the determinant of a matrix is zero, the matrix is said to be singular.

III RULES OF MATRIX OPERATIONS

Addition. Two matrices having the same number of rows and columns may be added by summing the corresponding elements.

Example:

$$\begin{bmatrix} 1 & 3 & 2 \\ 4 & 1 & 1 \end{bmatrix} + \begin{bmatrix} 2 & 0 & 4 \\ 1 & -2 & -3 \end{bmatrix} = \begin{bmatrix} 3 & 3 & 6 \\ 5 & -1 & -2 \end{bmatrix}$$

Multiplication. The product of two matrices A and B is another matrix C.

$$AB = C$$

The element C_{ij} of C is determined by multiplying the elements of the ith row in A by the elements of the jth column in B according to the rule

$$c_{ij} = \sum_k a_{ik} b_{kj}$$

Example:

$$\text{Let } A = \begin{bmatrix} 1 & 1 & 1 \\ 1 & 2 & 2 \\ 1 & 2 & 3 \end{bmatrix} \qquad B = \begin{bmatrix} 2 & 0 \\ 0 & 1 \\ 3 & -1 \end{bmatrix}$$

$$AB = \begin{bmatrix} 1 & 1 & 1 \\ 1 & 2 & 2 \\ 1 & 2 & 3 \end{bmatrix} \begin{bmatrix} 2 & 0 \\ 0 & 1 \\ 3 & -1 \end{bmatrix} = \begin{bmatrix} 5 & 0 \\ 8 & 0 \\ 11 & -1 \end{bmatrix} = C$$

ie., $$c_{21} = 1 \times 2 + 2 \times 0 + 2 \times 3 = 8$$

It is evident that the number of columns in A must equal the number of rows in B, or that the matrices must be conformable. We also note that $AB \neq BA$.

The post-multiplication of a matrix by a column matrix results in a column matrix

Example:

$$\begin{bmatrix} 1 & 1 & 1 \\ 1 & 5 & 2 \\ 2 & 1 & 3 \end{bmatrix} \begin{Bmatrix} 1 \\ 3 \\ 2 \end{Bmatrix} = \begin{Bmatrix} 6 \\ 20 \\ 11 \end{Bmatrix}$$

Pre-multiplication of a matrix by a row matrix (or transpose of a column matrix) results in a row matrix

Example:

$$[1 \quad 3 \quad 2]\begin{bmatrix}1 & 1 & 1\\ 1 & 5 & 2\\ 2 & 1 & 3\end{bmatrix} = [8 \quad 18 \quad 13]$$

The transpose of a product $AB = C$ is $C' = B'A'$

Example:

$$\text{Let } A = \begin{bmatrix}1 & 1\\ 2 & 3\end{bmatrix} \qquad B = \begin{bmatrix}2 & 1\\ 1 & 1\end{bmatrix}$$

$$C = AB = \begin{bmatrix}3 & 2\\ 7 & 5\end{bmatrix} \qquad C' = B'A' = \begin{bmatrix}2 & 1\\ 1 & 1\end{bmatrix}\begin{bmatrix}1 & 2\\ 1 & 3\end{bmatrix} = \begin{bmatrix}3 & 7\\ 2 & 5\end{bmatrix}$$

Inversion of a Matrix. Consider a set of equations

$$\begin{aligned} a_{11}x_1 + a_{12}x_2 + a_{13}x_3 &= y_1\\ a_{21}x_1 + a_{22}x_2 + a_{23}x_3 &= y_2\\ a_{31}x_1 + a_{32}x_2 + a_{33}x_3 &= y_3 \end{aligned} \tag{1}$$

which can be expressed in the matrix form

$$AX = Y \tag{2}$$

Pre-multiplying by the inverse A^{-1}, we obtain the solution

$$X = A^{-1}Y \tag{3}$$

We can identify the term A^{-1} by Cramer's rule as follows. The solution for x_1 is

$$\begin{aligned} x_1 &= \frac{1}{|A|}\begin{vmatrix}y_1 & a_{12} & a_{13}\\ y_2 & a_{22} & a_{23}\\ y_3 & a_{32} & a_{33}\end{vmatrix}\\ &= \frac{1}{|A|}\left\{y_1\begin{vmatrix}a_{22} & a_{23}\\ a_{32} & a_{33}\end{vmatrix} - y_2\begin{vmatrix}a_{12} & a_{13}\\ a_{32} & a_{33}\end{vmatrix} + y_3\begin{vmatrix}a_{12} & a_{13}\\ a_{22} & a_{23}\end{vmatrix}\right\}\\ &= \frac{1}{|A|}\{y_1C_{11} + y_2C_{21} + y_3C_{31}\} \end{aligned}$$

where A is the determinant of the coefficient matrix A, and C_{11}, C_{21}, and C_{31} are the cofactors of A corresponding to elements 11, 21, and 31. We can also write similar expressions for x_2 and x_3 by replacing the second and third columns by the y column respectively. Thus, the complete solution can be written in the matrix form

$$\begin{Bmatrix} x_1 \\ x_2 \\ x_3 \end{Bmatrix} = \frac{1}{|A|}\begin{bmatrix} C_{11} & C_{21} & C_{31} \\ C_{12} & C_{22} & C_{32} \\ C_{13} & C_{23} & C_{33} \end{bmatrix}\begin{Bmatrix} y_1 \\ y_2 \\ y_3 \end{Bmatrix} \tag{4}$$

or

$$\{x\} = \frac{1}{|A|}[C_{ji}]\{y\} = \frac{1}{|A|}[adj\ A]\{y\}$$

Thus, by comparison with Eq (3) we arrive at the result

$$A^{-1} = \frac{1}{|A|} adj\ A \tag{5}$$

Example: Find the inverse of the matrix

$$A = \begin{bmatrix} 1 & 1 & 1 \\ 1 & 2 & 2 \\ 1 & 0 & 3 \end{bmatrix}$$

(a) The determinant of A is $|A| = 3$

(b) The minors of A are

$$M_{11} = \begin{vmatrix} 2 & 2 \\ 0 & 3 \end{vmatrix} = 6,\ M_{12} = \begin{vmatrix} 1 & 2 \\ 1 & 3 \end{vmatrix} = 1, \text{ etc.}$$

(c) Supply the signs $(-1)^{i+j}$ to the minors to form the cofactors

$$[C_{ij}] = \begin{vmatrix} 6 & -1 & -2 \\ -3 & 2 & 1 \\ 0 & -1 & 1 \end{vmatrix}$$

(d) The adjoint matrix is the transpose of the cofactor matrix, or $[C_{ij}]' = [C_{ji}]$. Thus, the inverse A^{-1} is found to be

$$A^{-1} = \frac{1}{|A|} adj\ A = \frac{1}{3}\begin{bmatrix} 6 & -3 & 0 \\ -1 & 2 & -1 \\ -2 & 1 & 1 \end{bmatrix}$$

(e) The result can be checked as follows

$$A^{-1}A = \frac{1}{3}\begin{bmatrix} 6 & -3 & 0 \\ -1 & 2 & -1 \\ -2 & 1 & 1 \end{bmatrix}\begin{bmatrix} 1 & 1 & 1 \\ 1 & 2 & 2 \\ 1 & 0 & 3 \end{bmatrix}$$

$$= \frac{1}{3}\begin{bmatrix} 3 & 0 & 0 \\ 0 & 3 & 0 \\ 0 & 0 & 3 \end{bmatrix} = \begin{bmatrix} 1 & 0 & 0 \\ 0 & 1 & 0 \\ 0 & 0 & 1 \end{bmatrix}$$

It should be noted that for an inverse to exist, the determinant $|A|$ must not be zero.

Equation (5) for the inverse offers another means of evaluating a determinant. Premultiply Eq. (5) by A

$$AA^{-1} = \frac{A}{|A|} adj\ A = I$$

Thus we obtain the expression

$$|A|I = A\ adj\ A \tag{6}$$

Transpose of a Product

The following operations are given without proof.

$$(AB)' = B'A'$$
$$(A + B)' = A' + B' \tag{7}$$

Orthogonal Transformation. A matrix P is orthogonal if

$$P^{-1} = P'$$

The determinant of an orthogonal matrix is equal to ± 1. If A = symmetric matrix, then

$$P^{-1}AP = D = P'AP \text{ a diagonal matrix} \tag{8}$$

If A = symmetric matrix, then

$$P'A = AP$$
$$\{x\}'A = A\{x\} \tag{9}$$

Partitioned Matrices

A matrix may be partitioned into submatrices by horizontal and vertical lines as shown by the example below

$$\left[\begin{array}{cc|c} 2 & 4 & -1 \\ 0 & -3 & 4 \\ 1 & 2 & 2 \\ \hline 3 & -1 & -5 \end{array}\right] = \left[\begin{array}{c|c} [A] & [B] \\ \hline [C] & [D] \end{array}\right]$$

where the submatrices are

$$A = \begin{bmatrix} 2 & 4 \\ 0 & -3 \\ 1 & 2 \end{bmatrix} \qquad B = \begin{Bmatrix} -1 \\ 4 \\ 2 \end{Bmatrix}$$

$$C = [3 \quad -1] \qquad D = [-5]$$

Partitioned matrices obey the normal rules of matrix algebra and can be added, subtracted, and multiplied as though the submatrices were ordinary matrix elements. Thus

$$\left[\begin{array}{c|c} A & B \\ \hline C & D \end{array}\right] \left\{\begin{array}{c} x \\ \hline y \end{array}\right\} = \begin{bmatrix} A\{x\} + B\{y\} \\ C\{x\} + D\{y\} \end{bmatrix}$$

$$\left[\begin{array}{c|c} A & B \\ \hline C & D \end{array}\right] \left[\begin{array}{c|c} E & F \\ \hline G & H \end{array}\right] = \left[\begin{array}{c|c} AE + BG & AF + BH \\ \hline CE + DG & CF + DH \end{array}\right]$$

IV DETERMINATION OF EIGENVECTORS

The eigenvector X_i corresponding to the eigenvalue λ_i can be found from the cofactors of any row of the characteristic equation.

Let $[A - \lambda_i I]X_i = 0$ be written out for a third order system as

$$\begin{bmatrix} (a_{11} - \lambda_i) & a_{12} & a_{13} \\ a_{21} & (a_{22} - \lambda_i) & a_{23} \\ a_{31} & a_{32} & (a_{33} - \lambda_i) \end{bmatrix} \begin{Bmatrix} x_1 \\ x_2 \\ x_3 \end{Bmatrix}_i = 0 \tag{1}$$

Its characteristic equation $|A - \lambda_i I| = 0$ written out in determinant form is

$$\begin{vmatrix} (a_{11} - \lambda_i) & a_{12} & a_{13} \\ a_{21} & (a_{22} - \lambda_i) & a_{23} \\ a_{31} & a_{32} & (a_{33} - \lambda_i) \end{vmatrix} = 0 \tag{2}$$

The determinant expanded in terms of the cofactors of the first row is

$$(a_{11} - \lambda_i)C_{11} + a_{12}C_{12} + a_{13}C_{13} = 0 \tag{3}$$

Next replace the first row of the determinant by the second row, leaving the other two rows unchanged. The value of the determinant is still zero because of two identical rows

$$\begin{vmatrix} a_{21} & (a_{22} - \lambda_i) & a_{23} \\ a_{21} & (a_{22} - \lambda_i) & a_{23} \\ a_{31} & a_{32} & (a_{33} - \lambda_i) \end{vmatrix} = 0 \tag{4}$$

Again expand in terms of the cofactors of the first row, which are identical to the cofactors of the previous determinant.

$$a_{21}C_{11} + (a_{22} - \lambda_i)C_{12} + a_{23}C_{13} = 0 \tag{5}$$

Finally replace the first row by the third row and expand in terms of the first row of the new determinant.

$$\begin{vmatrix} a_{31} & a_{32} & (a_{33} - \lambda_i) \\ a_{21} & (a_{22} - \lambda_i) & a_{23} \\ a_{31} & a_{32} & (a_{33} - \lambda_i) \end{vmatrix} = 0 \tag{6}$$

$$a_{31}C_{11} + a_{32}C_{12} + (a_{33} - \lambda_i)C_{13} = 0 \tag{7}$$

Equations (3) (5) and (7) can now be assembled in a single matrix equation

$$\begin{bmatrix} (a_{11} - \lambda_i) & a_{12} & a_{13} \\ a_{21} & (a_{22} - \lambda_i) & a_{23} \\ a_{31} & a_{32} & (a_{33} - \lambda_i) \end{bmatrix} \begin{Bmatrix} C_{11} \\ C_{12} \\ C_{13} \end{Bmatrix} = 0 \tag{8}$$

Comparison of Eqs. (1) and (8) indicates that the eigenvector X_i may be determined from the cofactors of the characteristic equation with $\lambda = \lambda_i$. Since the eigenvectors are relative to a normalized coordinate, the column of cofactors may differ by a multiplying factor.

$$\begin{Bmatrix} x_1 \\ x_2 \\ x_3 \end{Bmatrix}_i = \alpha \begin{Bmatrix} C_{11} \\ C_{12} \\ C_{13} \end{Bmatrix}$$

Instead of the first row, any other row may have been used for the determination of the cofactors.

V CHOLESKY'S METHOD OF SOLUTION*

The matrix equation

$$[A]\{X\} = \{C\} \tag{1}$$

*Salvadori and Baron. *Numerical Methods in Engineering*. Englewood Cliffs, N.J.: Prentice-Hall, Inc., 1952, pp. 23–28.

may be solved for $\{X\}$ by premultiplying the equation by the inverse of $[A]$

$$\{X\} = [A]^{-1}\{C\}$$

Cholesky's method avoids the necessity of inverting the matrix $[A]$, the elements of $\{X\}$ being available by successive algebraic steps.

Cholesky's method depends on converting the original equation, Eq. (1), to the form

$$[T]\{X\} = \{k\} \tag{2}$$

where (for a 3×3 matrix)

$$[T] = \begin{bmatrix} 1 & t_{12} & t_{13} \\ 0 & 1 & t_{23} \\ 0 & 0 & 1 \end{bmatrix} \tag{3}$$

is an upper traiangular matrix with unit diagonal elements. For example, consider a 3×3 matrix

$$\begin{bmatrix} 0 & t_{12} & t_{13} \\ 0 & 1 & t_{23} \\ 0 & 0 & 1 \end{bmatrix} \begin{Bmatrix} x_1 \\ x_2 \\ x_3 \end{Bmatrix} = \begin{Bmatrix} k_1 \\ k_2 \\ k_3 \end{Bmatrix}$$

The elements of $\{X\}$ from the above equation are simply found by a backward substitution as follows

$$\begin{aligned} x_3 &= k_3 \quad \therefore \quad x_3 = k_3 \\ x_2 + t_{23}x_3 &= k_2 \quad \therefore \quad x_2 = k_2 - t_{23}x_3 \\ x_1 + t_{12}x_2 + t_{13}x_3 &= k_3 \quad \therefore \quad x_1 = k_3 - t_{12}x_2 - t_{13}x_3 \end{aligned}$$

Thus if $[T]$ and $\{k\}$ are known, the solution for $\{X\}$ in Eq. (1) is available.

To determine $[T]$ and $\{k\}$ multiply Eq. (2) by a lower triangular matrix

$$[L] = \begin{bmatrix} l_{11} & 0 & 0 \\ l_{21} & l_{22} & 0 \\ l_{31} & l_{32} & l_{33} \end{bmatrix} \tag{4}$$

as follows

$$[L][T]\{X\} = [L]\{k\} \tag{5}$$

For this equation to equal the original equation, the following relationships must exist

$$[A] = [L][T] \tag{6}$$

$$\{C\} = [L]\{k\} \tag{7}$$

Writing out the above equations in terms of their elements, we have

$$\begin{bmatrix} a_{11} & a_{12} & a_{13} \\ a_{21} & a_{22} & a_{23} \\ a_{31} & a_{32} & a_{33} \end{bmatrix} = \begin{bmatrix} l_{11} & l_{11}t_{12} & l_{11}t_{13} \\ l_{21} & (l_{21}t_{12} + l_{22}) & (l_{21}t_{13} + l_{22}t_{23}) \\ l_{31} & (l_{31}t_{12} + l_{32}) & (l_{31}t_{13} + l_{32}t_{23} + l_{33}) \end{bmatrix}$$

$$\begin{Bmatrix} c_1 \\ c_2 \\ c_3 \end{Bmatrix} = \begin{Bmatrix} l_{11}k_1 \\ l_{21}k_1 + l_{22}k_2 \\ l_{31}k_1 + l_{32}k_2 + l_{33}k_3 \end{Bmatrix}$$

By equating the elements in these equations we have

$$\begin{aligned}
a_{11} &= l_{11} & & \\
a_{21} &= l_{21} & & \\
a_{31} &= l_{31} & & \\
a_{12} &= l_{11}t_{12} & \therefore\quad t_{12} &= \frac{a_{12}}{l_{11}} \\
a_{13} &= l_{11}t_{13} & \therefore\quad t_{13} &= \frac{a_{13}}{l_{11}} \\
a_{22} &= (l_{21}t_{12} + l_{22}) & \therefore\quad l_{22} &= a_{22} - l_{21}t_{12} \\
a_{23} &= (l_{21}t_{13} + l_{22}t_{23}) & \therefore\quad t_{23} &= \frac{1}{l_{22}}(a_{23} - l_{21}t_{13}) \\
a_{32} &= (l_{31}t_{12} + l_{32}) & \therefore\quad l_{32} &= a_{32} - l_{31}t_{12} \\
a_{33} &= (l_{31}t_{13} + l_{32}t_{23} + l_{33}) & \therefore\quad l_{33} &= a_{33} - l_{31}t_{13} - l_{32}t_{23} \\
c_1 &= l_{11}k_1 & \therefore\quad k_1 &= \frac{c_1}{l_{11}} \\
c_2 &= l_{21}k_1 + l_{32}k_2 & \therefore\quad k_2 &= \frac{1}{l_{22}}(c_2 - l_{21}k_1) \\
c_3 &= l_{31}k_1 + l_{32}k_2 + l_{33}k_3 & \therefore\quad k_3 &= \frac{1}{l_{33}}(c_3 - l_{31}k_1 - l_{32}k_2)
\end{aligned}$$

Thus the elements of the matrices $[L]$, $[T]$ and $\{k\}$ are now available in terms of the known elements of $[A]$ and $[C]$ and Eq. (1) may be solved without inverting the matrix $[A]$.

D

Normal Modes of Uniform Beams

We assume the free vibrations of a uniform beam to be governed by Euler's differential equation.

$$EI\frac{\partial^4 y}{\partial x^4} + m\frac{\partial^2 y}{\partial t^2} = 0 \tag{1}$$

To determine the normal modes of vibration, the solution in the form

$$y(x, t) = \phi_n(x)e^{i\omega_n t} \tag{2}$$

is substituted into Eq. (1) to obtain the equation

$$\frac{d^4\phi_n(x)}{dx^4} - \beta_n^4\phi_n(x) = 0 \tag{3}$$

where:

$\phi_n(x)$ = characteristic function describing the deflection of the nth mode
m = mass density per unit length
$\beta_n^4 = m\omega_n^2/EI$
$\omega_n = (\beta_n l)^2\sqrt{EI/ml^4}$ = natural frequency of the nth mode.

The characteristic functions $\phi_n(x)$ and the normal-mode frequencies ω_n depend on the boundary conditions, and have been tabulated by Young and Felgar. An abbreviated summary taken from this work is presented here.

REFERENCE

Young, D. and R. P. Felgar Jr., *Tables of Characteristic Functions Representing Normal Modes of Vibration of a Beam.* The University of Texas Publication No. 4913, July 1, 1949.

Table 1.
Characteristic Functions and Derivatives
Clamped-Clamped Beam
First Mode

$\frac{x}{l}$	ϕ_1	$\phi_1' = \frac{1}{\beta_1}\frac{d\phi_1}{dx}$	$\phi_1'' = \frac{1}{\beta_1^2}\frac{d^2\phi_1}{dx^2}$	$\phi_1''' = \frac{1}{\beta_1^3}\frac{d^3\phi_1}{dx^3}$
0.00	0.00000	0.00000	2.00000	−1.96500
0.04	0.03358	0.34324	1.62832	−1.96285
0.08	0.12545	0.61624	1.25802	−1.94862
0.12	0.26237	0.81956	0.89234	−1.91254
0.16	0.43126	0.95451	0.53615	−1.84732
0.20	0.61939	1.02342	0.19545	−1.74814
0.24	0.81459	1.02986	−0.12305	−1.61250
0.28	1.00546	0.97870	−0.41240	−1.44017
0.32	1.18168	0.87608	−0.66581	−1.23296
0.36	1.33419	0.72992	−0.87699	−0.99452
0.40	1.45545	0.54723	−1.04050	−0.73007
0.44	1.53962	0.33897	−1.15202	−0.44611
0.48	1.58271	0.11478	−1.20854	−0.15007
0.52	1.58271	−0.11478	−1.20854	0.15007
0.56	1.53962	−0.33897	−1.15202	0.44611
0.60	1.45545	−0.54723	−1.04050	0.73007
0.64	1.33419	−0.72992	−0.87699	0.99452
0.68	1.18168	−0.87608	−0.66581	1.23296
0.72	1.00546	−0.97870	−0.41240	1.44017
0.76	0.81459	−1.02986	−0.12305	1.61250
0.80	0.61939	−1.02342	0.19545	1.74814
0.84	0.43126	−0.95451	0.53615	1.84732
0.88	0.26237	−0.81956	0.89234	1.91254
0.92	0.12545	−0.61624	1.25802	1.94862
0.96	0.03358	−0.34324	1.62832	1.96285
1.00	0.00000	0.00000	2.00000	1.96500

1 CLAMPED-CLAMPED BEAM

n	$\beta_n l$	$(\beta_n l)^2$	ω_n/ω_1
1	4.7300	22.3733	1.0000
2	7.8532	61.6728	2.7565
3	10.9956	120.9034	5.4039

2 FREE-FREE BEAM

The natural frequencies of the free-free beam are equal to those of the clamped-clamped beam. The characteristic functions of the free-free beam are related to those of the clamped-clamped beam as follows.

free-free		clamped-clamped
ϕ_n	=	ϕ_n''
ϕ_n'	=	ϕ_n'''
ϕ_n''	=	ϕ_n
ϕ_n'''	=	ϕ_n'

3 CLAMPED-FREE BEAM

n	$\beta_n l$	$(\beta_n l)^2$	ω_n/ω_1
1	1.8751	3.5160	1.0000
2	4.6941	22.0345	6.2669
3	7.8548	61.6972	17.5475

4 CLAMPED-PINNED BEAM

n	$\beta_n l$	$(\beta_n l)^2$	ω_n/ω_1
1	3.9266	15.4182	1.0000
2	7.0686	49.9645	3.2406
3	10.2102	104.2477	6.7613

5 FREE-PINNED BEAM

The natural frequencies of the free-pinned beam are equal to those of the clamped-pinned beam. The characteristic functions of the free-pinned beam are related to those of the clamped-pinned beam as follows.

free-pinned		clamped-pinned
ϕ_n	=	ϕ_n''
ϕ_n'	=	ϕ_n'''
ϕ_n''	=	ϕ_n
ϕ_n'''	=	ϕ_n'

Table 1.
Characteristic Functions and Derivatives
Clamped-Clamped Beam
Second Mode

$\frac{x}{l}$	ϕ_2	$\phi_2' = \frac{1}{\beta_2}\frac{d\phi_2}{dx}$	$\phi_2'' = \frac{1}{\beta_2^2}\frac{d^2\phi_2}{dx^2}$	$\phi_2''' = \frac{1}{\beta_2^3}\frac{d^3\phi_2}{dx^3}$
0.00	0.00000	0.00000	2.00000	−2.00155
0.04	0.08834	0.52955	1.37202	−1.99205
0.08	0.31214	0.86296	0.75386	−1.93186
0.12	0.61058	1.00644	0.16713	−1.78813
0.16	0.92602	0.97427	−0.35923	−1.54652
0.20	1.20674	0.79030	−0.79450	−1.21002
0.24	1.41005	0.48755	−1.11133	−0.79651
0.28	1.50485	0.10660	−1.28991	−0.33555
0.32	1.47357	−0.30736	−1.32106	0.13566
0.36	1.31314	−0.70819	−1.20786	0.57665
0.40	1.03457	−1.05271	−0.96605	0.94823
0.44	0.66150	−1.30448	−0.62296	1.21670
0.48	0.22751	−1.43728	−0.21508	1.35744
0.52	−0.22751	−1.43728	0.21508	1.35744
0.56	−0.66150	−1.30448	0.62296	1.21670
0.60	−1.03457	−1.05271	0.96605	0.94823
0.64	−1.31314	−0.70819	1.20786	0.57665
0.68	−1.47357	−0.30736	1.32106	0.13566
0.72	−1.50485	0.10660	1.28991	−0.33555
0.76	−1.41005	0.48755	1.11133	−0.79651
0.80	−1.20674	0.79030	0.79450	−1.21002
0.84	−0.92602	0.97427	0.35923	−1.54652
0.88	−0.61058	1.00644	−0.16713	−1.78813
0.92	−0.31214	0.86296	−0.75386	−1.93186
0.96	−0.08834	0.52955	−1.37202	−1.99205
1.00	0.00000	0.00000	−2.00000	−2.00155

Table 1.
Characteristic Functions and Derivatives
Clamped-Clamped Beam
Third Mode

$\frac{x}{l}$	ϕ_3	$\phi_3' = \frac{1}{\beta_3}\frac{d\phi_3}{dx}$	$\phi_3'' = \frac{1}{\beta_3^2}\frac{d^2\phi_3}{dx^2}$	$\phi_3''' = \frac{1}{\beta_3^3}\frac{d^3\phi_3}{dx^3}$
0.00	0.00000	0.00000	2.00000	−1.99993
0.04	0.16510	0.68646	1.12323	−1.97469
0.08	0.54804	0.99303	0.28189	−1.82280
0.12	0.98720	0.95006	−0.45252	−1.48447
0.16	1.34190	0.62285	−0.99738	−0.96698
0.20	1.50782	0.11050	−1.28572	−0.33199
0.24	1.42971	−0.46573	−1.28637	0.32333
0.28	1.10719	−0.98087	−1.01443	0.88956
0.32	0.59186	−1.32694	−0.53145	1.26880
0.36	−0.02445	−1.43171	0.06438	1.39529
0.40	−0.62837	−1.27099	0.65569	1.24912
0.44	−1.10739	−0.87257	1.12747	0.86096
0.48	−1.37174	−0.31031	1.38852	0.30669
0.52	−1.37174	0.31031	1.38852	−0.30669
0.56	−1.10739	0.87257	1.12747	−0.86096
0.60	−0.62837	1.27099	0.65569	−1.24912
0.64	−0.02445	1.43171	0.06438	−1.39529
0.68	0.59186	1.32694	−0.53145	−1.26880
0.72	1.10719	0.98087	−1.01443	−0.88956
0.76	1.42971	0.46573	−1.28637	−0.32333
0.80	1.50782	−0.11050	−1.28572	0.33199
0.84	1.34190	−0.62285	−0.99738	0.96698
0.88	0.98720	−0.95006	−0.45252	1.48447
0.92	0.54804	−0.99303	0.28189	1.82280
0.96	0.16510	−0.68646	1.12323	1.97469
1.00	0.00000	0.00000	2.00000	1.99993

Table 2.
Characteristic Functions and Derivatives
Clamped-Free Beam
First Mode

$\frac{x}{l}$	ϕ_1	$\phi_1' = \frac{1}{\beta_1}\frac{d\phi_1}{dx}$	$\phi_1'' = \frac{1}{\beta_1^2}\frac{d^2\phi_1}{dx^2}$	$\phi_1''' = \frac{1}{\beta_1^3}\frac{d^3\phi_1}{dx^3}$
0.00	0.00000	0.00000	2.00000	−1.46819
0.04	0.00552	0.14588	1.88988	−1.46805
0.08	0.02168	0.28350	1.77980	−1.46710
0.12	0.04784	0.41286	1.66985	−1.46455
0.16	0.08340	0.53400	1.56016	−1.45968
0.20	0.12774	0.64692	1.45096	−1.45182
0.24	0.18024	0.75167	1.34247	−1.44032
0.28	0.24030	0.84832	1.23500	−1.42459
0.32	0.30730	0.93696	1.12889	−1.40410
0.36	0.38065	1.01771	1.02451	−1.37834
0.40	0.45977	1.09070	0.92227	−1.34685
0.44	0.54408	1.15612	0.82262	−1.30924
0.48	0.63301	1.21418	0.72603	−1.26512
0.52	0.72603	1.26512	0.63301	−1.21418
0.56	0.82262	1.30924	0.54408	−1.15612
0.60	0.92227	1.34685	0.45977	−1.09070
0.64	1.02451	1.37834	0.38065	−1.01771
0.68	1.12889	1.40410	0.30730	−0.93696
0.72	1.23500	1.42459	0.24030	−0.84832
0.76	1.34247	1.44032	0.18024	−0.75167
0.80	1.45096	1.45182	0.12774	−0.64692
0.84	1.56016	1.45968	0.08340	−0.53400
0.88	1.66985	1.46455	0.04784	−0.41286
0.92	1.77980	1.46710	0.02168	−0.28350
0.96	1.88988	1.46805	0.00552	−0.14588
1.00	2.00000	1.46819	0.00000	0.00000

Table 2.
Characteristic Functions and Derivatives
Clamped-Free Beam
Second Mode

$\frac{x}{l}$	ϕ_2	$\phi_2' = \frac{1}{\beta_2}\frac{d\phi_2}{dx}$	$\phi_2'' = \frac{1}{\beta_2^2}\frac{d^2\phi_2}{dx^2}$	$\phi_2''' = \frac{1}{\beta_2^3}\frac{d^3\phi_2}{dx^3}$
0.00	0.00000	0.00000	2.00000	−2.03693
0.04	0.03301	0.33962	1.61764	−2.03483
0.08	0.12305	0.60754	1.23660	−2.02097
0.12	0.25670	0.80428	0.86004	−1.98590
0.16	0.42070	0.93108	0.49261	−1.92267
0.20	0.60211	0.99020	0.14007	−1.82682
0.24	0.78852	0.98502	−0.19123	−1.69625
0.28	0.96827	0.92013	−0.49475	−1.53113
0.32	1.13068	0.80136	−0.76419	−1.33373
0.36	1.26626	0.63565	−0.99384	−1.10821
0.40	1.36694	0.43094	−1.17895	−0.86040
0.44	1.42619	0.19593	−1.31600	−0.59748
0.48	1.43920	−0.06012	−1.40289	−0.32772
0.52	1.40289	−0.32772	−1.43920	−0.06012
0.56	1.31600	−0.59748	−1.42619	0.19593
0.60	1.17895	−0.86040	−1.36694	0.43094
0.64	0.99384	−1.10821	−1.26626	0.63565
0.68	0.76419	−1.33373	−1.13068	0.80136
0.72	0.49475	−1.53113	−0.96827	0.92013
0.76	0.19123	−1.69625	−0.78852	0.98502
0.80	−0.14007	−1.82682	−0.60211	0.99020
0.84	−0.49261	−1.92267	−0.42070	0.93108
0.88	−0.86004	−1.98590	−0.25670	0.80428
0.92	−1.23660	−2.02097	−0.12305	0.60754
0.96	−1.61764	−2.03483	−0.03301	0.33962
1.00	−2.00000	−2.03693	0.00000	0.00000

Table 2.
Characteristic Functions and Derivatives
Clamped-Free Beam
Third Mode

$\frac{x}{l}$	ϕ_3	$\phi_3' = \frac{1}{\beta_3}\frac{d\phi_3}{dx}$	$\phi_3'' = \frac{1}{\beta_3^2}\frac{d^2\phi_3}{dx^2}$	$\phi_3''' = \frac{1}{\beta_3^3}\frac{d^3\phi_3}{dx^3}$
0.00	0.00000	0.00000	2.00000	−1.99845
0.04	0.08839	0.52979	1.37287	−1.98892
0.08	0.31238	0.86367	0.75558	−1.92871
0.12	0.61120	1.00785	0.16974	−1.78480
0.16	0.92728	0.97665	−0.35563	−1.54286
0.20	1.20901	0.79394	−0.78975	−1.20575
0.24	1.41376	0.49285	−1.10515	−0.79124
0.28	1.51056	0.11405	−1.28189	−0.32872
0.32	1.48203	−0.29711	−1.31055	0.14479
0.36	1.32534	−0.69422	−1.19398	0.58908
0.40	1.05185	−1.03374	−0.94753	0.96533
0.44	0.68568	−1.27881	−0.59802	1.24030
0.48	0.26103	−1.40247	−0.18130	1.39004
0.52	−0.18130	−1.39004	0.26103	1.40247
0.56	−0.59802	−1.24030	0.68568	1.27881
0.60	−0.94753	−0.96533	1.05185	1.03374
0.64	−1.19398	−0.58908	1.32534	0.69422
0.68	−1.31055	−0.14479	1.48203	0.29711
0.72	−1.28189	0.32872	1.51056	−0.11405
0.76	−1.10515	0.79124	1.41376	−0.49285
0.80	−0.78975	1.20575	1.20901	−0.79394
0.84	−0.35563	1.54236	0.92728	−0.97665
0.88	0.16974	1.78480	0.61120	−1.00785
0.92	0.75558	1.92871	0.31238	−0.86367
0.96	1.37287	1.98892	0.08829	−0.52979
1.00	2.00000	1.99845	0.00000	0.00000

Table 3.
Characteristic Functions and Derivatives
Clamped-Pinned Beam
First Mode

$\frac{x}{l}$	ϕ_1	$\phi_1' = \frac{1}{\beta_1} \frac{d\phi_1}{dx}$	$\phi_1'' = \frac{1}{\beta_1^2} \frac{d^2\phi_1}{dx^2}$	$\phi_1''' = \frac{1}{\beta_1^3} \frac{d^3\phi_1}{dx^3}$
0.00	0.00000	0.00000	2.00000	−2.00155
0.04	0.02338	0.28944	1.68568	−2.00031
0.08	0.08834	0.52955	1.37202	−1.99203
0.12	0.18715	0.72055	1.06060	−1.97079
0.16	0.31214	0.86296	0.75386	−1.93187
0.20	0.45574	0.95776	0.45486	−1.87177
0.24	0.61058	1.00643	0.16712	−1.78812
0.28	0.76958	1.01105	−0.10554	−1.67975
0.32	0.92601	0.97427	−0.35923	−1.54652
0.36	1.07363	0.89940	−0.59009	−1.38932
0.40	1.20675	0.79029	−0.79450	−1.21002
0.44	1.32032	0.65138	−0.96918	−1.01128
0.48	1.41006	0.48755	−1.11133	−0.79652
0.52	1.47245	0.30410	−1.21875	−0.56977
0.56	1.50485	0.10661	−1.28992	−0.33555
0.60	1.50550	−0.09916	−1.32402	−0.09872
0.64	1.47357	−0.30736	−1.32106	0.13566
0.68	1.40913	−0.51224	−1.28180	0.36247
0.72	1.31313	−0.70820	−1.20786	0.57666
0.76	1.18741	−0.88996	−1.10157	0.77340
0.80	1.03457	−1.05270	−0.96606	0.94823
0.84	0.85795	−1.19210	−0.80507	1.09714
0.88	0.66151	−1.30448	−0.62295	1.21670
0.92	0.44974	−1.38693	−0.42455	1.30414
0.96	0.22752	−1.43727	−0.21507	1.35743
1.00	0.00000	−1.45420	0.00000	1.37533

Table 3.
Characteristic Functions and Derivatives
Clamped-Pinned Beam
Second Mode

$\frac{x}{l}$	ϕ_2	$\phi_2' = \frac{1}{\beta_2}\frac{d\phi_2}{dx}$	$\phi_2'' = \frac{1}{\beta_2^2}\frac{d^2\phi_2}{dx^2}$	$\phi_2''' = \frac{1}{\beta_2^3}\frac{d^3\phi_2}{dx^3}$
0.00	0.00000	0.00000	2.00000	−2.00000
0.04	0.07241	0.48557	1.43502	−1.99300
0.08	0.25958	0.81207	0.87658	−1.94824
0.12	0.51697	0.98325	0.33937	−1.83960
0.16	0.80176	1.00789	−0.15633	−1.65333
0.20	1.07449	0.90088	−0.58802	−1.38736
0.24	1.30078	0.68345	−0.93412	−1.05012
0.28	1.45308	0.38242	−1.17673	−0.65879
0.32	1.51208	0.02894	−1.30380	−0.23724
0.36	1.46765	−0.34350	−1.31068	0.18649
0.40	1.31923	−0.70122	−1.20092	0.58286
0.44	1.07550	−1.01270	−0.98634	0.92349
0.48	0.75348	−1.25090	−0.68631	1.18364
0.52	0.37700	−1.39515	−0.32640	1.34442
0.56	−0.02536	−1.43265	0.06348	1.39438
0.60	−0.42268	−1.35944	0.45136	1.33056
0.64	−0.78413	−1.18058	0.80569	1.15876
0.68	−1.08158	−0.90972	1.09776	0.89319
0.72	−1.29186	−0.56793	1.30395	0.55537
0.76	−1.39858	−0.18205	1.40755	0.17245
0.80	−1.39351	0.21752	1.40010	−0.22494
0.84	−1.27726	0.59923	1.28198	−0.60506
0.88	−1.05919	0.93288	1.06244	−0.93759
0.92	−0.75676	1.19208	0.75879	−1.19604
0.96	−0.39406	1.35629	0.39504	−1.35983
1.00	0.00000	1.41251	0.00000	−1.41592

Table 3.
Characteristic Functions and Derivatives
Clamped-Pinned Beam
Third Mode

$\frac{x}{l}$	ϕ_3	$\phi_3' = \frac{1}{\beta_3}\frac{d\phi_3}{dx}$	$\phi_3'' = \frac{1}{\beta_3^2}\frac{d^2\phi_3}{dx^2}$	$\phi_3''' = \frac{1}{\beta_3^3}\frac{d^3\phi_3}{dx^3}$
0.00	0.00000	0.00000	2.00000	−2.00000
0.04	0.14410	0.65020	1.18532	−1.97961
0.08	0.48626	0.97168	0.39742	−1.85535
0.12	0.89584	0.98593	−0.30845	−1.57331
0.16	1.25604	0.74002	−0.86560	−1.13046
0.20	1.47476	0.30725	−1.21523	−0.56678
0.24	1.49419	−0.21934	−1.32168	0.04683
0.28	1.29662	−0.73864	−1.18195	0.62397
0.32	0.90489	−1.15556	−0.82867	1.07934
0.36	0.37703	−1.39512	−0.32637	1.34445
0.40	−0.20439	−1.41364	0.23807	1.37996
0.44	−0.74658	−1.20525	0.76897	1.18287
0.48	−1.16223	−0.80234	1.17711	0.78746
0.52	−1.38422	−0.26994	1.39411	0.26005
0.56	−1.37687	0.30522	1.38344	−0.31179
0.60	−1.14194	0.82907	1.14631	−0.83344
0.64	−0.71844	1.21582	0.72134	−1.21873
0.68	−0.17628	1.40210	0.17821	−1.40403
0.72	0.39519	1.35742	−0.39391	−1.35870
0.76	0.90188	1.08924	−0.90103	−1.09010
0.80	1.26035	0.64175	−1.25980	−0.64233
0.84	1.41160	0.08860	−1.41124	−0.08900
0.88	1.33072	−0.47918	−1.33049	0.47891
0.92	1.03098	−0.96820	−1.03085	0.96800
0.96	0.56168	−1.29798	−0.56162	1.29782
1.00	0.00000	−1.41429	0.00000	1.41414

E

Specifications of Vibration Bounds

Specifications for vibrations are often based on harmonic motion.

$$x = x_0 \sin \omega t$$

The velocity and acceleration are then available from differentiation and the following relationships for the peak values can be written.

$$\dot{x}_0 = 2\pi f x_0$$

$$\ddot{x}_0 = -4\pi^2 f^2 x_0 = -2\pi f \dot{x}_0$$

These equations can be represented on the log-log paper by rewriting them in the form

$$\ln \dot{x}_0 = \ln x_0 + \ln 2\pi f$$

$$\ln \dot{x}_0 = -\ln \ddot{x}_0 - \ln 2\pi f$$

By letting x_0 = constant, the plot of $\ln \dot{x}_0$ against $\ln 2\pi f$ is a straight line of slope equal to $+1$. By letting $\ddot{x}_0$ = constant, the plot of $\ln \dot{x}_0$ versus $\ln 2\pi f$ is again a straight line of slope -1. These lines are shown graphically in

Fig. E-1. The graph is often used to specify bounds for the vibration. Shown in heavy lines are bounds for a maximum acceleration of 10 g, minimum and maximum frequencies of 5 and 500 c.p.s., and an upper limit for the displacement of 0.30 inch.

Figure E-1.

Answers to Selected Problems

Chapter 1

1-1 $\dot{x}_{max} = 8.38$ in./sec
$\ddot{x}_{max} = 351$ in./sec^2

1-3 $x_{max} = 2.86$ in, $\tau = 0.10$ sec, $\ddot{x}_{max} = 11300$ in./sec^2,

1-5 $z = 5e^{0.6435i}$

1-8 $R = 5.76$, $\theta = 13°20'$

1-9 $x(t) = 4/\pi(\sin \omega t + \frac{1}{3} \sin 3\omega t + \frac{1}{5} \sin 5\omega t + \cdots)$

1-11 $x(t) = 1/2 + 4/\pi^2(\cos \omega_1 t + 1/3^2 \cos 3\omega_1 t + 1/5^2 \cos 5\omega_1 t + \cdots)$
where $\omega_1 = 2\pi$

1-13 $\sqrt{\overline{x^2}} = A/2$

1-14 $\overline{x^2} = 1/3$

1-16 $b_n = \frac{1}{n\pi}(1-\cos kn\pi)$, $a_n = \frac{1}{n\pi} \sin kn\pi$, $a_o = k$

Chapter 2

2-1 5.62 cps

2-3 0.159 sec

2-5 $x = \frac{W_2}{k}\left\{\sqrt{\frac{2kh}{W_1 + W_2}} \sin \omega t + (1 - \cos \omega t)\right\} \qquad \omega = \sqrt{\frac{kg}{W_1 + W_2}}$

2-7 $J_0 = 9.27$ lb in. sec^2

2-9 8.0 in.

2-11 $\omega = \sqrt{\frac{k}{m + J_0/r^2}}$

2-13 $\tau = 1.98$ sec

2-15 $\tau = \frac{2\pi L}{a}\sqrt{\frac{h}{3g}}$

2-17 $f = \frac{1}{2\pi}\sqrt{\frac{gab}{h\kappa^2}} \qquad \kappa =$ rad. of gyr.

2-20 $m_{\text{eff}} = (m/n^2)[1/12 + (1/2 - n)^2]$

2-22 $m_{\text{eff}} = 0.257(wl/g)$

2-24 $T = (\dot{x}^2)\frac{1}{2}[J_1/r_1^2 + m_0 + m_2 + J_2/r_2^2]$

2-26 $c_{cr} = 0.288$

2-27 $\zeta = 1.45$

2-29 (a) $\omega_d = 27.8$ rad/sec, (b) $\delta = 0.0202$
(c) $\zeta = 0.00322$ (d) $c = 0.00232$

2-32 (a) $c_{cr} = 2\frac{l}{a}\sqrt{km}$, (b) $\omega_d = \omega_n\sqrt{1 - \zeta^2}$
$\omega_n = \frac{a}{l}\sqrt{\frac{k}{m}}$, $\zeta = \frac{ca}{2l\sqrt{km}}$

2-34 $\zeta = 0.00818$

2-38 $x_{\max} = 4.96$ in., $t = 0.0179$ sec

2-39 $\zeta_1 = 0.59$, $x = 0.378$

2-41 $k_{eq} = \frac{(k_1 + k_2)k_3}{k_1 + k_2 + k_3}$

2-43 $k_{\text{eff}} = k_1 + (a/b)^2 k_2$

2-45 Equal amplitudes at $n = 3$

Chapter 3

3-2 144%

3-3 $X = 1.42$ in., $\phi = 59°15'$

3-5 $\zeta = 0.143$

3-7 $\frac{1}{3}w$ required (down at left, up at right) in Fig. 3-7

3-9 (a) $f_n = 900$ rpm, (b) $\zeta = 0.0117$
(c) $X = 0.0457$ in. (d) $178°$

3-11 319 lb total for 1″ shaft
54.4 lb total for $\frac{3}{4}$″ shaft.

3-13 $t = 0.0104$ sec

3-15 $X = 7.703$ in.

3-17 $\ddot{x} + \frac{g}{l}x = \frac{g}{l}X_0 \sin \omega t,\ h = \frac{l}{1 - (X_0/X)}$ from m to node

3-19 $Y = 0.0488$ in.

3-22 $k = 18.8$ lb/in. each

3-24 $X = 0.0065$ in.

3-26 $\eta = c\omega/k$

3-28 $(\tau_m/\tau_d)^2 + \zeta^2 = 1$

3-30 $c_d = \frac{1}{2\omega W} \pm \sqrt{\left(\frac{1}{2\omega W}\right)^2 - \left(\frac{k - m\omega^2}{\omega}\right)^2}$
$W = W_d/\pi F^2$, W_d = energy dissipated/cycle

3-32 δ independent of amplitude when $W_d/U \neq f(X)$.

3-34 $X = \frac{1}{1 - (\omega/\omega_n)^2}\sqrt{\left(\frac{F}{k}\right)^2 - \left(\frac{4D}{\pi k}\right)^2},\quad c_{eq} = \frac{4D}{\pi\omega X}$

Chapter 4

4-5 $x = F_0/k(1 - \cos \omega_n t) - F_0/k[1 - \cos \omega_n(t - t_0)],\ t > t_0$

4-9
$$\bar{x}(s) = \frac{(s + 2\zeta\omega_n)x(0)}{s^2 + 2\zeta\omega_n s + \omega_n^2} + \frac{\dot{x}(0)}{s^2 + 2\zeta\omega_n s + \omega_n^2}$$
$$= e^{-\zeta\omega_n t}\left\{\frac{\dot{x}(0) + \zeta\omega_n x(0)}{\omega_n\sqrt{1 - \zeta^2}} \sin \sqrt{1 - \zeta^2}\,\omega_n t + x(0) \cos \sqrt{1 - \zeta^2}\,\omega_n t\right\}$$

4-10
$$z = \frac{100}{\omega_n^2}(1 - \cos \omega_n t) - \frac{20}{\omega_n} \sin \omega_n t$$
$$z_{\max} = \frac{100}{\omega_n^2}\left(1 - \frac{5}{\sqrt{5^2 + \omega_n^2}}\right) - \frac{20}{\sqrt{5^2 + \omega_n^2}}$$

4-13 $\tan \omega_n t = \frac{\sqrt{2mgs/k}}{s - mg/2k}$

4-14 $\omega t_0 = 4.52 = 259°$, From Eq. 4.4-14, $X_1 = 0.546''$

4-15 $x_{\max} = 12.08''$, $t = 0.392$ sec

4-22 $x_{\max} = 11.65''$

4-33 $\ddot{x} = -2\zeta\omega_n(\dot{x} - \dot{y}) - \omega_n^2(x - y)$
Integrate with initial conditions = 0
$$\dot{x} = -2\zeta\omega_n(x - y) - \omega_n^2 \int (x - y)\, dt$$

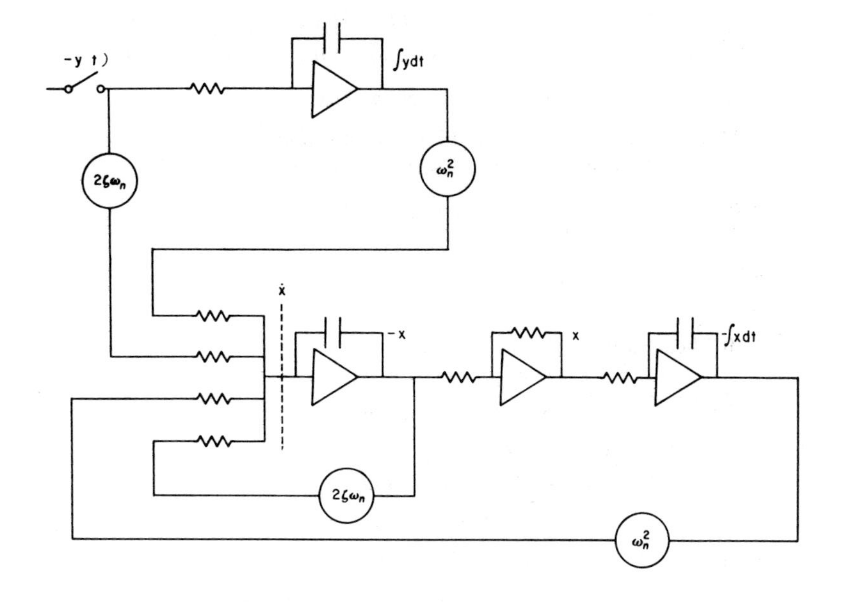

4-34 $x = \dfrac{F_0 e^{-\zeta\omega_n t}}{c\omega_n\sqrt{1-\zeta^2}} \sin(\sqrt{1-\zeta^2}\,\omega_n t + \sin^{-1}\sqrt{1-\zeta^2}) - \dfrac{F_0 cos\omega_n t}{c\omega_n}$

Chapter 5

5-2 $\omega_1^2 = k/m \qquad (X_1/X_2)_1 = 1$
$\omega_1^2 = 3k/m \qquad (X_1/X_2)_2 = -1$

5-4 $\omega_1^2 = 0.570\,k/m \qquad (X_1/X_2)_1 = 3.43$
$\omega_2^2 = 4.096\,k/m \qquad (X_1/X_2)_2 = -0.096$

5-6 $\omega_n = \sqrt{\dfrac{K_2(J_1 + J_2)}{J_1 J_2}}$

5-8 $\omega_n = 15.7$ rad/sec

5-10 $2\ddot{\theta}_1 + \ddot{\theta}_2 + 2g/l\,\theta_1 = 0$
$\ddot{\theta}_1 + \ddot{\theta}_2 + g/l\,\theta_2 = 0$

5-11 $\begin{bmatrix} m_1 & 0 \\ 0 & m_2 \end{bmatrix}\begin{Bmatrix} \ddot{y}_1 \\ \ddot{y}_2 \end{Bmatrix} + \dfrac{T}{l}\begin{bmatrix} 2 & -1 \\ -1 & 2 \end{bmatrix}\begin{Bmatrix} y_1 \\ y_2 \end{Bmatrix} = 0$

5-13 $\omega_1 = 0.796\sqrt{T/ml},\ (y_1/y_2)_1 = 1.365$
$\omega_2 = 1.536\sqrt{T/ml},\ (y_1/y_2)_2 = -0.365$

5-15 $\omega = \sqrt{\dfrac{g}{l} + \dfrac{k}{ml^2}(1 \pm 1)}$ beat period = 52.3 sec

5-17 $x_1 = X(0.500 \cos \omega_1 t + 0.500 \cos \omega_2 t) \qquad \omega_1^2 = 0.586\,g/l$
$x_2 = X(1.207 \cos \omega_1 t - 0.207 \cos \omega_2 t) \qquad \omega_2^2 = 3.414\,g/l$
$(x_1/x_2)_1 = 0.414/1.0$
$(x_1/x_2)_2 = -2.414/1.0$

5-20 $\begin{bmatrix} m & 0 \\ 0 & J \end{bmatrix}\begin{Bmatrix} \ddot{x} \\ \ddot{\theta} \end{Bmatrix} + \begin{bmatrix} 2k & -kl/4 \\ -kl/4 & (5/16)kl^2 \end{bmatrix}\begin{Bmatrix} x \\ \theta \end{Bmatrix} = \begin{Bmatrix} 0 \\ 0 \end{Bmatrix}$ $\quad \downarrow x \quad \circlearrowleft \theta$

5-22 both static and dynamic coupling exist

5-24 $f_1 = 0.963$ cps node 10.9 ft ahead of cg.
$f_2 = 1.33$ cps node 1.48 ft behind cg.

5-26 $\omega_1 = 31.6$ rad/sec $(X_1/X_2)_1 = 0.50$
$\omega_2 = 63.4$ rad/sec $(X_1/X_2)_2 = -1$

5-29 $x_1 = \frac{8}{9}\cos\omega_1 t + \frac{1}{9}\cos\omega_2 t$
$x_2 = \frac{4}{9}\cos\omega_1 t - \frac{1}{9}\cos\omega_2 t$

5-30 Shear ratio 1st/2nd story = 2

5-32 $x_1 = \dfrac{2k_1^2 X_g \sin\omega t}{(\omega^2 - \frac{1}{2}k_1/m_1)(\omega^2 - 2k_1/m_1)\, m_1^2}$,

$x_2 = \dfrac{2k_1(k_1 - \omega^2 m_1)X_g \sin\omega t}{(\omega^2 - \frac{1}{2}k_1/m_1)(\omega^2 - 2k_1/m_1)\, m_1^2}$

5-34 $(\omega/\omega_h)_2 = 2.73$, $(Y_1/Y_0)_2 = -0.74$

5-36 $V_1 = 43.3$ ft/sec, $V_2 = 60.3$ ft/sec

5-38 $X_1 = \dfrac{m\omega^2 e[k_2 - M_2\omega^2 + i\omega c]}{[(k_1 + k_2) - M_1\omega^2 + i\omega c][k_2 - M_2\omega^2 + i\omega c] - [k_2 + i\omega c]^2}$

5-41 $d_2 = 1/2''$

5-43 $w = 11.4$ lb $\quad k = 17.9$ lb/in.

5-45 $\zeta_0 = 0.423 \quad \omega/\omega_n = 0.943$

5-48 $y = \delta_F F + \delta_M M,\quad \delta_F = \dfrac{4l^3}{3 \times 81EI},\quad \delta_M = \dfrac{2l^2}{81EI}$

$\theta = \alpha_F F + \alpha_M M,\quad \alpha_F = \dfrac{2l^2}{81EI},\quad \alpha_M = \dfrac{l}{9EI}$

$F = m\omega_1^2 y(l/3),\ M = -(J_p - J_d)\omega\omega_1\theta$

5-50 $\ddot{x} = -40x + 40y$ where $x = x_1$
$\ddot{y} = 40x - 100y + 60z$ $\quad y = x_2$
$Z = 10\sin\pi t$

Use Eq. 4.7-6 to initiate calculations

$x(2) = \dfrac{40}{6}\cdot\dfrac{60}{6}\dfrac{Z(2)\Delta t^4}{[1 + (40/6)\Delta t^2][1 + (100/6)\Delta t^2]}$

$y(2) = \dfrac{1}{[1 + (100/6)\Delta t^2]}\left[\dfrac{40}{6}x(2) + \dfrac{60}{6}Z(2)\right]\Delta t^2$

Chapter 6

6-1 $[K] = \begin{bmatrix} (K_1 + K_2) & -K_2 & 0 \\ -K_2 & (K_2 + K_3) & -K_3 \\ 0 & -K_3 & K_3 \end{bmatrix}$

$[a] = [K]^{-1} = \begin{bmatrix} 1/K_1 & 1/K_1 & 1/K_1 \\ 1/K_1 & (1/K_1 + 1/K_2) & (1/K_1 + 1/K_2) \\ 1/K_1 & (1/K_1 + 1/K_2) & (1/K_1 + 1/K_2 + 1/K_3) \end{bmatrix}$

6-3 $a_{11} = 0.0114\, l^3/EI$
$a_{12} = a_{21} = 0.01448\, l^3/EI$
$a_{22} = 0.01920\, l^3/EI$

6-5
$$[K] = \begin{bmatrix} (k_1+k_2) & -k_2 & 0 & 0 & 0 & \\ -k_2 & (k_2+k_3) & -k_3 & 0 & 0 & \\ 0 & -k_3 & (k_3+k_4) & -k_4 & 0 & \\ 0 & 0 & -k_4 & (k_4+k_5) & -k_5 \cdots & \\ \vdots & & & & \vdots & \\ & & & & & k_n \end{bmatrix}$$

6-6
$$[a] = \begin{bmatrix} 1/k_1 & 1/k_1 & 1/k_1 & \cdots \\ 1/k_1 & 1/k_1 + 1/k_2 & 1/k_1 + 1/k_2 + 1/k_3 & \\ 1/k_1 & 1/k_1 + 1/k_2 & \cdot & \\ \vdots & \vdots & \vdots & \end{bmatrix}$$

6-9
$$\begin{bmatrix} J_1 & 0 \\ 0 & J_2 \end{bmatrix}\begin{Bmatrix} \ddot{\theta}_1 \\ \ddot{\theta}_2 \end{Bmatrix} + l^2\begin{bmatrix} \frac{9}{16}k_1 & -\frac{9}{16}k_1 \\ -\frac{9}{16}k_1 & (\frac{9}{16}k_1 + \frac{1}{4}k_2) \end{bmatrix}\begin{Bmatrix} \theta_1 \\ \theta_2 \end{Bmatrix} = \begin{Bmatrix} 0 \\ 0 \end{Bmatrix}$$

where θ_1 ↻ and θ_2 ↻
$J_1 = ml^2/3 \qquad J_2 = 7/48ml^2$

6-14 x = coord. of c.g. and θ ↻

$$\lambda_1 = \frac{\omega_1^2 m}{k} = 1.64 \qquad \begin{Bmatrix} \theta \\ x \end{Bmatrix}_1 = \begin{Bmatrix} 1.44/l \\ 1.00 \end{Bmatrix}$$

$$\lambda_2 = \frac{\omega_2^2 m}{k} = 4.10 \qquad \begin{Bmatrix} \theta \\ x \end{Bmatrix}_2 = \begin{Bmatrix} -8.40/l \\ 1.00 \end{Bmatrix}$$

$$P'[K]P = \begin{bmatrix} 1.92 & 0 \\ 0 & 28.3 \end{bmatrix} \qquad [K] = \begin{bmatrix} \frac{5}{16}l^2 & -\frac{1}{4}l \\ -\frac{1}{4} & 2 \end{bmatrix}$$

6-17
$$[K] = k\begin{bmatrix} 2 & -1 \\ -1 & 2 \end{bmatrix} \qquad [P] = \begin{bmatrix} 0.732 & -2.732 \\ 1.00 & 1.00 \end{bmatrix}$$

$$\ddot{q}_1 + 0.634q_1 = 0 \qquad \ddot{q}_2 + 2.366q_2 = 0 \qquad [\tilde{P}] = \begin{bmatrix} 0.732/\sqrt{2.535} & -2.732/\sqrt{9.48} \\ 1/\sqrt{2.535} & 1/\sqrt{9.48} \end{bmatrix}\frac{1}{\sqrt{m}}$$

6-20
$$\begin{bmatrix} (ms^2+2k) & -k \\ -k & (ms^2+2k) \end{bmatrix}\begin{Bmatrix} \bar{x}_1(s) \\ \bar{x}_2(s) \end{Bmatrix} = \begin{Bmatrix} \frac{F_0\omega}{s^2+\omega^2} + msx_1(0) + m\dot{x}_1(0) \\ msx_2(0) + m\dot{x}_2(0) \end{Bmatrix}$$

$$\bar{x}_1(s) = \frac{\begin{vmatrix} \left(\frac{F_0\omega}{s^2+\omega^2} + msx_1(0) + m\dot{x}_1(0)\right) & -k \\ (msx_2(0) + m\dot{x}_2(0) & (ms^2+2k) \end{vmatrix}}{m(s^2+k/m)(s^2+3k/m)}$$

6-21
$$[K] = k\begin{bmatrix} 2 & -1 \\ -1 & 2 \end{bmatrix}, \quad [c] = c\begin{bmatrix} 2 & -1 \\ -1 & 1 \end{bmatrix} \qquad \therefore \text{ not proportional}$$

6-24 $[A - \lambda I] = \begin{bmatrix} -(\lambda + 4) & 0 & 1 \\ 0 & -\lambda & 1 \\ -10 & -100 & -\lambda \end{bmatrix}$

6-30 $\begin{Bmatrix} \dot{z}_1 \\ \dot{z}_2 \end{Bmatrix} = -\begin{bmatrix} 0 & 1 \\ k/m & c/m \end{bmatrix}\begin{Bmatrix} z_1 \\ z_2 \end{Bmatrix} + \begin{Bmatrix} 0 \\ F/m \end{Bmatrix} \sin \omega t$

Chapter 7

7-2 $a_{11} = \dfrac{k_2 + k_3}{\sum k_i k_j}, \quad a_{21} = a_{12} = \dfrac{k_2}{\sum k_i k_j}, \quad a_{22} = \dfrac{k_1 + k_2}{\sum k_i k_j}$

7-3 $\omega^2 = \dfrac{k}{m}\left(\dfrac{4 - 6n + 5n^2}{1 + 2n^2}\right); n = \dfrac{X_2}{X_1}$

7-6 $a_{11} = a_{21} = a_{12} = a_{31} = a_{13} = \dfrac{l_1}{(m_1 + m_2 + m_3)g}$

$a_{22} = a_{23} = a_{32} = \dfrac{l_1}{(m_1 + m_2 + m_3)g} + \dfrac{l_2}{(m_2 + m_3)g}$

$a_{33} = \dfrac{l_1}{(m_1 + m_2 + m_3)g} + \dfrac{l_2}{(m_2 + m_3)g} + \dfrac{l_3}{m_3 g}$

7-11 $\omega = \sqrt{\dfrac{6EI}{Ml^3}\left(1 + \dfrac{n}{2}\right)}; \dfrac{y_1}{y_2} = -\dfrac{n}{2}$

7-13 $\omega_1 = 0.00340\sqrt{EI}$

7-15 $\omega_1 = 0.00374\sqrt{EI}$

7-18 $\omega_1 = 2.90\sqrt{\dfrac{gEI}{Wl^3}}$

7-20 $f_{11} = 498$ c.p.s.

7-23 $\begin{Bmatrix} x_1 \\ x_2 \\ x_3 \end{Bmatrix} = \dfrac{m\omega^2}{3k}\begin{bmatrix} 4 & 2 & 1 \\ 4 & 8 & 4 \\ 4 & 8 & 7 \end{bmatrix}\begin{Bmatrix} x_1 \\ x_2 \\ x_3 \end{Bmatrix}$

7-25 $\omega_1 = 4.93\sqrt{\dfrac{gEI}{Wl^3}}$

7-27 $\omega_1^2 = 0.705 \times 10^5 \qquad \{\theta\}_1 = \begin{Bmatrix} 1.00 \\ 0.534 \\ -0.181 \\ -0.654 \end{Bmatrix}$

$\omega_2^2 = 3.00 \times 10^5 \qquad \{\theta\}_2 = \begin{Bmatrix} 1.00 \\ -1.00 \\ -1.00 \\ 0.50 \end{Bmatrix}$

7-49 Start with Eq. 7.12-14

$$\begin{bmatrix} y \\ \theta \\ M \\ V \\ \varphi \\ T \end{bmatrix} \overset{=}{\rightarrow} \begin{bmatrix} & \\ & \\ & \\ & \\ & \\ & \end{bmatrix} \begin{bmatrix} 0 \\ 0 \\ M \\ V \\ 0 \\ T \end{bmatrix}_0 = \begin{bmatrix} \frac{l^2}{2EI} M_0 + \frac{l^3}{6EI} V_0 \\ \frac{l}{EI} M_0 + \frac{l^2}{2EI} V_0 \\ M_0 + V_0 l \\ V_0 \\ hT_0 \\ T_0 \end{bmatrix}$$

7-58 $\omega_k = 2\sqrt{\frac{K}{J}} \sin \frac{(2k-1)\pi}{2(2N+1)} \qquad k = 1, 2, 3, \cdots N$

7-59 $Y_{r+1} - 2\left(1 - \frac{\omega^2 ml}{2T}\right) Y_r + Y_{r-1} = 0$

$$\omega_k = 2\sqrt{\frac{T}{ml}} \sin \frac{k\pi}{2(N+1)}, \quad k = 1, 2, 3, \cdots N$$

7-63 β is found from $2 \cos \beta\left(N + \frac{1}{2}\right) \sin \frac{\beta}{2} = -\frac{K_N}{k} \sin \beta N$

and substituted into $\qquad \omega = 2\sqrt{\frac{k}{m}} \sin \frac{\beta}{2}$

7-65 $m\ddot{y}_n = k(y_{n+1} - 2y_n + y_{n-1})$ boundary eqs.

$$m\ddot{y}_N = -k(y_N - y_{N-1} + h\theta)$$

$$\sum_{n=1}^{N} nh(m\ddot{y}_n) - K_\theta \theta = (N+1) m\rho^2 \ddot{\theta}$$

ρ = radius of gyration of each floor about its center of mass.

7-29 $\theta_1 = 0.00618$ rad, $\theta_2 = -0.000778$ rad, $\theta_3 = -0.00338$ rad

7-31 $\theta_1 = 0.0798$ rad, 0°; $\theta_2 = 0.0558$ rad, 0°; $\theta_2 = 0.0204$ rad, 42°; $\theta_4 = 0.0442$ rad, 128°8′

7-33 22.8 cps

7-34 22.5 cps; 52.5 cps

7-38 $\omega_1 = 0.537\sqrt{\frac{k}{J}} \qquad \{\theta\}_1 = \begin{Bmatrix} 1.00 \\ 0.714 \\ 0.239 \\ -0.326 \end{Bmatrix}$

$$\omega_2 = 1.27\sqrt{\frac{k}{J}} \qquad \{\theta\}_2 = \begin{Bmatrix} 1.00 \\ -0.614 \\ -1.237 \\ 0.1416 \end{Bmatrix}$$

$$\omega_3 = 1.81\sqrt{\frac{k}{J}} \qquad \{\theta\}_3 = \begin{Bmatrix} 1.00 \\ -2.27 \\ 1.87 \\ -0.10 \end{Bmatrix}$$

7-40 $\omega_1 = 0.585\sqrt{\frac{EI}{ml^3}}$; $\omega_2 = 3.88\sqrt{\frac{EI}{ml^3}}$

7-41 $\omega = \sqrt{\frac{3EI}{ml^3} + \frac{3}{2}\Omega^2}$

7-43 $\begin{vmatrix} u_{12} & u_{14} \\ u_{32} & u_{34} \end{vmatrix} = 0$

7-46 $\begin{vmatrix} u_{32} & u_{34} \\ u_{42} & u_{44} \end{vmatrix} = 0$

Chapter 8

8-2 $f = \frac{n}{2l}\sqrt{\frac{T}{l}}, \quad n = 1, 2, 3, \ldots$

8-3 $\tan\frac{\omega l}{c} = \frac{(T/kl)(l/c)}{(mc^2/kl^2)(l/c)^2 - 1}$

8-5 16,600 ft./sec.

8-11 $\omega_n = (2n + 1)\frac{\pi}{l}\sqrt{\frac{Gg}{\rho}}, n = 0, 1, 2, \ldots$

8-12 $\tan\frac{\omega l}{c} = \frac{2(J_0/J_s)(\omega\, l/c)}{(J_0/J_s)^2\,(\omega\, l/c)^2 - 1}$

8-14 $b = 2/\pi$, node at $x = 0.219l$ from ends

8-16 Same f_n as free-free bar

8-18 $\omega = n^2\sqrt{\frac{gEIl}{W_b}}$ where n is determined from

$$(1 + \cosh nl \cdot \cos nl) = nl\frac{W_0}{W_c}(\sinh nl \cdot \cos nl - \cosh nl \cdot \sin nl)$$

8-20 $\tan\omega\sqrt{\frac{\rho}{Gg}}l = -\omega\frac{I_pG}{K}\sqrt{\frac{\rho}{Gg}}$

Chapter 9

9-3 $K_i = \frac{p_0}{l}\int_0^l \phi_i(x)\,dx$

9-8 $y(x, t) = \frac{4p_0 l}{\pi M\omega_2^2}\sin\frac{2x\pi}{l}(1 - \cos\omega_2 t)$

9-10 Modes absent are 2*nd*, 5*th*, 8*th*, etc.

9-11 $K = \sqrt{2}\cos(2n - 1)\pi/6$, $D_n = (1 - \cos\omega_n t)$

$$u = \frac{2F_0 l}{AE}\left[\frac{\cos(\pi/6)\cos(\pi/2)(x/l)}{(\pi/2)^2}D_1 + \frac{\cos(5\pi/6)\cos(5\pi/2)(x/l)}{(5\pi/2)^2}D_2 + \cdots\right]$$

9-14 $K_1 = \frac{1}{l}\int_0^l \phi_1(x)\,dx = 0.784$

$K_2 = \frac{1}{l}\int_0^l \phi_2(x)\,dx = 0.434$

$$K_3 = \frac{1}{l}\int_0^l \phi_3(x)\,dx = 0.254$$

9-19 $$\left\{1 + \frac{K\varphi_2'^2(0)}{M\omega_1^2[1-(\omega/\omega_1)^2]}\right\}\left\{1 + \frac{K\varphi_2'^2(0)}{M\omega_2^2[1-(\omega/\omega_2)^2]}\right\} = \left\{\frac{K\varphi_1'(0)\varphi_2'(0)}{M\omega_1^2[1-(\omega/\omega_1)^2]}\right\}\left\{\frac{K\varphi_1'(0)\varphi_2'(0)}{M\omega_2^2[1-(\omega/\omega_2)^2]}\right\}$$

$\varphi_1 = \sqrt{2}\sin\frac{\pi x}{l}, \quad \varphi_1' = \frac{\pi}{l}\sqrt{2}\cos\frac{\pi x}{l}$, etc.

One mode approximation gives

$$\left(\frac{\omega}{\omega_1}\right)^2 = 1 + \frac{2K}{M\omega_1^2}\left(\frac{\pi}{l}\right)^2, \quad \omega_1 = \pi^2\sqrt{\frac{EI}{Ml^3}}$$

9-20 One mode approximation

$$\left(\frac{\omega}{\omega_1}\right)^2 = 1 + \frac{4K}{M\omega_1^2}\left(\frac{\pi}{l}\right)^2$$

9-21 Using one free-free mode and translation mode of M_0

$$\left(\frac{\omega}{\omega_1}\right)^2 = \frac{M_1}{M_1 + M_0\varphi_1^2(0) - [M_0^2\varphi_1^2(0)/(M_0 + 2ml)]}$$

where $M_1 = \int \varphi_1^2(x)m\,dx = 2ml, \quad \omega_1 = 22.4\sqrt{\frac{EI}{m(2l)^4}}$

Chapter 10

10-5 $\bar{x} = 0.50,\ \overline{x^2} = 0.333$

10-6 $\bar{x} = A_0,\ \overline{x^2} = A_0^2 + \frac{1}{2}A_1^2,$

10-8 $f = f_n\sqrt{1 \pm (1/Q)}$

10-10 $$\bar{x}(s) = \frac{1}{ms^2 + k(1 + i\gamma)}$$

10-14 $$f(t) = \sum_{n=1}^{\infty} b_n \sin\frac{2\pi}{\tau}nt, \quad b_n = \frac{2}{n\pi} \text{ for } n \text{ odd}$$
$$= -\frac{2}{n\pi} \text{ for } n \text{ even}$$

10-16 $$f(t) = \frac{2A}{i\pi}\sum_{n=-\infty}^{\infty}\frac{1}{n}e^{in\omega_0 t} \quad n \text{ odd}$$

10-20

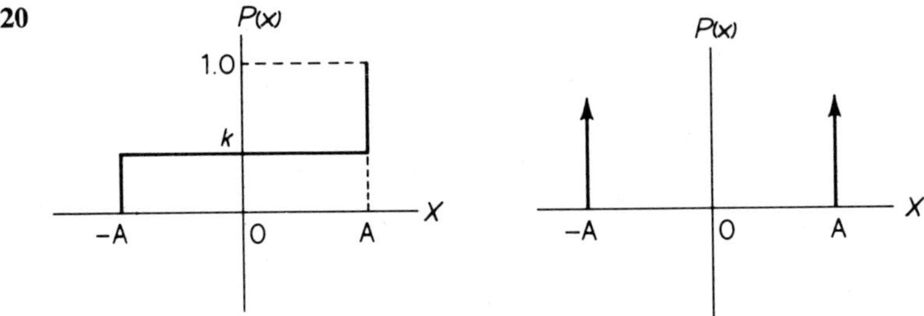

10-21 $\sigma = 1.99$ rms $= 3.60$

10-22 $P[x > 2\sigma] = 4.6\%$ $P[X > 2\sigma] = 13.5\%$

10-24

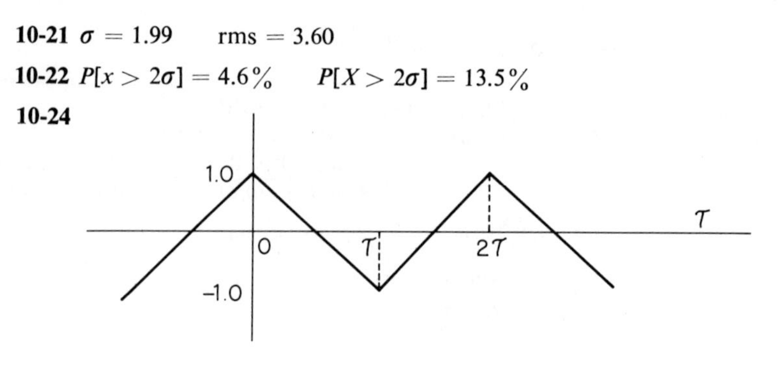

10-27

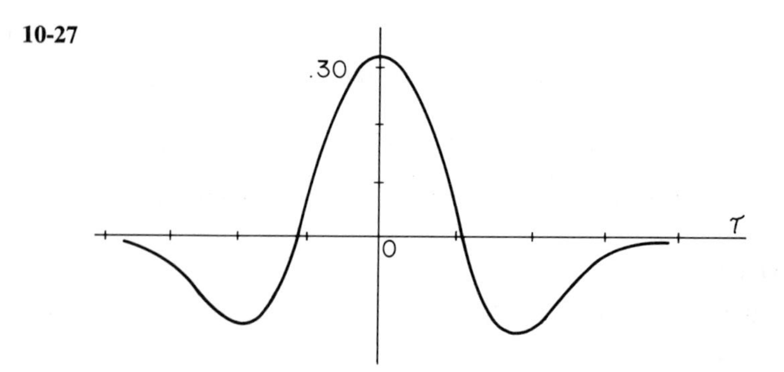

Chapter 11

11-2 $m\ddot{x} + \frac{2T_0}{l_0}x\left[1 + \frac{1}{2}\left(\frac{EA}{T_0} - 1\right)\left(\frac{x}{l}\right)^2\right] = 0$

11-3 $m\ddot{x} + \frac{\pi r_0^2 \rho}{3(h - x_0)^2}[(h - x)^3 - (h - x_0)^3]$

r_0 = radius of circle at water line

$\rho = wt/\text{vol}$ of water

11-8 $V = \sqrt{y^2 + \omega_n^4 x^2}$ $x = y = 0$ for equilibrium

11-11 Shift origin of phase plane to π in Fig. 11.5-2

11-13 $\lambda_{1,2} = 3, 4$ $\begin{Bmatrix} x \\ y \end{Bmatrix}_1 = \begin{Bmatrix} .50 \\ 1.00 \end{Bmatrix}_1$ $\begin{Bmatrix} x \\ y \end{Bmatrix}_2 = \begin{Bmatrix} 1 \\ 1 \end{Bmatrix}$

11-14 $\begin{Bmatrix} x \\ y \end{Bmatrix} = \begin{bmatrix} 0.5 & 1 \\ 1 & 1 \end{bmatrix}\begin{Bmatrix} \xi \\ \eta \end{Bmatrix}$

11-25 $y = \frac{-x(\omega_n^2 + \mu x^2)}{c}$ $c = \frac{dy}{dx}$

11-27 $\dfrac{dy}{dx} = \dfrac{-(x + \delta)}{y}$ where $\delta = \left(\dfrac{c}{m}y + \dfrac{\mu}{\omega_n^2 m}x^3\right)$

$$\omega_n^2 = k/m$$

$$\tau = \omega_n t, \quad y = \frac{dx}{d\tau}$$

11-30 $\tau = 4\sqrt{\dfrac{l}{g}}\displaystyle\int_0^{60^\circ} \dfrac{d\theta}{\sqrt{2(\cos\theta - \cos\theta_0)}} = 4\sqrt{\dfrac{l}{g}}\displaystyle\int_0^{\pi/2} \dfrac{d\theta}{\sqrt{1 - k^2\sin^2\phi}}$

where $k = \sin\dfrac{\theta_0}{2}$, $\sin\dfrac{\theta}{2} = k\sin\phi$

11-40
11-41 $\tau \cong \sqrt{\dfrac{l}{g}}\left(1 + \dfrac{1}{16}\theta_0^2\right)$

11-45 $\ddot{\theta} + \left(-\dfrac{g}{l} + \dfrac{4}{l}\omega^2 y_0 \cos 2\omega t\right)\theta = 0$

Index

E

F

G

H

I

J

K

L

Q

R

S

T

U

V

W

Y